Closed Loop Neuroscience

Closed Loop Neuroscience

Ahmed El Hady
Max Planck Institute for Biophysical Chemistry, Goettingen, Germany

AMSTERDAM • BOSTON • HEIDELBERG • LONDON
NEW YORK • OXFORD • PARIS • SAN DIEGO
SAN FRANCISCO • SINGAPORE • SYDNEY • TOKYO
Academic Press is an imprint of Elsevier

Academic Press is an imprint of Elsevier
125 London Wall, London EC2Y 5AS, United Kingdom
525 B Street, Suite 1800, San Diego, CA 92101-4495, United States
50 Hampshire Street, 5th Floor, Cambridge, MA 02139, United States
The Boulevard, Langford Lane, Kidlington, Oxford OX5 1GB, United Kingdom

Notices

Knowledge and best practice in this field are constantly changing. As new research and experience broaden our understanding, changes in research methods, professional practices, or medical treatment may become necessary.

Practitioners and researchers must always rely on their own experience and knowledge in evaluating and using any information, methods, compounds, or experiments described herein. In using such information or methods they should be mindful of their own safety and the safety of others, including parties for whom they have a professional responsibility.

To the fullest extent of the law, neither the Publisher nor the authors, contributors, or editors, assume any liability for any injury and/or damage to persons or property as a matter of products liability, negligence or otherwise, or from any use or operation of any methods, products, instructions, or ideas contained in the material herein.

Library of Congress Cataloging-in-Publication Data
A catalog record for this book is available from the Library of Congress

British Library Cataloguing-in-Publication Data
A catalogue record for this book is available from the British Library

ISBN 978-0-12-802452-2

For information on all Academic Press publications
visit our website at https://www.elsevier.com/

Working together
to grow libraries in
developing countries

www.elsevier.com • www.bookaid.org

Publisher: Mara Conner
Acquisitions Editor: Mara Conner
Editorial Project Manager: Kathy Padilla
Production Project Manager: Chris Wortley
Cover Designer: Victoria Pearson

Typeset by SPi Global, India

Dedication

Dedicated to those who question established knowledge.

Contents

4. Testing the Theory of Practopoiesis Using Closed Loops

D. Nikolić

5. Local Field Potential Analysis for Closed-Loop Neuromodulation

N. Maling and C. McIntyre

6. Online Event Detection Requirements in Closed-Loop Neuroscience

P. Varona, D. Arroyo, F.B. Rodríguez and T. Nowotny

7. Closing Dewey's Circuit

A. Wallach, S. Marom and E. Ahissar

8. Stochastic Optimal Control of Spike Times in Single Neurons

A. Iolov, S. Ditlevsen and A. Longtin

9. Hybrid Systems Neuroscience

*E.M. Navarro-López, U. Çelikok and
N.S. Şengör*

10. Computational Complexity and the Function-Structure-Environment Loop of the Brain

B. Juba

11. Subjective Physics

R. Brette

List of Contributors

Numbers in parenthesis indicate the pages on which the authors' contributions begin.

E. Ahissar (93), The Weizmann Institute of Science, Rehovot, Israel

D. Arroyo (81), Autonomous University of Madrid, Madrid, Spain

P. beim Graben (171), Humboldt-University of Berlin, Berlin, Germany

M. Beudel (213), University of Groningen, Groningen, The Netherlands

F. Boi (201), Italian Institute of Technology, Rovereto, Italy

R. Brette (145), Sorbonne Universités, Paris, France; INSERM, Paris, France; CNRS, Paris, France

U. Çelikok (113), Boğaziçi University, Istanbul, Turkey

S. Ching (35), Washington University, St. Louis, MO, United States

J. Couto (187), University of Antwerp, Antwerpen, Belgium; University of Sheffield, Sheffield, United Kingdom

S. Ditlevsen (101), University of Copenhagen, Copenhagen, Denmark

A. Gharabaghi (223), Eberhard Karls University, Tuebingen, Germany

M. Giugliano (187), University of Antwerp, Antwerpen, Belgium; Neuro-Electronics Research Flanders, Leuven, Belgium; University of Sheffield, Sheffield, United Kingdom; Brain Mind Institute, EPFL, Lausanne, Switzerland

W. Glannon (259), University of Calgary, Calgary, AB, Canada

C. Ineichen (259), University of Zurich, Zurich, Switzerland

A. Iolov (101), University of Ottawa, Ottawa, Canada; University of Copenhagen, Copenhagen, Denmark

B. Juba (131), Washington University, St. Louis, MO, United States

G. Kumar (35), Washington University, St. Louis, MO, United States

D. Linaro (187), University of Antwerp, Antwerpen, Belgium; University of Sheffield, Sheffield, United Kingdom

A. Longtin (101), University of Ottawa, Ottawa, Canada

C.J. Maley (271), University of Kansas, Lawrence, KS, United States

N. Maling (67), Case Western Reserve University, Cleveland, OH, United States

S. Marom (93), Technion—Israel Institute of Technology, Haifa, Israel

C. McIntyre (67), Case Western Reserve University, Cleveland, OH, United States; Louis Stokes Cleveland Veterans Affairs Medical Center, Cleveland, OH, United States

C.T. Moritz (229), University of Washington, Seattle, WA, United States

E.M. Navarro-López (113), The University of Manchester, Manchester, United Kingdom

D. Nikolić (53), Max Planck Institute for Brain Research, Frankfurt/M, Germany; Frankfurt Institute for Advanced Studies (FIAS), Frankfurt/M, Germany; Ernst Strüngmann Institute (ESI) for Neuroscience in Cooperation with Max Planck Society, Frankfurt/M, Germany; University of Zagreb, Zagreb, Croatia

T. Nowotny (81), University of Sussex, Brighton, United Kingdom

M. Park (3), University College London, London, United Kingdom

G. Piccinini (271), University of Missouri—St. Louis, St. Louis, MO, United States

J.W. Pillow (3), Princeton University, Princeton, NJ, United States

R. Pulizzi (187), University of Antwerp, Antwerpen, Belgium

J.T. Ritt (35), Boston University, Boston, MA, United States

F.B. Rodríguez (81), Autonomous University of Madrid, Madrid, Spain

G. Ruffini (241), Starlab, Barcelona, Spain; Neuroelectrics Corporation, Cambridge, MA, United States

M. Semprini (201), Italian Institute of Technology, Rovereto, Italy

N.S. Şengör (113), Istanbul Technical University, Istanbul, Turkey

M. Tatsuno (19), The University of Lethbridge, Lethbridge, AB, Canada

P. Varona (81), Autonomous University of Madrid, Madrid, Spain

A. Vato (201), Italian Institute of Technology, Rovereto, Italy

A. Wallach (93), University of Ottawa, Ottawa, ON, Canada

A.S. Widge (229), Massachusetts General Hospital, Charlestown, MA, United States; Picower Institute for Learning & Memory, Massachusetts Institute of Technology, Cambridge, MA, United States

Foreword and Introduction

There is a strong current in contemporary culture advocating 'holistic' views as some sort of cure-all. Reductionism implies attention to a lower level while holistic implies attention to higher level. These are intertwined in any satisfactory description: and each entails some loss relative to our cognitive preferences, as well as some gain... there is no whole system without an interconnection of its parts and there is no whole system without an environment.

Varela, 1977. On being autonomous: the lessons of natural history for systems theory. In: George Klir (Ed.), Applied Systems Research. Plenum Press, New York. pp. 77–85

Since childhood, I was obsessed with drawing circles inside circles. I would have never imagined this would end up being the manner I abstract the brain, as loops within loops. My imagined abstraction of the brain as a hierarchy of loops was heavily influenced by my readings of the works of Francesco Varela, who had a significant influence on my thinking. His work on Autopoeisis can help us, experimentalists and theorists, formulate self-organization and systems structure in a global interdependent manner. I drew the following sketch in 2008 as part of a thought experiment that I was going through asking the basic question "What does it mean to theorize about the brain?"

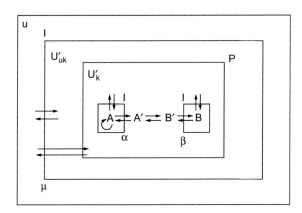

In the above drawn sketch:

A is the internal structure of agent A.
B is the internal structure of agent B.
A' is agent A external state emergent of the internal structure A.

B' is agent B external state emergent of the internal structure B.
$I\alpha$ is the interface between the internal structure of A and other components of the universe. This interface allows bidirectional interaction between the internal structure of A and other components of the universe (the same applies for agent B but through the $I\beta$ interface).
P is the paradigm which defines the set of epistemological tools developed making accessible a certain knowledge of the universe and the other agents.
U'_k is the external state emergent from the internal state of the universe. For the agents U'_k is knowable using the set of epistemological tools (P).
U'_{uk} is the external state emergent from the internal state of the universe. For the agents U'_{uk} is unknowable using the set of epistemological tools (P).
U is the internal state of the universe that is inaccessible to the agents.
$I\mu$ is the interface between the set of the emergent states of the universe (both knowable and unknowable) and the internal state of the universe. This interface allows bidirectional interaction between the universe itself and other components of the universe.

In the above sketch, agents A and B decide to develop a physical theory about their internal structures using P. This act of theorizing in conjunction with other interactions in the universe can be sketched as follows:

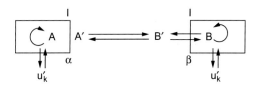

The above sketch shows the different interactions experienced by an agent. The agent interacts with itself. Self-interaction embodies the process in which the agent is engaged in theorizing about its internal structure. Agents have a bidirectional interaction between states of the internal structure and emergent external states. These bidirectional interactions occur through an interface $I\alpha$ (Agent A) or $I\beta$ (Agent B). Moreover, there are bidirectional interactions between external emergent states of both agents in the universe. In addition, there are bidirectional interactions between agents and the knowable partition of the universe. These interactions are facilitated through the interface Iα (Agent A) or Iβ (Agent B).

The thought experiment assumes that there is no a priori knowledge for either agents or for an external observer of the universe about the internal states of any component of the proposed universe and no a priori knowledge of what is happening on the interface. It also assumes that the agents do not have a priori mind-brain distinction. What is assumed is that P as an epistemological space includes the formal/empirical methods used to abstract knowable perceived universe; subsequently agents develop physical theories about U'_k and by definition, about their own internal structure using P.

Based on the above assumptions and supposing an external observer is observing the proposed universe, this observer would ask the following questions:

- Can an agent theorize about its internal structure?
- Can an agent theorize about another agent's internal structure?
- Can an agent theorize about the universe in which it is embedded?
- Can an agent develop new epistemological tools that does not belong to the epistemological space P? How?
- Is P as an epistemological space suitable to theorize about the internal structure?

The aforementioned questions arise from the pattern of interactions of our agents in the universe and they question the possibility of theorizing about the internal structure. This is equivalent to the attempt of *Homo sapiens* to theorize about their brains which can be defined to a great extent as an internal structure that allows the interaction of agents dynamically with each other and with the universe in which they exist.

This thought experiment has kept me both anxious and busy thinking deeply about the challenging questions that it had generated. Most importantly, the prospects of limitations on our endeavors to theorize about the brain. One crucial issue, which the aforementioned thought experiment suggests, is that we need to define the epistemological space within which the "Brain" is viewed and probed. This has led me to closed-loop approaches in neuroscience that I strongly think will expand our epistemological space, shifting the conceptual framework through which we view the brain. In such a shift of framework, a new brain ontology

can be defined where the brain is a complex set of loops within loops, instead of assumed loops that we interact with it in an open loop form. In this case, Brain ontology is a relational one meaning that it defines the loops, not as isolated entities, but as entities that are in continuous dynamical interaction with their environments.

In order for this endeavor to succeed, a highly interdisciplinary approach is needed. I have been actively working towards creating a community of neuroscientists that transcends the peculiarities of their specific sub-fields aiming to benefit from unconventional cross-fertilizations. To this end, I organized a workshop on closed-loop methods in neuroscience during the 2014 FENS conference in Milan. In parallel, I have edited along with Eberhard Fetz and Steve Potter a special topic issue of frontiers in neural circuits on closed-loop neuroscience. It has become evident for me that closed-loop approaches in neuroscience are ripe to be used for a better understanding of the brain. Nevertheless, I thought of closed loops not only as an experimental paradigm (in terms of just software and hardware) but, as I previously eluded to, as a framework within which foundations of contemporary neuroscience can be redefined. Therefore, I came up with the idea of this book that have successfully brought scientists from mathematics, computer science, physics, linguistics, philosophy, engineering, and medical backgrounds. In an interdisciplinary tour de force, the book aims to establish "Closed-Loop Neuroscience" as a standalone field. In the era of big data driven neuroscience, it is important to revisit foundational questions on whether pilling large amounts of data will help us understand the brain. The conceptual framework within which we view the brain (that I term "Brain Ontology") is crucial, as from it stems the manner we abstract the brain. We have to then question whether our contemporary view of the brain will help us understand it or not. I remain agnostic on the question but I feel the compulsion to ask these conceptual questions that shape a field that I think is currently losing sight of its aims. The book might appear fragmented or even incoherent, a sign of a nascent burgeoning field that is taking shape. I am hoping it will motivate more efforts in this direction of creating truly interdisciplinary work that aims to question, define and redefine the conceptual framework within which we view the brain.

Ahmed El Hady
Princeton, January 2016

Acknowledgments

I would like to thank my graduate advisor, Fred Wolf and my postdoctoral advisor, Carlos Brody for giving me the freedom to explore new ideas and for helping me to nurture my career. I am greatly indebted to them. Throughout the course of my career, I have met scientists who have foundationally changed the way I think about the brain. The largest influence came from Shimon Marom whom I have the greatest respect for his approach to neuroscience and his openness to interdisciplinarity. He has been a great host for me whenever I visited the Technion in Israel. Steve Potter and Eberhard Fetz have been very generous in accepting that I co-edit with them a special topic of frontiers in neural circuits on closed-loop methodologies in neuroscience. In addition, I got to interact with many great scientists who are changing the manner in which experiments in neuroscience are done: Elon Vaadia, Hagai Bergman, Jose Carmena, and Michele Giuaglano. I would like to specially thank Avner Wallach, a talented young neuroscientist who has co-organized with me the closed-loop methods in neuroscience workshop at FENS Milan 2014.

I would like to acknowledge the many inspiring discussions and continuous encouragement that I got from my friends Darashana Naranayan and Joel Finkelstein, who have widened my horizons on the applicability of closed-loops to living systems design. I would like to thank Athena Akrami, Ryan Low, and Sam Lewallen for the continued discussions on concepts in neuroscience that are still ongoing, and will hopefully unravel deep principles about the brain organization.

I would like to acknowledge several indepth discussions with my friend and collaborator Pepe Alcami who has always encouraged me to do daring experiments and think of alternative hypotheses. I would also like to thank Megan deBettencourt for interesting discussions about the application of closed-loop methods to cognitive neuroscience and her dedication to move the field forward.

This book would not have appeared in this great form without the dedicated effort of the editorial team at Academic Press who have been very supportive, specifically Mica Haley, Kathy Padilla, and Mara Conner.

Part I

Theoretical Axis

Chapter 1

Adaptive Bayesian Methods for Closed-Loop Neurophysiology

J.W. Pillow* and M. Park†

*Princeton University, Princeton, NJ, United States, †University College London, London, United Kingdom

1 INTRODUCTION

A primary goal in systems neuroscience is to characterize the statistical relationship between environmental stimuli and neural responses. This is commonly known as the *neural coding problem*, which is the problem of characterizing what aspects of neural activity convey information about the external world (Bialek et al., 1991; Rieke et al., 1996; Aguera y Arcas and Fairhall, 2003; Schwartz et al., 2006; Victor and Nirenberg, 2008). This problem is challenging because the relevant stimulus space is often high-dimensional and neural responses are stochastic, meaning that repeated presentations of a single stimulus elicit variable responses. Moreover, neural datasets are limited by the finite length of neurophysiological recordings. In many cases, experimenters wish to rapidly characterize neural tuning as a precursor to other experiments, or to track dynamic changes in tuning properties over time. Adaptive Bayesian methods for stimulus selection, which seek to choose stimuli according to an optimality criterion, provide a natural framework for addressing these challenges.

In classic "fixed design" experiments, stimuli are selected prior to the start of the experiment, or are selected randomly from a fixed distribution, without regard to the observed responses. By contrast, adaptive or closed-loop designs take account of the responses as they are observed during the experiment in order to select future stimuli. Thus, if one observes that a neuron is insensitive to one dimension or region of stimulus space, one can spend more time exploring how the response varies across others. A variety of studies have developed adaptive methods for closed-loop experiments, with applications to linear receptive field (RF) estimation (Földiák, 2001; Lewi et al., 2009a, 2011; Park and Pillow, 2012), color processing in macaque V1 (Horwitz and Hass, 2012; Park et al., 2011, 2014), sound processing in auditory neurons (Nelken et al., 1994; deCharms et al., 1998; Machens et al., 2005; O'Connor et al., 2005), and nonlinear stimulus integration (DiMattina and Zhang, 2011; Bölinger and Gollisch, 2012; Gollisch

and Herz, 2012). See Benda et al. (2007) and DiMattina and Zhang (2013) for recent reviews.

Here we focus on Bayesian methods for adaptive experimental design, known in machine learning as *Bayesian active learning*. The basic idea is to define a statistical model of the neural response, then design stimuli or experiments to estimate the model parameters as efficiently as possible. The learning goal is specified by a utility function that determines the "most useful" stimulus given posterior uncertainty (Mackay, 1992; Chaloner and Verdinelli, 1995; Cohn et al., 1996; Watson and Pelli, 1983; Paninski, 2003, 2005). In the following, we introduce the basics of Bayesian active learning and describe two specific applications to closed-loop neurophysiology.

2 BAYESIAN ACTIVE LEARNING

A Bayesian active learning method or Bayesian adaptive design has three basic ingredients:

1. An *encoding model* $p(r|\mathbf{x},\theta)$, which describes the conditional probability of a neural response r given a stimulus $\mathbf{x}$ and model parameters θ.
2. A *prior distribution* $p(\theta)$ over the model parameters.
3. A *utility function* $u(\theta,r,\mathbf{x}|\mathcal{D}_t)$, which quantifies the usefulness of stimulus-response pair $(\mathbf{x},r)$ for learning about θ, given the data recorded so far in the experiment, $\mathcal{D}_t = \{(\mathbf{x}_1,r_1),...,(\mathbf{x}_t,r_t)\}$.

Here we will consider the stimulus $\mathbf{x}$ to be a vector (eg, the position, orientation, and spatial frequency of a sinusoidal grating, or the vector formed from a binary white noise image). We will consider the elicited response r to be a scalar (eg, the spike count in some time window), although extending this framework to multivariate responses represents an important avenue for future research. Taken together, these ingredients fully specify both the uncertainty about the parameters given the observed data and the optimal stimulus at any point in the experiment.

Closed Loop Neuroscience. http://dx.doi.org/10.1016/B978-0-12-802452-2.00001-9

2.1 Posterior and Predictive Distributions

The encoding model $p(r|\mathbf{x},\theta)$ captures our assumptions about the encoding relationship between stimulus and response, that is, the noisy process that takes stimuli and transforms them into spike responses. When considered as a function of the parameters θ, the encoding model provides the likelihood function. The prior distribution $p(\theta)$, in turn, characterizes our uncertainty about the parameters before the beginning of the experiment. These two ingredients combine to specify the *posterior distribution* over the parameters given data according to Bayes' rule:

$$p(\theta|\mathcal{D}_t) \propto p(\theta) \prod_{i=1}^{t} p(r_i|\mathbf{x}_i,\theta), \qquad (1)$$

where the model provides a likelihood term $p(r_i|\mathbf{x}_i,\theta)$ for each stimulus-response pair. The product of likelihood terms arises from the assumption that responses are conditionally independent given the stimulus in each time step. (Later, we will discuss relaxing this assumption.)

Another important distribution that we can obtain from these ingredients is the *predictive distribution* of the response, $p(r|\mathbf{x},\mathcal{D}_t)$ which is the conditional distribution of the response given the stimulus and all previously observed data $\mathcal{D}_t$, with uncertainty about the parameters θ integrated out. This is given by

$$p(r|\mathbf{x},\mathcal{D}_t) = \int p(r|\mathbf{x},\theta)\, p(\theta|\mathcal{D}_t)\, d\theta. \qquad (2)$$

2.2 Utility Functions and Optimal Stimulus Selection

In sequential optimal designs, the experimenter selects a stimulus on each trial according to an optimality criterion known as the *expected utility*. This quantity is the average of the utility function over $p(r,\theta|x,\mathcal{D})$, the joint distribution of r and θ given a stimulus and the observed data so far in the experiment:

$$U(\mathbf{x}|\mathcal{D}_t) \triangleq \mathbb{E}_{r,\theta}[u(\theta,r,\mathbf{x}|\mathcal{D}_t)] = \iint u(\theta,r,\mathbf{x}|\mathcal{D}_t) \\ p(r,\theta|\mathbf{x},\mathcal{D}_t)\, d\theta\, dr, \qquad (3)$$

or if responses are integer spike counts,

$$= \int \sum_{r=0}^{\infty} u(\theta,r,\mathbf{x}|\mathcal{D}_t)\, p(r,\theta|\mathbf{x},\mathcal{D}_t)\, d\theta. \qquad (4)$$

The experimenter selects the stimulus with maximal expected utility for the next trial:

$$\mathbf{x}_{t+1} = \arg \max_{\mathbf{x}^* \in \Omega} U(\mathbf{x}^*|\mathcal{D}_t), \qquad (5)$$

where Ω is some set of candidate stimuli. Fig. 1 shows a schematic of the three iterative steps in Bayesian active

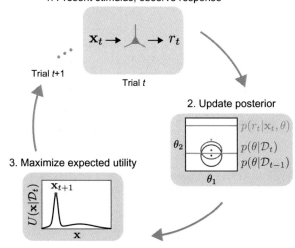

FIG. 1 Schematic of Bayesian active learning for closed-loop neurophysiology experiments. At time step t of the experiment, we present stimulus $\mathbf{x}_t$ and record neural response r_t. Then, we update the posterior $p(\theta|\mathcal{D}_t)$ by combining the likelihood $p(r_t|\mathbf{x}_t,\theta)$ with the prior $p(\theta|\mathcal{D}_{t-1})$, which is the posterior from the previous time step. Finally, we search for the stimulus $\mathbf{x}_{t+1}$ that maximizes the expected utility $U(\mathbf{x}|\mathcal{D}_t)$, which quantifies the learning objective in terms of a utility function integrated over the joint distribution of r and θ given the data. These steps are repeated until some stopping criterion.

learning: (1) present stimulus and observe response; (2) update posterior; and (3) maximize expected utility to select a new stimulus.

The choice of utility function determines the notion of optimality for a Bayesian active learning paradigm. Intuitively, the utility function can be understood as providing a precise specification of what kinds of posterior distributions are considered "good." We will now review several choices for utility.

2.2.1 Maximum Mutual Information (Infomax)

The most popular Bayesian active learning method seeks to maximize the gain in information about model parameters, an approach commonly known as *infomax learning* (Lindley, 1956; Bernardo, 1979; Luttrell, 1985; Mackay, 1992; Paninski, 2005; Lewi et al., 2009a; Kontsevich and Tyler, 1999). This approach selects stimuli that maximize the mutual information between response r and the parameters θ, which is equivalent to minimizing the expected entropy of the posterior distribution.

Formally, infomax learning arises from a utility function given by the log ratio of the posterior to the prior,

$$u(\theta,r,\mathbf{x}|\mathcal{D}_t) = \log \frac{p(\theta|r,\mathbf{x},\mathcal{D}_t)}{p(\theta|\mathcal{D}_t)}, \qquad (6)$$

where the numerator is the updated posterior after observing a new stimulus-response pair $(\mathbf{x},r)$, and the denominator is the prior, given by the posterior at trial t.

The expected utility is therefore the mutual information between r and θ:

$$U_{\text{infomax}}(\mathbf{x}|\mathcal{D}_t) = \mathbb{E}_{r,\theta}\left[\log\frac{p(\theta|r,\mathbf{x},\mathcal{D}_t)}{p(\theta|\mathcal{D}_t)}\right] = H(\theta|\mathcal{D}_t)$$
$$-H(\theta;r|\mathbf{x},\mathcal{D}_t) \triangleq I(\theta,r|\mathbf{x},\mathcal{D}_t), \tag{7}$$

where we use $H(\theta;r|\mathbf{x},\mathcal{D}_t)$ to denote the conditional entropy of θ given r for fixed $\mathbf{x}$ and $\mathcal{D}_t$, and $H(\theta|\mathcal{D}_t)$ is the entropy of the posterior after the previous trial. Note that we can perform infomax learning by selecting the stimulus that minimizes $H(\theta;r|\mathbf{x},\mathcal{D}_t)$, since $H(\theta|\mathcal{D}_t)$ is independent of the stimulus and response on the current trial. The mutual information utility function is also commonly referred to as the *expected information gain*, since it is the expected change in the posterior entropy from a single stimulus-response pair (Paninski, 2005; Lewi et al., 2009a).

It is worth noting that the mutual information can also be written as

$$I(\theta,r|\mathbf{x},\mathcal{D}_t) = H(r|\mathcal{D}_t) - H(r;\theta|\mathbf{x},\mathcal{D}_t), \tag{8}$$

which is the difference between the marginal entropy of r (under the predictive distribution) and the conditional entropy of r given θ. This expression is sometimes easier to compute than the expected utility as given in Eq. (7), as we will show in our first application below.

2.2.2 Minimum Mean Squared Error (MMSE)

Another possible approach is to select stimuli that will allow for optimal least-squares estimation of θ, a paradigm commonly known as *MMSE* learning (Kuck et al., 2006; Müller and Parmigiani, 1996; Park et al., 2014). We can formalize MMSE learning with a utility function given by the negative mean squared error (since we wish to maximize utility):

$$u = -\|\theta - \hat{\theta}_{(r,\mathbf{x},\mathcal{D}_t)}\|^2, \tag{9}$$

where $\hat{\theta}_{(r,\mathbf{x},\mathcal{D}_t)} = \mathbb{E}_\theta[\theta|r,\mathbf{x},\mathcal{D}_t]$ is the mean of the posterior $p(\theta|r,\mathbf{x},\mathcal{D}_t)$, also known as the *Bayes' least squares estimate*. The expected utility is therefore given by

$$U_{\text{MMSE}}(\mathbf{x}|\mathcal{D}_t) = -\mathbb{E}_{r,\theta}\left[(\theta - \hat{\theta}_{(r,\mathbf{x},\mathcal{D}_t)})^\top(\theta - \hat{\theta}_{(r,\mathbf{x},\mathcal{D}_t)})\right]$$
$$= -\mathbb{E}_r[\text{Tr}[\text{cov}(\theta|r,\mathbf{x},\mathcal{D})]], \tag{10}$$

which is equal to the negative trace of the posterior covariance given the response, averaged over $p(r|\mathbf{x},\mathcal{D}_t)$. Because trace and expectation can be exchanged, this can also be written as the sum of the expected marginal posterior variances:

$$U_{\text{MMSE}}(\mathbf{x}|\mathcal{D}_t) = -\sum_i \mathbb{E}_r[\text{var}(\theta_i|r,\mathbf{x},\mathcal{D}_t)]. \tag{11}$$

One might therefore refer to MMSE learning as *minimum-variance* learning because it seeks to minimize the sum of posterior variances for each parameter (Park et al., 2014).

2.2.3 Other Utility Functions

A variety of other utility functions have been proposed in the literature, including prediction error on test data (Roy and Mccallum, 2001; Cohn et al., 1996); misclassification error (Kapoor and Basu, 2007); and the mutual information between function values at tested and untested locations (Krause et al., 2008). Other "nongreedy" active learning methods do not attempt to maximize expected utility in each time step, but define optimality in terms of the stimulus that will best allow one to achieve some learning or prediction goal in 2 or n trials after the current trial (King-Smith et al., 1994).

For the remainder of this chapter, we will focus on infomax learning, due to its popularity and relative computational tractability. But readers should be aware that the choice of utility function can have significant effects on learning, due to the fact that different utility functions imply different notions of "goodness" of a posterior distribution. For example, if we consider an uncorrelated Gaussian posterior, entropy depends on the product of variances, whereas the MSE depends on the sum of variances. Thus, infomax learning would strongly prefer a posterior with variances $(\sigma_1^2 = 1, \sigma_2^2 = 100)$ to one with variances $(\sigma_1^2 = 20, \sigma_2^2 = 20)$ since $(1 \cdot 100) < (20 \cdot 20)$. MMSE learning, however, would have the opposite preference because $(1 + 100) > (20 + 20)$. These differences may have practical effects on learning by determining which stimuli are selected. (See Park et al. (2014) for an detailed exploration in the context of tuning curve estimation.)

2.2.4 Uncertainty Sampling

Before moving on to applications, it is worth mentioning methods that are not fully Bayesian but rely on some of the basic ingredients of Bayesian active learning. A prominent example is *uncertainty sampling*, which was introduced for training probabilistic classifiers (Lewis and Gale, 1994). The original idea was to select stimuli for which the current classifier (given the labeled data available so far) is maximally uncertain. This employs the entirely reasonable heuristic that a classifier should be able to learn more from examples with uncertain labels. However, uncertainty sampling is not necessarily a Bayesian active learning method because the selection rule does not necessarily maximize an expected utility function.

Uncertainty sampling can take at least two different forms in a neurophysiology setting. The first, which we will call *response uncertainty sampling* and involves picking the stimulus for which the response has maximal entropy,

$$\mathbf{x}_{t+1} = \arg\max_{\mathbf{x}^* \in \Omega} H(r|\mathbf{x}^*,\mathcal{D}_t) \tag{12}$$

which involves maximizing entropy of the predictive distribution (Eq. 2). For a binary neuron, this would correspond to selecting a stimulus for which the spike probability is closest to 0.5.

An alternative approach is to select stimuli for which we have maximal uncertainty about the function underlying the response, which we will refer to as *parameter uncertainty sampling*. This method can be used when the parameters are in a one-to-one correspondence with the stimulus space. If we consider a tuning curve $f(\mathbf{x})$ that specifies the mean response to stimulus $\mathbf{x}$, parameter uncertainty sampling would correspond to selection rule:

$$\mathbf{x}_{t+1} = \arg \max_{\mathbf{x}^* \in \Omega} H(f(\mathbf{x}^*)|\mathcal{D}_t), \qquad (13)$$

where $p(f(\mathbf{x})|\mathcal{D}_t)$ is the posterior distribution over the value of the tuning curve at $\mathbf{x}$. This differs from infomax learning because it fails to take into account the *amount* of information that the response r is likely to provide about f. For example, higher-mean regions of the tuning curve will in general provide less information than lower-mean regions under Poisson noise, because noise variance grows with the mean. In the next two sections, we will explore active learning methods for two specific applications of interest to neurophysiologists.

3 APPLICATION: TUNING CURVE ESTIMATION

First, we consider the problem of estimating a neuron's tuning curve in a parametric stimulus space. The object of interest is a nonlinear function f that describes how a neuron's firing rate changes as a function of some stimulus parameters (eg, orientation, spatial frequency, position). Canonical examples would include orientation tuning curves, spatial frequency tuning curves, speed tuning curves, hippocampal place fields, entorhinal grid cell fields, and absorption spectra in photoreceptors. We might equally call such functions "firing rate maps" or "response surfaces." Our goal is to select stimuli in order to characterize these functions using the smallest number of measurements. We will separately discuss methods for parametric and nonparametric tuning curves under Poisson noise.

3.1 Poisson Encoding Model

We will model a neuron's average response to a stimulus $\mathbf{x}$ by a nonlinear function $f(\mathbf{x})$, known as the tuning curve, and assume that the response is corrupted by Poisson noise. The resulting encoding model is given by:

$$\lambda = f(\mathbf{x}) \qquad (14)$$

$$p(r|\mathbf{x}) = \frac{1}{r!} \lambda^r e^{-\lambda}. \qquad (15)$$

Let $\mathcal{D}_t = \{(\mathbf{x}_i, r_i)\}_{i=1}^t$ denote the data collected up to time t in an experiment. Then we have log-likelihood function

$$\mathcal{L}(\lambda_t|\mathcal{D}_t) = \log p(R_t|\lambda_t) = R_t^\top \log \lambda_t - \mathbf{1}^\top \lambda_t \qquad (16)$$

where $R_t = (r_1,...,r_t)^\top$ is the vector of responses, $\lambda_t = (f(\mathbf{x}_1),...,f(\mathbf{x}_t))^\top$ is the vector of spike rates for the stimuli presented, and $\mathbf{1}$ is a vector of one. We have ignored a constant that does not depend on f.

3.2 Parametric Tuning Curves

In many settings, the experimenter has a particular parametric tuning curve in mind (eg, a von-Mises function for a V1 orientation tuning curve, or a Naka-Rushton function for a contrast-response function). This approach confers advantages in terms of speed: by making strong assumptions about the form of the tuning curve, active learning algorithms can rule out many functions a priori and more quickly identify regions of stimulus space that are informative about the parameters. For example, if the desired tuning curve is a Gaussian bump, Bayesian active learning will not waste time trying to determine if there are other bumps in unexplored regions of parameter space once a single bump has been identified. The potential disadvantage of this approach is that it may fail when tuning curves violate the assumptions of the parametric model. If a neuron has a bimodal tuning curve, for example, an active learning algorithm designed for unimodal function may never discover the second mode.

Here we describe a simple approach for infomax learning of a parametric tuning curve $f(\mathbf{x}; \theta)$, which describes the mean response to a stimulus $\mathbf{x}$ and is described by parameters θ. In general, the log-likelihood (Eq. 16) is not convex as a function of θ, and gradient ascent methods may therefore not find the global maximum of the likelihood. We therefore use Markov Chain Monte Carlo (MCMC) sampling to obtain samples from the posterior distribution over θ, an approach used previously for Bayesian tuning curve inference in a fixed design setting (Cronin et al., 2010).

We can use a standard MCMC sampling method (eg, Metropolis-Hastings or slice sampling) to obtain a set of m samples $\{\theta^{(i)}\} \sim p(\theta|\mathcal{D}_t)$ from the posterior given the data so far in the experiment (Eq. 1). We can then evaluate the expected information gain for any candidate stimulus $\mathbf{x}^*$ using a grid over spike counts $r \in \{0,1,...,r_{max}\}$ to compute the marginal and conditional response entropies. We set r_{max} to some suitably high value (eg, 200 spikes) based on the current posterior over spike rates. Mutual information is given by the difference of these entropies:

$$\begin{aligned} I(\theta, r|\mathbf{x}^*, \mathcal{D}_t) \approx &-\sum_{r=0}^{r_{max}} p(r|\mathbf{x}^*) \log p(r|\mathbf{x}^*) \\ &+ \frac{1}{m} \sum_{i=1}^m \sum_{r=0}^{r_{max}} p(r|\mathbf{x}^*, \theta^{(i)}) \log p(r|\mathbf{x}^*, \theta^{(i)}), \end{aligned} \qquad (17)$$

where the marginal response distribution is given by the mean over MCMC samples:

$$p(r|\mathbf{x}^*) = \frac{1}{m}\sum_{i=1}^{m} p(r|\mathbf{x}^*, \theta^{(i)}) \qquad (18)$$

for each response value r. The resulting algorithm is remarkably simple, and may be implemented with fewer than 150 lines of code in Matlab using a 2D grid over stimulus locations and spike counts r (code available from http://pillowlab.princeton.edu/code_activelearningTCs.html).

Fig. 2 shows an illustration of this algorithm in a simulated experiment for estimating a 1D Gaussian tuning curve with baseline, parameterized as:

$$f(x; \theta) = b + A\exp\left(-\frac{1}{2\sigma^2}(x-\mu)^2\right), \qquad (19)$$

where the parameters θ include a preferred stimulus μ, tuning width σ, amplitude A, and baseline firing rate b. We obtained samples from the posterior with a standard implementation of slice sampling.

The first column of Fig. 2 shows a snapshot of the algorithm after three trials of the experiment. The top plot shows the three observed stimulus-response pairs (black dots), the true tuning curve (black trace), and $m=100$ posterior samples $f^{(i)}$ (gray traces), given by $f(x|\theta^{(i)})$ for each sample $\theta^{(i)}$ from the posterior (Eq. 1). The posterior mean $\hat{f}$ (red trace) is the mean of the samples $\frac{1}{m}\sum_i f^{(i)}$. The bottom plot

shows the expected information gain, computed using (Eq. 17), for each stimulus x on a grid over the stimulus range $[-10, 10]$. Intuitively, the expected information gain for each stimulus is related to the spread of the sample tuning functions at that location (gray traces in the top plots); the more the sample tuning curves disagree, the higher the information gain from that stimulus. A black asterisk marks the maximum of the expected utility function, which determines the stimulus selected for the next trial. Subsequent columns show analogous snapshots after 4, 5, and 10 trials. The top-right plot shows the estimate after 50 trials, along with the stimuli-response pairs obtained (black dots). Note that the stimuli are far from uniformly distributed, with most dots clustered at the peak and sides of the tuning curve. This indicates that these stimulus locations tend to provide maximal information about the tuning curve under this parameterization.

The bottom right plot (Fig. 2) shows a comparison of the average error in the tuning curve estimate $|f(x) - \hat{f}(x)|$ under infomax learning and nonadaptive learning (uniform *iid* stimulus sampling), averaged over 250 runs of each algorithm. On average, the nonadaptive sampling method requires 25 trials to achieve the same error as infomax learning after only 10 trials. Longer runs reveal that, even asymptotically, the nonadaptive method requires 50% more trials to achieve the same error as infomax learning for this 1D example. Substantially greater improvements can be obtained in higher dimensions.

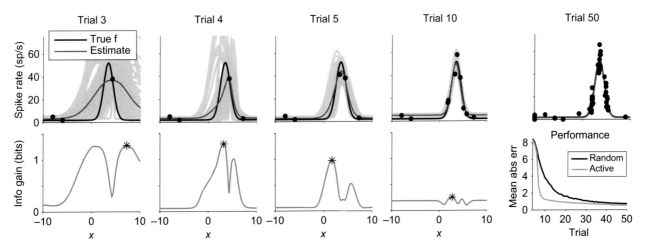

FIG. 2 Active learning of 1D parametric tuning curve using MCMC sampling. The true tuning curve (*black*) has preferred stimulus $\mu = 3.4$, tuning width $\sigma = 1$, amplitude $A = 50$, and baseline firing rate $b = 2$. We placed a uniform prior on each of these parameters: $\mu \in [-10, 10]$, $\sigma \in [0.1, 20]$, $A \in [1, 200]$, and $b \in [0.1, 50]$. Top row: True tuning curve (*black*) and Bayes least-squares estimate (*red*), shown along with 50 samples from the posterior (*gray traces*) after 3, 4, 5, 10, and 50 trials (50-trial figure at right generated from an independent run). Bottom row: Expected information gain for each candidate stimulus given the data so far in the experiment. *Black asterisk* indicates location of the selected stimulus, which is presented on the next trial. Bottom right: Comparison of mean absolute error between true tuning curve and BLS estimate under random (*black*) and Bayesian active learning stimulus election (*yellow*), averaged over 250 runs of each method. The active method achieves a maximal speedup factor of 2.5, with an error after 10 trials approximately equal to the random sampling error after 25 trials.

3.3 Nonparametric Tuning Curves With Gaussian Process Priors

In some settings, the experimenter may not wish to make a strong parametric assumption about the form of the tuning curve, and instead prefer an approach that will converge for any possible f. This motivates a nonparametric approach, which allows the number of degrees of freedom in the tuning curve to grow flexibly with the amount of data. Here we discuss an approach based on transformed Gaussian process (GP) priors, described previously in Park et al. (2011, 2014).

3.3.1 Gaussian Processes

GPs provide a flexible and computationally tractable family of prior distributions over smooth functions. They have been used for nonparametric tuning curve estimation (Rad and Paninski, 2010), and for a variety of other neuroscience applications, including spike rate estimation (Cunningham et al., 2008), factor analysis (Yu et al., 2009), and estimation of cortical maps (Macke et al., 2011). A GP can be understood as an extension of a multivariate Gaussian distribution to the continuum, so that each value of the function has a Gaussian distribution. Formally, a GP is characterized by a *mean function* $\mu(\cdot)$ and *covariance function* $K(\cdot,\cdot)$ which specify the mean and covariance of the Gaussian distribution over function values at any collection of locations. For a d-dimensional function $\phi : \mathbb{R}^d \to I\!R$ distributed according to a GP, we write $\phi \sim \mathcal{GP}(\mu, K)$. This means that for any pair of locations $\mathbf{x}_1, \mathbf{x}_2 \in \mathbb{R}^d$, the values of the function at these locations have a bivariate Gaussian distribution:

$$\begin{bmatrix} \phi(\mathbf{x}_1) \\ \phi(\mathbf{x}_2) \end{bmatrix} \sim \mathcal{N}\left(\begin{bmatrix} \mu(\mathbf{x}_1) \\ \mu(\mathbf{x}_2) \end{bmatrix}, \begin{bmatrix} K(\mathbf{x}_1, \mathbf{x}_1) & K(\mathbf{x}_1, \mathbf{x}_2) \\ K(\mathbf{x}_1, \mathbf{x}_2) & K(\mathbf{x}_2, \mathbf{x}_2) \end{bmatrix} \right). \quad (20)$$

Similarly, for any set of N input locations $\mathbf{x}_{1:N} = \{ \mathbf{x}_i \in \mathbb{R}^d \}_{i=1}^{N}$, the vector of function values $\boldsymbol{\phi}_{1:N} = \{ \phi(\mathbf{x}_i) \in \mathbb{R} \}_{i=1}^{N}$ has a multivariate Gaussian distribution with mean vector whose i'th element is $\mu(\mathbf{x}_i)$ and covariance matrix whose i,j'th element is $K(\mathbf{x}_i, \mathbf{x}_j)$. A common approach is to fix the mean function to a constant μ and select a covariance function that has desired degree of smoothness or differentiability (see Rasmussen and Williams (2006)). Here we will use the popular Gaussian or "squared exponential" covariance function (inaptly named because it takes the form of an exponentiated square, not a squared exponential), which is given by:

$$K(\mathbf{x}_i, \mathbf{x}_j) = \rho \exp\left(- \| \mathbf{x}_i - \mathbf{x}_j \|^2 / (2\tau^2) \right). \quad (21)$$

This covariance function has a pair of hyperparameters: a marginal variance ρ, which determines how far function values deviate from the mean; and length-scale τ, which determines function smoothness. We can consider the mean μ as a third hyperparameter, so that the GP is determined by the hyperparameter vector $\psi = (\rho, \tau, \mu)$.

If we wished to model a neuron's response as having fixed variance Gaussian noise, then we could place a GP prior directly over the tuning curve f, and use the standard formulas to compute the posterior over f after each trial (Rasmussen and Williams, 2006). In this case, there's no need for adaptive stimulus selection: it turns out that the posterior covariance over f and the expected utility depend only on the stimuli $\{ \mathbf{x}_i \}$, and are independent of the observed responses $\{ r_i \}$. This means that we can plan out a maximally informative set of stimuli before the experiment begins. (Note however, that this is true only if we consider the GP hyperparameters ψ and Gaussian noise variance σ^2 to be fixed; if they are to be estimated or integrated over during the experiment, then active learning becomes worthwhile even in this setting.)

3.3.2 Transformed Gaussian Processes

If we wish to model neurons with Poisson noise, the tuning curve must be nonnegative. This rules out the use of standard GP priors because they place probability mass on negative as well as positive function values. A straightforward solution is to transform a GP by a nonlinear function g with range. This suggests we parameterize the tuning curve as

$$f(\mathbf{x}) = g(\phi(\mathbf{x})), \quad (22)$$

where $g(\cdot)$ is an invertible nonlinear function with nonnegative output, and ϕ is a real-valued function governed by GP prior. We refer to the resulting model as the GP-Poisson model (see Fig. 3).

This model lends itself to infomax learning because information gain about f is the same as the information gain about ϕ. An invertible nonlinearity g can neither create nor destroy information. Moreover, if g is convex and logconcave (meaning g grows at least linearly and at most exponentially), then the posterior over ϕ under a GP prior

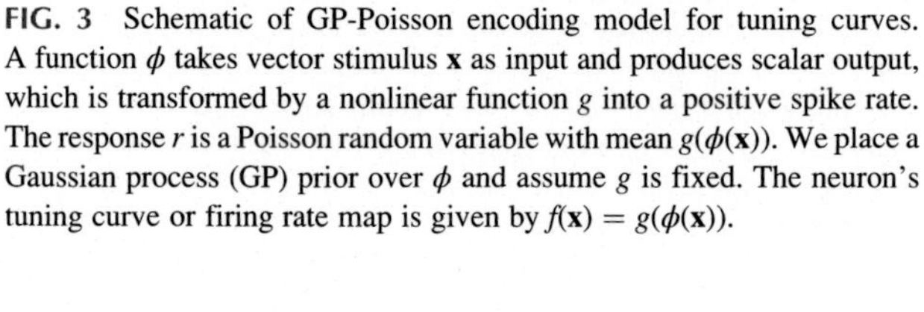

FIG. 3 Schematic of GP-Poisson encoding model for tuning curves. A function ϕ takes vector stimulus $\mathbf{x}$ as input and produces scalar output, which is transformed by a nonlinear function g into a positive spike rate. The response r is a Poisson random variable with mean $g(\phi(\mathbf{x}))$. We place a Gaussian process (GP) prior over ϕ and assume g is fixed. The neuron's tuning curve or firing rate map is given by $f(\mathbf{x}) = g(\phi(\mathbf{x}))$.

will be strictly log-concave (by an argument similar to that given in Paninski (2004)). This ensures that the posterior over ϕ, written $p(\phi|\mathcal{D}_t)$, has a single mode and can be reasonably well approximated by a GP (Park et al., 2014).

3.3.3 Posterior Updating

We can approximate the posterior over ϕ using the Laplace approximation, a Gaussian approximation that results from finding the posterior mode and using the Hessian (second derivative matrix) at the mode to approximate the covariance (Kass and Raftery, 1995; Pillow et al., 2011). Given data $\mathcal{D}_t$, we find the *maximum a posteriori* (MAP) estimate of ϕ at the set of stimulus locations presented so far in the experiment:

$$\hat{\boldsymbol{\phi}}_{\mathrm{map}} \triangleq \arg\max_{\boldsymbol{\phi}_t} \; \mathcal{L}(\boldsymbol{\phi}_t|\mathcal{D}_t) + \log p(\boldsymbol{\phi}_t|\mu\mathbf{1}, \mathbf{K}) \quad (23)$$

$$= \arg\max_{\boldsymbol{\phi}_t} \; R_t^\top \log(g(\boldsymbol{\phi}_t)) - \mathbf{1}^\top g(\boldsymbol{\phi}_t) \\ - \frac{1}{2}(\boldsymbol{\phi}_t - \mu\mathbf{1})^\top \mathbf{K}^{-1}(\boldsymbol{\phi}_t - \mu\mathbf{1}), \quad (24)$$

where $\mathcal{L}(\boldsymbol{\phi}_t|\mathcal{D}_t)$ denotes the Poisson log-likelihood (Eq. 16) and $\log p(\boldsymbol{\phi}_t|\mu, \mathbf{K})$ is the log probability of the vector of function values $\boldsymbol{\phi}_t = (\phi(\mathbf{x}_1), \ldots \phi(\mathbf{x}_t))$ under the GP prior. This latter term is given by a multivariate normal density with $t \times t$ covariance matrix $\mathbf{K}_{i,j} = K(\mathbf{x}_i, \mathbf{x}_j)$, and $\mathbf{1}$ is a length-t vector of ones.

The Laplace approximation to the posterior at the stimulus locations $X_t = (\mathbf{x}_1, \ldots, \mathbf{x}_t)$ is a multivariate Gaussian with mean $\hat{\boldsymbol{\phi}}_{\mathrm{map}}$ and covariance equal to the negative inverse Hessian of the log-posterior:

$$p(\boldsymbol{\phi}_t|\mathcal{D}_t) \approx \mathcal{N}(\hat{\boldsymbol{\phi}}_{\mathrm{map}}, \Sigma), \quad \Sigma = (L + \mathbf{K}^{-1})^{-1}, \quad (25)$$

where L is the Hessian of the negative log-likelihood, which is a diagonal matrix with diagonal elements given by:

$$L_{ii} = -\frac{\partial^2}{\partial \phi_i^2} \mathcal{L}(\boldsymbol{\phi}_t|\mathcal{D}_t) = r_i \frac{g(\phi_i)g''(\phi_i) - g'(\phi_i)^2}{g(\phi_i)^2} + g''(\phi_i) \quad (26)$$

evaluated at $\boldsymbol{\phi}_t = \hat{\boldsymbol{\phi}}_{\mathrm{map}}$. When $g(\cdot)$ is exponential, the first term vanishes and we have simply $L = \mathrm{diag}(\exp(\boldsymbol{\phi}_t))$. The full posterior GP is then determined by the Gaussian approximation at the points in X_t: the likelihood enters only at these points, and its effects on the posterior at all other points are mediated entirely by the smoothing properties of the GP prior.

For any set of possible next stimuli $\mathbf{x}^* = \{\mathbf{x}_i^*\}_{i=1}^N$, let $\boldsymbol{\phi}_t^* = \phi(\mathbf{x}^*)$ denote the vector of associated function values. The posterior distribution over these function values under the Laplace approximation is given by:

$$p(\boldsymbol{\phi}_t^*|\mathcal{D}_t) \approx \mathcal{N}(\boldsymbol{\mu}_t, \Lambda_t), \quad (27)$$

with mean and covariance given by

$$\boldsymbol{\mu}_t = \mu\mathbf{1} + \mathbf{K}^*\mathbf{K}^{-1}(\hat{\boldsymbol{\phi}}_{\mathrm{map}} - \mu\mathbf{1}), \quad (28)$$

$$\Lambda_t = \mathbf{K}^{**} - \mathbf{K}^*(L^{-1} + \mathbf{K})^{-1}\mathbf{K}^{*\top}, \quad (29)$$

where $\mathbf{K}_{ij}^* = K(\mathbf{x}_i^*, \mathbf{x}_j)$ and $\mathbf{K}_{ij}^{**} = K(\mathbf{x}_i^*, \mathbf{x}_j^*)$ (see Rasmussen and Williams (2006, Section 3.4.2)). To perform uncertainty sampling (Eq. 13), we could simply select the stimulus for which the corresponding diagonal element of Λ_t is largest. However, this will not take account of the fact that the expected information gain under Poisson noise depends on the posterior mean as well as the variance, as we will see next.

3.3.4 Infomax Learning

Because mutual information is not altered by invertible transformations, the information gain about the tuning curve f is the same as about ϕ. We can therefore perform infomax learning by maximizing mutual information between r and $\boldsymbol{\phi}_t$, or equivalently, minimizing the conditional entropy of $\boldsymbol{\phi}_t$ given r (Eq. 7). Because $\boldsymbol{\phi}_t$ is a function instead of a vector, this entropy is not formally well defined, but for practical purposes we can consider the vector of function values $\boldsymbol{\phi}_t^*$ defined on some grid of points $\mathbf{x}^*$, which has a tractable, approximately Gaussian distribution (Eq. 27).

The expected utility of a stimulus $\mathbf{x}^*$ can therefore be computed using the formula for negative entropy of a Gaussian:

$$U_{\mathrm{infomax}}(\mathbf{x}^*|\mathcal{D}_t) = -\frac{1}{2}\mathbb{E}_{[r^*,|\mathbf{x}]}\left[\log|\Lambda_t^{-1}\Lambda_t^*|\right], \quad (30)$$

where Λ_t is the posterior covariance at time t (Eq. 29) and Λ_t^* is the posterior covariance after updating with observation $(\mathbf{x}^*, r^*)$. The expectation is taken with respect to $p(r^*|\mathbf{x}^*, \mathcal{D}_t)$, the predictive distribution at $\mathbf{x}^*$ (Eq. 2). If we use exponential g, then the Λ_t^* lacks explicit dependence on r^* because the Hessian depends only on the value of ϕ at $\mathbf{x}^*$, that is:

$$\Lambda_t^* = \left(\Lambda_t^{-1} + \delta_{(\mathbf{x}^*,\mathbf{x}^*)}\, e^{\hat{\phi}_{\mathrm{map}}(\mathbf{x}^*)}\right)^{-1}, \quad (31)$$

where $\delta_{(\mathbf{x}^*,\mathbf{x}^*)}$ is a matrix with a single 1 in the diagonal position corresponding to stimulus $\mathbf{x}^*$ and zeros elsewhere, and $\hat{\boldsymbol{\phi}}_{\mathrm{map}}(\mathbf{x}^*)$ is the MAP estimate of ϕ after observing r^*. Using the matrix-determinant lemma to compute the determinant of Λ_t^*, we can simplify the expected utility to:

$$U_{\mathrm{infomax}}(\mathbf{x}^*|\mathcal{D}_t) = \frac{1}{2}\mathbb{E}_{[r^*,|\mathbf{x},*]}\left[\log\left(1 + \sigma_t^2(\mathbf{x}^*)e^{\hat{\phi}_{\mathrm{map}}(\mathbf{x}^*)}\right)\right] \\ \approx \frac{1}{2}\mathbb{E}_{r^*|\mathbf{x}^*}\left[\sigma_t^2(\mathbf{x}^*)e^{\hat{\phi}_{\mathrm{map}}(\mathbf{x}^*)}\right], \quad (32)$$

where $\sigma_t^2(\mathbf{x}^*) = \Lambda_t(\mathbf{x}^*, \mathbf{x}^*)$ is the marginal variance of the posterior over the value of ϕ at $\mathbf{x}^*$ (Eq. 29), and the approximation on the right results from a first-order Taylor expansion of $\log(x)$. The information gain therefore depends on r^* only via its influence on the MAP estimate of $\phi(\mathbf{x}^*)$. Because there is no analytical form for this expectation, it is reasonable to use the expectation of $e^{\hat{\phi}_{map}(\mathbf{x}^*)}$ over the current posterior, which follows from the formula for the mean of a log-normal distribution. This allows us to evaluate the information gain for each candidate stimulus $\mathbf{x}^*$:

$$U_{\text{infomax}}(\mathbf{x}^*|\mathcal{D}_t) \approx \frac{1}{2}\sigma_t^2(\mathbf{x}^*)e^{\mu_t(\mathbf{x}^*) + \frac{1}{2}\sigma_t^2(\mathbf{x}^*)}. \quad (33)$$

We take the next stimulus $\mathbf{x}_{t+1}$ to be the maximizer of this function, which depends on both the mean and variance of the posterior over ϕ. (Note that this differs from uncertainty sampling, which uses only the posterior variance to select stimuli).

The form of the mean dependence observed in Eq. (33) depends critically on the choice of the nonlinear transformation g. For exponential g, as assumed here, the information gain is an increasing function of the mean, meaning that the algorithm will explore high-firing rate regions of the tuning curve first. For a soft-rectifying g, however, information gain turns out to be a *decreasing* function of the posterior mean (Park et al., 2014), meaning the algorithm prefers to explore moderate-to-low firing-rate regions of the tuning curve first. See Park et al. (2014) for a more thorough discussion of this issue, and along with other choices for expected utility.

3.3.5 Simulations

Fig. 4 shows a comparison of several stimulus selection methods using data simulated from a Poisson neuron with a two-dimensional tuning curve. We compared infomax stimulus selection to random sampling and uncertainty sampling, which selects the stimulus for which the posterior variance of firing rate map is maximal. We also tested the performance of a method based on a model with Gaussian (instead of Poisson) noise, for which information gain can be computed analytically. For this model, the utility does not depend on the observations, meaning that the entire sequence of stimuli can be planned out before the

experiment. Fig. 4B shows the performance of each method in terms of posterior entropy and mean squared error (average over 100 independent repetitions). Uncertainty sampling, in this case, focuses on picking stimuli only around the high peak areas of the true map, which results in slower decrease in MSE than random sampling.

4 APPLICATION: LINEAR RECEPTIVE FIELD ESTIMATION

Another important class of problems in electrophysiology experiments is linear RF estimation. This differs from tuning curve estimation in that the estimation problem is high dimensional, since RF dimensionality is equal to the number spatio-temporal elements or "pixels" in the relevant stimulus driving the neuron. This high-dimensional characterization problem is simplified, however, by the fact that the assumed function is linear. This assumption is justified by the fact that many sensory neurons have approximately linear response properties in a suitably defined input space (which may involve a nonlinear transformation of the raw stimulus (David and Gallant, 2005)). The active learning problem is to select stimuli in this high-dimensional space that are maximally informative about the neuron's weighting function. This poses unique challenges because, unlike the tuning curve problem, we cannot grid up the input space and compute the expected utility for each stimulus.

4.1 Generalized Linear Model

A popular model for linear RFs arises from the generalized linear model (GLM) (Nelder and Wedderburn, 1972; Truccolo et al., 2005; Pillow et al., 2008). This model (Fig. 5) consists of a linear filter $\mathbf{k}$, which describes how the neuron integrates the stimulus over time and space, followed by a point nonlinearity $g(\cdot)$, which transforms filter output into the neuron's response range, and exponential-family noise that captures stochasticity in the response. Typically, one regards the nonlinearity in a

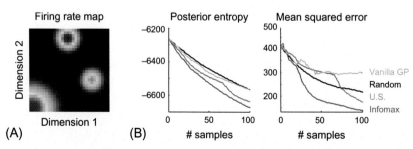

FIG. 4 Comparison of stimulus selection methods for 2D tuning curve. (A) True tuning curve in a 2D input space, with maximum of 90 sp/s and minimum of 1 sp/s. (B) The posterior entropy (left) and mean squared error (right), as a function of the number of experimental trials, for each of four methods: (1) stimulus selection under a standard or "vanilla" Gaussian process model with Gaussian noise (where stimulus selection does not depend on the observed responses); (2) random *iid* stimulus selection; (3) uncertainty sampling; and (4) infomax learning. Infomax learning exhibits the best performance in terms of both information and MSE.

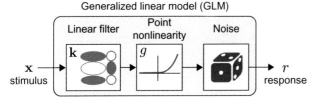

Generalized linear model (GLM)

FIG. 5 Schematic of generalized linear encoding model (GLM) for RF characterization. The model contains a weight vector **k** that linearly filters the stimulus, a nonlinear function g, and noise from an exponential family distribution.

GLM as fixed, which simplifies estimation and stimulus selection problems (Paninski, 2004).

The Poisson GLM, also known as the linear-nonlinear-Poisson (LNP) model, is given by

$$\lambda = g(\mathbf{k}^\top \mathbf{x}), \quad r|\mathbf{x} \sim \text{Poiss}(\Delta \lambda), \quad (34)$$

where $\mathbf{k}^\top \mathbf{x}$ is the dot product between the filter **k** and the stimulus **x**, the nonlinearity g ensures the spike rate λ is nonnegative, and Δ is a time bin size. The model can be extended to incorporate linear dependencies on spike-history and other covariates like the responses from other neurons (Truccolo et al., 2005; Pillow et al., 2008; Koepsell and Sommer, 2008; Kelly et al., 2010), but we will focus here on the simple case where the stimulus **x** is the only input.

4.2 Infomax Stimulus Selection for Poisson GLM

Lewi et al. (2009a) developed an infomax learning method for RF characterization under a Poisson GLM, which we will refer to as "Lewi-09." This method assumes an isotropic Gaussian prior over **k**, which leads to a posterior formed by the product of a Gaussian prior and Poisson likelihood, just as in the GP-Poisson tuning curve model considered in the previous section. We will omit details of the derivation, but the method developed in Lewi et al. (2009a) is closely related to the infomax learning for the GP-Poisson model (and in fact, was a direct source of inspiration for our work). In brief, the Lewi-09 method performs approximate MAP inference for **k** after each response and computes a Gaussian "Laplace" approximation to the posterior. When the nonlinearity g is exponential, this leads to a concise formula for the expected information gain for a candidate stimulus $\mathbf{x}^*$ (which closely resembles Eq. (33)):

$$I(r, \mathbf{k}|\mathbf{x}^*, \mathcal{D}_t) \approx \frac{1}{2} \sigma_\rho^2 e^{\mu_\rho + \frac{1}{2}\sigma_\rho^2}, \quad (35)$$

where μ_ρ and σ_ρ^2 are projections of $\mathbf{x}^*$ onto the posterior mean and covariance of **k**:

$$\mu_\rho = \boldsymbol{\mu}_t^\top \mathbf{x}^*, \quad \sigma_\rho^2 = \mathbf{x}^{*\top} \Lambda_t \mathbf{x}^* \quad (36)$$

(see Eqs. 4.14 and 4.20 in Lewi et al. (2009a) for details). An interesting consequence of this formula is that, although

the stimulus space is high-dimensional, the informativeness of a candidate stimulus depends on its linear projection onto the current posterior mean and covariance matrix.

The key contributions of Lewi et al. (2009a) include fast methods for updating the posterior mean and covariance, and efficient methods for selecting maximally informative stimuli subject to a "power" constraint (ie, $\|\mathbf{x}^*\|^2 < \text{const}$). The Lewi-09 algorithm yields substantial improvements relative to randomized *iid* (ie, "white noise") stimulus selection. Intriguingly, the authors show that in high-dimensional settings, two methods that might have been expected to work well, in fact do not: (1) the maximal eigenvector of the posterior covariance matrix; and (2) the most informative stimulus from a limited set of *iid*-sampled stimuli. Both of these methods turn out to perform approximately as poorly as random stimulus selection.

4.3 Infomax for Hierarchical RF Models

One shortcoming of the Lewi-09 method is it does not exploit prior information about the structure of neural RFs. It uses an isotropic Gaussian prior, also known as the *ridge* prior,

$$\mathbf{k} \sim \mathcal{N}(0, \rho I), \quad (37)$$

where ρ is the common prior variance of all RF coefficients. This assumes the RF elements are *a priori* independent, with prior probability mass spread out equally in all directions (see Fig. 6). By contrast, we know that RFs tend to be structured, eg, smooth and sparse in space and time. An active learning method that incorporates such structure can concentrate prior probability mass closer to the manifold of likely RF shapes and spend less time exploring regions of stimulus space that are unlikely to be informative about the RF.

To put this in Bayesian terms, the goal of infomax learning is to obtain a posterior distribution with minimal entropy. This can be achieved by either: (1) selecting informative stimuli, that is, stimuli that lead to maximal narrowing via new likelihood terms; or (2) using a prior with minimal entropy, so that less learning is required in the first place. Although this second point might seem trivial or uninteresting, we will show that it is not, as priors that encourage the forms of structure found in biological systems are not straightforward to construct, and are more difficult to use for active learning than Gaussian priors. Here we will discuss a method that uses this second strategy. Our method uses a hierarchical prior, formulated as a covariance mixture of Gaussians, that flexibly encodes statistical regularities in RF structure in a way that speeds up learning when such structure is present (eg, smoothness, sparsity, locality), but defaults to an uninformative prior when it is not.

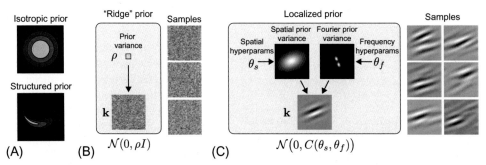

FIG. 6 Effects of the prior distributions on active learning of RFs. (A) An isotropic Gaussian prior (top) has comparatively large prior entropy, and will thus require a lot of data to achieve a concentrated posterior. A "structured" prior (*below*), however, concentrates prior probability mass closer to the region of likely RF shapes, so less data is required to concentrate the posterior if the true RF lies close to this region. (B) Graphical model for ridge regression prior, which models all RF coefficients as *iid* Gaussian. RF samples from this prior (*right*) are simply Gaussian white noise. (C) Graphical model for localized prior from Park and Pillow (2011). The prior simultaneously encourages localized support in space (left) and in the Fourier domain (*right*), with support controlled by hyperparameters θ_s and θ_f for each domain, respectively. The support depicted in this diagram assigns high prior probability to a Gabor filter (*bottom*), and samples from the prior conditioned on these hyperparameters exhibit similar spatial location, frequency content, and orientation (*right*). The full hierarchical prior consists of a mixture of these conditional distributions for hyperparameters covering all locations, frequencies, and orientations, and includes the ridge prior as a special case.

For simplicity, we use a linear-Gaussian encoding model, which can be viewed as a GLM with "identity" nonlinearity and Gaussian noise:

$$r = \mathbf{k}^\top \mathbf{x} + \epsilon, \quad \epsilon \sim \mathcal{N}(0, \sigma^2), \qquad (38)$$

where ϵ is zero-mean Gaussian noise with variance σ^2. While this provides a less accurate statistical model of spike responses than the Poisson GLM used in Lewi et al. (2009a), it simplifies the problem of computing and optimizing the posterior expectations needed for active learning.

We consider a "structured" hierarchical, conditionally Gaussian prior of the form:

$$\mathbf{k} \mid \theta \sim \mathcal{N}(0, C_\theta), \quad \theta \sim p_\theta, \qquad (39)$$

where C_θ is a prior covariance matrix that depends on hyperparameters θ, and p_θ is a hyperprior over θ. The effective prior over $\mathbf{k}$ is a mixture-of-Gaussians, also known as a *covariance mixture of Gaussians* because the component distributions are zero-mean Gaussians with difference covariances:

$$p(\mathbf{k}) = \int p(\mathbf{k}|\theta)p(\theta)d\theta = \int \mathcal{N}(0, C_\theta)\, p_\theta(\theta)d\theta. \qquad (40)$$

4.3.1 Posterior Distribution

For this model, the posterior distribution is also a mixture of Gaussians:

$$p(\mathbf{k}|\mathcal{D}_t) = \int p(\mathbf{k}|\mathcal{D}_t, \theta\, p(\theta|\mathcal{D}_t)d\theta = \int \mathcal{N}(\mu_\theta, \Lambda_\theta)\, p(\theta|X, Y)d\theta. \qquad (41)$$

with conditional mean and covariance

$$\mu_\theta = \frac{1}{\sigma^2}\Lambda_\theta X^\top Y, \quad \Lambda_\theta = \left(\frac{1}{\sigma^2}X^\top X + C_\theta^{-1}\right)^{-1}, \qquad (42)$$

and mixing distribution given by the marginal posterior:

$$p(\theta|\mathcal{D}_t) \propto p(R_t|X_t, \theta)\, p_\theta(\theta). \qquad (43)$$

The marginal posterior is proportional to the product of the conditional marginal likelihood or "evidence" $p(R_t|X_t, \theta)$ and the hyperprior $p_\theta(\theta)$. For the linear-Gaussian encoding model, the conditional evidence has an analytic form:

$$p(R_t|X_t, \theta) = (\sqrt{2\pi}\sigma)^{-t}|\Lambda_\theta|^{\frac{1}{2}}|C_\theta|^{-\frac{1}{2}}\exp\left[\frac{1}{2}\left(\mu_\theta^\top \Lambda_\theta^{-1}\mu_\theta - m^\top L^{-1}m\right)\right],$$
$$L = \sigma^2(X_t^\top X_t)^{-1}, \quad m = \frac{1}{\sigma^2}LX_t^\top R_t, \qquad (44)$$

where $X = [\mathbf{x}_1, \ldots, \mathbf{x}_t]^\top$ is the $t \times d$ design matrix and $R = (r_1, \ldots, r_t)^\top$ is the vector of observed responses.

As mentioned earlier, active learning confers no benefits over nonadaptive stimulus selection when the response is governed by a linear-Gaussian model and a Gaussian prior, due to the fact that the posterior covariance is response-independent (Eq. 42). However, this response-independence does not hold for models with mixture-of-Gaussian priors, as the response R_t affects the posterior via the conditional evidence (Eq. 43). Intuitively, as the responses come in, the conditional evidence tells us what hyperparameter settings (ie, which prior distributions) are best. As we will show, active learning under a structured hierarchical prior can confer substantial advantages over methods based on an isotropic prior.

4.3.2 Localized RF Prior

We illustrate this approach using a flexible prior designed to capture *localized* structure in neural RFs (Park and Pillow, 2011, 2012). The prior, introduced as part of an empirical Bayesian RF estimation method called *automatic locality determination* (*ALD*), seeks to exploit the observation that neural RFs tend to be localized in both space-time and spatio-temporal frequency. Locality in space-time means that neurons typically integrate sensory input over a limited region of space and time. Locality in frequency, on the other hand, means that neurons tend to respond to a restricted range of frequencies and orientations, or equivalently, that the Fourier transform of a neural RF tends to be band-limited.

The ALD prior captures localized structure using a covariance matrix $C(\theta_s, \theta_f)$ controlled by two sets of hyperparameters: (1) spatial hyperparameters θ_s, which define an elliptical region of space-time where the RF has nonzero prior variance; and (2) frequency hyperparameters θ_f, which define a pair of elliptical regions in Fourier space where the Fourier transform of the RF has nonzero prior variance. The full covariance matrix C is given by a product of diagonal covariance matrices with a Fourier change-of-basis operator sandwiched between them (Park and Pillow, 2011). Fig. 6C shows a graphical illustration of this prior for one setting of the hyperparameters. The spatial support (left) covers a large region in the middle of a two-dimensional stimulus image, while the frequency support (right) includes two small regions with a particular orientation in the 2D Fourier plane, giving rise to a small range of oriented, smooth, band-pass RFs. Six samples from the conditional prior $p(\mathbf{k}|\theta_s, \theta_f)$ all exhibit this common tendency (Fig. 6C, right).

The full prior distribution $p(\mathbf{k})$ is a continuous mixture of Gaussians (Eq. 40), where each mixing component is a zero-mean Gaussian with covariance $C(\theta_s, \theta_f)$, and the mixing distribution $p(\theta_s, \theta_f)$ assigns prior probability to a broad range of possible hyperparameter values, including settings of θ_s and θ_f for which there is no localized structure in either space-time or frequency. The prior does not rule out any RFs a priori because it includes the ridge prior as one of its components (Fig. 6B). However, many of the mixing components *do* have restricted spatial or Fourier support, meaning that probability mass is more concentrated on the manifold of likely RFs (as depicted schematically in Fig. 6A). The key virtue of this prior in the context of active learning is that, for neurons with localized, smooth, oriented, band-pass, or other relevant structure, the support of the marginal posterior $p(\theta|\mathcal{D}_t)$ shrinks down relatively quickly to a restricted region of hyperparameter space, ruling out vast swaths of the underlying parameter space. This allows the posterior entropy over $\mathbf{k}$ to shrink far more quickly than could be achieved under an isotropic prior.

4.3.3 Active Learning With Localized Priors

To develop a practical method for active learning under the localized prior, we still need two ingredients: an efficient way to update the posterior distribution $p(\mathbf{k}|\mathcal{D}_t)$ after each trial, and a tractable method for computing and maximizing the expected information gain $I(r, \mathbf{k}|\mathbf{x}^*, \mathcal{D}_t)$. The posterior distribution contains an intractable integral over the hyperparameters (Eq. 41) and lacks the log-concavity property that motivated the Gaussian approximation-based methods developed in Lewi et al. (2009a). However, we can instead exploit the conditionally Gaussian structure of the posterior

to develop a fully Bayesian approach using Markov Chain Monte Carlo (MCMC) sampling. We approximate the posterior using a set of hyperparameter samples $\{\theta^{(i)}\}$, $i \in \{1,...,m\}$, drawn from the marginal posterior $p(\theta|\mathcal{D}_t)$ (Eq. 43) via a standard MCMC sampling technique. Here each sample represents a full set of space-time and frequency hyperparameters for the localized prior, $\theta^{(i)} = \{\theta_s^{(i)}, \theta_f^{(i)}\}$. The posterior can then be approximated as:

$$p(\mathbf{k}|\mathcal{D}_t) \approx \frac{1}{m}\sum_i p(\mathbf{k}|\mathcal{D}_t, \theta^{(i)}) = \frac{1}{m}\sum_i \mathcal{N}(\mu^{(i)}, \Lambda^{(i)}) \quad (45)$$

with mean and covariance of each Gaussian as given in Eq. (42). To update the posterior rapidly after each trial, we use a version of the *resample-move particle filter*, which involves resampling a full set of "particles" $\{\theta^{(i)}\}$ using the new data from each trial and then performing a small number of additional MCMC steps (Gilks and Berzuini, 2001). The main computational bottleneck is the cost of updating the conditional posterior mean $\mu^{(i)}$ and covariance $\Lambda^{(i)}$ for each particle $\theta^{(i)}$, which requires inverting of a $d \times d$ matrix. However, this cost is independent of the amount of data, and particle updates can be performed efficiently in parallel, since the particles do not interact, except for the evaluation of mutual information. Full details are provided in Park and Pillow (2012).

We can select the most informative stimulus for the next trial using a moment-based approximation to the posterior. Although the entropy of a mixture-of-Gaussians has no tractable analytic form, the marginal mean and covariance of the (sample-based) posterior is given by:

$$\tilde{\mu}_t = \frac{1}{m}\sum \mu_t^{(i)}, \qquad \tilde{\Lambda}_t = \frac{1}{m}\sum_{i=1}^m \left(\Lambda_t^{(i)} + \mu_t^{(i)}\mu_t^{(i)\top}\right) - \tilde{\mu}_t\tilde{\mu}_t^\top.$$

$$(46)$$

The entropy of the posterior is upper-bounded by $\frac{1}{2}|2\pi e\tilde{\Lambda}_t|$ due to the fact that Gaussians are maximum-entropy distributions for given covariance. We select the next stimulus to be proportional to the maximum-variance eigenvector of $\tilde{\Lambda}_t$, which is the most informative stimulus for a linear Gaussian model and a power constraint (Lewi et al., 2009a). This selection criterion is not guaranteed to maximize mutual information under the true posterior, but it has the heuristic justification that it selects stimulus directions for which the posterior has maximal variance, making it a form of parameter uncertainty sampling.

It is worth noting from Eq. (46) that directions of large posterior variance can arise in at least two different ways: (1) they can be directions of large variance for all conditional covariances $\Lambda^{(i)}$, meaning that all hyperparameter samples assign high posterior uncertainty over the component of $\mathbf{k}$ in this direction of stimulus space, or (2) they can be directions in which the conditional means $\mu^{(i)}$ are highly dispersed, meaning the conditional posteriors from different hyperparameter samples disagree about the mean of **k** along this direction. In either scenario, it seems intuitively reasonable that presenting a stimulus along this direction will reduce variance in the marginal posterior, but a more careful theoretical treatment is certainly warranted. In practice, we find that the method performs well both for simulated data from a Poisson GLM and for neural data from primate V1.

4.4 Comparison of Methods for RF Estimation

4.4.1 Simulated Data

We compared our method and the Lewi-09 method using simulated data from a Poisson-GLM with exponential nonlinearity and RF given by a Gabor filter (20 × 20 pixels, shown in Fig. 7). This is the encoding model assumed by the Lewi-09 method, whereas our method (which assumes linear-Gaussian encoding) exhibits model mismatch. For the Lewi-09 method, we used a ridge prior over **k**, with prior variance set by maximizing marginal likelihood for a small dataset. We tested implementations with two different numbers of particles (10 and 100), to examine the tradeoff between computational complexity and accuracy, and used the angular difference (in degrees) between the true and estimated RF as a performance measure.

Fig. 7A shows estimation error as a function of the number of trials. The localized-prior estimate exhibits faster reduction in error. Moreover, the localized algorithm performed better, suggesting that accurately preserving uncertainty over the hyperparameters aided performance. Fig. 7B shows the estimates obtained by each method after a different number of trials. Notice that the estimate with 100 hyperparameter particles after 200 trials is almost identical to the true filter, where the number of trials (200) is substantially lower than the dimensionality of the filter itself ($d = 400$).

4.4.2 Application to Neural Data

We also compared methods using an off-line analysis of real neural data (Rust et al., 2005) from a simple cell in primate V1. The stimuli were 1D white noise "flickering bars," with 16 spatial bars aligned with the cell's preferred orientation. We used 16 time bins to model the temporal RF, resulting in an a 256-dimensional stimulus space. We performed the off-line active learning by extracting the stimuli from the entire 46 minutes of recording of the experimental data. We computed the expected information gain on each trial from presenting each of the remaining stimuli, without access to the neuron's actual response. We tested our ALD-based active learning with 10 hyperparameter samples, and selected the most informative stimuli from ≈276,000 possible stimuli on each trial.

Fig. 8B shows the average angular difference between the maximum likelihood estimate (computed with the entire dataset) and the estimate obtained by each active learning method, as a function of trial number. The ALD-based method decreased the angular difference by 45 degrees with only 160 stimuli, while the Lewi-09 method required four times more data to achieve the same accuracy.

The most computationally expensive step in the algorithm is the eigendecomposition of the 256 × 256 posterior covariance matrix $\tilde{\Lambda}$, which took 30 ms on a circa-2012 quad-core Mac Pro. In total, it took less than 60 ms to compute the optimal stimulus in each trial using a simple implementation of our algorithm, which we expect to be fast enough for use in real-time neurophysiology experiments.

5 DISCUSSION

We have discussed methods for stimulus selection in closed-loop neurophysiology experiments based on Bayesian active learning, also known as Bayesian adaptive experimental design. The key ingredients of any such method are: (1) a response model; (2) a prior distribution

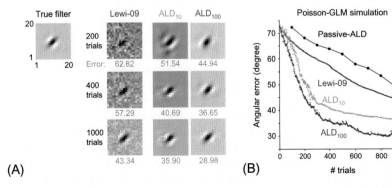

(A) (B)

FIG. 7 Simulated RF estimation experiment using a Poisson neuron with a Gabor filter RF and exponential nonlinearity. (A) Left: true RF is a 20 × 20 pixel Gabor filter, yielding a 400-dimensional stimulus space. Right: RF estimates obtained by Lewi-09 method (*blue*) and hierarchical model with localized ("ALD") prior using 10 (*orange*) or 100 (*red*) samples to represent the posterior, with posterior mean estimate shown after 200, 400, and 1000 trials of active learning. Grey numbers below indicate angular error in estimate (degree). (B) Angular error vs. number of stimuli for three active learning methods, along with ALD inference using "passive" random *iid* stimulus selection (*black*). Traces show average error over 20 repetitions.

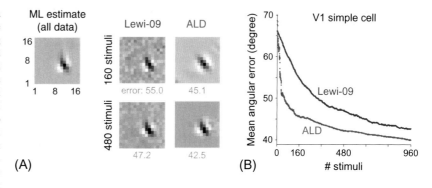

FIG. 8 Active learning of a linear RF using data from a primate V1 simple cell. Original data were recorded in response to a 1D white noise "flickering bars" stimulus aligned with the neuron's preferred orientation (see Rust et al. (2005)). We simulated active learning experiments via an offline analysis of a fixed dataset, where active learning methods had access to the set of stimuli but not the responses, and could reorder stimulus presentation according to expected utility. (A) Left: maximum likelihood estimate computed from entire 46-min dataset (166K stimuli sampled at 100 Hz). Right: RF estimates after 10 and 30 s of data selected using the Lewi-09 and localized (ALD) active learning methods. (B) Mean angular error between active learning estimate and all-data maximum likelihood estimate as a function of the number of stimuli.

over model parameters; and (3) a utility function. The first two ingredients define the posterior distribution over parameters and the predictive distribution over future responses. The expectation of the loss function over the joint distribution over parameters and future responses defines the expected utility of a stimulus; the methods we have considered operate by greedily selecting the stimulus with maximal expected utility on each trial, given the data collected so far in the experiment. Finally, we discussed the details for two prominent application domains: tuning curves (or "firing rate maps"), and linear RFs. For tuning curves, we discussed methods for both parametric and nonparametric (Park et al., 2014) models; for linear RFs, we examined methods based on Poisson models with simple priors (Lewi et al., 2009a) and Gaussian models with structured hierarchical priors (Park and Pillow, 2012). While these methods hold great promise for improving the efficiency of neurophysiology experiments, there remain a variety of challenges to overcome. We review several of these challenges below.

5.1 Adaptation

The methods we have described all assumed that the response on each trial was conditionally independent of responses on previous trials (Eqs. 15 and 34). Clearly, this assumption is violated by the fact that neurons adapt: the response after a series of large-response trials may differ from the response to the same stimulus after a series of weak-response trials. An active learning method that models responses as conditionally independent may misattribute the effect of adaptation to a reduced response to certain stimuli.

Fortunately, it is straightforward to incorporate a simple form of response adaptation under the conditionally Poisson models discussed in this chapter (Truccolo et al., 2005). We can augment the model with a set of weights $\mathbf{h}$ that capture dependence of spike rate λ on the recent response history: $\lambda = g(\mathbf{k}^\top \mathbf{x} + \mathbf{h}^\top \mathbf{r}_{hist})$ for the RF model (Lewi et al.,

2009a), or $\lambda = g(\phi(\mathbf{x}) + \mathbf{h}^\top \mathbf{r}_{hist})$ for the tuning curve model (Park and Pillow, 2011), where $\mathbf{r}_{hist}$ is some vector representation of response history at the current time.

This approach may not be sufficient to capture *stimulus specific* adaptation effects (eg, differential adaptation to inputs that would ordinarily elicit similar spike rates), which suggests one key avenue for future research. However, it is worth noting that infomax learning and related Bayesian active learning methods typically interleave stimuli that elicit a wide range of spike rates (see Fig. 2), which stands in contrast to staircase methods, which tend move slowly across stimulus space, or methods that seek a maximal response (Nelken et al., 1994; deCharms et al., 1998; Földiák, 2001) or to find a particular isoresponse contour (Gollisch and Herz, 2012; Horwitz and Hass, 2012). Thus, Bayesian methods may in general be less susceptible to systematic biases induced by adaptation than other adaptive selection methods.

5.2 Greediness

A second possible problem with the methods we have discussed is that they are "greedy": they select the stimulus that maximizes expected utility *on each trial*. This may be sub-optimal compared to strategies that select stimuli with an eye to maximizing expected utility some finite number of trials in the future (King-Smith et al., 1994). The computational difficulty of computing and maximizing expected utility multiple trials into the future makes this a challenging problem to undertake. Moreover, a technical result in Paninski (2005) shows that greedy infomax methods are still provably better than standard methods under certain consistency conditions.

A different and more unavoidable greediness problem arises in the setting of linear RF estimation (Section 4). Specifically, if the filter $\mathbf{k}$ contains a temporal component, such as one might characterize using reverse correlation (Chichilnisky, 2001), then the methods described in

Section 4 will neglect a large fraction of the available information. In this setting, the response depends on a temporal convolution of the filter with a stimulus movie. The effective stimulus at each time is a shifted copy of the previous stimulus plus a single new stimulus frame, so one must consider the response at multiple lags in order to accurately quantify the information each stimulus provides about the RF. Lewi and colleagues addressed this problem by extending the Lewi-09 method to the selection of maximally informative stimulus *sequences*, taking into account the mutual information between the model RF and an extended sequence of responses (Lewi et al., 2011). A comparable extension of our method based on structured priors (Park and Pillow, 2012) has not yet been undertaken, and represents an opportunity for future work.

5.3 Model Specification

A third important concern for Bayesian active learning methods is the specification or selection of the neural response model. Although it is possible for methods based on a misspecified model to outperform well-specified models (eg, as we showed in Section 4, using a method based on Gaussian noise for a Poisson neuron), there are few theoretical guarantees, and it is possible to find cases of model mismatch where adaptive methods are inferior to random *iid* methods (Paninski, 2005). It is therefore important to consider the accuracy and robustness of neural response models used for active learning.

One simple model selection problem involves the setting of hyperparameters governing the prior or response model, such the GP covariance function for tuning curve estimation (Section 3.3). In Park et al. (2014), we set hyperparameters by maximizing marginal likelihood after each trial using the data collected so far in the experiment, an approach that common in Bayesian optimization and active learning. This is not strictly justified in a Bayesian framework, but empirically it performs better than fixing hyperparameters governing smoothness and marginal variance a priori. A more principled approach would be to place a prior distribution over hyperparameters and use sampling-based inference so that the expected utility incorporates uncertainty about the model (Roy and McCallum, 2001). Lewi et al. explored several specific forms of model mismatch for RF estimation (Lewi et al., 2009a), and DiMattina and Zhang proposed methods aimed at the dual objective of parameter estimation and model comparison (DiMattina and Zhang, 2011, 2013).

5.4 Future Directions

One natural direction for future research will be to combine the Poisson-GLM response model from Lewi et al. (2009a) with the localized RF priors from Park and Pillow (2012) to build improved methods for RF estimation. This synthesis would combine the benefits of a more accurate neural model with a more informative and lower-entropy prior, provided that computational challenges can be overcome. A related idea was proposed in Lewi et al. (2009b), which altered the utility function to maximize information along particular manifolds defined by parametric RF shapes (eg, the manifold of Gabor filters within the larger space of all possible RFs). Both methods offer a desirable tradeoff between efficiency and robustness: they are set up to learn quickly when the true RF exhibits some form of low-dimensional structure, but still yield consistent estimates for arbitrary RFs. A promising opportunity for future work will be to design flexible priors to incorporate other forms of RF structure, such as space-time separability (Park and Pillow, 2013), sparsity (Seeger et al., 2007), or structured sparsity (Wu et al., 2014).

Another direction for improved active learning is the design of more flexible and accurate neural response models. Preliminary work in this direction has focussed on models with nonlinear input transformations (Lewi et al., 2009a), and "overdispersed" response noise (Park et al., 2013; Goris et al., 2014). Other recent work has focussed on hierarchical models for population responses, in which the responses from all previously recorded neurons are used to help determine the most informative stimuli for characterizing each subsequent neuron (Kim et al., 2014). In future work, we hope to extend these methods to simultaneous multineuron recordings so that stimuli provide maximal information about an entire population of neurons, including their correlations. Taken together, we believe these methods will greatly improve the speed and accuracy of neural characterization and allow for ambitious, high-throughput neurophysiology experiments that are not possible with standard, nonadaptive methods. We feel these methods will be especially useful in higher cortical areas, where neurons exhibit nonlinear "mixed" selectivity in high-dimensional stimulus spaces where tuning is poorly understood.

ACKNOWLEDGMENTS

This work was supported by the Sloan Foundation (JP), McKnight Foundation (JP), Simons Collaboration on the Global Brain Award (JP), NSF CAREER Award IIS-1150186 (JP), the Gatsby Charitable Trust (MP), and a grant from the NIH (NIMH grant MH099611 JP). We thank N. Rust and T. Movshon for V1 data, and N. Roy and C. DiMattina for helpful comments on this manuscript.

REFERENCES

Aguera y Arcas, B., Fairhall, A.L., 2003. What causes a neuron to spike? Neural Comput. 15 (8), 1789–1807.

Benda, J., Gollisch, T., Machens, C.K., Herz, A.V., 2007. From response to stimulus: adaptive sampling in sensory physiology. Curr. Opin. Neurobiol. 17 (4), 430–436.

Bernardo, J.M., 1979. Expected information as expected utility. Ann. Stat. 7 (3), 686–690.

Bialek, W., Rieke, F., de Ruyter van Steveninck, R.R., Warland, D., 1991. Reading a neural code. Science 252, 1854–1857.

Bölinger, D., Gollisch, T., 2012. Closed-loop measurements of iso-response stimuli reveal dynamic nonlinear stimulus integration in the retina. Neuron 73 (2), 333–346.

Chaloner, K., Verdinelli, I., 1995. Bayesian experimental design: a review. Stat. Sci. 10 (3), 273–304.

Chichilnisky, E.J., 2001. A simple white noise analysis of neuronal light responses. Netw. Comput. Neural Syst. 12, 199–213.

Cohn, D.A., Ghahramani, Z., Jordan, M.I., 1996. Active learning with statistical models. J. Artif. Intell. Res. 4, 129–145.

Cronin, B., Stevenson, I.H., Sur, M., Körding, K.P., 2010. Hierarchical Bayesian modeling and Markov chain Monte Carlo sampling for tuning-curve analysis. J. Neurophysiol. 103 (1), 591–602. http://dx.doi.org/10.1152/jn.00379.2009. http://jn.physiology.org/content/103/1/591.short.

Cunningham, J.P., Yu, B.M., Shenoy, K.V., Sahani, M., 2008. Inferring neural firing rates from spike trains using Gaussian processes. In: Advances in Neural Information Processing Systems 20, pp. 329–336.

David, S.V., Gallant, J.L., 2005. Predicting neuronal responses during natural vision. Netw. Comput. Neural Syst. 16 (2), 239–260.

deCharms, R.C., Blake, D.T., Merzenich, M.M., 1998. Optimizing sound features for cortical neurons. Science 280 (5368), 1439–1443.

DiMattina, C., Zhang, K., 2011. Active data collection for efficient estimation and comparison of nonlinear neural models. Neural Comput. 23 (9), 2242–2288.

DiMattina, C., Zhang, K., 2013. Adaptive stimulus optimization for sensory systems neuroscience. Front. Neural Circuits 7, 101.

Földiák, P., 2001. Stimulus optimisation in primary visual cortex. Neurocomputing 38, 1217–1222.

Gilks, W.R., Berzuini, C., 2001. Following a moving target—Monte Carlo inference for dynamic Bayesian models. J. R. Stat. Soc. Ser. B 63 (1), 127–146.

Gollisch, T., Herz, A.V., 2012. The iso-response method: measuring neuronal stimulus integration with closed-loop experiments. Front. Neural Circuits 6, 104.

Goris, R.L.T., Movshon, J.A., Simoncelli, E.P., 2014. Partitioning neuronal variability. Nat. Neurosci. 17 (6), 858–865. http://dx.doi.org/10.1038/nn.3711.

Horwitz, G.D., Hass, C.A., 2012, 06. Nonlinear analysis of macaque v1 color tuning reveals cardinal directions for cortical color processing. Nat. Neurosci. 15 (6), 913–919.

Kapoor, A., Horvitz, E., Basu, S., 2007. Selective supervision: guiding supervised learning with decision-theoretic active learning. In: International Joint Conference on Artificial Intelligence.

Kass, R., Raftery, A., 1995. Bayes factors. J. Am. Stat. Assoc. 90, 773–795.

Kelly, R.C., Smith, M.A., Kass, R.E., Lee, T.S., 2010. Local field potentials indicate network state and account for neuronal response variability. J. Comput. Neurosci. 29 (3), 567–579.

Kim, W., Pitt, M.A., Lu, Z.L., Steyvers, M., Myung, J.I., 2014, 2015/02/09. A hierarchical adaptive approach to optimal experimental design. Neural Comput. 26 (11), 2465–2492. http://dx.doi.org/10.1162/NECO_a_00654.

King-Smith, P.E., Grigsby, S.S., Vingrys, A.J., Benes, S.C., Supowit, A., 1994. Efficient and unbiased modifications of the quest threshold method: theory, simulations, experimental evaluation and practical implementation. Vis. Res. 34 (7), 885–912.

Koepsell, K., Sommer, F.T., 2008. Information transmission in oscillatory neural activity. Biol. Cybernet. 99 (4), 403–416.

Kontsevich, L.L., Tyler, C.W., 1999. Bayesian adaptive estimation of psychometric slope and threshold. Vision Res. 39, 2729–12737.

Krause, A., Singh, A., Guestrin, C., 2008. Near-optimal sensor placements in Gaussian processes: theory, efficient algorithms and empirical studies. J. Mach. Learn. Res. 1532-4435. 9, 235–284. http://portal.acm.org/citation.cfm?id=1390689.

Kuck, H., de Freitas, N., Doucet, A., 2006. SMC samplers for Bayesian optimal nonlinear design. In: Proceedings of the IEEE Nonlinear Statistical Signal Processing Workshop. IEEE, pp. 99–102.

Lewi, J., Butera, R., Paninski, L., 2009a. Sequential optimal design of neurophysiology experiments. Neural Comput. 21 (3), 619–687.

Lewi, J., Butera, R., Schneider, D.M., Woolley, S., Paninski, L., 2009b. Designing neurophysiology experiments to optimally constrain receptive field models along parametric submanifolds. In: Koller, D., Schuurmans, D., Bengio, Y., Bottou, L. (Eds.), Advances in Neural Information Processing Systems 21. Curran Associates, Inc., Red Hook, NY, pp. 945–952.

Lewi, J., Schneider, D.M., Woolley, S.M.N., Paninski, L., 2011, Feb. Automating the design of informative sequences of sensory stimuli. J. Comput. Neurosci. 30 (1), 181–200. http://dx.doi.org/10.1007/s10827-010-0248-1.

Lewis, D.D., Gale, W.A., 1994. A sequential algorithm for training text classifiers. In: Proceedings of the ACM SIGIR Conference on Research and Development in Information Retrieval. Springer-Verlag, pp. 3–12.

Lindley, D., 1956. On a measure of the information provided an experiment. Ann. Math. Stat. 27, 986–1005.

Luttrell, S.P., 1985. The use of transinformation in the design of data sampling scheme for inverse problems. Inverse Prob. 1, 199–218.

Machens, C.K., Gollisch, T., Kolesnikova, O., Herz, A.V.M., 2005. Testing the efficiency of sensory coding with optimal stimulus ensembles. Neuron 47 (3), 447–456. http://dx.doi.org/10.1016/j.neuron.2005.06.015.

Mackay, D., 1992. Information-based objective functions for active data selection. Neural Comput. 4, 589–603.

Macke, J.H., Gerwinn, S., White, L.E., Kaschube, M., Bethge, M., 2011. Gaussian process methods for estimating cortical maps. Neuroimage 56 (2), 570–581. http://dx.doi.org/10.1016/j.neuroimage.2010.04.272.

Müller, P., Parmigiani, G., 1996. Bayesian Analysis in Statistics and Econometrics: Essays in Honor of Arnold Zellner. Wiley, New York, pp. 397–406.

Nelder, J.A., Wedderburn, R.W., 1972. Generalized linear models. J. R. Stat. Soc. Ser. A 135 (3), 370–384.

Nelken, I., Prut, Y., Vaadia, E., Abeles, M., 1994. In search of the best stimulus: an optimization procedure for finding efficient stimuli in the cat auditory cortex. Hear. Res. 72, 237–253.

O'Connor, K.N., Petkov, C.I., Sutter, M.L., 2005. Adaptive stimulus optimization for auditory cortical neurons. J. Neurophysiol. 94 (6), 4051–4067.

Paninski, L., 2003. Design of experiments via information theory. Adv. Neural Inform. Process. Syst. 16, 1319–1326.

Paninski, L., 2004. Maximum likelihood estimation of cascade point-process neural encoding models. Netw. Comput. Neural Syst. 15, 243–262.

Paninski, L., 2005. Asymptotic theory of information-theoretic experimental design. Neural Comput. 17 (7), 1480–1507.

Park, M., Horwitz, G., Pillow, J.W., 2011. Active learning of neural response functions with Gaussian processes. In: Shawe-Taylor, J., Zemel, R., Bartlett, P., Pereira, F., Weinberger, K. (Eds.), Advances in Neural Information Processing Systems 24. pp. 2043–2051.

Park, M., Pillow, J.W., 2011. Receptive field inference with localized priors. PLoS Comput. Biol. 7 (10), e1002219. http://dx.doi.org/10.1371%2Fjournal.pcbi.1002219.

Park, M., Pillow, J.W., 2012. Bayesian active learning with localized priors for fast receptive field characterization. In: Bartlett, P., Pereira, F., Burges, C., Bottou, L., Weinberger, K. (Eds.), Advances in Neural Information Processing Systems 25. pp. 2357–2365. http://books.nips.cc/papers/files/nips25/NIPS2012_1144.pdf.

Park, M., Pillow, J.W., 2013. Bayesian inference for low rank spatiotemporal neural receptive fields. Adv. Neural Inform. Process. Syst. 26, 2688–2696. http://media.nips.cc/nipsbooks/nipspapers/paper_files/nips26/1259.pdf.

Park, M., Weller, J.P., Horwitz, G.D., Pillow, J.W., 2013. Adaptive estimation of firing rate maps under super-Poisson variability. In: Computational and Systems Neuroscience (CoSyNe) Annual Meeting.

Park, M., Weller, J.P., Horwitz, G.D., Pillow, J.W., 2014. Bayesian active learning of neural firing rate maps with transformed Gaussian process priors. Neural Comput. 26 (8), 1519–1541. http://www.mitpressjournals.org/doi/abs/10.1162/NECO_a_00615.

Pillow, J.W., Ahmadian, Y., Paninski, L., 2011, Jan. Model-based decoding, information estimation, and change-point detection techniques for multineuron spike trains. Neural Comput. 23 (1), 1–45. http://dx.doi.org/10.1162/NECO_a_00058.

Pillow, J.W., Shlens, J., Paninski, L., Sher, A., Litke, A.M., Chichilnisky, E.J., Simoncelli, E.J., 2008. Spatio-temporal correlations and visual signaling in a complete neuronal population. Nature 454, 995–999.

Rad, K.R., Paninski, L., 2010. Efficient, adaptive estimation of two-dimensional firing rate surfaces via Gaussian process methods. Netw. Comput. Neural Syst. 21 (3–4), 142–168.

Rasmussen, C., Williams, C., 2006. Gaussian Processes for Machine Learning. MIT Press, Cambridge, MA.

Rieke, F., Warland, D., de Ruyter van Steveninck, R., Bialek, W., 1996. Spikes: Exploring the Neural Code. MIT Press, Cambridge, MA. ISBN 0-262-18174-6.

Roy, N., Mccallum, A., 2001. Toward optimal active learning through sampling estimation of error reduction. In: Proceedings of the 18th International Conference on Machine Learning. Morgan Kaufmann, San Francisco, CA, pp. 441–448.

Rust, N.C., Schwartz, O., Movshon, J.A., Simoncelli, E.P., 2005. Spatio-temporal elements of macaque V1 receptive fields. Neuron 46 (6), 945–956. http://dx.doi.org/10.1016/j.neuron.2005.05.021.

Schwartz, O., Pillow, J.W., Rust, N.C., Simoncelli, E.P., 2006. Spike-triggered neural characterization. J. Vis. 1534-73626 (4), 484–507. http://journalofvision.org/6/4/13/.

Seeger, M., Gerwinn, S., Bethge, M., 2007. Bayesian inference for sparse generalized linear models. In: Machine Learning: ECML. Springer, New York.

Truccolo, W., Eden, U.T., Fellows, M.R., Donoghue, J.P., Brown, E.N., 2005. A point process framework for relating neural spiking activity to spiking history, neural ensemble and extrinsic covariate effects. J. Neurophysiol. 93 (2), 1074–1089.

Victor, J.D., Nirenberg, S., 2008. Indices for testing neural codes. Neural Comput. 20 (12), 2895–2936. http://dx.doi.org/10.1162/neco.2008.10-07-633.

Watson, A., Pelli, D., 1983. QUEST: a Bayesian adaptive psychophysical method. Percept. Psychophys. 33, 113–120.

Wu, A., Park, M., Koyejo, O.O., Pillow, J.W., 2014. Sparse Bayesian structure learning with dependent relevance determination priors. In: Ghahramani, Z., Welling, M., Cortes, C., Lawrence, N., Weinberger, K. (Eds.), Advances in Neural Information Processing Systems 27. Curran Associates, Inc., pp. 1628–1636. http://papers.nips.cc/paper/5233-sparse-bayesian-structure-learning-with-dependent-relevance-determination-priors.pdf

Yu, B.M., Cunningham, J.P., Santhanam, G., Ryu, S.I., Shenoy, K.V., Sahani, M., 2009. Gaussian-process factor analysis for low-dimensional single-trial analysis of neural population activity. J. Neurophysiol. 102 (1), 614.

Chapter 2

Information Geometric Analysis of Neurophysiological Data

M. Tatsuno

The University of Lethbridge, Lethbridge, AB, Canada

1 INTRODUCTION

Understanding how information is represented and processed in the brain is one of the most important questions in neuroscience. Recent advances in experimental technology, such as multielectrode recording and calcium imaging, have enabled recording of a large number of neurons simultaneously from a freely behaving animal (Buzsaki, 2004; Chapin et al., 1999; Davidson et al., 2009; Dragoi and Tonegawa, 2013; Euston et al., 2007; Hoffman and McNaughton, 2002; Kudrimoti et al., 1999; Laubach et al., 2000; Peyrache et al., 2009; Tatsuno, 2015; Tatsuno et al., 2006; Wilson and McNaughton, 1993). A number of statistical methods have been developed and have been applied to analyze the neurophysiological data. These methods can be broadly categorized into a single-neuron, two-neuron, and multineuron methods. Briefly, the mean firing rate is the most commonly used single-neuron measure (Maass and Bishop, 1998). It can be calculated as a temporal average of the spike counts in the single trial, r,

$$r = \frac{C_{sp}}{T}, \tag{1}$$

where C_{sp} is the spike count and T is an interval. It can be also calculated as an average over multiple trials, $\eta(t)$,

$$\eta(t) = \frac{1}{\Delta t} \frac{C(t; t + \Delta t)}{n}, \tag{2}$$

where n is the number of repetition and $C(t; t + \Delta t)$ is the spike count between time t and $t + \Delta t$. When the firing rate $\eta(t)$ is calculated in response to stimulus or event, it is reported as a peristimulus time histogram or a perievent time histogram. Historically both r and $\eta(t)$ provide the basis of the single-neuron doctrine that spike frequency corresponds to certainty (Adrian, 1928; Barlow, 1972). In the study of memory consolidation, the firing rate r was the first to be used for the detection of memory reactivation during sleep (Pavlides and Winson, 1989). Other useful single-neuron measures are an interspike intervals (ISI) histogram and auto-correlations (AC). ISI is represented as a histogram of the time intervals of all successive spikes. AC is a special case of cross-correlations (CCs) in which a signal is cross-correlated to itself.

The most popular two-neuron measure would be the CC. It is defined as

$$CC(\tau) = \sum_{n=-\infty}^{\infty} f(n + \tau) g(n), \tag{3}$$

where $f(n + \tau)$ and $g(n)$ are the spike count at the $(n + \tau)$th and nth bins, respectively. Essentially, CC is a plot that shows how one neuron fires with reference to the other neuron. CC has been applied to spike train analysis and provided evidence that experience modifies the temporal relationship of spike trains between two neurons (Euston et al., 2007; Skaggs and McNaughton, 1996; Wilson and McNaughton, 1994).

Neural interactions are not necessarily limited to two neurons. In fact, multineuronal sequential activity has been reported in various brain areas (Tatsuno, 2015). One of the most straightforward measures to detect multineuronal sequences would be template matching, in which the sequences that occur more frequently are detected by matching the template patterns (Euston et al., 2007; Ikegaya et al., 2004; Lee and Wilson, 2002; Louie and Wilson, 2001; Tatsuno et al., 2006). In addition, it has been conjectured that the statistical structure of firing patterns distributed across multiple neurons plays a key role in neural information processing (Macke et al., 2011). Recently, the approaches based on the maximum entropy model (Jaynes, 1957) have attracted attention (Cocco et al., 2009; Schneidman et al., 2006; Shlens et al., 2006, 2009; Tang et al., 2008). They suggested that pairwise interactions are sufficient to describe the statistical structure of firing patterns. In contrast, more recent studies have suggested that there were significant higher-order interactions

Closed Loop Neuroscience. http://dx.doi.org/10.1016/B978-0-12-802452-2.00002-0

beyond second-order interactions in neural activity (Ganmor et al., 2011; Montani et al., 2009; Ohiorhenuan et al., 2010; Shimazaki et al., 2012; Yu et al., 2011).

An information-geometric (IG) method for neural spikes is another promising multineuronal approach that is closely related to the maximum entropy approach (Nakahara and Amari, 2002). It is based on a theory that was developed for investigating a space of probability distributions (Amari, 1985; Amari and Nagaoka, 2000). It has been demonstrated that IG measures have desirable properties for spike train analysis and provide powerful statistical tools for neuroscience (Amari, 2001; Nakahara and Amari, 2002). In this chapter, we provide an introduction of the IG method and review how it has been applied to analyze neurophysiological data. In Section 2, we introduce the essential features of information geometry. In Section 3, we describe IG analysis of multineuronal spike data. In Section 4, we describe the relationship between the IG measures and neural parameters such as connection weights and the external inputs. In Section 5, we discuss the extension of IG method to neural signals that are influenced by correlated inputs and that are nonstationary. In Section 6, we also discuss IG analysis of closed-loop neuroscience, such as brain-computer interface (BCI). We then conclude the chapter in Section 7.

2 INTRODUCTION OF THE IG METHOD

In this section, we provide an intuitive introduction to the IG method. A more detailed and mathematically rigorous description can be found elsewhere (Amari, 1985; Amari and Nagaoka, 2000; Murray and Rice, 1993). Information geometry is a subfield of probability theory that has emerged from investigations of the geometrical structures of the parameter space of probability distributions. It initially aimed at providing geometrical interpretation of statistical inference theory, but it has been applied to a wide range of scientific and engineering fields, including machine learning, image processing, signal processing, optimization problems, and neuroscience.

The starting point of the IG approach is to describe the space of probability distributions as a manifold M. Intuitively, an N-dimensional manifold is a topological space in which each point in the space has a neighborhood that can be represented by an N-dimensional coordinate $\boldsymbol{\xi} = (\xi_1, ..., \xi_n)$. For example, a probability density function of a one-dimensional Gaussian distribution is given as

$$p\left(x; \mu, \sigma^2\right) = \frac{1}{\sqrt{2\pi}\sigma} e^{-\frac{(x-\mu)^2}{2\sigma^2}}, \qquad (4)$$

where x is a random variable, μ is the mean, and σ is the standard deviation. Because the Gaussian distribution is determined by a pair of parameters (μ, σ), the space of

$\boldsymbol{\xi} = (\mu, \sigma)$ provides a two-dimensional manifold of the distribution. However, this is not a Euclidean space, but rather a Riemannian space, and this is why the IG approach employs a method of differential geometry.

We now consider a measure of distance between two probability distributions. In statistics, the distance is often called the divergence. For the space of probability distributions, a commonly used measure is Kullback-Leibler divergence (KL divergence),

$$D[p(x) : q(x)] = \int p(x) \log \frac{p(x)}{q(x)} dx, \qquad (5)$$

where $p(x)$ and $q(x)$ are probability density functions (Kullback and Leibler, 1951). For a discrete distribution $P(x)$ and $Q(x)$, the integral is replaced by the sum and the KL divergence is given as,

$$D[P(x) : Q(x)] = \sum_i P(x_i) \log \frac{P(x_i)}{Q(x_i)}. \qquad (6)$$

Similar to the Euclidean distance, the KL divergence satisfies

$$D[P(x) : Q(x)] \geq 0, \qquad (7)$$

$$D[P(x) : Q(x)] = 0 \text{ if and only if } P(x) = Q(x). \qquad (8)$$

However, the KL divergence is asymmetric,

$$D[P(x) : Q(x)] \neq D[Q(x) : P(x)]. \qquad (9)$$

Therefore, the KL divergence is a measure for separation of two probability distributions, but is not the distance in the ordinary Euclidean sense.

Now let us consider three probability distributions, $p(x)$, $q(x)$, and $r(x)$. From Eq. (5), it is easy to show that

$$\begin{aligned} &D[p : q] - D[p : r] - D[r : q] \\ &= \int [p(x) - r(x)][\log r(x) - \log q(x)] dx. \end{aligned} \qquad (10)$$

If the right-hand side vanishes, the equation reduces to the Pythagoras theorem

$$D[p : q] = D[p : r] + D[r : q]. \qquad (11)$$

This relationship can be interpreted that two vectors $p(x) - r(x)$ and $\log r(x) - \log q(x)$ are orthogonal if the inner product of two vectors is defined by the right-hand side of Eq. (10). We will see that the Pythagoras relationship plays a crucial role in IG analysis of neural signals later.

It is also educational to consider the relationship between the mutual information and the KL divergence. The mutual information between two random variables X and Y is defined as,

$$I(X; Y) = H(X) - H(X|Y) = H(Y) - H(Y|X), \qquad (12)$$

where $H(X)$ and $H(Y)$ are the marginal entropies and $H(X|Y)$ and $H(Y|X)$ are the conditional entropies given as,

$$H(X) = -\int p(x)\log p(x)dx = -\int\int p(x,y)\log p(x)dxdy,$$

(13)

$$H(Y) = -\int p(y)\log p(y)dy = -\int\int p(x,y)\log p(y)dxdy,$$

(14)

$$H(X|Y) = -\int\int p(y)p(x|y)\log p(x|y)dxdy$$
$$= \int\int p(x,y)\log\frac{p(y)}{p(x,y)}dxdy,$$

(15)

$$H(Y|X) = -\int\int p(x)p(y|x)\log p(y|x)dxdy$$
$$= \int\int p(x,y)\log\frac{p(x)}{p(x,y)}dxdy.$$

(16)

Here, $p(x)$ and $p(y)$ are probability density functions of X and Y, respectively. $p(x,y)$ is a joint probability density function. $p(x|y)$ and $p(y|x)$ are conditional probability density functions of X given Y, and Y given X, respectively. As in Eq. (12), the mutual information estimates the uncertainty of a variable X, that still remains after knowing the other variable Y, and vice versa. Using Eqs. (13)–(16), Eq. (12) can be written as,

$$I(X;Y) = \int\int p(x,y)\log\frac{p(x,y)}{p(x)p(y)}dxdy$$
$$= D(p(x,y);p(x)p(y)).$$

(17)

Therefore, the mutual information $I(X;Y)$ is equivalent to the KL divergence from the joint distribution $p(x,y)$ to the marginal independent distribution $p(x)p(y)$.

Unlike the Euclidean space, the space of probability distributions is the curved Riemannian space. It has been shown that two types of straight lines (geodesics) can be found in the space; they are called a mixture geodesic (m-geodesic) and an exponential geodesic (e-geodesic), respectively (Amari, 1985; Amari and Nagaoka, 2000). The m-geodesic is defined as a set of points dividing two probability density functions as,

$$r(x;t) = (1-t)p(x) + tq(x), \ 0 \leq t \leq 1,$$

(18)

where t is a parameter. Note that $r(x;t)$ is again a probability density function. Similarly, the e-geodesic is defined as a set of points dividing logarithmic probability density functions as,

$$\log r(x;t) = (1-t)\log p(x) + t\log q(x) - \phi(t), \ 0 \leq t \leq 1,$$

(19)

where $\phi(t)$ is given as

$$\phi(t) = \log\int p(x)^{1-t}q(x)^t dx.$$

(20)

$\phi(t)$ is derived from the constraint that $r(x;t)$ be a probability density function,

$$\int r(x;t)dx = 1.$$

(21)

By extending a line to a surface, we can also define a flat surface. The surface defined by

$$M_m = \left\{ r(x;t) = \sum_{i=1}^n t_i p_i(x), \ t_i > 0, \ \sum_{i=1}^n t_i = 1 \right\},$$

(22)

is said to be mixture-flat (m-flat). The m-geodesic connecting any two probability distributions in M_m is again included in M_m. Similarly, the surface defined by

$$M_e = \left\{ r(x;t) = \exp\left[\sum_{i=1}^n t_i \log p_i(x) - \phi(t)\right], \ t_i > 0, \ \sum_{i=1}^n t_i = 1 \right\},$$

(23)

is said to be exponential-flat (e-flat). The e-geodesic connecting any two probability distributions in M_e is also included in M_e. Using the manifolds that are m-flat and e-flat, it is possible to fill the space of probability distributions. Specifically, if m-flat and e-flat manifolds are orthogonal to each other, the m-geodesic and e-geodesic also intersects orthogonally. Suppose $p(x)$ and $q(x)$ are on m-geodesic and e-geodesic, respectively, and if they intersect at $r(x)$, this is the case in which the Pythagoras theorem in Eq. (11) holds.

We now investigate the property of a specific family of probability distributions, called an exponential family. In general, the exponential family can be expressed as

$$p(x,\theta) = \exp\left[\sum_{i=1}^n \theta^i k_i(x) + r(x) - \psi(\theta)\right],$$

(24)

where x is a random variable, $\theta = (\theta^1, ..., \theta^n)$ is a parameter of probability distributions and $k_1(x), ..., k_n(x)$ and $r(x)$ are functions of x. Using $\boldsymbol{x} = (x_1, ..., x_n)$ where $x_i = k_i(x)$ and letting $r(x) = 0$, Eq. (24) is written as a standard form,

$$p(\boldsymbol{x}, \boldsymbol{\theta}) = \exp[\boldsymbol{\theta}\cdot\boldsymbol{x} - \psi(\boldsymbol{\theta})].$$

(25)

The parameter $\boldsymbol{\theta}$ is called the natural parameter. $\psi(\boldsymbol{\theta})$ is a normalization function, assuring $\int p(\boldsymbol{x}, \theta)d\boldsymbol{x} = 1$, which is given by,

$$\psi(\boldsymbol{\theta}) = \log\int \exp(\boldsymbol{\theta}\cdot\boldsymbol{x})d\boldsymbol{x}.$$

(26)

The exponential family is important because it includes the discrete distribution that is used to represent neural spiking activity. It also includes a wide variety of important distributions, such as the Normal distribution, Poisson distribution, gamma distribution, and log-normal distribution. In addition, it has been shown that the Pythagoras theorem in Eq. (11) holds true for the exponential family (Amari, 1985; Amari and Nagaoka, 2000). Here, let us briefly demonstrate that the Normal distribution and the discrete distribution belong to the exponential family. For the Normal distribution in Eq. (4), we convert the variables as,

$$x_1 = x, x_2 = x^2, \theta^1 = -\frac{1}{2\sigma^2}, \theta^2 = \frac{\mu^2}{\sigma^2}. \qquad (27)$$

We can see that the Normal distribution can be written in the standard form

$$P(x, \boldsymbol{\theta}) = \exp\left[\sum_{i=1}^{2} \theta^i x_i - \psi(\boldsymbol{\theta})\right]. \qquad (28)$$

A normalization function $\psi(\boldsymbol{\theta})$ is given as

$$\psi(\boldsymbol{\theta}) = \frac{\mu^2}{2\sigma^2} + \log\sqrt{2\pi}\sigma = \frac{(\theta^1)^2}{4\theta^2} - \frac{1}{2}\log(-\theta^2) + \frac{1}{2}\log\pi. \quad (29)$$

For the discrete distribution of a stochastic variable $x = \{0, 1, \ldots, n\}$ which takes $n+1$ values, a probability $p_i = \text{Prob}\{x = i\}, (i = 0, 1, \ldots, n)$ can be written as

$$p(x) = \sum_{i=0}^{n} p_i \delta_i(x), \qquad (30)$$

where $\delta_i(x)$ is given as

$$\delta_i(x) = \begin{cases} 1, x = i, \\ 0, x \neq i. \end{cases} \qquad (31)$$

Here we assume $p_i > 0$ for all i. Due to a constraint that x must satisfy $\sum p_i = 1$, it is sufficient to consider an n-dimensional vector $\boldsymbol{p} = (p_1, \ldots, p_n)$. By introducing

$$x_i = \delta_i(x), \ \theta^i = \log\frac{p_i}{p_0}, \qquad (32)$$

the discrete distribution can be written in the normal form,

$$P(x, \boldsymbol{\theta}) = \exp\left[\sum_{i=1}^{n} \theta^i x_i - \psi(\boldsymbol{\theta})\right], \qquad (33)$$

where a normalization function $\psi(\boldsymbol{\theta})$ is given as

$$\psi(\boldsymbol{\theta}) = -\log p_0 = \log\left[1 + \sum_{i=1}^{n} \exp(\theta^i)\right]. \qquad (34)$$

Note that $p_0 = 1 - \sum_{i=1}^{n} p_i$. The fact that the discrete distribution belongs to an exponential family is important because neural spiking activity can be represented using the discrete distribution. We will see this in the next section.

Lastly, let us briefly comment on some basic properties of the exponential family of probability distributions. Firstly, $\psi(\boldsymbol{\theta})$ is a convex function and $\boldsymbol{\theta} = (\theta^1, \ldots, \theta^n)$ provides an affine coordinate system that forms the e-flat manifold. There exists a dually convex function $\varphi(\boldsymbol{\eta})$ for a dually affine coordinate system $\boldsymbol{\eta} = (\eta_1, \ldots, \eta_n)$ that forms the m-flat manifold. An important consequence is that $\boldsymbol{\theta}$ and $\boldsymbol{\eta}$ are dually flat. That is, they form mutually orthogonal coordinate systems and the Pythagoras theorem in Eq. (11)

holds true. In the next section, we will see that this orthogonal property plays an important role in the IG analysis. Secondly, $\varphi(\boldsymbol{\eta})$ and $\psi(\boldsymbol{\theta})$ are connected by the Legendre transform,

$$\varphi(\boldsymbol{\eta}) = \max_{\theta}(\boldsymbol{\theta} \cdot \boldsymbol{\eta} - \psi(\boldsymbol{\theta})). \qquad (35)$$

In fact, $\varphi(\boldsymbol{\eta})$ takes the form of the negative of entropy,

$$\varphi(\boldsymbol{\eta}) = \int p(\boldsymbol{x}, \boldsymbol{\eta})\log p(\boldsymbol{x}, \boldsymbol{\eta})d\boldsymbol{x}. \qquad (36)$$

Furthermore, $\boldsymbol{\theta}$ and $\boldsymbol{\eta}$ are obtained by calculating the gradient of $\varphi(\boldsymbol{\eta})$ and $\psi(\boldsymbol{\theta})$ as

$$\boldsymbol{\theta} = \nabla\varphi(\boldsymbol{\eta}), \boldsymbol{\eta} = \nabla\psi(\boldsymbol{\theta}). \qquad (37)$$

By computing $\boldsymbol{\eta} = \nabla\psi(\boldsymbol{\theta})$, we have

$$\boldsymbol{\eta} = \nabla\psi(\boldsymbol{\theta}) = \int \boldsymbol{x}p(\boldsymbol{x}, \boldsymbol{\theta})d\boldsymbol{x} = E[\boldsymbol{x}], \qquad (38)$$

where $E[\boldsymbol{x}]$ represents an expectation of $\boldsymbol{x}$. Therefore, $\boldsymbol{\eta}$ is called an expectation parameter.

3 IG ANALYSIS OF NEUROPHYSIOLOGICAL DATA

As we saw in the previous section, the discrete probability distribution belongs to an exponential family. First, let us describe how neural firing can be treated in the framework of information geometry. As the simplest example, we consider a network of two neurons. Let x_1 and x_2 be two binary variables where $x_i = \{1, 0\}$ $(i = 1, 2)$ represents neuronal firing or silence, respectively. Following Eq. (33), the joint probability $p(x_1, x_2; \boldsymbol{\theta})$ is written as

$$p(x_1, x_2; \boldsymbol{\theta}) = \exp\left[\theta_1^{(2,2)} x_1 + \theta_2^{(2,2)} x_2 + \theta_{12}^{(2,2)} x_1 x_2 - \psi(\boldsymbol{\theta})^{(2,2)}\right], \qquad (39)$$

where the first and second part of superscript (2,2) represent the order of the log-linear model and the number of neurons in the network, respectively. Rewriting $p(x_1, x_2; \boldsymbol{\theta})$ as $p_{x_1 x_2}$ and taking the logarithm, it is expanded by a polynomial of x_1 and x_2,

$$\log p_{x_1 x_2} = \theta_1^{(2,2)} x_1 + \theta_2^{(2,2)} x_2 + \theta_{12}^{(2,2)} x_1 x_2 - \psi(\boldsymbol{\theta})^{(2,2)}. \qquad (40)$$

Note that we have a constraint $\sum p_{ij} = 1$ and $p_{ij} > 0$. Eq. (40) is called a log-linear model. The $\boldsymbol{\theta}^{(2,2)} = \left(\theta_1^{(2,2)}, \theta_2^{(2,2)}, \theta_{12}^{(2,2)}\right)$ and $\psi(\boldsymbol{\theta})^{(2,2)}$ can be obtained by putting p_{00}, p_{01}, p_{10} and p_{11} into Eq. (40), which yield,

$$\log p_{00} = -\psi(\boldsymbol{\theta})^{(2,2)}, \qquad (41)$$

$$\log p_{01} = \theta_2^{(2,2)} - \psi(\boldsymbol{\theta})^{(2,2)}, \qquad (42)$$

$$\log p_{10} = \theta_1^{(2,2)} - \psi(\boldsymbol{\theta})^{(2,2)}, \qquad (43)$$

$$\log p_{11} = \theta_1^{(2,2)} + \theta_2^{(2,2)} + \theta_{12}^{(2,2)} - \psi(\boldsymbol{\theta})^{(2,2)}. \quad (44)$$

By solving Eqs. (41)–(44) simultaneously, we obtain,

$$\theta_1^{(2,2)} = \log\frac{p_{10}}{p_{00}}, \ \theta_2^{(2,2)} = \log\frac{p_{01}}{p_{00}}, \ \theta_{12}^{(2,2)} = \log\frac{p_{11}p_{00}}{p_{01}p_{10}}, \quad (45)$$

$$\psi(\boldsymbol{\theta})^{(2,2)} = -\log p_{00}. \quad (46)$$

$\boldsymbol{\theta}^{(2,2)} = \left(\theta_1^{(2,2)}, \theta_2^{(2,2)}; \theta_{12}^{(2,2)}\right)$ are natural parameters and provide an e-flat affine coordinate system. We also call $\boldsymbol{\theta}^{(2,2)}$ the IG measures and they represent the first- and second-order interactions from the IG point of view. We call θ_i the one-neuron IG measure and θ_{ij} the two-neuron IG measure, rather than the first-order and second-order IG measures in order to avoid confusion between the order of IG interaction and the order of log-linear model. The corresponding dually flat (m-flat) affine coordinate system $\boldsymbol{\eta} = (\eta_1, \eta_2; \eta_{12})$ are calculated as

$$\eta_1 = E[x_1] = \mathrm{Prob}\{x_1 = 1\}, \ \eta_2 = E[x_2] = \mathrm{Prob}\{x_2 = 1\},$$
$$\eta_{12} = E[x_1 x_2] = \mathrm{Prob}\{x_1 = 1, x_2 = 1\}.$$
$$(47)$$

Here η_i and η_{12} represent firing probability of a single neuron and their joint firing probability, respectively.

There are several coordinate systems that can be used for representing two-neuron firings. For example, the single-neuron firing probability and joint firing probability $\boldsymbol{\eta} = (\eta_1, \eta_2; \eta_{12})$ provide a coordinate system. Similarly, the one-neuron and two-neuron IG measures $\boldsymbol{\theta}^{(2,2)} = \left(\theta_1^{(2,2)}, \theta_2^{(2,2)}; \theta_{12}^{(2,2)}\right)$ provide another coordinate system. However, one of the significant advantages of the IG method is the fact that the IG measures $\boldsymbol{\theta}$ and firing probabilities $\boldsymbol{\eta}$ can be decomposed orthogonally. That is, the two-neuron IG measure $\theta_{12}^{(2,2)}$ is orthogonal to the firing probability η_1 and η_2. Information geometry therefore proposes to use a mixed coordinate system $\boldsymbol{\zeta} = \left(\eta_1, \eta_2; \theta_{12}^{(2,2)}\right)$. In this coordinate system, a pairwise neural interaction $\theta_{12}^{(2,2)}$ can be assessed independently from the change in the firing probability of neurons η_1 and η_2.

In information geometry, the orthogonality is defined by the fact that a small change of $\log p(x_1, x_2)$ in the direction of η_i ($i = 1$ or 2) is not correlated with a small change of $\log p(x_1, x_2)$ in the direction of $\theta_{12}^{(2,2)}$. The small change in the direction of η_i ($i = 1$ or 2) is given by

$$\frac{\partial}{\partial \eta_i} \log p(x_1, x_2), \quad (48)$$

and the small change in the direction of $\theta_{12}^{(2,2)}$ is given by

$$\frac{\partial}{\partial \theta_{12}^{(2,2)}} \log p(x_1, x_2). \quad (49)$$

Then, if

$$E\left[\frac{\partial}{\partial \eta_i} \log p(x_1, x_2) \frac{\partial}{\partial \theta_{12}^{(2,2)}} \log p(x_1, x_2)\right] = 0, \quad (50)$$

holds true where E is the expectation with respect to $p(x_1, x_2)$, these directions are said to be orthogonal to each other (Nakahara and Amari, 2002). Importantly, the orthogonal property does not hold if a pairwise neural interaction is measured differently (Amari, 2009). For example, if the pairwise interaction is measured by covariance,

$$cov = E[x_1 x_2] - E[x_1]E[x_2] = \eta_{12} - \eta_1 \eta_2, \quad (51)$$

or by correlation coefficient,

$$\rho = \frac{cov}{\sqrt{\eta_1(1 - \eta_1)\eta_2(1 - \eta_2)}} = \frac{\eta_{12} - \eta_1 \eta_2}{\sqrt{\eta_1(1 - \eta_1)\eta_2(1 - \eta_2)}}, \quad (52)$$

The coordinate systems $(\eta_1, \eta_2; cov)$ and $(\eta_1, \eta_2; \rho)$ do not satisfy the orthogonal property except for the point that $cov = 0$ or $\rho = 0$.

Another advantage of the IG method is that neural interactions in different orders can be calculated in a straightforward manner. For example, let us consider a network consisting of three neurons. The joint firing probability $p_{x_1 x_2 x_3}$ is expanded as,

$$\log p_{x_1 x_2 x_3} = \theta_1^{(3,3)} x_1 + \theta_2^{(3,3)} x_2 + \theta_3^{(3,3)} x_3 + \theta_{12}^{(3,3)} x_1 x_2$$
$$+ \theta_{13}^{(3,3)} x_1 x_3 + \theta_{23}^{(3,3)} x_2 x_3 + \theta_{123}^{(3,3)} x_1 x_2 x_3 - \psi^{(3,3)}. \quad (53)$$

Eight equations are obtained by putting $p_{000}, p_{001}, p_{010}, p_{011}, p_{100}, p_{101}, p_{110}$ and p_{111} into Eq. (53). By solving them simultaneously, we have

$$\theta_1^{(3,3)} = \log\frac{p_{100}}{p_{000}}, \ \theta_2^{(3,3)} = \log\frac{p_{010}}{p_{000}}, \ \theta_3^{(3,3)} = \log\frac{p_{001}}{p_{000}}, \quad (54)$$

$$\theta_{12}^{(3,3)} = \log\frac{p_{110}p_{000}}{p_{100}p_{010}}, \ \theta_{23}^{(3,3)} = \log\frac{p_{011}p_{000}}{p_{010}p_{001}},$$
$$\theta_{13}^{(3,3)} = \log\frac{p_{101}p_{000}}{p_{100}p_{001}}, \quad (55)$$

$$\theta_{123}^{(3,3)} = \log\frac{p_{111}p_{100}p_{010}p_{001}}{p_{110}p_{101}p_{011}p_{000}}, \quad (56)$$

$$\psi^{(3,3)} = -\log p_{000}. \quad (57)$$

$\boldsymbol{\theta}^{(3,3)} = \left(\theta_i^{(3,3)}; \theta_{ij}^{(3,3)}; \theta_{ijk}^{(3,3)}\right)$ represents the single, pairwise, and triple-wise IG interactions among neurons x_1, x_2, and x_3, respectively. Note that other commonly used correlation measures, such as cov and ρ cannot be expanded to multiple neurons easily.

Similar to the two-neuron case, we can form a mixed coordinate system using the firing probabilities and IG measures. It has been shown that two kinds of mixed coordinates $\boldsymbol{\zeta}_1 = \left(\eta_1, \eta_2, \eta_3; \theta_{12}^{(3,3)}, \theta_{23}^{(3,3)}, \theta_{31}^{(3,3)}, \theta_{123}^{(3,3)}\right)$ and $\boldsymbol{\zeta}_2 = \left(\eta_1, \eta_2, \eta_3, \eta_{12}, \eta_{23}, \eta_{31}; \theta_{123}^{(3,3)}\right)$ are available (Nakahara and Amari, 2002). ζ_1 is a coordinate system in which $\eta_1 = (\eta_1, \eta_2, \eta_3)$ and $\boldsymbol{\theta}_{1+}^{(3,3)} = \left(\theta_{12}^{(3,3)}, \theta_{23}^{(3,3)}, \theta_{31}^{(3,3)}, \theta_{123}^{(3,3)} \right)$ are orthogonal to each other. In other words, $\boldsymbol{\theta}_{1+}^{(3,3)}$ represents the neural interactions that are independent from the changes in a single-neuron firing η_1. ζ_2 is a coordinate system in which $\eta_2 = (\eta_1, \eta_2, \eta_3, \eta_{12}, \eta_{23}, \eta_{31})$ and $\boldsymbol{\theta}_{2+}^{(3,3)} = \theta_{123}^{(3,3)}$ are orthogonal to each other. Here, $\boldsymbol{\theta}_{2+}^{(3,3)}$ represents the neural interaction that is independent from the changes in both single-neuron firing and joint firing η_2. In this way, the IG neural interactions at different orders can be constructed in a hierarchical manner.

For a network of two neurons, we saw the relationship $\theta_{12}^{(2,2)} = \log \frac{p_{11}p_{00}}{p_{01}p_{10}}$ in Eq. (45). For a network of three neurons, we have the relationship $\theta_{12}^{(3,3)} = \log \frac{p_{110}p_{000}}{p_{100}p_{010}}$ in Eq. (55). Both represent a pairwise neural interaction between the neuron 1 and neuron 2 in a two-neuron system and three-neuron system, respectively. Suppose we are not able to record the activity of the third neuron in the network of three neurons. In this case, we need to use a partially expanded log-linear model,

$$\log p_{x_1 x_2 *} = \theta_1^{(2,3)} x_1 + \theta_2^{(2,3)} x_2 + \theta_{12}^{(2,3)} x_1 x_2 - \psi^{(2,3)}, \quad (58)$$

where "$*$" represents the marginalization over the third neuron. $p_{x_1 x_2 *}$ represents the marginal distribution of x_1 and x_2. The two-neuron IG measure, $\theta_{12}^{(2,3)}$ is then given as,

$$\theta_{12}^{(2,3)} = \log \frac{p_{11*}p_{00*}}{p_{01*}p_{10*}}. \quad (59)$$

We will discuss how $\theta_{12}^{(2,2)}$, $\theta_{12}^{(3,3)}$, and $\theta_{12}^{(2,3)}$ are related to the parameters of networks such as connection weights in the next section.

In general, the full log-linear expansion for an N-neuron system is given by

$$\log p_{x_1 x_2 \ldots x_N} = \sum_i \theta_i^{(N,N)} x_i + \sum_{i<j} \theta_{ij}^{(N,N)} x_i x_j + \sum_{i<j<k} \theta_{ijk}^{(N,N)} x_i x_j x_k$$
$$+ \ldots + \ldots + \theta_{12,\ldots N}^{(N,N)} x_1 x_2 \ldots x_N - \psi(\boldsymbol{\theta})^{(N,N)}, \quad (60)$$

where $\theta_{12,\ldots N}^{(N,N)}$ is the Nth-order neuronal correlation. The first few IG measures and the normalization function are expressed as

$$\theta_i^{(N,N)} = \log \frac{p_{x_1=0,\ldots,x_i=1,\ldots,x_N=0}}{p_{x_1=0,\ldots,x_N=0}},$$

$$\theta_{ij}^{(N,N)} = \log \frac{p_{x_1=0,\ldots,x_i=1,\ldots,x_j=1,\ldots,x_N=0} \, p_{x_1=0,\ldots,x_N=0}}{p_{x_1=0,\ldots,x_i=1,\ldots,x_j=0,\ldots,x_N=0} \, p_{x_1=0,\ldots,x_i=0,\ldots,x_j=1,\ldots,x_N=0}},$$

$$\theta_{ijk}^{(N,N)} = \log \frac{p_{x_i=1,x_j=1,x_k=1;x_{N-i,j,k}=0} \, p_{x_i=1;x_{N-i}=0} \, p_{x_j=1;x_{N-j}=0} \, p_{x_k=1;x_{N-k}=0}}{p_{x_i=1,x_j=1;x_{N-i,j}=0} \, p_{x_j=1,x_k=1;x_{N-j,k}=0} \, p_{x_i=1,x_k=1;x_{N-i,k}=0} \, p_{x_{1:N}=0}},$$

$$\theta_{ijkl}^{(N,N)} = \log \left(\frac{p_{x_i=1,x_j=1,x_k=1,x_l=1;x_{N-i,j,k,l}=0} \, p_{x_i=1,x_j=1;x_{N-i,j}=0} \, p_{x_i=1,x_k=1;x_{N-i,k}=0}}{p_{x_i=1,x_j=1,x_k=1;x_{N-i,j,k}=0} \, p_{x_i=1,x_j=1,x_l=1;x_{N-i,j,l}=0} \, p_{x_i=1,x_k=1,x_l=1;x_{N-i,k,l}=0}} \right.$$
$$\times \frac{p_{x_i=1,x_l=1;x_{N-i,l}=0} \, p_{x_j=1,x_k=1;x_{N-j,k}=0} \, p_{x_j=1,x_l=1;x_{N-j,l}=0} \, p_{x_k=1,x_l=1;x_{N-k,l}=0}}{p_{x_j=1,x_k=1,x_l=1;x_{N-j,k,l}=0} \, p_{x_i=1;x_{N-i}=0} \, p_{x_j=1;x_{N-j}=0} \, p_{x_k=1;x_{N-k}=0}}$$
$$\left. \times \frac{p_{x_{1:N}=0}}{p_{x_l=1;x_{N-l}=0}} \right), \quad (61)$$
$$\ldots \quad \ldots \quad \ldots$$
$$\psi(\boldsymbol{\theta})^{(N,N)} = -\log p_{x_1=0,\ldots,x_N=0},$$

where $1 \leq i < j < k < l \leq N$. Note that for $\theta_{ijk}^{(N,N)}$ and $\theta_{ijkl}^{(N,N)}$, we used the following form of representation:

$$p_{x_i=1,x_j=1,x_k=1;x_{N-i,j,k}=0}$$
$$= p_{x_1=0,\ldots,x_i=1,\ldots,x_j=1,\ldots,x_k=1,\ldots,x_N=0} \quad (62)$$
$$p_{x_i=1,x_j=1,x_k=1,x_l=1;x_{N-i,j,k,l}=0}$$
$$= p_{x_1=0,\ldots,x_i=1,\ldots,x_j=1,\ldots,x_k=1,\ldots,x_l=1,\ldots,x_N=0}.$$

In practice, the estimates of $\boldsymbol{\theta}$s are obtained by maximum likelihood estimates of $p_{x_1 x_2,\ldots,x_N}$, given by

$$p_{x_1 x_2,\ldots,x_N} = \frac{c_{x_1 x_2,\ldots,x_N}}{\sum_{x_1 x_2,\ldots,x_N} c_{x_1 x_2,\ldots,x_N}}, \quad (63)$$

where $c_{x_1 x_2,\ldots,x_N}$ is the number of counts in which the event $(X_1 = x_1, X_2 = x_2, \ldots, X_N = x_N)$ occurs. If the number of observations is rather small, the specific event may have never occurred. This is specifically true for the event that includes the high-order interactions. A possible treatment in the framework of maximum a posteriori estimate has been discussed elsewhere (Shimazaki et al., 2012).

In real experiments, it is not possible to specify the total number of neurons in the network. Furthermore, because the number of $\boldsymbol{\theta}^{(N,N)}$ parameters in the log-linear model increases as $2^N - 1$, it is not realistic to obtain a robust estimation of all the parameters for a large N. To overcome this difficulty, the previous study (Tatsuno et al., 2009) proposed to use the second-order log-linear model regardless of the number of neurons N in the network,

$$\log p_{x_1 x_2 *...*} = \theta_1^{(2,N)} x_1 + \theta_2^{(2,N)} x_2 + \theta_{12}^{(2,N)} x_1 x_2 - \psi^{(2,N)},$$

(64)

where the superscript (2, N) represents that the second-order log-linear model was used for an N-neuron system. Note that interactions with other $N-2$ neurons were included in the equation implicitly. The IG measures are then given by

$$\theta_1^{(2,N)} = \log \frac{p_{10*...*}}{p_{00*...*}}, \theta_2^{(2)} = \log \frac{p_{01*...*}}{p_{00*...*}},$$

$$\theta_{12}^{(2)} = \log \frac{p_{11*...*} p_{00*...*}}{p_{01*...*} p_{10*...*}}, \psi^{(2,N)} = -\log p_{00*...*},$$

(65)

where "$* ... *$" represents the marginalization over the other N-2 neurons. In general, the IG measures partly expanded up to the kth order in an N-neuron network are obtained from

$$\log p_{x_1 x_2 ... x_{k*}...*} = \sum_i \theta_i^{(k,N)} x_i + \sum_{i<j} \theta_{ij}^{(k,N)} x_i x_j + ... +$$

$$\sum_{i<j<...<m} \theta_{ij,...,m}^{(k,N)} x_i x_j ... x_m + \cdots + \theta_{12,...k}^{(k,N)} x_1 x_2 ... x_k - \psi(\boldsymbol{\theta})^{(k,N)},$$

(66)

where $1 \le m \le k \le N$. The first few terms and normalizing factor are given as follows

$$\theta_i^{(k,N)} = \log \frac{p_{x_1=0,...,x_i=1,...,x_k=0,*...*}}{p_{x_1=0,...,x_k=0,*...*}},$$

$$\theta_{ij}^{(k,N)} = \log \frac{p_{x_1=0,...,x_i=1,...,x_j=1,...,x_k=0,*...*} \cdot p_{x_1=0,...,x_k=0,*...*}}{p_{x_1=0,...,x_i=1,...,x_j=0,...,x_k=0,*...*} \cdot p_{x_1=0,...,x_i=0,...,x_j=1,...,x_k=0,*...*}},$$

$$\theta_{ijq}^{(k,N)} = \log \frac{p_{x_i=1,x_j=1,x_q=1;x_{k-i,j,q}=0;*} \cdot p_{x_i=1;x_{k-i}=0;*}}{p_{x_j=1;x_{k-j}=0;*} \cdot p_{x_q=1;x_{k-q}=0;*}}{p_{x_i=1,x_j=1;x_{k-i,j}=0;*} \cdot p_{x_j=1,x_q=1;x_{k-j,q}=0;*}},$$

(67)

$$\theta_{ijqr}^{(k,N)} = \log \left(\frac{p_{x_i=1,x_q=1;x_{k-i,q}=0;*} \cdot p_{x_{1:k}=0;*}}{p_{x_i=1,x_j=1,x_q=1,x_r=1;x_{k-i,j,q,r}=0;*} \cdot p_{x_i=1,x_j=1;x_{k-i,j}=0;*} \cdot p_{x_i=1,x_q=1;x_{k-i,q}=0;*}}{p_{x_i=1,x_j=1,x_q=1;x_{k-i,j,q}=0;*} \cdot p_{x_i=1,x_j=1,x_r=1;x_{k-i,j,r}=0;*} \cdot p_{x_i=1,x_q=1,x_r=1;x_{k-i,q,r}=0;*}}\right.$$

$$\times \frac{p_{x_i=1,x_r=1;x_{k-i,r}=0;*} \cdot p_{x_j=1,x_q=1;x_{k-j,q}=0;*} \cdot p_{x_j=1,x_r=1;x_{k-j,r}=0;*} \cdot p_{x_q=1,x_r=1;x_{k-q,r}=0;*}}{p_{x_j=1,x_q=1,x_r=1;x_{k-j,q,r}=0;*} \cdot p_{x_i=1;x_{k-i}=0;*} \cdot p_{x_j=1;x_{k-j}=0;*} \cdot p_{x_q=1;x_{k-q}=0;*}}$$

$$\left. \times \frac{p_{x_{1:k}=0;*}}{p_{x_r=1;x_{k-r}=0;*}} \right),$$

$$... \quad ... \quad ...$$

$$\psi(\boldsymbol{\theta})^{(k,N)} = -\log p_{x_1=0,...,x_k=0,*...*},$$

where "$* ... *$" represents the marginalization over the $(N-k)$ neurons. Also note that for $\theta_{ijq}^{(k,N)}$ and $\theta_{ijqr}^{(k,N)}$, we used the following form of representation:

$$p_{x_i=1,x_j=1,x_q=1;x_{k-i,j,q}=0;*}$$
$$= p_{x_1=0,...,x_i=1,...,x_j=1,...,x_q=1,...,x_k=0,*...*}$$
$$p_{x_i=1,x_j=1,x_q=1,x_r=1;x_{k-i,j,q,r}=0;*}$$
$$= p_{x_1=0,...,x_i=1,...,x_j=1,...,x_q=1,...,x_r=1,...,x_k=0,*...*}$$

(68)

Both $\theta_{ij,...m}^{(N,N)}$ (the IG measure from the full model in Eq. (60)) and $\theta_{ij,...m}^{(k,N)}$ (the IG measure from the kth order partial model in Eq. (66)) represent the m-neuron IG interactions. However, note the difference between them; $\theta_{ij,...m}^{(N,N)}$ is calculated from the full information of all N neurons.

By contrast, $\theta_{ij,...m}^{(k,N)}$ is calculated from the partial information of k neurons by marginalizing $(N-k)$ neurons. It has been shown that $\theta_{ij,...N}^{(N,N)}$ is statistically orthogonal to any η_i. On the other hand, $\theta_{ij,...k}^{(k,N)}$ is orthogonal to η_i for i that is included in the k neurons (Amari, 2001; Nakahara and Amari, 2002).

We have so far treated the case for binary random variables. The extension to the case in which a random variable takes a finite set of discrete values has been worked out elsewhere (Amari, 2001). However, the extension to continuous variables is a challenging problem and requires further study (Cena and Pistone, 2007; Newton, 2012).

4 IG MEASURES AND THE UNDERLYING NETWORK PARAMETERS

In the previous section, we have reviewed the framework of IG method and several advantages to the conventional correlation measures, such as covariance and correlation coefficient. In neuroscience, one of the important questions is to find the possible relationships between correlation measures and the underlying neural architectures. This is a difficult problem because multiple network architectures could produce the same correlation values. In other words, determination of the underlying neural architectures can be considered an ill-posed inverse problem, such as reconstruction of an original three-dimensional image from a two-dimensional image on a retina, whereas extracting the neural interaction can be considered a forward problem.

Here, we investigate the possible relationship between the IG measures and the underlying neural architectures by assuming that spikes are generated by a network of stochastic model neurons (Ginzburg and Sompolinsky, 1994). The simplicity of the model allows us to study the relationship analytically. Briefly, let $x_i(t)$ be the state of each neuron at time t and take the binary values, 0 and 1, corresponding to a quiescent and active state, respectively. The total input to the ith neuron at time t is written as

$$u_i(t) = \sum_j J_{ij}x_j(t) + h_i, \qquad (69)$$

where J_{ij} denotes the connection strength from the jth presynaptic neuron to the ith postsynaptic neuron, and h_i represents the external input to the ith neuron. We assume that there is no self-coupling (ie, $J_{ii} = 0$), but the effect of self-coupling is negligible in the analysis with a large N limit (Ginzburg and Sompolinsky, 1994; Tatsuno and Okada, 2004). The neuron dynamics is determined by the transition rate w between the binary states as,

$$w(S_i \rightarrow (1-x_i)) = \frac{1}{2\tau_0}\{1 - (2x_i - 1)[2g(h_i) - 1]\}, \quad (70)$$

where τ_0 is a microscopic characteristic time, and $g(u_i)$ is a sigmoidal activation function whose value is bounded in the interval [0, 1] such as

$$g(u_i) = \frac{1 + \tanh(\beta(u_i - m))}{2}, \qquad (71)$$

where $\beta > 0$ and m are constants determining the slope and offset-threshold of the sigmoid. The probability of finding the system in a state $\{x^{(N)}\} = (x_1, ..., x_N, t)$ at time t is characterized by the following master equation

$$\frac{d}{dt}P\left(\{x^{(N)}\};t\right) = -\sum_i w(x_i \rightarrow (1-x_i))P\left(\{x^{(N)}\};t\right) + \sum_i w((1-x_i) \rightarrow x_i)P\left(\{x_i^{(N)}\};t\right), \quad (72)$$

where $\{x_i^{(N)}\} = (x_1, ...(1-x_i), ..., x_N, t)$. For instance, the probability of finding the system in a state $\{x^{(2)}\} = (x_1, x_2, t)$ for a network of two neurons is given as

$$\frac{d}{dt}P\left(\{x^{(2)}\};t\right) = -w(x_1 \rightarrow (1-x_1))P\left(\{x^{(2)}\};t\right)$$
$$-w(x_2 \rightarrow (1-x_2))P\left(\{x^{(2)}\};t\right) + w((1-x_1) \rightarrow x_1) \quad (73)$$
$$P\left(\{x_1^{(2)}\};t\right) + w((1-x_2) \rightarrow x_2)P\left(\{x_2^{(2)}\};t\right),$$

where $\{x_1^{(2)}\} = ((1-x_1), x_2, t)$ and $\{x_2^{(2)}\} = (x_1, (1-x_2), t)$.

Once $P(x^{(N)}; t)$ is given, the mean firing rate $\langle x_i(t)\rangle$ is written by

$$\langle x_i(t)\rangle = \sum_{\{x^{(N)}\}} x_i P\left(\{x^{(N)}\};t\right), \qquad (74)$$

where a summation is taken over all possible configurations. Note that $\langle x_i(t)\rangle$ is equivalent to $\eta(t)$ in Eq. (2). The time evolution is then given as,

$$\frac{d}{dt}\langle x_i(t)\rangle = \sum_{\{x^{(N)}\}} x_i \frac{d}{dt}P\left(\{x^{(N)}\};t\right). \qquad (75)$$

Substituting Eq. (72) into Eq. (75) yields

$$\frac{d}{dt}\langle x_i(t)\rangle = \sum_{\{x^{(N)}\}} (1 - 2x_i)w(x_i \rightarrow (1-x_i))P\left(\{x^{(N)}\};t\right)$$
$$= \langle(1 - 2x_i)w(x_i \rightarrow (1-x_i))\rangle. \qquad (76)$$

Further substitution of Eq. (70) into Eq. (76) yields the equation that the mean firing rate $\langle x_i(t)\rangle$ obeys:

$$\tau_0 \frac{d}{dt}\langle x_i(t)\rangle = -\langle x_i(t)\rangle + \langle g(u_i(t))\rangle. \qquad (77)$$

For the evolution of the mean coincident firing rate $\langle x_i(t)x_j(t)\rangle$, we start with

$$\frac{d}{dt}\langle x_i(t)x_j(t)\rangle = \sum_{\{x^{(N)}\}} x_i x_j \frac{d}{dt}P\left(\{x^{(N)}\};t\right). \qquad (78)$$

A similar calculation yields the equation that the mean coincident firing rate $\langle x_i(t)x_j(t)\rangle$ obeys:

$$\tau_0 \frac{d}{dt}\langle x_i(t)x_j(t)\rangle = -2\langle x_i(t)x_j(t)\rangle + \langle x_i(t)g(u_j(t))\rangle + \langle x_j(t)g(u_i(t))\rangle. \qquad (79)$$

In general, the equation that the coincident firing of N neurons $\langle x_1(t)x_2(t)...x_N(t)\rangle$ obeys is written as:

$$\tau_0 \frac{d}{dt}\langle x_1(t)x_2(t)\ldots x_N(t)\rangle = -N\langle x_1,(t),x_2,(t),\ldots,x_N,(t)\rangle$$
$$+\langle x_2(t)x_3(t)\ldots x_{N-1}(t)x_N(t)g(u_1(t))\rangle$$
$$+\langle x_1(t)x_3(t)\ldots x_{N-1}(t)x_N(t)g(u_2(t))\rangle$$
$$+\ldots + \langle x_1(t)x_2(t)\ldots x_{N-2}(t)x_{N-1}(t)g(u_N(t))\rangle. \tag{80}$$

For mathematical clarity, we investigate the relationship between the IG measures and the neural architectures when the network is in the equilibrium state. By letting $\tau_0 \frac{d}{dt}\langle x_i(t)\rangle = 0$ in Eq. (77), $\tau_0 \frac{d}{dt}\langle x_i(t)x_j(t)\rangle = 0$ in Eq. (79), and $\tau_0 \frac{d}{dt}\langle x_1(t)x_2(t)\ldots x_N(t)\rangle = 0$ in Eq. (80), we obtain

$$\langle x_i\rangle = \langle g(u_i)\rangle, \tag{81}$$

$$\langle x_i x_j\rangle = \frac{1}{2}\{\langle x_i g(u_j)\rangle + \langle x_j g(u_i)\rangle\}, \tag{82}$$

$$\langle x_1 x_2 \ldots x_N\rangle = \frac{1}{N}(\langle x_2 x_3 \ldots x_{N-1}x_N g(u_1)\rangle$$
$$+ \langle x_1 x_3 \ldots x_{N-1}x_N g(u_2)\rangle$$
$$+\ldots + \langle x_1 x_2 \ldots x_{N-2}x_{N-1} g(u_N)\rangle). \tag{83}$$

In this article, we focus on $\langle x_i\rangle$ and $\langle x_i x_j\rangle$. More general cases can be found elsewhere (Nie et al., 2014a).

To have a better insight into the relationship between the IG measures and the underlying neural architecture, first, let's consider a system of two neurons. Eq. (81) becomes,

$$\langle x_1\rangle = \langle g(u_1)\rangle = \langle g(J_{12}x_2 + h_1)\rangle, \langle x_2\rangle = \langle g(u_2)\rangle$$
$$= \langle g(J_{21}x_1 + h_2)\rangle. \tag{84}$$

Because x_i takes 1 or 0, $\langle g(J_{ij}x_j + h_i)\rangle$ can be separated into $\langle x_j\rangle$ and $g(\cdot)$ as,

$$\langle g(J_{ij}x_j + h_i)\rangle = \langle x_j g(J_{ij} + h_i) + (1 - x_j)g(h_i)\rangle$$
$$= \langle x_j g(J_{ij} + h_i)\rangle + \langle (1 - x_j)g(h_i)\rangle$$
$$= \langle x_j\rangle g(J_{ij} + h_i) + g(h_i) - \langle x_j\rangle g(h_i)$$
$$= \langle x_j\rangle\{g(J_{ij} + h_i) - g(h_i)\} + g(h_i). \tag{85}$$

Using this trick, Eq. (84) is rewritten as,

$$\langle x_1\rangle = \langle x_2\rangle\{g(J_{12} + h_1) - g(h_1)\} + g(h_1),$$
$$\langle x_2\rangle = \langle x_1\rangle\{g(J_{21} + h_2) - g(h_2)\} + g(h_2). \tag{86}$$

Hence, we have,

$$\langle x_1\rangle = \frac{g(h_1) + \triangle g_1 g(h_2)}{1 - \triangle g_1 \triangle g_2}, \quad \langle x_2\rangle = \frac{g(h_2) + \triangle g_2 g(h_1)}{1 - \triangle g_1 \triangle g_2} \tag{87}$$

where $\triangle g_1 = g(J_{12} + h_1) - g(h_1)$ and $\triangle g_1 = g(J_{12} + h_1) - g(h_1)$, respectively. Eq. (87) provides the relationship between the mean firing rate $\langle x_i\rangle$ (left-hand-side) and the network parameters J_{ij} and h_i (right-hand side). Using $\langle x_1\rangle$ and $\langle x_2\rangle$, the mean coincident firing rate $\langle x_1 x_2\rangle$ is also obtained as

$$\langle x_1 x_2\rangle = \frac{1}{2}\{x_1 g(J_{21} + h_2) + x_2 g(J_{12} + h_1)\}. \tag{88}$$

To convert Eqs. (87) and (88) into the relationship between the IG measures and the network parameters, we first express the joint probability distribution p_{ij} in terms of $\langle x_i\rangle$ and $\langle x_i x_j\rangle$, which is given as

$$p_{00} = 1 - \langle x_1\rangle - \langle x_2\rangle + \langle x_1 x_2\rangle,$$
$$p_{01} = \langle x_2\rangle - \langle x_1 x_2\rangle,$$
$$p_{10} = \langle x_1\rangle - \langle x_1 x_2\rangle,$$
$$p_{11} = \langle x_1 x_2\rangle. \tag{89}$$

Plugging these expressions into Eq. (45) and using Eq. (71) yields,

$$\theta_1^{(2,2)} = 2\beta(h_1 - m)$$
$$+ \log\left(\frac{2\exp[2\beta(h_1 + h_2 + J_{12})] + 2\exp[4\beta m] + 2\exp[2\beta(h_1 + J_{12} + m)]}{\exp[2\beta(h_1 + h_2 + J_{12})] + \exp[2\beta(h_1 + h_2 + J_{21})] + 2\exp[4\beta m]}\right.$$
$$\times \left.\frac{+\exp[2\beta(h_2 + J_{12} + m)] + \exp[2\beta(h_2 + J_{21} + m)]}{+2\exp[2\beta(h_1 + J_{12} + m)] + 2\exp[2\beta(h_2 + J_{21} + m)]}\right),$$

$$\theta_2^{(2,2)} = 2\beta(h_2 - m)$$
$$+ \log\left(\frac{2\exp[2\beta(h_1 + h_2 + J_{21})] + 2\exp[4\beta m] + 2\exp[2\beta(h_2 + J_{21} + m)]}{\exp[2\beta(h_1 + h_2 + J_{12})] + \exp[2\beta(h_1 + h_2 + J_{21})] + 2\exp[4\beta m]}\right.$$
$$\times \left.\frac{+\exp[2\beta(h_1 + J_{12} + m)] + \exp[2\beta(h_1 + J_{21} + m)]}{+2\exp[2\beta(h_1 + J_{12} + m)] + 2\exp[2\beta(h_2 + J_{21} + m)]}\right), \tag{90}$$

$$\theta_{12}^{(2,2)} = \beta(J_{12} + J_{21})$$
$$+ \log\left(\frac{(\exp[2\beta(h_1 + h_2 + J_{12})] + \exp[2\beta(h_1 + h_2 + J_{21})] + 2\exp[2\beta m]}{(2\exp[2\beta(h_1 + h_2 + J_{12})] + 2\exp[4\beta m] + 2\exp[2\beta(h_1 + J_{12} + m)]}\right.$$
$$\times \frac{+2\exp[2\beta(h_1 + J_{12} + m)] + 2\exp[2\beta(h_2 + J_{21} + m)])(2\exp[\beta(2h_1 + 2h_2 + J_{12} + J_{21})]}{+\exp[2\beta(h_2 + J_{12} + m)] + \exp[2\beta(h_2 + J_{21} + m)])(2\exp[2\beta(h_1 + h_2 + J_{21})]}$$
$$\times \frac{+2\exp[\beta(2h_1 + J_{12} + J_{21} + 2m)] + 2\exp[2h_2 + J_{12} + J_{21} + 2m]}{+2\exp[4\beta m] + \exp[2\beta(h_1 + J_{12} + m)] + \exp[2\beta(h_1 + J_{21} + m)]}$$
$$\times \left.\frac{+\exp[\beta(J_{12} - J_{21} + 4m)] + \exp[\beta(J_{21} - J_{12} + 4m)])}{+2\exp[2\beta(h_2 + J_{21} + m)])}\right).$$

The result demonstrates that $(\theta_1^{(2,2)}, \theta_2^{(2,2)})$ are expressed by the term corresponding to the external input h_i and an additional logarithmic bias term. In contrast, $\theta_{12}^{(2,2)}$ has the first term corresponding to the sum of the connection strength $(J_{12} + J_{21})$ and an additional logarithmic bias term.

If the connection is symmetric $(J_{12} = J_{21})$, Eq. (90) is significantly simplified to

$$\theta_1^{(2,2)} = 2\beta(h_1 - m), \quad \theta_2^{(2,2)} = 2\beta(h_2 - m), \quad \theta_{12}^{(2,2)} = 2\beta J_{12}. \tag{91}$$

These equations show that $(\theta_1^{(2,2)}, \theta_2^{(2,2)})$ depend on the external inputs h_i only and $\theta_{12}^{(2,2)}$ on the connection strength $J_{12}(=J_{21})$ only (Tatsuno and Okada, 2004). However, for the general asymmetric connection $(J_{12} \neq J_{21})$, the relationship in Eq. (90) is very complex, even for a simple two-neuron-network.

In the previous section, we discussed the differences among $\theta_{12}^{(2,2)}$, $\theta_{12}^{(2,3)}$ and $\theta_{12}^{(3,3)}$. Let's consider how they are related to the network parameters. If the order of the log-linear model (eg, the first superscript 2 in $\theta_i^{(2,3)}$ and $\theta_{12}^{(2,3)}$) and the number of neurons in the network (eg, the second superscript 3 in $\theta_i^{(2,3)}$ and $\theta_{12}^{(2,3)}$) are different, the simple relationships in Eq. (91) do not hold anymore, even for symmetric networks. Suppose that we recorded the activity of neuron 1 and neuron 2 in the network of 3 neurons. We calculate the IG measures as $\theta_1^{(2,3)} = \log \frac{p_{10*}}{p_{00*}}, \theta_2^{(2,3)} = \log \frac{p_{01*}}{p_{00*}}$, and $\theta_{12}^{(2,3)} = \log \frac{p_{11*}p_{00*}}{p_{01*}p_{10*}}$. $p_{x_1 x_2 *}$ is the marginal distribution of x_1 and x_2. In this case, even the connections are symmetric $(J_{ij} = J_{ji})$, we find the IG measures have additional bias terms as

$$\theta_1^{(2,3)} = 2\beta(h_1 - m) + \log \frac{1 + \exp[2\beta((h_3 - m) + J_{13})]}{1 + \exp[2\beta(h_3 - m)]},$$

$$\theta_2^{(2,3)} = 2\beta(h_2 - m) + \log \frac{1 + \exp[2\beta((h_3 - m) + J_{23})]}{1 + \exp[2\beta(h_3 - m)]},$$

$$\theta_{12}^{(2,3)} = 2\beta J_{12} + \log \frac{(1 + \exp[2\beta(h_3 - m)]) (1 + \exp[2\beta((h_3 - m) + J_{13} + J_{23})])}{(1 + \exp[2\beta((h_3 - m) + J_{13})]) (1 + \exp[2\beta((h_3 - m) + J_{23})])}. \tag{92}$$

However, if we calculate $\theta_1^{(3,3)} = \log \frac{p_{100}}{p_{000}}$, $\theta_2^{(3,3)} = \log \frac{p_{010}}{p_{000}}$, and $\theta_{12}^{(3,3)} = \log \frac{p_{110}p_{000}}{p_{100}p_{010}}$ in which the activity of all three neurons in the network are recorded, the bias terms vanish and we can recover,

$$\theta_1^{(3,3)} = 2\beta(h_1 - m), \quad \theta_2^{(3,3)} = 2\beta(h_2 - m), \quad \theta_{12}^{(3,3)} = 2\beta J_{12} \tag{93}$$

In reality, it is not plausible to assume that we know the number of neurons N in the network. However, we can assume that N in the brain is relatively large, eg, in the order of 10^3 to 10^4. Then, an important question is how the logarithmic bias term in $\theta_i^{(2,N)}$ and $\theta_{ij}^{(2,N)}$ scale when N gets large. If the logarithmic term does not increase as $N \to \infty$, the IG measures are expected to provide a good estimation of connection strengths and external inputs. This is the question we investigated in our previous study (Tatsuno et al., 2009). After some algebra, we found the following relationship,

$$\theta_i^{(2,N)} \propto h_i + O\left(\frac{1}{N}\right),$$

$$\theta_{ij}^{(2,N)} \propto (J_{ij} + J_{ji}) + O\left(\frac{1}{N}\right), \tag{94}$$

where $O(1/N)$ represents the term in the order $1/N$. Eq. (94) suggests that the logarithmic bias terms decrease as $O(1/N)$ and the IG measures $\theta_i^{(2,N)}$ and $\theta_{ij}^{(2,N)}$ provide a robust estimate of an external input and connection strength separately.

5 EXTENSION OF THE IG METHOD

We have so far reviewed the basic property of IG method (Section 2), application of IG method to multineuronal spiking data (Section 3), and the relationship between the IG measures and network parameters such as connection weights and external inputs (Section 4). In this section, let us review several recent extensions of the IG method.

5.1 IG Measures Under Correlated Inputs

We have seen that IG measure $\theta_{ij}^{(2,N)}$ is specifically useful to infer the connection weight between two neurons. However, this property would hold only when external inputs to neurons h_i are not correlated. Since neurons in the brain often receive correlated inputs, this would hinder the application of the IG method to real data. In order to overcome this limitation, $\theta_{ij}^{(2,N)}$ has been extended to $\theta_{ij}^{(k,N)}$ with the kth order log-linear model (Nie and Tatsuno, 2012). The goal was to find the order k such that $\theta_{ij}^{(k,N)}$ provides robust estimation of $(J_{ij} + J_{ji})$ when the neurons receive correlated inputs and when the size of network is relatively large (10^3 to 10^4).

Briefly, a simple neural network consisting of a layer of recurrently connected N neurons n_i ($i = 1, ..., N$) was considered. Each neuron in the layer receives a common input W from a single neuron x_0 whose activity can be modulated by an input h. The total inputs to the ith neuron in the layer and to x_0 are written

$$u_i = \sum_{j \neq i} J_{ij} x_j(t) + W x_0, \tag{95}$$

$$u_0 = h. \tag{96}$$

First, in order to obtain explicit analytical results and insights into a more general situation, a simple case in which all recurrent connections are uniformly connected ($J_{ij} = J$) was treated. Following the approach in the previous section and assuming equilibrium, the relationship between firing activity and the network parameters under the correlated inputs can be obtained. For example, if the layer contains only two neurons, the relationship between the neural activity $(\langle x_1 \rangle, \langle x_2 \rangle, \langle x_1 x_2 \rangle, \langle x_1 x_0 \rangle, \langle x_2 x_0 \rangle, \langle x_1 x_2 x_0 \rangle)$ and the network parameters J, W, and h are given as

$$\langle x_i \rangle = \langle x_j x_0 \rangle [g(J + W) - g(J) - g(W) + g(0)],$$
$$\langle x_0 \rangle = g(h),$$
$$\langle x_1 x_2 \rangle = \frac{1}{2} \{ (\langle x_1 x_0 \rangle + \langle x_2 x_0 \rangle)[g(J + W) - g(J)] + (\langle x_1 \rangle + \langle x_2 \rangle)g(J) \},$$
$$\langle x_i x_0 \rangle = \frac{1}{2} \{ \langle x_j x_0 \rangle [g(J + W) - g(W)] + \langle x_i \rangle g(h) + g(W)g(h) \},$$
$$\langle x_1 x_2 x_0 \rangle = \frac{1}{3} \{ \langle x_1 x_2 \rangle g(h) + 2\langle x_1 x_0 \rangle g(J + W) \}. \tag{97}$$

Note that we have up to a three-neuron interaction $\langle x_1 x_2 x_0 \rangle$. The relationship was then converted to the analytical relationship between the IG measures $\theta_{ij}^{(2,2)}$ in the layer and the network parameters using Eqs. (45), (71), and (89).

This procedure was systematically extended to the case in which the layer contains more neurons, eg, up to $N = 10$. We found that $\theta_{ij}^{(2,N)}$ was strongly influenced by the correlated input W, suggesting that $\theta_{ij}^{(2,N)}$ failed to estimate the connection weights for a small network. However, when the order k of log-linear model was increased, $\theta_{ij}^{(k,N)}$ with $k \geq 3$ performed increasingly better. In addition, we also found that when the number of neurons N was increased, $\theta_{ij}^{(k,N)}$ performed increasingly better. Based on these insights from the analytical investigation, we performed computer simulation with more realistic conditions; neurons are asymmetrically connected ($J_{ij} \neq J_{ji}$) and network size is realistically large ($N = 10^3$). We found that $\theta_{ij}^{(4,N)}$ provides an accurate estimation of connection strength within approximately a 10% error. This result suggests that if the IG measure $\theta_{ij}^{(4,N)}$ between neurons i and j is calculated together with the firing activity of two additional neurons, it provides a robust estimator of connection weight even under correlated inputs (Nie and Tatsuno, 2012).

5.2 Correlated Inputs and Higher-Order Interactions

In their seminal theoretical work published in 2003, Amari and his colleagues investigated the mechanisms of how higher-order interactions can arise when they receive common inputs (Amari et al., 2003). They considered a pool of N homogeneous neurons $\boldsymbol{x} = (x_1, x_2, \ldots, x_N)$ and set recurrent connections at zero ($J_{ij} = 0$). Each neuron receives M common inputs $\boldsymbol{s} = (s_1, s_2, \ldots, s_M)$ as

$$u_i = \sum_{j=1}^{M} W_{ij} s_j(t) - h, \tag{98}$$

where W_{ij} is a random binary connection; $W_{ij} = 1$ with probability c and $W_{ij} = 0$ with probability $1 - c$. The output of neurons is determined by a thresholding mechanisms; $x_i = 1$ for $u_i > 0$ and $x_0 = 0$ for $u_i \leq 0$. The common inputs $\boldsymbol{s}$ are assumed to be Gaussian and thus the internal states $\boldsymbol{u} = (u_1, u_2, \ldots, u_N)$ are subject to a joint Gaussian distribution. For simplicity, they normalized the variance of u_i to 1 and their covariance to α. Note that this holds true for any neuron pairs because of homogeneous assumption. Using this model, they investigated the distribution of mean firing activity $q(r)$ in which r is given as $r = \frac{1}{N} \sum x_i$. The $q(r)$ is the fraction of firing neurons in the pool and is a function of the covariance α (in Theorem 1 $q(r)$ is explicitly written as $q(r, \alpha)$). Using the mathematical trick that correlated Gaussian random variables u_i can be represented by independent standard Gaussian random variables v_i and ε as,

$$u_i = \sqrt{1 - \alpha} v_i + \sqrt{\alpha} \varepsilon - h, \tag{99}$$

they derived the following theorem (Amari et al., 2003).

Theorem 1 *The activity distribution* q(r, α) *is given, in the limit* $N \to \infty$, *by*

$$q(r, \alpha) = C \exp \left[\frac{2\alpha - 1}{2(1 - \alpha)} \left\{ F^{-1}(r) - \frac{\sqrt{\alpha}}{2\alpha - 1} h \right\}^2 \right], \tag{100}$$

where C *is the normalization constant and*

$$F(\varepsilon; \alpha, h) = \frac{1}{\sqrt{2\pi}} \int_{\frac{h - \sqrt{\alpha}\varepsilon}{\sqrt{1 - \alpha}}}^{\infty} \exp \left[-\frac{u^2}{2} \right] du. \tag{101}$$

Theorem 1 describes how the activity distribution $q(r, \alpha)$ changes the shape depending on α. Briefly, when $\alpha \sim 0$, $q(r, 0)$ is the delta function and is concentrated. In the range $0 < \alpha < \frac{1}{2}$, $q(r, \alpha)$ has a continuous unimodal shape where peak is found at $r = F\left(\frac{\sqrt{\alpha}}{2\alpha - 1} h \right)$. When $\alpha = \frac{1}{2}$, $q(r, \alpha)$ is a uniform distribution. For $\frac{1}{2} < \alpha \leq 1$, $q(r, \alpha)$ is a bimodal continuous distribution. By IG analysis, they demonstrated that the concentration of distribution was not resolved even when pairwise or third-order interactions are considered. In fact, weak, but all orders of interactions were necessary for generation of a smooth distribution with bimodal peaks.

Interestingly, the bimodal distribution corresponds to synchronous firing of neuron pools. Thus, this study elucidated an important intrinsic mechanism of synchronous firing. It is interesting to note that the distribution of $\boldsymbol{u}$ has the freedom of $\frac{n(n+3)}{2}$ only, but the distribution of $\boldsymbol{x}$, which was generated by thresholding of $\boldsymbol{u}$, has all orders of higher interactions. Further insights have also been obtained by other studies (Hamaguchi et al., 2005; Macke et al., 2011; Montani et al., 2009; Yu et al., 2011).

5.3 Relationship Between Higher-Order IG Measures and Network Parameters

In typical multielectrode recordings, activity of multiple neurons $k\,(\geq 3)$ is recorded simultaneously. Therefore, we can in principle calculate up to the kth order IG measure $\theta_{12,\ldots k}^{(k,N)} x_1 x_2 \ldots x_k$, as in Eq. (67). However, relatively little was known about the property of IG measures involving neuronal interactions ≥ 4. We therefore analytically investigated the influence of external inputs and the level of asymmetry of connections on the IG measures in cases ranging from one-neuron to ten-neuron interactions (Nie et al., 2014a). The system in Eqs. (95) and (96) was extended as follows.

$$u_i(t) = \sum_{j \neq i} J_{ij} x_j(t) + W_{i0} x_0(t) + h_i(t), \qquad (102)$$

$$u_0(t) = h_0(t), \qquad (103)$$

where each neuron in the layer receives a correlated input from a single upstream neuron x_0 with a connection strength represented by W_{i0}. It also receives a background input $h_i(t)$. The upstream neuron x_0 receives a background input $h_0(t)$. A background input is assumed to be a random variable $h_i \sim N(m_i, \sigma_i)$, where $N(m_i, \sigma_i)$ is the normal distribution with the mean (m_i) and variance (σ_i^2). The evolution of the coincident firing of N neurons $x_1(t)x_2(t)\ldots x_N(t)$ is given by Eq. (80) and its equilibrium version is Eq. (83). Assuming uniform connections $(J_{ij} = J$ and $W_{i0} = W)$, we analytically expressed the relationship between the IG measures up to 10-neuron interactions and the network parameters $(J, W,$ and $h)$. We found that all orders of IG measures were strongly influenced by external inputs in a highly nonlinear manner.

Next, we extended our analytical results to asymmetric connections $(J_{ij} \neq J$ and $W_{i0} \neq W)$. By computer simulation, we found that when the network size was increased to 10^3, the nonlinear influence of external inputs almost vanished. The result suggests that all IG measures from one-neuron to ten-neuron interactions are robust against the influence of external inputs. We also investigated how the level of asymmetry of connection weights influenced the IG measures.

Computer simulation revealed that all the IG measures were robust against the modulation of the asymmetry of connections. These results provide additional support that an IG approach provides useful tools for analysis of multineuronal spiking data.

5.4 IG Measures and Oscillatory Brain States

The investigation described so far assumed that the network is in the equilibrium state. Although the equilibrium assumption is useful in theoretical studies, in reality, neural activity is highly nonstationary. For example, the brain exhibits various oscillations depending on cognitive demands or when an animal is asleep. The investigation of the IG measures during oscillatory network states is important for testing if the IG method can be applied to real neural data. Using model networks of binary neurons described above and more realistic spiking neurons (Izhikevich, 2003), we studied how $\theta_i^{(4,N)}$ and $\theta_{ij}^{(4,N)}$ were influenced by oscillatory neural activity (Nie et al., 2014b). Two general oscillatory mechanisms, externally driven oscillations and internally induced oscillations, were considered. In both mechanisms, we found that $\theta_i^{(4,N)}$ was linearly related to the magnitude of the external input, and that $\theta_{ij}^{(4,N)}$ was linearly related to the sum of connection strengths between two neurons. We also observed that $\theta_{ij}^{(4,N)}$ was not dependent on the oscillation frequency. These results are consistent with the previous findings that were obtained under the equilibrium conditions. It was demonstrated that the IG method provides useful insights into neural interactions under the oscillatory conditions that are often observed in the real brain.

5.5 State-Space Analysis of Time-Varying IG Measures

In a recent important extension of the IG method to nonstationary neural spike train data, Shimazaki and his colleagues introduced a state-space approach (Shimazaki et al., 2012). In this study, they developed a recursive Bayesian filter/smoother for the extraction of the IG measures $\boldsymbol{\theta}$. Briefly, they considered neurophysiological experiments in which spiking activity of N neurons are repeatedly obtained under identical n experimental trials. After discretizing neural activity into T small bins, parallel spike sequences in the tth bin, and the lth trial can be represented as a binary vector $\boldsymbol{X}^{t,l} = \left(X_1^{t,l}, X_2^{t,l}, \ldots, X_N^{t,l}\right)$. The key assumption is that $\boldsymbol{X}^{t,l}$ is nonstationary but n samples at a specific time bin t, $(\boldsymbol{X}^{t,1}, \boldsymbol{X}^{t,2}, \ldots, \boldsymbol{X}^{t,n})$ are considered stationary. In order to write the results compactly, they introduced the notation Ω_k that represents a collection of all the k-element subsets of a set with N elements; eg, $\Omega_1 = \{1, 2, ..., N\}$, $\Omega_2 = \{12, 13, ..., (N-1)(N)\}$, etc. They also represented x_i, $x_i x_j, ...,$ and $x_i x_j ... x_N$ as $f_i(\boldsymbol{x}) = x_i$, $f_{ij}(\boldsymbol{x}) = x_i x_j, ...,$ and $f_{i...N}(\boldsymbol{x}) = x_i \cdots x_N$, respectively. Using $f_I(\boldsymbol{x})$ ($I \in \{\Omega_1, ..., \Omega_N\}$), the log-linear model and the expectation parameters can be written compactly as $\log p(\boldsymbol{x}) = \sum_I \theta_I f_I(\boldsymbol{x}) - \psi(\boldsymbol{\theta})$ and $\eta_I = E[f_I(\boldsymbol{x})]$, respectively.

Then an efficient estimator of the observed spike synchrony rate for $I \in \{\Omega_1, ..., \Omega_r\}$ at the tth bin is given as

$$y_I^t = \frac{1}{n} \sum_{l=1}^n f_I\left(\boldsymbol{X}^{t,l}\right). \tag{104}$$

The likelihood of the observed parallel spike sequences is given as,

$$p(\boldsymbol{y}_{1:T}|\boldsymbol{\theta}_{1:T})$$

$$= \prod_{l=1}^n \prod_{t=1}^T \exp\left[\sum_{I \in \{\Omega_1, ..., \Omega_r\}} \theta_I^t f_I\left(\boldsymbol{X}^{t,l}\right) - \psi(\boldsymbol{\theta}^t)\right]$$

$$= \prod_{t=1}^T \exp\left[n \sum_{I \in \{\Omega_1, ..., \Omega_r\}} \theta_I^t y_I^t - \psi(\boldsymbol{\theta}^t)\right], \tag{105}$$

where $\boldsymbol{y}_{1:T} = [\boldsymbol{y}_1, \boldsymbol{y}_2, ..., \boldsymbol{y}_T]$ and $\boldsymbol{\theta}_{1:T} = [\boldsymbol{\theta}_1, \boldsymbol{\theta}_2, ..., \boldsymbol{\theta}_T]$. The time-dependent IG parameters $\boldsymbol{\theta}^t$ are assumed to follow the state-space equation as,

$$\boldsymbol{\theta}^t = \boldsymbol{F}\boldsymbol{\theta}^{t-1} + \boldsymbol{\xi}^t, \tag{106}$$

where a matrix $\boldsymbol{F}$ is the first order autoregressive parameters and $\boldsymbol{\xi}^t$ is a random vector drawn from a zero-mean multivariate Gaussian distribution with covariance matrix $\boldsymbol{Q}$. The initial condition $\boldsymbol{\theta}^1$ is assumed to follow the Gaussian distribution as $\boldsymbol{\theta}^1 \sim N(\boldsymbol{\mu}, \boldsymbol{\Sigma})$. Here $\boldsymbol{w} = (\boldsymbol{F}, \boldsymbol{Q}, \boldsymbol{\mu}, \boldsymbol{\Sigma})$ are called hyperparameters. Given Eqs. (105) and (106), the goal is to estimate the posterior distribution of the IG parameters given the spike data. This can be achieved by Bayes theorem,

$$p(\boldsymbol{\theta}_{1:T}|\boldsymbol{y}_{1:T}, \boldsymbol{w}) = \frac{p(\boldsymbol{y}_{1:T}|\boldsymbol{\theta}_{1:T})p(\boldsymbol{\theta}_{1:T}|\boldsymbol{w})}{\int p(\boldsymbol{y}_{1:T}|\boldsymbol{\theta}_{1:T})p(\boldsymbol{\theta}_{1:T}|\boldsymbol{w})d\boldsymbol{\theta}_{1:T}}$$
$$= \frac{p(\boldsymbol{y}_{1:T}|\boldsymbol{\theta}_{1:T})p(\boldsymbol{\theta}_{1:T}|\boldsymbol{w})}{p(\boldsymbol{y}_{1:T}|\boldsymbol{w})}. \tag{107}$$

The hyperparameters can be optimized by the principle of maximizing the logarithm of the marginal likelihood $l(\boldsymbol{w})$, which is given as,

$$l(\boldsymbol{w}) = \log p(\boldsymbol{y}_{1:T}|\boldsymbol{w}). \tag{108}$$

This is the logarithm of the denominator in Eq. (107). In practice, however, the calculation of Eq. (107) is not easy. Instead, the solution can be obtained by the expectation-maximization algorithm; construction of a posterior density $p(\boldsymbol{\theta}_{1:T}|\boldsymbol{y}_{1:T}, \boldsymbol{w})$ with the given hyperparameters $\boldsymbol{w}$ (E-step) and optimization of the hyperparameters $\boldsymbol{w}$ (M-step) are performed iteratively. Shimazaki et al. (2012) validated the method using simulated spike data with known underlying correlation dynamics. They also applied the method to multineuronal data from the monkey motor cortex (Riehle et al., 1997) and demonstrated that the higher-order spike correlation did organize dynamically in relation to a behavioral demand.

5.6 Estimation of Spiking Irregularities for Nonstationary Firing Rate

Another important extension of the IG method to nonstationary spiking data is the estimation of spike irregularities when firing rate is nonstationary (Miura et al., 2006). It has been demonstrated that ISI histograms of cortical neurons are well described by a gamma distribution (Baker and Lemon, 2000). The gamma distribution is given as

$$q(T; \eta, \kappa) = \frac{(\eta\kappa)^2}{\Gamma(\kappa)} T^{\kappa-1} \exp[-\eta\kappa T], \tag{109}$$

where a random variable T represents an ISI, η is a mean firing rate and κ is a shape parameter that characterizes spiking irregularities. For $\kappa > 1$ an ISI becomes more regular and for $\kappa < 1$ it becomes more irregular. $\kappa = 1$ corresponds to a Poisson process. Therefore, the gamma distribution can be considered as an extension of the Poisson distribution by taking a neuron's refractory period into account. Previous studies suggest that the shape parameter κ is unique to individual neurons (Shinomoto et al., 2005). If this is the case, κ would provide useful information about neuron types and possibly about the function of neurons. Estimation of κ is, however, difficult because the firing rate η is nonstationary and is, in fact, an unknown function that changes with the environment. A statistical model with an unknown function is called a semiparametric model. Assuming that the time-dependent firing rate $\eta(t)$ is stationary over the m consecutive ISIs, Miura and his colleagues applied a novel IG approach and derived the statistical measure S_I for estimating κ without estimating the unknown firing rate $\eta(t)$ (Miura et al., 2006). For example, the measure S_I for $m = 2$ is given as,

$$S_I = -\frac{1}{n-1} \sum_{i=1}^{n-1} \frac{1}{2} \log\left(\frac{4T_i T_{i+1}}{(T_i + T_{i+1})^2}\right), \tag{110}$$

where n is the number of independent observations. Note that S_I is similar to the measure of local variation L_V, which is given as

$$L_V = 3 - \frac{12}{N-1} \sum_{i=1}^{N-1} \frac{T_i T_{i+1}}{(T_i + T_{i+1})^2}. \tag{111}$$

L_V has been shown to be useful for cell classification (Shinomoto et al., 2005, 2003).

6 INFORMATION GEOMETRY AND CLOSED-LOOP NEUROSCIENCE

Information geometry was developed as a branch of mathematics that investigates the space of probability distributions by means of differential geometry (Amari, 1985; Amari and Nagaoka, 2000). It has been primarily used for offline analysis of neural data, as described above. Recently, however, several attempts to extend the IG method to online closed-loop neuroscience, specifically in the framework of BCI, have been reported. In BCI, brain signals such as EEG and single unit activity are translated into control signals for external devices. A key step is to extract relevant information from high-dimensional neurophysiological data. In principle, this can be achieved by directly analyzing the data, but it becomes unfeasible when the size of data gets larger. By taking advantage of the fact that higher-order neural interactions can arise from a thresholding mechanism of Gaussian variables (Amari et al., 2003; Macke et al., 2011) (see Section 5.2), Yu et al. (2011) proposed to an IG method for extracting relevant information efficiently for a BCI application (Yu et al., 2011).

They considered that Gaussian variables correspond to the membrane potentials of neurons and that the thresholding mechanism corresponds to action potential generation. Note that the Gaussian variables contain up to pairwise interactions but action potentials exhibit higher-order interactions. This model, also called the dichotomized Gaussian (DG) model, requires only neurons' firing rates and pairwise correlations to approximate the probabilities of the patterns in the data. It allows efficient characterization of neural activity with a relatively small amount of data, which is desirable in BCI applications. Yu and his colleagues applied the DG model to electrophysiological data from the motor cortex of awake monkeys and from the visual cortex of anesthetized cats (Yu et al., 2011). They reported that the DG model based on information geometry could successfully extract relevant information from relatively a short duration ($\sim$15 min).

In the framework of BCI, characterization of (averaged) signal covariance matrices is often a crucial step in designing BCI. One of the most popular approaches for classifying EEG signals such as those during motor imagery is the common spatial pattern (CSP) algorithm (Ramoser et al., 2000). In short, the CSP tries to find a relevant low-dimensional space by linearly transforming the EEG signals. It aims at maximizing the variance of signals in one class and minimizes the variance of signals in the other class simultaneously. The divergence of covariance matrices has been often treated in Euclidean space but the covariance matrices are symmetric positive-definite (SPD). Therefore, it is possible that the CSP algorithm is better implemented in Riemannian geometry. In fact, it has been shown that consideration taking into account Riemannian structure of the SPD matrices can handle such data sets effectively (Barachant et al., 2012). Investigation of how the Riemannian-geometric and IG approaches promote BCI has become one of the active research areas in closed-loop neuroscience (Samek and Muller, 2014; Yger et al., 2015).

7 CONCLUSION

In this chapter, we have reviewed how the method based on information geometry provides useful toolsets to the analysis of neurophysiological data. We learned that the heart of the IG approach resides in dually orthogonal coordinate systems of exponential family of distributions; one corresponds to neural firings and the other to neural interactions. Information geometry provides hierarchical decomposition of neural interactions such that they are orthogonal to the changes in firing probabilities and this is very useful to assess the amount of higher-order interactions in multineuronal spike data. In addition, using a network of simple stochastic neurons in the equilibrium condition, it has been shown that the single-neuron and two-neuron IG measures correspond to the amount of uncorrelated external inputs and the sum of connection weights between neurons, respectively. Promising attempts have also been made toward the analysis of nonstationary neural signals. In addition, development of novel IG approach toward closed-loop neuroscience, such as BCI by taking Riemannian geometrical structure into account is underway.

We have focused on IG analysis of neural signals, but information geometry has much wider applications, including statistical inference, signal processing, optimization problems, and machine learning. In the machine learning field, deep learning has attracted much attention recently due to its superb performance in image recognition (Hinton et al., 2006; Hinton and Salakhutdinov, 2006). Deep learning can be considered an extension of the brain-inspired information processing, such as the Perceptron by Rosenblatt (Rosenblatt, 1962) in 1960s and the connectionist approach, such as Back Propagation in 1980s (Rumelhart et al., 1986). As a layered network in deep learning and the neocortex share certain common features, understanding the mechanism of deep learning may shed a light on how the neocortex processes information.

In fact, the manifold of connection weights of layered neural networks lies in Riemannian space and therefore it can be treated in the framework of information geometry. Although it has been shown that this is challenging due to intrinsic singularities of the system (Wei et al., 2008), information geometry is expected to provide a novel insight into how the deep learning works and hence, how the brain process probabilistic information efficiently.

REFERENCES

Adrian, E.D., 1928. The Basis of Sensation. Christophers, London.

Amari, S., 1985. Differential Geometrical Methods in Statistics. Springer-Verlag, Berlin.

Amari, S., 2001. Information geometry on hierarchy of probability distributions. IEEE Trans. Inf. Theory 47 (5), 1701–1711.

Amari, S., 2009. Measure of correlation orthogonal to change in firing rate. Neural Comput. 21 (4), 960–972. http://dx.doi.org/10.1162/neco.2008.03-08-729.

Amari, S., Nagaoka, H., 2000. Methods of Information Geometry. Oxford University Press, New York, NY.

Amari, S., Nakahara, H., Wu, S., Sakai, Y., 2003. Synchronous firing and higher-order interactions in neuron pool. Neural Comput. 15 (1), 127–142.

Baker, S.N., Lemon, R.N., 2000. Precise spatiotemporal repeating patterns in monkey primary and supplementary motor areas occur at chance levels (Research Support, Non-U.S. Gov't). J. Neurophysiol. 84 (4), 1770–1780.

Barachant, A., Bonnet, S., Congedo, M., Jutten, C., 2012. Multiclass brain-computer interface classification by Riemannian geometry. (Research Support, Non-U.S. Gov't). IEEE Trans. Biomed. Eng. 59 (4), 920–928. http://dx.doi.org/10.1109/TBME.2011.2172210.

Barlow, H.B., 1972. Single units and sensation: a neuron doctrine for perceptual psychology? Perception 1 (4), 371–394.

Buzsaki, G., 2004. Large-scale recording of neuronal ensembles. Nat. Neurosci. 7 (5), 446–451.

Cena, A., Pistone, G., 2007. Exponential statistical manifold. Ann. Inst. Stat. Math. 59 (1), 27–56. http://dx.doi.org/10.1007/s10463-006-0096-y.

Chapin, J.K., Moxon, K.A., Markowitz, R.S., Nicolelis, M.A., 1999. Real-time control of a robot arm using simultaneously recorded neurons in the motor cortex. Nat. Neurosci. 2 (7), 664–670.

Cocco, S., Leibler, S., Monasson, R., 2009. Neuronal couplings between retinal ganglion cells inferred by efficient inverse statistical physics methods. Proc. Natl. Acad. Sci. U. S. A. 106 (33), 14058–14062. http://dx.doi.org/10.1073/pnas.0906705106.

Davidson, T.J., Kloosterman, F., Wilson, M.A., 2009. Hippocampal replay of extended experience. Neuron 63 (4), 497–507. http://dx.doi.org/10.1016/j.neuron.2009.07.027. pii:S0896-6273(09)00582-0.

Dragoi, G., Tonegawa, S., 2013. Distinct preplay of multiple novel spatial experiences in the rat. (Research Support, N.I.H., Extramural Research Support, Non-U.S. Gov't). Proc. Natl. Acad. Sci. U. S. A. 110 (22), 9100–9105. http://dx.doi.org/10.1073/pnas.1306031110.

Euston, D.R., Tatsuno, M., McNaughton, B.L., 2007. Fast-forward playback of recent memory sequences in prefrontal cortex during sleep. Science 318 (5853), 1147–1150.

Ganmor, E., Segev, R., Schneidman, E., 2011. Sparse low-order interaction network underlies a highly correlated and learnable neural population code. Proc. Natl. Acad. Sci. U. S. A. 108 (23), 9679–9684. http://dx.doi.org/10.1073/pnas.1019641108. pii:1019641108.

Ginzburg, I.I., Sompolinsky, H., 1994. Theory of correlations in stochastic neural networks. Phys. Rev. E Stat. Phys. Plasmas Fluids Relat. Interdiscip. Top. 50 (4), 3171–3191.

Hamaguchi, K., Okada, M., Yamana, M., Aihara, K., 2005. Correlated firing in a feedforward network with Mexican-hat-type connectivity (Research Support, Non-U.S. Gov't). Neural Comput. 17 (9), 2034–2059. http://dx.doi.org/10.1162/0899766054322937.

Hinton, G.E., Salakhutdinov, R.R., 2006. Reducing the dimensionality of data with neural networks. Science 313 (5786), 504–507. http://dx.doi.org/10.1126/science.1127647.

Hinton, G.E., Osindero, S., Teh, Y.W., 2006. A fast learning algorithm for deep belief nets. (Research Support, Non-U.S. Gov't). Neural Comput. 18 (7), 1527–1554. http://dx.doi.org/10.1162/neco.2006.18.7.1527.

Hoffman, K.L., McNaughton, B.L., 2002. Coordinated reactivation of distributed memory traces in primate neocortex. Science 297 (5589), 2070–2073.

Ikegaya, Y., Aaron, G., Cossart, R., Aronov, D., Lampl, I., Ferster, D., Yuste, R., 2004. Synfire chains and cortical songs: temporal modules of cortical activity. Science 304 (5670), 559–564.

Izhikevich, E.M., 2003. Simple model of spiking neurons. IEEE Trans. Neural Netw. 14 (6), 1569–1572. http://dx.doi.org/10.1109/TNN.2003.820440.

Jaynes, E.T., 1957. Information theory and statistical mechanics. Phys. Rev. E Stat. Phys. Plasmas Fluids Relat. Interdiscip. Top. 106, 620–630.

Kudrimoti, H.S., Barnes, C.A., McNaughton, B.L., 1999. Reactivation of hippocampal cell assemblies: effects of behavioral state, experience, and EEG dynamics. J. Neurosci. 19 (10), 4090–4101.

Kullback, S., Leibler, R.A., 1951. On information and sufficiency. Ann. Math. Stat. 22 (1), 79–86.

Laubach, M., Wessberg, J., Nicolelis, M.A., 2000. Cortical ensemble activity increasingly predicts behaviour outcomes during learning of a motor task. Nature 405 (6786), 567–571. http://dx.doi.org/10.1038/35014604.

Lee, A.K., Wilson, M.A., 2002. Memory of sequential experience in the hippocampus during slow wave sleep. Neuron 36 (6), 1183–1194.

Louie, K., Wilson, M.A., 2001. Temporally structured replay of awake hippocampal ensemble activity during rapid eye movement sleep. Neuron 29 (1), 145–156.

Maass, W., Bishop, C., 1998. Pulsed Neural Networks. MIT Press, Cambridge, MA.

Macke, J.H., Opper, M., Bethge, M., 2011. Common input explains higher-order correlations and entropy in a simple model of neural population activity. (Research Support, Non-U.S. Gov't). Phys. Rev. Lett. 106 (20), 208102.

Miura, K., Okada, M., Amari, S., 2006. Estimating spiking irregularities under changing environments. Neural Comput. 18 (10), 2359–2386.

Montani, F., Ince, R.A., Senatore, R., Arabzadeh, E., Diamond, M.E., Panzeri, S., 2009. The impact of high-order interactions on the rate of synchronous discharge and information transmission in somatosensory cortex. (Research Support, Non-U.S. Gov't). Philos. Trans. A Math. Phys. Eng. Sci. 367 (1901), 3297–3310. http://dx.doi.org/10.1098/rsta.2009.0082.

Murray, M.K., Rice, J.W., 1993. Differential Geometry and Statistics, vol. 48 Chapman and Hall, London.

Nakahara, H., Amari, S., 2002. Information-geometric measure for neural spikes. Neural Comput. 14 (10), 2269–2316.

Newton, N.J., 2012. An infinite-dimensional statistical manifold modelled on Hilbert space. J. Funct. Anal. 263 (6), 1661–1681. http://dx.doi.org/10.1016/j.jfa.2012.06.007.

Nie, Y., Tatsuno, M., 2012. Information-geometric measures for estimation of connection weight under correlated inputs. Neural Comput. http://dx.doi.org/10.1162/NECO_a_00367.

Nie, Y., Fellous, J.M., Tatsuno, M., 2014a. Information-geometric measures estimate neural interactions during oscillatory brain states. (Research Support, Non-U.S. Gov't). Front. Neural Circuits 8, 11. http://dx.doi.org/10.3389/fncir.2014.00011

Nie, Y., Fellous, J.M., Tatsuno, M., 2014b. Influence of external inputs and asymmetry of connections on information-geometric measures involving up to ten neuronal interactions. Neural Comput. 26 (10), 2247–2293. http://dx.doi.org/10.1162/NECO_a_00633.

Ohiorhenuan, I.E., Mechler, F., Purpura, K.P., Schmid, A.M., Hu, Q., Victor, J.D., 2010. Sparse coding and high-order correlations in fine-scale cortical networks. Nature 466 (7306), 617–621. http://dx.doi.org/10.1038/nature09178. pii:nature09178.

Pavlides, C., Winson, J., 1989. Influences of hippocampal place cell firing in the awake state on the activity of these cells during subsequent sleep episodes. J. Neurosci. 9 (8), 2907–2918.

Peyrache, A., Khamassi, M., Benchenane, K., Wiener, S.I., Battaglia, F.P., 2009. Replay of rule-learning related neural patterns in the prefrontal cortex during sleep. Nat. Neurosci. 12 (7), 919–926. http://dx.doi.org/10.1038/nn.2337 pii:nn.2337.

Ramoser, H., Muller-Gerking, J., Pfurtscheller, G., 2000. Optimal spatial filtering of single trial EEG during imagined hand movement. (Clinical Trial Randomized Controlled Trial Research Support, Non-U.S. Gov't). IEEE Trans. Rehabil. Eng. 8 (4), 441–446.

Riehle, A., Grun, S., Diesmann, M., Aertsen, A., 1997. Spike synchronization and rate modulation differentially involved in motor cortical function. Science 278 (5345), 1950–1953.

Rosenblatt, F., 1962. Principles of Neurodynamics. Spartan Books, Washington, DC.

Rumelhart, D.E., Hinton, G.E., Williams, R.J., 1986. Learning representations by back-propagating errors. Nature 323 (6088), 533–536.

Samek, W., Muller, K.R., 2014. Information geometry meets BCI: Spatial filtering using divergences. In: Paper Presented at the 2014 International Winter Workshop on Brain-Computer Interface, Jeongsun-kun.

Schneidman, E., Berry 2nd, M.J., Segev, R., Bialek, W., 2006. Weak pairwise correlations imply strongly correlated network states in a neural population. Nature 440 (7087), 1007–1012.

Shimazaki, H., Amari, S., Brown, E.N., Grun, S., 2012. State-space analysis of time-varying higher-order spike correlation for multiple neural spike train data. (Research Support, N.I.H., Extramural Research Support, Non-U.S. Gov't). PLoS Comput. Biol. 8 (3), e1002385. http://dx.doi.org/10.1371/journal.pcbi.1002385.

Shinomoto, S., Shima, K., Tanji, J., 2003. Differences in spiking patterns among cortical neurons. (Research Support, Non-U.S. Gov't). Neural Comput. 15 (12), 2823–2842. http://dx.doi.org/10.1162/089976603322518759.

Shinomoto, S., Miyazaki, Y., Tamura, H., Fujita, I., 2005. Regional and laminar differences in in vivo firing patterns of primate cortical neurons. (Comparative Study Research Support, Non-U.S. Gov't). J. Neurophysiol. 94 (1), 567–575. http://dx.doi.org/10.1152/jn.00896.2004.

Shlens, J., Field, G.D., Gauthier, J.L., Grivich, M.I., Petrusca, D., Sher, A., Litke, A.M., Chichilnisky, E.J., 2006. The structure of multi-neuron firing patterns in primate retina. J. Neurosci. 26 (32), 8254–8266.

Shlens, J., Field, G.D., Gauthier, J.L., Greschner, M., Sher, A., Litke, A.M., Chichilnisky, E.J., 2009. The structure of large-scale synchronized firing in primate retina. (Comparative Study Research Support, N.I.H., Extramural Research Support, Non-U.S. Gov't Research Support, U.S. Gov't, Non-P.H.S.). J. Neurosci. Off. J. Soc. Neurosci. 29 (15), 5022–5031. http://dx.doi.org/10.1523/JNEUROSCI.5187-08.2009.

Skaggs, W.E., McNaughton, B.L., 1996. Replay of neuronal firing sequences in rat hippocampus during sleep following spatial experience. Science 271 (5257), 1870–1873.

Tang, A., Jackson, D., Hobbs, J., Chen, W., Smith, J.L., Patel, H., Prieto, A., Petrusca, D., Grivich, M.I., Sher, A., Hottowy, P., Dabrowski, W., Litke, A.M., Beggs, J.M., 2008. A maximum entropy model applied to spatial and temporal correlations from cortical networks in vitro. J. Neurosci. 28 (2), 505–518. http://dx.doi.org/10.1523/JNEUROSCI.3359-07.2008 pii:28/2/505.

Tatsuno, M., 2015. Analysis and Modeling of Coordinated Multi-Neuronal Activity. Springer, New York, NY.

Tatsuno, M., Okada, M., 2004. Investigation of possible neural architectures underlying information-geometric measures. Neural Comput. 16 (4), 737–765.

Tatsuno, M., Lipa, P., McNaughton, B.L., 2006. Methodological considerations on the use of template matching to study long-lasting memory trace replay. J. Neurosci. 26 (42), 10727–10742.

Tatsuno, M., Fellous, J.M., Amari, S.I., 2009. Information-geometric measures as robust estimators of connection strengths and external inputs. Neural Comput. 21 (8), 2309–2335. http://dx.doi.org/10.1162/neco.2009.04-08-748.

Wei, H., Zhang, J., Cousseau, F., Ozeki, T., Amari, S., 2008. Dynamics of learning near singularities in layered networks. Neural Comput. 20 (3), 813–843. http://dx.doi.org/10.1162/neco.2007.12-06-414.

Wilson, M.A., McNaughton, B.L., 1993. Dynamics of the hippocampal ensemble code for space. Science 261 (5124), 1055–1058.

Wilson, M.A., McNaughton, B.L., 1994. Reactivation of hippocampal ensemble memories during sleep. Science 265 (5172), 676–679.

Yger, F., Lotte, F., Sugiyama, M., 2015. Averaging covariance matrices for EEG signal classification based on the CSP: an empirical study. In: Paper Presented at the EUSIPCO 2015, Nice, France.

Yu, S., Yang, H., Nakahara, H., Santos, G.S., Nikolic, D., Plenz, D., 2011. Higher-order interactions characterized in cortical activity. (Research Support, N.I.H., Intramural Research Support, Non-U.S. Gov't). J. Neurosci. Off. J. Soc. Neurosci. 31 (48), 17514–17526. http://dx.doi.org/10.1523/JNEUROSCI.3127-11.2011.

Chapter 3

Control Theory for Closed-Loop Neurophysiology

G. Kumar*, J.T. Ritt[†] and S. Ching*

Washington University, St. Louis, MO, United States, [†]Boston University, Boston, MA, United States

1 INTRODUCTION

Neuroscience and neurology are experiencing rapid development of new tools and technology for modifying neural activity. These tools hold extraordinary promise as probes with which to elucidate basic mechanisms in both the normal and pathological brain, and potentially as new therapeutic options in neurological disease. A major goal in the development of these tools is to eventually enable closed-loop instantiations, wherein brain activity is modified in near real-time as a function of observed activity. Many outstanding problems remain, however, in the pursuit of this goal including the development of methods to robustly model, measure, and alter the behavior of highly complex and nonlinear neuronal circuits.

Electrophysiological measurement of nerve activity extends as far back as the modern understanding of electricity, for example, in Galvani's pioneering experiments (Piccolino, 1997). Examples of crude forms of stimulating the nervous systems can be found in antiquity (Rossi, 2003). In modern times, structured stimulation has been instrumental in uncovering some aspects of neural organization (Penfield and Boldrey, 1937). Neurostimulation has revealed a great deal about visual motion discrimination and control of eye motions and integration of tactile inputs (Romo et al., 1993). Clinical successes include deep brain stimulation (DBS) as a treatment for Parkinson's disease and an ever widening list of other conditions (Perlmutter and Mink, 2006), and cochlear implants to restore hearing (Rauschecker and Shannon, 2002). However, most of these applications take advantage of fortuitous details of neural organization (such as a spatial organization of frequency tuning in the cochlea, or a localized area that can be "transiently ablated" in DBS). In particular, typically stimulation is open loop and follows a binary design (on or off), even if initial parameters are calibrated by a clinician or scientist.

To fill these gaps, interest has continued to grow in approaching neurostimulation from the perspective of systems and control theory—the study of dynamical systems subject to input (and, specifically, closed-loop feedback) (Schiff, 2012; Agarwal and Sarma, 2012a; Santaniello et al., 2012, 2011)—in attempts to develop more effective outcomes through principled design of stimulation parameters.

In this chapter, we discuss some of the principal research directions that may help to enable control-theoretic approaches in closed-loop neuroscience. We offer a survey of existing paradigms and efforts in the modeling and control of neural dynamics; and how emerging research challenges may influence the evolution of these previous efforts into new theoretical and experimental lines of investigation. This chapter is mostly focussed on closed-loop neuroscience in the context of stimulation as an actuation modality within the closed-loop paradigm. Thus, we emphasize this chapter is not exhaustive. For instance, we will only briefly discuss the extensive line of research on design of sensory stimuli in a closed-loop fashion and the more recent use of control theory for study of the intrinsic feedback mechanisms in neural circuits, absent overt stimulation.

This chapter is written for both engineering and scientific audiences, at a depth that is intended to introduce concepts at a relatively high level. For control systems engineers and theorists who are interested in entering these areas in more depth, several excellent texts are available. The books of Dayan and Abbott (2005), Izhikevich (2007), and Ermentrout and Terman (2010) provide rigorous, but highly accessible overviews of basic and computational neuroscience. The recent monograph by Schiff (2012) offers a cross section of some of the key domain areas within neuroscience and brain medicine where control theory can have an immediate impact.

The remainder of our paper is organized thus: in Section 2 we provide an overview of basic neuroscience concepts and review several of the most well-established paradigms for modeling neuronal activity. In Section 3 we proceed to discuss existing problems in neurostimulation, and present existing and emerging methods for

Closed Loop Neuroscience. http://dx.doi.org/10.1016/B978-0-12-802452-2.00003-2

bringing principled control-theoretic approaches to such problems. In Section 4, we discuss the many paradigms that have been developed for observing, estimating, and inferring parameters for these models from neuroscience data. Finally, in Section 6 we discuss wider challenges in neurocontrol, including the definition of objectives that go beyond the space of neural activity and into harder to define functional and behavioral metrics.

2 DYNAMICS OF NEURAL PHYSIOLOGY AND MODELING PARADIGMS

2.1 Fundamentals of Neural Function

For readers with a background in systems and control theory, we provide here a brief section overviewing basic neurophysiology. The below summary includes the major ideas making up a consensus view among neuroscientists, but it should be noted that neural systems show tremendous diversity, and almost every rule admits some exception. Our intent is to lay out a partially schematic view that still captures the salient features relevant to control problems. Readers interested in further detail are pointed to the monographs noted in Section 1, as well as the comprehensive neuroscience and neurobiology textbooks (Kandel and Schwartz, 2013; Nicholls, 2012; Bear et al., 2007; Johnston and Wu, 1995).

The basic cellular unit of the nervous system is the neuron (Johnston and Wu, 1995; Kandel and Schwartz, 2013) (see Fig. 1). A typical neuron is comprised of a cell body (or soma); one or more dendritic trees, which are thin, branching processes eminating from the soma and permeating the region near the cell; and an axon, which is another thin process that can extend large distances from the soma, often branching along the way. Like all cells, neurons are encapsulated by a cellular membrane that separates the extracellular solution outside the cell from the intracellular solution (or cytoplasm). An essential feature of neurons is that, through various mechanisms that control transport of charged molecules across their membrane, they maintain a voltage difference of around 60–90 mV, comparing the cytoplasm to the outside fluid. This voltage difference is connected to a concentration gradient of various species (eg, there is a substantially higher concentration of sodium ions outside than inside a neuron, and the reverse is true for potassium ions). The most significant behavior of a neuron is the action potential, or less formally, "spike," which is a fast (around 1 ms) large (around 100 mV) depolarization and repolarization of the cell membrane, that usually starts where the axon emerges from the soma (called the axon hillock), and propagates down the axon to its terminals.

Neurons connect to each other through synapses, of which there are many types. The most common are chemical synapses, which are usually located at the termination of axon branches, juxtaposed against another neuron's dendrite or soma. The axon is said to come from a presynaptic neuron, while the cell on the other side of the synaptic junction is said to be postsynaptic. When an action potential arrives at a synapse, chemicals known as neurotransmitters are released into the synaptic cleft, where they bind with receptors in the membrane of the postsynaptic cell, causing effects that depend on the transmitter and receptor. At excitatory synapses, the postsynaptic neuron is depolarized (the membrane potential becomes less negative); at inhibitory synapses the opposite occurs. There are many variations on this theme, including neuromodulators that can have long-lasting and

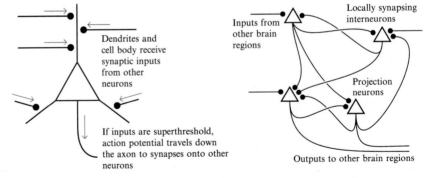

FIG. 1 Schematic depictions of major features of neural function. Left: Neurons consist of a cell body, dendrites, and axons. Dendrites are small diameter processes that are a principal termination point for synapses (*black circles*) from other neurons that provide excitatory or inhibitory inputs. If the cell membrane potential reaches threshold, an action potential is generated at the axon hillock and travels down the axon; the neuron's axon typically makes multiple branches, which ultimately terminate in synapses onto other neurons. Gray arrows depict the convention understanding of information flow. Right: Networks of neurons interact to produce the complex dynamics that underlie brain function. A typical brain region will contain local interneurons that synapse only onto nearby neurons, and projection neurons that send axons to other brain regions, possibly in addition to making local connections. Some subset of neurons will also receive synaptic input from other regions. There is an extraordinary degree of diversity in neuron structure, membrane dynamics, biochemical makeup, and projection pattern found throughout the brain.

Dendrites and cell body receive synaptic inputs from other neurons

If inputs are superthreshold, action potential travels down the axon to synapses onto other neurons

Inputs from other brain regions

Locally synapsing interneurons

Projection neurons

Outputs to other brain regions

indirect effects, rather than directly influencing membrane potential. The overall picture at the single neuron level is that it receives synaptic input to its dendrite and soma from other neurons (a typical neuron in mammalian neocortex may receive 10,000 synapses), and if the net effect of all those inputs at a given time is to raise the soma membrane potential a sufficient amount, the neuron will send an action potential down its own axon, inducing synaptic inputs onto other neurons. For a typical neuron, the threshold amount of depolarization required to initiate an action potential is around 5–20 mV.

The initiation and propagation of action potentials are generally considered the central factor of brain and neural function. When a neuron spikes, it can be thought of as a signal being sent down the axon to other neurons. The aggregate signaling of all the neurons in a nervous system comprises the information processing and behavioral capabilities of that system. At a systems level, the interesting unit of computation is a (local) network of neurons (Fig. 1) which may have an extensive pattern of connectivity, including interneurons whose synapses are primarily on other nearby neurons. At a still larger scale, local brain regions appear to subserve specific functions (eg, processing of visual input, or planning of actions), and are themselves interconnected by "projection neurons" that have long axons spanning multiple brain regions. In formulating a control objective, it is important to determine the scale at which control is required (single cell, local network, or multiregion), and the associated structure of synaptic connectivity and the dynamical properties of membrane fluctuations in response to inputs. We give further details in the context of standard models of neural activity in the following section.

We note that at a coarse scale, the nervous system is divided into peripheral and central systems. The central nervous system, consisting of the spine and brain, is considered the core of computational capability of the nervous system. Peripheral nerves (primarily bundles of axons) serve mainly to convey information to or from the many different parts of the body (muscles, skin, vital, and sensory organs). Although there are many clinical applications of peripheral nerve stimulation, and some important remaining problems, we focus here on technologies and approaches with primary applicability to the central nervous system, which presents far more complicated structures and neural interactions, and hence also substantial challenges requiring development of novel ideas in control theory.

2.2 Dynamical Systems Models: Neuron-Level Models

As one may surmise, the biophysics through which neurons produce and transmit their activity admits a fundamental mathematical characterization. Indeed, the field of mathematical modeling in neuroscience dates to the seminal work of Hodgkin and Huxley (1952), who provided a dynamical systems-based description of action potential production in neurons of the giant squid. The key insight of Hodgkin and Huxley was that the production of each spike involved both positive and negative feedback mechanisms to modulate the flow of charged ions across a cell membrane; leading, respectively to rapid depolarization (ie, the spike upstroke), followed by repolarization (ie, the spike downstroke).

2.2.1 Voltage-Gated Conductance Equations

Essential in the formulation of Hodgkin and Huxley was the mechanism that mediated this positive and negative feedback: voltage gating of ionic conductance. The Hodgkin-Huxley (HH) model contains three trans-membrane currents: sodium, potassium, and a nonspecific "Leak" current (see Fig. 2 for the equivalent circuit). The leak current is a strictly passive current, in that its conductance is fixed. In contrast, the trans-membrane currents due to sodium and potassium are mediated by conductances that are themselves gated by the membrane potential (see Fig. 3).[1] The nonlinear dynamics of this gating are the substrate through which spikes are generated. For the purposes of our discussion, assume that the current due to sodium serves to depolarize the membrane; while that due to Potassium serves to hyperpolarize. In the HH model, a slight depolarization in the membrane potential of a neuron above its rest (equilibrium) potential leads to a rapid increase in sodium conductance and further depolarization, that is, a positive feedback. As the membrane becomes progressively more depolarized, however, the sodium conductance becomes actively attenuated and potassium conductance increases, that is, negative feedback. In concert, these effects lead to an equally rapid repolarization, usually involving a certain amount of hyperpolarization, that is, undershoot, giving the spike its characteristic morphology (see Fig. 3).

Hodgkin and Huxley formulated their model with a total of four state variables: one for the membrane potential, two for sodium, and one for potassium. Since its inception, significant effort has gone into developing models for the

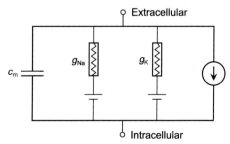

FIG. 2 Hodgkin-Huxley equivalent circuit model.

---

1. Note that each transmembrane current depends intimately on both electrical and diffusive forces, that is, concentration gradients.

FIG. 3 Voltage gating of the membrane conductance is the key mechanism underlying the generation of action potentials.

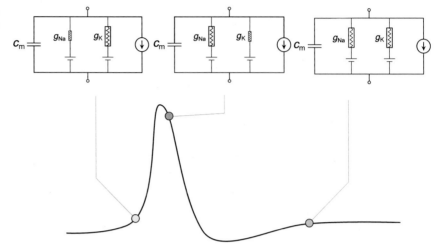

myriad of other ion channels that exist in different types of neurons. In this context, the basic modeling paradigm of Hodgkin and Huxley persists, wherein the dynamics of a neuron are described through equations of the form:

$$c_m \frac{dV}{dt} = -\sum I_{ion} - \sum I_{syn} + I_{app}, \qquad (1)$$

where I_{ion}, I_{syn}, and I_{app} denote, respectively, the ionic, synaptic, and external applied currents specific to each cell, and c_m is the membrane capacitance (see, again, Fig. 2). Ionic currents are described by equations of the form

$$I_x(V) = \bar{g}_x m^p h^q (V - E_x), \qquad (2)$$

where

$$\dot{m} = \frac{m_\infty(V) - m}{\tau_m(V)}, \qquad \dot{h} = \frac{h_\infty(V) - h}{\tau_h(V)}. \qquad (3)$$

Note, importantly, that these currents are themselves functions of the membrane potential V. The parameters $\bar{g}_x$ and E_x are constant, while p and q are nonnegative integers. The term $\bar{g}_x m^p h^q$ is the ionic conductance and determines the behavior of the given current as the voltage deviates from equilibrium.

Connectivity between cells is established through the synaptic currents I_{syn}, which are described by an equation of the form

$$I_{syn} = \bar{g} s(v_{pre})(V - E_s), \qquad (4)$$

where s is an activation variable that depends on the voltage of the presynaptic cell v_{pre} and E_s is a constant.

For sheer versatility of dynamical regimes, the HH/voltage-gated conductance paradigm cannot be matched by any other class of dynamical systems neuronal model. Generalizing the model to include different channels and spatial details has been successfully carried out toward describing a wide range of experimentally observed neural

phenomena (Ermentrout and Terman, 2010). This explanatory power, of course, comes at the cost of analytical and numerical tractability, especially when formulating and studying larger networks of neurons. Since the original formulation of Hodgkin and Huxley, several models have been formulated that preserve many of its dynamical features, but with lower dimension (Izhikevich, 2004, 2003) and overall equation complexity.

Later, we will discuss how specific reductions and approximations of these spiking neuron models (eg, integrate and fire models; and phase oscillator-based models) have been used to enable control design for regime-specific objectives such as manipulating spike timing and synchronization.

2.3 Statistical Models

Two pivotal problems in the analysis of neural data pertain to encoding (predicting the firing rate of a population from an arbitrary stimulus) and decoding (estimating the stimulus based on neural response data). A highly successful approach to these problems has been through the development of statistical models, often embedded in formal state space methods (Chen et al., 2014; Chen, 2014), to describe observed patterns of neuronal activation (Paninski et al., 2007). In statistical models, such as those embedded in the point process generalized linear model paradigm (Paninski, 2004), the probability of a neuron producing a spike at a given time t is described by a rate $\lambda(t)$, akin to the rate parameter of a renewal stochastic process. The time evolution of $\lambda(t)$ is determined through a prior state model, for example, of the autoregressive type. The fitting and parameterization of such models offer mechanistic insight into the neuronal network interactions that might mediate observed phenomenology. Later, we will discuss how similar methods have been used for the specific problem of inferring connectivity between neurons and brain regions.

2.4 Mean Field Models

Motivated by mean field theory, a class of models has been developed that seeks to create low-dimensional descriptions of the collective activity of large populations of neurons. Several popular examples of mean field neuronal models exist, most notably the well-established Wilson-Cowan model (Wilson and Cowan, 1972, 1973), wherein the dynamics in a cortical macrocolumn are reduced to a two-dimensional descriptions of the form

$$\dot{e}_j = -e_j + (k_e - r_e e_j)\mathcal{F}[c_1 e_j - c_2 i_j + P + \phi(t)] + W(t), \quad (5)$$

$$\dot{i}_j = -i_j + (k_e - r_i e_j)\mathcal{F}[c_3 e_j - c_4 i_j + Q], \quad (6)$$

where e_j and i_j represent the overall activity in the excitatory and inhibitory populations. The function $\mathcal{F}$ is a logistic sigmoid of the form

$$\mathcal{F}(x) = \frac{1}{1 + \exp[-a(x - \theta)]} - \frac{1}{1 + \exp(a\theta)}, \quad (7)$$

where a and θ are free parameters. The constants P and Q determine the level of excitation present in the system. Depending on these values, the system may exhibit either a stable equilibrium or periodic limit cycle behavior. Indeed, from a dynamical-systems perspective, this model is not qualitatively different from many low-dimensional single neuron models.

The continuum mean field model by Liley et al. (2002) similarly describes dynamics at the macrocolumn level and has been successfully been used to derive mechanistic interpretations for observed electroencephalogram patterns associated with anesthesia and other drug effects (Foster et al., 2011, 2008).

3 FROM NEUROSTIMULATION TO NEUROCONTROL

Controlling the dynamics of neurons and populations thereof is a highly nontrivial problem because complex dynamics, nonlinearities, and irregularities appear in individual cells and the couplings between them. The *underactuated* nature of most current brain stimulation modalities adds to this complexity, because a single actuator (eg, an optical fiber) typically affects thousands of neurons so that single units cannot be "addressed" individually. Finding a tractable yet meaningful mathematical formulation for such control problems is a fundamental challenge.

3.1 Actuating the Brain: Technologies for Neurostimulation

Complementing a wide range of technologies for measuring neural activity is a wide and rapidly expanding range of stimulation methods. These can be classified among several overlapping categories. At the coarsest level, stimulation can be invasive, requiring penetration of the device into tissue, or noninvasive, in which devices external to the body modulate neural activity by imposing electrical or magnetic fields. Examples of the latter include transcranial magnetic stimulation (TMS) (Walsh et al., 2003) and transcranial direct or alternating current stimulation (tDCS or tACS) (Gandiga et al., 2006). While these technologies are under active research and hold substantial clinical potential, they usually operate at large spatial scales and in the absence of detailed knowledge of their effects at the single neuron or local network level, and we focus on control at a finer scale using invasive technologies.

Invasive stimulation includes electrical, optical, pharmacological, and thermal methods. Again there is a difference in scale, with pharmacological (Devor and Zalkind, 2001) and thermal (Nakayama et al., 1963) typically providing modulation on long time scales and/or large spatial scales, rather than moment to moment control within specified networks. Other hybrid methods with similar modulatory properties are available, such as optical alteration of gene expression. However, by far the majority of applications considered from a control perspective employ electrical stimulation, including all currently available clinical devices. Optical methods hold immense promise and are under rapid development, though so far without a certain path to human application (Bernstein and Boyden, 2011; Deisseroth, 2011). We will focus on those electrical and optical approaches that provide the greatest interest for control engineering at time scales of order 100 ms or below.

As discussed above, neurons typically maintain a -60 to -90 mV potential across their membrane (inside minus outside) and have action potential thresholds roughly 10–20 mV above their resting potential. In most applications, the control objective is to drive this threshold level depolarization in a selected subset of neurons. For simplicity, we will assume stimulation is moderately fast-acting, and ignore dynamical effects such as depolarization block, in which too slow a rise in membrane potential can result in a "superthreshold" potential that nevertheless does not initiate an action potential. We now summarize the principal mechanisms by which electrical and optical approaches achieve this depolarization.

In most electrical approaches, a distant electrode supplies the current return path and ground potential. There are then two common tools to drive the potential changes, intra- or extracellular electrodes. Intracellular stimulation requires an electrode to physically penetrate the cell membrane and make contact with the (approximately) equipotential cytoplasm inside the cell. Current between stimulating and reference electrodes must thus cross the cell membrane, producing a change in potential. Intracellular electrodes include fluid-filled sharpened glass pipettes ("sharps") that puncture the membrane, and "whole cell" and related techniques where a pipette tip is brought into contact with the membrane, and suction is applied until the membrane breaks inside the tip, resulting in continuity of the cytoplasm and electrode solution. Whole cell interfaces generally result in higher quality electrical connections, but are technically more demanding. While intracellular methods provide the ability to observe subthreshold changes in membrane potential, and drive nearly arbitrary changes in that potential, they are not suitable for most applications of systems level or clinical relevance. Since physical contact is required between the neuron and electrode, only a small number of cells can be controlled at a time, and even very small tissue movement can disrupt the quality of connection.

Therefore almost all applications involve the use of extracellular electrodes, placed near, but not in contact with the neurons of interest, and without direct connection to the internal cytoplasm. How can the potential across the membrane, separating inside and outside of the cell, be altered by electrodes outside the cell? The answer is that the volume conductor resistivity inside and outside the cell creates voltage gradients along the direction of current flow, which if aligned appropriately with the cell structure, induce flows of current across the (resistive) membrane. A similar process in reverse results in loop currents across a neuron (outward flow through one patch of membrane returning through inward flow through another patch) driving a potential change in the extracellular matrix, which can be recorded by a nearby electrode. General discussion of extracellular recording and stimulation can be found in Plonsey and Barr (2007), and an example of the complicated interaction between imposed field shape, cell morphology, and stimulation pattern can be found in McIntyre et al. (2004).

However, an essential fact of extracellular recordings is that they generally are unable to resolve subthreshold changes in membrane potential, and are limited to the detection of the timing of action potentials. Also, it should be clear that an extracellular electrode must perform some form of source separation if multiple neurons are nearby, resulting in a persistent ambiguity as to what is being recorded. During stimulation, the electrode imposes a nonuniform electric field that influences all neurons in the neighborhood of its tip, making selective stimulation of subsets of neurons difficult to impossible (techniques of current steering with multiple contacts provide some improvement (Butson and McIntyre, 2008), but do not fundamentally alter the electrode-neuron coupling mechanism). Thus, while extracellular electrodes form the only currently viable hardware for most applications, the significant limitations inherent in their mechanisms of recording and stimulation create major challenges to the control engineer, which we discuss more fully in Section 3.2.

As an alternative to electrical interfaces, the last decade has seen an explosive growth in optogenetic stimulation and recording (Bernstein and Boyden, 2011; Deisseroth, 2011; Fenno et al., 2011). In all optical methods, applied light directly alters some facet of the membrane to induce a transmembrane current, which results in the shift of membrane potential. Under the right conditions, application of light can drive activity in any neuron, possibly by opening micropores that allow passive ionic flow, or through alteration of thermal and capacitive properties. However, most common optical stimulation methods introduce some alteration of selected neurons that becomes the mechanism of light sensitivity. The canonical example of optogenetics is the genetic manipulation of target neurons to express the blue light sensitive ion channel Channelrhodopsin-2 (ChR2), originally isolated from algae, which when open generates an inward (depolarizing) current. There are now many optogenetic constructs, that differ in current capacity or kinetics, wavelength sensitivity, and other properties. There are also classes of optogenetic constructs that hyperpolarize cells through effective outward currents, thereby inhibiting activity rather than exciting it, which is difficult or impossible to do with electrical stimulation.

In addition to the bidirectional stimulation possible with optogenetic stimulation, the key advantage is the ability to tie stimulation specificity to genetic parameters, such as promoters that are correlated with cell type. For example, in a typical application, only a certain subset of excitatory neurons may be activated by a light source that illuminates a much larger volume of cells, most of which do not express the optogenetic construct; in particular, neighboring cells may have differential responses to the light. This enhanced specificity is in contrast to electrical stimulation, which will influence all cells in the electrode neighborhood. Moreover, new experimental designs become possible using molecular biological tools to guide expression. For example, viral injections in one brain region can be used to express ChR2 in another "upstream" region, but only in those neurons that project to the original region. In this way, illuminating the upstream region induces activity only in those neurons whose output goes to a particular location, regardless of where they sit physically within the illuminate network. The wide range of available tools to guide expression supports "circuit cracking" or reverse engineering of complex neural networks (Packer et al., 2013) and the ability to associate specific neuron classes with behavioral outcomes (O'Connor et al., 2013).

There are limitations to general application of optogenetic stimulation. Brain tissue scatters and absorbs light such that sufficiently targeted illumination can be difficult to achieve. While in principle, illuminators could be left outside the brain, in practice for all but the most shallow regions, some degree of penetrating instrumentation, and associated tissue damage, is required. In some cases the implanted devices can be larger and do more damage than electrical wires, which can be made very thin. In general, the kinetics of activation tend to be slower than electrical activation, and since they involve cycling of membrane spanning ion channels or pumps, introduce new complexities of time-dependent responsiveness. As with all molecular alterations of biological cells, there may be unintended effects related to overexpression, excitotoxicity, or other detriments to cell health. However, the largest obstacle to implementation in the clinic is that gene therapy methods are required to introduce the optogenetic tools, but remain experimental and unproven for human use (Chow and Boyden, 2013). The superior stimulation properties of optogenetic tools motivate their continual development, but it is an open question how widely such tools may be viable in future treatments.

3.2 Actuating the Brain: Characterization of Control Inputs

In the previous section we focussed on the mechanism of action of the most common methods for stimulating nervous tissue. Here we relate those methods to function in a control setting. By *actuator* we mean an independent channel of stimulation, for example, a single electrode contact whose voltage can be set independently of any other contacts, or the tip of an optical fiber driven by its own light source. The emphasis is on the channel of information rather than the physical layout; for example, in our nomenclature, multiple electrode contacts shorted together at a common potential would be classified as a single actuator. It is also not necessarily the case that a physical layout determines the effective dimension of actuation. For example, in a scanning laser system, the actuation density would be determined by the optics, namely, the number of locations that can effectively be illuminated independently of other regions. The optically determined density depends not only on the hardware employed, but also the properties (scattering and absorption) of the tissue. One major limitation of optical methods is that greater depths cannot be illuminated with the same specificity as shallow depths due to tissue scattering (although attempts to shift to longer wavelengths improve the effective depth).

The principal issue for control is having an actuator *density* that is sufficiently high, rather than the total number of channels. In a typical local network of neurons, such as a column of neocortex, there may be up to 40,000 neurons spanning a radius of 0.5 mm (Oberlaender et al., 2012). It is a challenging problem to build hardware capable of providing independent stimulation to so many cells in such a small volume. For example, microwire or MEMS fabricated arrays of electrodes typically span 1–2 mm. While there may be hundreds to thousands (Bernstein and Boyden, 2011) of channels of stimulation available, they cover a volume that may contain millions of neurons, and therefore the neuron-to-actuator ratio is still of order 1000 to 1 or more. Thus one central issue in formulating a control problem is recognition that currently available technology limits stimulation to highly underactuated input (Ching and Ritt, 2013).

We have written so far only of the direct effect from the stimulator on the neuron. Since neurons are coupled, even a neuron not directly affected by stimulation might be brought under control through its synaptically coupled neighbors. Several issues arise in this context. If both pre- and postsynaptic neurons must be controlled, there may not be an effective way to achieve independent activity patterns, since transmission of influence to the postsynaptic neuron occurs only following presynaptic action potentials. Moreover, indirect stimulation of a neuron through a network introduces delays due to the activation time of the presynaptic neuron and the speed of synaptic transmission. Also, while the effects of stimulation are typically somewhat temporally constrained in an isolated neuron (at least, on the order of 100 ms correlation time), once network perturbations are introduced, it is possible for the correlation time of the response to stimulation to extend to seconds or longer, as activity propagates through the larger structure. While control of a single neuron receiving a single input is a relatively straightforward problem, at the network level, the structure and dynamics can become highly complex and provide a formidable control problem. What the effective dynamical dimension of a given neural network in a brain region might be is a hard and largely still open question.

3.3 Probing Brain Circuits With Principled Objective Functions

Due to the advent of the aforementioned technologies increasingly sophisticated characterizations and understanding of basic neuronal functions are being uncovered. To date, however, the use of these methods has remained largely perturbative: using a stereotyped, "pulse" type input, a population of neurons is either activated or deactivated en masse (see Fig. 4). Any resulting change in the behavioral phenotype can then be attributed to the population of neurons in question. What such an approach lacks is precise *control* of neuronal dynamics, that is, a fine-grained manipulation of the patterning, sequencing and synchronization of spiking activity within the population. The next generation of neuroscience questions are likely to center on how such dynamics—and not simply activation/deactivation—are related to function and behavior.

FIG. 4 Neurocontrol technologies and optogenetics in particular, are being used to gain increasingly sophisticated insight into brain function. Left: This technology offers the ability to simultaneously modulate input intensity and record neuronal activity, potentially in a closed-loop implementation. Right: Currently, neurostimulation is typically used in a *perturbative* manner, in which pulse-type inputs are used to modulate population firing rates. We seek to understand and design precise neuronal *control*, to better probe the role of neuronal dynamics in function, and to gain insight into the ways that neurons "control" each other.

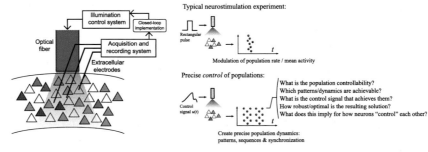

3.4 Control of Single Neurons

In recognition of the above objectives, efforts have been increasing toward simple neuronal control. The most straightforward (and yet, nevertheless, nontrivial) objective is the reliable control of spiking in single neurons.

In this context, classical control theory has be applied to modulate the interspike interval (ISI) of a neuron (Miranda-Dominguez et al., 2010). Here, the authors propose a proportional-integral (PI) controller to inject a current across a cell membrane using dynamic clamp,[2] so that the neuron can maintain a firing rate. The goal therein was to compensate for the slow change in neuronal ISI with respect to time due to "drift" induced by the neuron's phase response curve (PRC). Note that the PRC is an experimentally obtained characterization of the phase advance/shift in spiking due to application of an impulsive input to a periodically spiking neuron.

In Stigen et al. (2011), an algorithm was developed to control spike timing and synchrony of a neuron, also based on computation of the PRC. The PRC data, estimated in real time from the neuron, is first fitted to a sigmoid function. Then, inverting the functional relationship (in essence, via a feedback linearization), the control is calculated to achieve a spike at a desired interval.

Using optimal control theory applied to single phase-oscillator-based neurons (see also Section 3.5), (Nabi et al., 2013b) shows the design of minimum energy charge-balanced input waveforms which performs better than conventional monophasic pulse stimuli both in terms of energy and control.

In Ahmadian et al. (2011), the problem of single neuron spike control is framed in the context of a statistical model. There, finding the best time-dependent stimulus that maximizes the probability of a desired spike train is reduced to a maximum a posteriori (MAP) estimation problem. In this sense, the formulation of the problem reduces to calculating the stimulus that maximizes the likelihood function dependent on the target and observed spike train, under physiological and hardware implementation constraints.

In Iolov et al. (2014), a stochastic optimal control problem is solved for a single neuron leaky integrate-and-fire (LIF) model with the objective of precisely controlling the spike times of a neuron. No constraint has been imposed on noise intensity level. The problem is approached using dynamic programming in a closed-loop setting and using the Pontryagin maximum principle to synthesize the open loop control.

Clearly, controlling spiking in single neurons is a prerequisite for the eventual manipulation of spiking in larger populations and networks. However, standard control theoretic formulations and notions are likely to be intractable for such a problem at neurophysiologically relevant scales, subject to the technological constraints discussed above.

Consider, for instance, the classical notion of controllability, which refers to the ability of an input to steer a dynamical system exactly along a desired trajectory. For large neuronal populations, this implies exact specification of the membrane potential and ionic states of every neuron at any point in time. As described above, with present-day stimulation technology, in which a given input affects up to thousands of neurons, this notion is intractable. Moreover, while subthreshold activity is meaningful, many questions in neuroscience and neural coding consider spikes as the fundamental entities of neural processing.

With these constraints in mind, Ching and Ritt (the present authors) in Ching and Ritt (2013) have introduced controllability definitions and control strategies for neuronal populations in an underactuated stimulation environment. In particular, two controllability notions are defined. The first is spike *sequence controllability*, which concerns the ability of an external input to create an ordered progression of spikes without regard for timing. The second is spike *pattern controllability*, which, in addition to ordering, constrains when spikes occur.

---

2. In dynamic clamp, the current through an intracellular electrode is itself regulated via closed-loop feedback.

These properties were studied in the context of integrate-and-fire (IF) neuronal models. The IF model is the simplest dynamical systems model of neuronal spiking, wherein the membrane dynamics are linear except for a discontinuity associated with spiking. Specifically, an IF model was considered in the form

$$\dot{V} = -\alpha V + u(t)b(E - V), \tag{8}$$

where $u(t) \in \mathcal{U} \subset \mathbb{R}$ is the exogenous input. The neuron is said to spike when V evolves from $V = 0$ to, typically, $V = 1$.

In Ching and Ritt (2013), analytic conditions for pattern control in uncoupled IF populations of arbitrary size. These conditions amount to geometric constraints on the parameters α, b, and E. Motivated by the application domain of optogenetics, the input here is affine in the state variable V, rendering a bilinear control problem. To validate the robustness of the solution, the resulting inputs were applied to biophysically detailed HH models. The resulting HH output showed agreement with the IFX design.

3.5 Control of Neuronal Oscillator Networks: Synchronization

Phase model design paradigms: A significant thrust in the study of neuronal dynamics has centered on neural oscillations, owing to their prevalence in electrophysiologic recordings and the fact that the HH model, in a typical parameterization, is fundamentally a nonlinear oscillator. In context, a particular case of the general spike control objectives introduced above is oscillatory synchrony.

To study the dynamics associated with oscillations and synchrony in networks of neurons, phase reduction is a popular technique. Here, the dynamics of a neuron is reduced to a single phase variable, that is,

$$\dot{\phi} = \omega + Z(\phi)u(t), \tag{9}$$

where ϕ is the phase, which varies from 0 to 2π, ω is the neurons natural frequency, $Z(\phi)$ is the PRC, and $u(t)$ is an exogenous (eg, control) input. In the context of modeling neuronal activity, the model (9) is said to spike at every instance that ϕ traverses a multiple of 2π in the forward direction.

Models of the form (9) have been a popular medium for studying the patterns of synchrony that result from different types of neuronal coupling (Brown et al., 2003; Cohen et al., 1982; Ashwin and Swift, 1992; Ghigliazza and Holmes, 2004; Kopell and Ermentrout, 1990), and how the presence of external stimuli change the oscillatory dynamics of large groups of oscillators (Brown et al., 2004a, b; Tass, 1999; Lajoie and Shea-Brown, 2011).

Control designs embedded in phase models have also been proposed to drive neuron oscillators to form certain synchronized spiking patterns (Tass, 2000; Kiss et al.,

2007; Danzl et al., 2008; Danzl and Moehlis, 2007; Tass, 1999; Kano and Kinoshita, 2010; Moehlis et al., 2006) and complete network synchronization (Moehlis et al., 2006; Harada et al., 2010; Marella and Ermentrout, 2008; Hata et al., 2011; Abouzeid and Ermentrout, 2009). For instance, optimal synchronizing control of an ensemble of neurons (usually uncoupled) were investigated and formulated by way of the Pontryagin maximum principle as a boundary value problem (Dasanayake and Li, 2011a, b), which can be solved by methods such as pseudospectral approximations or homotopy perturbation (Li et al., 2013; Dasanayake et al., 2013). In addition, optimal control techniques have been applied to phase models to perform various tasks such as to alter firing frequency (Zlotnik and Li, 2011, 2012).

Control of nonlinear oscillator networks: Efforts to control synchronization high-dimensional nonlinear oscillator-based neuronal models are more limited (Parlitz et al., 1996; Pikovsky et al., 2001; Lian et al., 2004; Gu et al., 2014). This may be partially due to the fact that coupling in many of these networks tends to produce highly robust synchronization without additional control.

Because of their interesting dynamics and potential applications in optimization and information processing (Adachi and Aihara, 1997; Tokuda et al., 1997), chaotic synchronization for neuronal oscillators is receiving increasing attention (Pecora and Carroll, 1990; Grebogi and Yorke, 1990; Wang et al., 2006; Shrimali et al., 2007; He et al., 2003; Shrimali, 2009). Indeed, it is well known that neuronal networks can exhibit complicated and even chaotic dynamics (Zou and Nossek, 1993; Gilli, 1993; Chen et al., 2005). Different schemes have been developed for synchronizing such systems (Shen et al., 2012; Wu and Park, 2013; Jeong et al., 2013; Gan, 2012; Zhang et al., 2014), for example, adaptive control in delayed neural networks (Zhou et al., 2006; Cheng et al., 2005); synchronizing time varying neural systems (Cao and Lu, 2006; Cui and Lou, 2009); and stochastic nonlinear control (Sun and Cao, 2007; Sun et al., 2007; Yu and Cao, 2007; Li et al., 2010).

3.6 Control of Neuronal Oscillator Networks: Desynchronization

The active desynchronization of neuronal populations has received substantial recent interest owing to the observation that pathological synchronization neurons of the basal ganglia cortical loop is thought to be a factor contributing to Parkinsons disease (PD) (Agarwal and Sarma, 2012c, 2010). DBS therapy, in which a permanent, high-frequency, pulsatile signal is administered to points in this loop, has been shown to alleviate many of the motor symptoms (eg, tremor) associated with PD (Agarwal and Sarma,

2012b). At present, DBS therapy is applied in a strictly open loop fashion with pulse parameters tuned by hand. This has led to a significant push to develop principled methods for synthesizing optimal stimulation controls for desynchronizing these neuronal populations (Agarwal and Sarma, 2010, 2012b, c).

Optimal control: Optimal desynchronizing control objectives include minimizing desynchronization time and energy. For example, minimum time control (Danzl et al., 2009), energy-optimal control (Danzl et al., 2010; Wilson and Moehlis, 2014a; Nabi and Moehlis, 2011), and both minimum energy and time control (Nabi et al., 2013a). In Wilson and Moehlis (2014a), a procedure for finding an energy-optimal stimulus was proposed based on computation of Lyapunov exponents. Conditions for desynchronizing two neurons were derived and were extended to large neural population by maximizing the phase distribution. Unlike other proposed methods in Danzl et al. (2009) and Nabi et al. (2013a), the procedure does not need the full model of the dynamics. Later, the methodology was moved closer toward experimentation (adapted for extracellular neural stimulation as is the case of DBS) in Wilson and Moehlis (2014b).

Delayed feedback control: In contrast to the above approaches, which center on the synthesis of optimal open-loop desynchronizing controls, efforts have been directed at developing actual closed-loop feedback-based stimulation paradigms. Here, mild stimulation techniques can be computed to restore desynchronized dynamics in a network of oscillatory neurons (Tass, 1999). The idea is to shift the neurons to their asynchronous physiological mode, thereby suppressing disease symptoms. Feedback control techniques include linear single input (Rosenblum and Pikovsky, 2004a, b) or multiple inputs (Hauptmann et al., 2005a, b, c) and nonlinear delayed feedback control (Popovych et al., 2005, 2006).

3.7 Control of Bursting and Seizure Activity

In a similar spirit to the DBS scenario described above, there has been steady interest in using neurostimulation to mitigate or suppress the cortical dynamics associated with epileptic seizures. Seizures, in general, are associated with widespread cortical synchronization that may begin focally, or in a spatially diffuse manner (Kramer and Cash, 2012; Kramer et al., 2008).

Several putative designs have been proposed for seizure control based on mean field neuronal models. These include feedback schemes that alter the bifurcation structure via linear feedback (Kramer et al., 2006; Roberts and Robinson, 2008) and schemes that use spatially distributed proportional control to mitigate seizure spread (Ching et al., 2012). It is worth noting that several of the optimal control methods noted above for desynchronization of neural

population could equivalently be deployed in the context of seizure control. In particular, Wilson and Moehlis (2014a) noted in their design is the mitigation of burst-like fast slow dynamics associated with seizure kindling and onset.

4 IDENTIFICATION AND ESTIMATION OF NEURONAL DYNAMICS

In order to successfully implement any control strategy, it is necessary to have a suitable and accurately parameterized model of the neuronal networks (and dynamics thereof) in question. The identification of models in neuroscience and neural engineering research has generally centered on the question of network structure, that is, the connectivity, or statistical associations between brain regions as observed through different neural recording modalities. Below, we provide a detailed review of neural connectivity identification and inference; then proceed to discuss more recent efforts to identify both connectivity and dynamics, toward facilitating neuronal state estimation.

4.1 Inference of Neuronal Network Structure and Connectivity

In neuronal network identification, several notions of connectivity are considered: anatomical connectivity, functional connectivity, and effective connectivity. Anatomical connectivity involves the characterization of structural or physical topology of brain networks through synaptic or fiber tract (white matter) connections. Such connections are stable over short time scales (seconds to minutes), but may encounter morphological changes at longer time scales (hours to days). Conventional magnetic resonance (MR) imaging was used in early studies of the assessment of white matter fiber tracts in the brain to measure volumetric differences in white (and gray) matter between patients and controls (Shenton et al., 2001). Recent studies rely on diffusion tensor imaging (DTI) which has the ability to visualize anatomical connections between different parts of the brain noninvasively and on an individual basis (Burns et al., 2003).

Brain connectivity is usually represented in graph or matrix format. The graph may be weighted, with weights representing connection densities or efficacies, or binary, with binary elements indicating the presence or absence of a connection. A structural description of the human brain is the objective in the human connectome project (Van Essen et al., 2012; Sporns, 2011). In this study, it was revealed that network studies of structural connectivity obtained from noninvasive neuroimaging modalities have a number of highly nonrandom network properties, including high clustering and modularity combined with high efficiency and short path length (Sporns, 2011; Sporns et al., 2005).

Functional connectivity generally refers to the data-driven study of the statistical dependencies between different sites or nodes within the brain. An edge or a connection between two nodes in functional connectivity defines a statistical link between these two sites. Generally, functional brain connectivity metrics are bidirectional, yielding full symmetric matrices which may be thresholded to yield binary undirected graphs, with the setting of the threshold controlling the degree of edge sparsity.

Functional connectivity can be estimated in a variety of ways. A simple and common measure is the Pearson correlation (ie, the cross-correlation) which is usually obtained over a single trial (Hampson et al., 2002; Achard et al., 2006; Kafashan et al., 2014). Another measure of functional connectivity is coherence. Two signals are said to be coherent if they have constant relative phase, or, equivalently, if their power spectra correlate, for a given time and/or frequency window. Because this measure is defined based on the frequency content of the two signals, it is insensitive to a fixed lag between two time series. Coherence can be estimated either for a single (often narrow) frequency range, or estimated within multiple frequency ranges. Functional connectivity can be also be obtained by using an information theoretic measure called mutual information (MI). MI quantifies the shared information between two signals and unlike correlation can identify nonlinear dependencies (Jeong et al., 2001; Zhou et al., 2009). Nonlinear measures of functional connectivity can also be used, though they typically require long stationary segments of signal and can be very sensitive to noise (Pereda et al., 2005; Netoff et al., 2006; Quiroga et al., 2002).

Functional connectivity in neural data can result from stimulus-locked transients, evoked by a common afferent input, or reflect stimulus-induced oscillations. Hence a more difficult question is finding the causal influences of one neural element over another.

Effective brain connectivity yields a full nonsymmetric matrix. Applying a threshold to such matrices yields binary directed graphs. Causal interactions in a network can be measured by characterizing the effects of selective perturbations. In Tononi and Sporns (2003), the authors used an active stimulation of selected subsets of a network to interpret the interaction between the selected subset and the rest of the network as the measure of effective connectivity. Moreover, effective connectivity can be inferred through time series analysis. Structural equation modeling (SEM) is one such time series analysis technique which has been successfully adopted to functional brain imaging data (Büchel and Friston, 1997; Kim et al., 2007). In SEM, the directions of interactions among set of nodes or variables are hypothesized and correlation analysis is used to quantify the strength of connections. A generalization of this

approach is dynamic causal modeling (DCM), which is a Bayesian model comparison procedure that operates on comparing different directed network structures that best fit neural data. A challenge associated with SEM and DCM is that neither of these methods can efficiently search across all possible network structures and parameters, so that it is common for only a small set of potential networks to be hypothesized and compared (Smith et al., 2011).

Granger-Geweke causality (GGC) is another method to determine a causal influence of one time series on another. The basic concept of the GGC is based on the improvement of the prediction of one time series by incorporating the knowledge of the second one. Since Granger causality is a more data-driven method, it has been widely used in recent years. Nonlinear versions of GGC have been studied in Marinazzo et al. (2011) and Diks and Wolski (2013). In Zou and Feng (2009) GGC was compared to Bayesian networks to determine causal influence in synthesized networks. The authors stated that the Bayesian network technique outperforms the Granger causality approach when the data length is short, and vice versa.

Transfer entropy and partial correlation are two other approaches which are able to measure direct interaction to a certain extent and can be used to infer effective connectivity in a neuronal network (Marrelec et al., 2009). Transfer entropy from process X to another process Y is the amount of uncertainty reduced in future values of Y by knowing the past values of X given past values of Y. Partial correlation measures the degree of association between two random variables, with the effect of a set of controlling random variables removed. In Smith et al. (2011), connectivity analysis was conducted for different metrics and it is asserted that partial correlation has the best performance.

Other spectral measures for inferring causal influences include directed transfer function (DTF), partial directed coherence (PDC), and directed DTF (dDTF). In Fasoula et al. (2013), it has been demonstrated that the connectivity estimation performed by DTF is most robust in terms of low estimation error, while the PDC in terms of low fictitious causal density. The dDTF provides lower fictitious causal density and higher spectral selectivity as compared to DTF, at high enough SNR.

Graph theoretical approaches have been utilized recently as a means to characterize the networks constructed through the above formalisms. In general, graph theory can be applied to either functional or effective connectivity measures. However, neuroscience connectivity studies have been biased toward functional metrics (Bullmore and Sporns, 2009). There are several measures that can be used to quantify graph topology such as node degree, degree distribution, assortativity, clustering coefficient, motifs, connection density or cost, hubs, centrality, robustness, and modularity. It is not yet established which measures are most appropriate for the analysis of brain networks (Bullmore and Sporns, 2009), however, many important statistical properties underlying the topological organization of the human brain, including modularity, small-worldness, and the existence of highly connected network hubs have been investigated in He and Evans (2010), Bullmore and Sporns (2009), and Fallani et al. (2012).

4.2 State Estimation and Kalman Filtering

We move now from the identification of broad network structure, to the identification of local states and dynamics at smaller neuronal scales. Filtering is one of the most pervasive techniques in estimation of state from noisy sensor data for many engineering problems. The nonlinearities associated with spiking neuron models makes the filtering problem highly nontrivial. More generally, since Kalman's pioneering work (Kalman, 1960) on tracking and estimation, the extended Kalman filter has been developed for application to nonlinear systems. EKF simply linearizes all nonlinear model to facilitate use of Kalman filtering. However EKF faces stability problems on violation of local linearity assumptions and calculation of computationally expensive Jacobian matrices. To overcome the limitations, Julier and Uhlmann (1997) developed a new tool, the unscented Kalman filter (UKF), which is more efficient than EKF by obviating the need of linearization.

Single neuron filtering: In Voss et al. (2004), the authors used UKF to discuss the problem of estimating parameters and unobserved trajectory components from noisy time series measurements of continuous nonlinear dynamical systems. This provided the framework for how to estimate states applying UKF in nonlinear neuron models. In an extension of Voss et al. (2004), Schiff and Sauer (2008) estimated the state and parameters of spatiotemporal excitable system with an observer UKF and, moreover, showed how to optimally control the frequency of such a system. With the advent of UKF, Ullah and Schiff (2009) demonstrated that membrane potential measurements alone can predict the system trajectory defined by the HH equations, thus reconstructing ion channel dynamics. Prediction of the input stimulus from experimentally recorded single neuron membrane potential by UKF and designing an observer similar to the classical Luenberger observer to estimate the internal variables in an HH neuron, for example, gating variables, ion concentrations has been shown in Wei et al. (2011) and Sinha et al. (2013), respectively.

In Lankarany et al. (2014), dual and joint estimation strategies were employed to derive four Kalman filtering algorithm namely, joint UKF (JUKF), dual UKF (DUKF), joint EKF (JEKF), and dual EKF (DEKF) for estimating the state and parameters for HH neurons. It was argued therein that with fast computation, EKF can be preferred over UKF-based filtering in the context of single neuron dynamic clamp experiments.

Networks: Naturally, the filtering problem becomes progressively harder with scale. For networks of neurons, data assimilation techniques can be used to predict the unobservable state and parameters. In Sauer and Schiff (2009), a novel "Consensus State Method" is proposed which makes the prediction of dynamics for partially observed heterogeneous networks of neurons, tractable.

State estimation via statistical models: In an extension of the general case of linear Gaussian state equations and Gaussian observation process, where Kalman filtering (Kalman, 1960) is ubiquitously used to solve the problem, Smith and Brown (2003) developed an algorithm to estimate a state-space model where the observation model is a point-process (serving as a model of neural spike trains). They modeled the stimulus which modulates the neuron spiking activity as a Gaussian autoregressive model and formulated an approximate expectation-maximization (EM) algorithm (Dempster et al., 1977) to solve this simultaneous estimation problem for unobservable state-space and the parameters of both the point process and the state space.

5 CONTROL THEORETIC ANALYSIS OF NEURAL CIRCUITS

In parallel to overt closed-loop control using neurostimulation, an important line of emerging research involves performing control-theoretical analyses, such as assessment of controllability and observability, on neuronal networks across different spatial scales. Such fundamental analyses can provide key insights into the extent to which a considered network is controllable or observable, and thus serve as a backbone upon neurocontrol or neurostimulation solutions can be developed.

In this direction, the authors in Whalen et al. (2015) investigated how network topology and connectivity weights affect controllability and observability of nonlinear neuronal networks by deploying mathematical concepts from Lie algebra (a tool to analyze controllability of nonlinear systems) and group theory. In particular, they focussed on three neurons motifs subject to a common external input where the Fitzhugh-Nagumo neuron model with sigmoidal coupling was used to represent the neuronal dynamics. The authors highlight important insights into the role played by network symmetry in shaping these control-theoretic properties, with the key limitation being difficulty in scaling the analytical approach to larger network configurations.

One approach to obviating scalability issues is to consider higher-level descriptions of brain dynamics, such as in Gu et al. (2015), wherein linear networks were formulated to investigate how brain network structure affects controllability. In particular, they constructed structural brain networks with 254 large-scale cortical and subcortical regions using diffusion tensor imaging data and performed diffusion tractography to estimate the number of connections between these brain regions. By representing these estimates in a weighted adjacency matrix and assuming each of these brain regions as a node of the network, the authors performed a suite of detailed control-theoretic analyses using the controllability Gramian of discrete-time time-invariant linear systems. This work has provided a framework to study controllability of brain networks, under the assumption of the linear dynamics of the underlying model.

Developing control theoretic methods that reconcile scale, model complexity, and analytical tractability will be an important aspect of the overall problem of developing closed-loop solutions in the context of neurophysiology.

6 TOWARD CLOSED-LOOP MODULATION OF NEURAL CIRCUITS

6.1 New Formulations to Meet New Problems

As this chapter highlights, neurophysiology exhibits many complex and challenging dynamics that complicate the analysis and design of closed-loop paradigms. These dynamics span discrete and continuous formulations of systems ranging in scale from single neurons to continuum ensembles. We posit several lines of investigation, each connected to systems and control theory, that will be required in order to meet these challenges:

- The creation of alternative formulations and methods for systems-theoretic analysis, including controllability/observability, of neuronal dynamics at large scale.
- Establishing reliable model validation techniques to facilitate meaningful analysis and control-theoretic design.
- The design of rapid, but likely suboptimal, control synthesis methods for creating patterns of activation in vivo, subject to the constraints discussed herein.
- The development of efficient ways to estimate model parameters and perform meaningful observation of underlying dynamics from highly constrained real-time sensing.
- The development of objective functions that transcend neuronal activation per se, and move toward behavioral or cognitive endpoints.
- Explicit connection of the current neuroscience theories in cognition to control-theoretic formulations for scientific study or neuroengineering translation.

The latter, in particular, is a domain full of important applications including the design of DBS brain machine interfaces, and sensory prosthesis. Below, we briefly elaborate on recent control-theoretic developments in these areas.

6.2 Control Engineering in the Design of Brain-Machine Interfaces

Brain-machine interfaces (BMIs) or brain-computer interfaces (BCIs) (Nicolelis and Lebedev, 2009; Moritz et al., 2008) are artificially designed systems which can sense and interpret brain activity for the purpose of restoring impaired motor functions in patients (Nicolas-Alonso and Gomez-Gil, 2012). As shown in Fig. 5, central to these systems are devices that continuously measure the motor-relevant cortical neuronal activity, a decoder that extracts task-relevant motor information, and an encoder that feeds back the motor-relevant sensory information back to the brain.

In the last decade, most of the effort toward the development of clinically deployable BMIs has been centered on advancing devices such as multichannel electrodes to measure neuronal activity over a long period of time and developing complex decoding and learning algorithms (Gilja et al., 2012; Sussillo et al., 2012; Dangi et al., 2013; Ifft et al., 2013). In recognition of the importance of missing sensory feedback in BMIs and thus facilitating next-generation feedback-enabled BMIs, recent attempts have shown the incorporation of artificial texture (Doherty et al., 2011) and proprioception (Weber et al., 2011) information in experimental settings. In these attempts, the approaches are primarily based on the learning paradigm where the subject is trained to differentiate artificial sensory feedback in a task-dependent context.

In parallel, a systematic control-theoretic approach has been taken to optimize the learning process in BMIs, rigorously analyze the BMI system and design artificial sensory feedback optimally by developing a theoretical framework

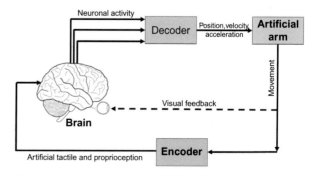

FIG. 5 A schematic of closed-loop brain-machine interface (BMI).

based on optimal control theory (Heliot et al., 2010; Kumar et al., 2011, 2013; Kumar and Kothare, 2013; Lagang and Srinivasan, 2013). For instance in Heliot et al. (2010), the authors have leveraged tools from control theory such as inverse models and feedback control and proposed a learning model to maximize the performance of closed-loop BMIs. In Kumar et al. (2011), the authors have used the framework of model predictive control to design an optimal stimulus for a single spiking neuron and analyzed a single joint movement task-based closed-loop BMI to elucidate the importance of sensory feedback. A systematic approach in designing optimal sensory feedback in closed-loop BMIs has been shown in Kumar et al. (2013) where the authors have proposed an optimal design of feedback enabled closed-loop BMI and designed optimal sensory feedback using a model predictive controller. In Kumar and Kothare (2013), the authors have used theoretical analysis approaches embedded in dynamical systems theory to investigate systems properties such as continuity and differentiability of inter-spike intervals in spiking recurrent neuronal networks. In Lagang and Srinivasan (2013), the authors have used the framework of stochastic optimal control to analyze the performance of open and closed-loop BMIs and explained key phenomenon in closed-loop BMI operation.

ACKNOWLEDGMENTS

Portions of this chapter were used as the basis of a tutorial workshop at the 2015 American Control Conference, held in Chicago, IL. We are grateful for the contributions of Mohammad Mehdi Kafashan, Aniban Nandi, and Sensen Liu to sections of this manuscript. Jason T. Ritt and ShiNung Ching hold Career Awards at the Scientific Interface from the Burroughs-Wellcome Fund.

REFERENCES

Abouzeid, A., Ermentrout, B., 2009. Type-II phase resetting curve is optimal for stochastic synchrony. Phys. Rev. E 80, 011911.

Achard, S., Salvador, R., Whitcher, B., Suckling, J., Bullmore, E., 2006. A resilient, low-frequency, small-world human brain functional network with highly connected association cortical hubs. J. Neurosci. 26 (1), 63–72.

Adachi, M., Aihara, K., 1997. Associative dynamics in a chaotic neural network. Neural Netw. 10, 83–89.

Agarwal, R., Sarma, S.V., 2010. Restoring the basal ganglia in Parkinson's disease to normal via multi-input phase-shifted deep brain stimulation. Conf. Proc. IEEE Eng. Med. Biol. Soc. 2010, 1539–1542.

Agarwal, R., Sarma, S.V., 2012a. The effects of DBS patterns on basal ganglia activity and thalamic relay: a computational study. J. Comput. Neurosci. 33 (1), 151–167. http://dx.doi.org/10.1007/s10827-011-0379-z.

Agarwal, R., Sarma, S.V., 2012b. The effects of DBS patterns on basal ganglia activity and thalamic relay: a computational study. J. Comput. Neurosci. 33 (1), 151–167. http://dx.doi.org/10.1007/s10827-011-0379-z.

Agarwal, R., Sarma, S.V., 2012c. Performance limitations of relay neurons. PLoS Comput. Biol. 8 (8), e1002626. http://dx.doi.org/10.1371/journal.pcbi.1002626.

Ahmadian, Y., Packer, A.M., Yuste, R., Paninski, L., 2011. Designing optimal stimuli to control neuronal spike timing. J. Neurophysiol. 106 (2), 1038–1053.

Ashwin, P., Swift, J., 1992. The dynamics of n weakly coupled identical oscillators. J. Nonlin. Sci. 2, 69–108.

Bear, M., Connors, B., Paradiso, M., 2007. Neuroscience: Exploring the Brain. Lippincott Williams & Wilkins, Philadelphia, PA. ISBN 9780781760034.

Bernstein, J.G., Boyden, E.S., 2011. Optogenetic tools for analyzing the neural circuits of behavior. Trends Cogn. Sci. 15 (12), 592–600.

Brown, E., Holmes, P., Moehlis, J., 2003. Globally Coupled Oscillator Networks. Springer-Verlag, Berlin, pp. 183–215.

Brown, E., Moehlis, J., Holmes, P., 2004a. On the phase reduction and response dynamics of neural oscillator populations. Neural Comp. 16, 673–715.

Brown, E., Moehlis, J., Holmes, P., Clayton, E., Rajkowski, J., Aston-Jones, G., 2004b. The influence of spike rate and stimulus duration on noradrenergic neurons. J. Comp. Neurosci. 17, 13–29.

Büchel, C., Friston, K., 1997. Modulation of connectivity in visual pathways by attention: cortical interactions evaluated with structural equation modelling and FMRI. Cereb. Cortex 7 (8), 768–778.

Bullmore, E., Sporns, O., 2009. Complex brain networks: graph theoretical analysis of structural and functional systems. Nat. Rev. Neurosci. 10 (3), 186–198.

Burns, J., Job, D., Bastin, M., Whalley, H., Macgillivray, T., Johnstone, E., Lawrie, S.M., 2003. Structural disconnectivity in schizophrenia: a diffusion tensor magnetic resonance imaging study. Br. J. Psychiatry 182 (5), 439–443.

Butson, C.R., McIntyre, C.C., 2008. Current steering to control the volume of tissue activated during deep brain stimulation. Brain Stimul. 1 (1), 7–15.

Cao, J., Lu, J., 2006. Adaptive synchronization of neural networks with or without time-varying delay. Chaos 16 (1), 013133.

Chen, Z., 2014. Advanced State Space Methods for Neural and Clinical Data. Cambridge University Press, Cambridge.

Chen, T., Lu, W., Chen, G., 2005. Dynamical behaviors of a large class of general delayed neural networks. Neural Comput. 17, 949–968.

Chen, Z., Gomperts, S.N., Yamamoto, J., Wilson, M.A., 2014. Neural representation of spatial topology in the rodent hippocampus. Neural Comput. 26 (1), 1–39.

Cheng, C.J., Liao, T.L., Hwang, C.C., 2005. Exponential synchronization of a class of chaotic neural networks. Chaos Solitons Fract. 24 (1), 197–206.

Ching, S., Ritt, J.T., 2013. Control strategies for underactuated neural ensembles driven by optogenetic stimulation. Front. Neural Circuits 7, 54. http://dx.doi.org/10.3389/fncir.2013.00054.

Ching, S., Brown, E.N., Kramer, M.A., 2012. Distributed control in a mean-field cortical network model: implications for seizure suppression. Phys. Rev. E 86 (2 Pt 1), 021920.

Chow, B.Y., Boyden, E.S., 2013. Optogenetics and translational medicine. Sci. Transl. Med. 5 (177). 177ps5–177ps5.

Cohen, A., Holmes, P., Rand, R.H., 1982. The nature of coupling between segmental oscillators of the lamprey spinal generator for locomotion: a model. J. Math. Biol. 13, 345–369.

Cui, B., Lou, X., 2009. Synchronization of chaotic recurrent neural networks with time-varying delays using nonlinear feedback control. Chaos Solitons Fract. 39 (1), 288–294.

Dangi, S., Orsborn, A.L., Moorman, H.G., Carmena, J.M., 2013. Design and analysis of closed-loop decoder adaptation algorithms for brain-machine interfaces. Neural Comput 25 (7), 1693–1731.

Danzl, P., Moehlis, J., 2007. Event-based feedback control of nonlinear oscillators using phase response curves. In: Proceedings of the 46th IEEE Conference on Decision and Control. pp. 5806–5811.

Danzl, P., Hansen, R., Bonnet, G., Moehlis, J., 2008. Partial phase synchronization of neural populations due to random poisson inputs. J. Comp. Neurosci. 25, 141–157.

Danzl, P., Hespanha, J., Moehlis, J., 2009. Event-based minimum-time control of oscillatory neuron models. Biol. Cybernet. 101, 387–399.

Danzl, P., Nabi, A., Moehlis, J., 2010. Charge-balanced spike timing control for phase models of spiking neurons. Discrete Contin. Dyn. Syst. 28, 1413–1435.

Dasanayake, I., Li, J.S., 2011a. Constrained minimum-power control of spiking neuron oscillators. In: Proceedings of the 50th IEEE Conference on Decision and Control. Orlando, FL, USA.

Dasanayake, I., Li, J.S., 2011b. Optimal design of minimum-power stimuli for phase models of neuron oscillators. Phy. Rev. E 83, 061916.

Dasanayake, I., Zlotnik, A., Zhang, W., Li Jr., S., 2013. Optimal control of neurons using the homotopy perturbation method. In: 2013 IEEE 52nd Annual Conference on Decision and Control (CDC), pp. 3385–3390.

Dayan, P., Abbott, L., 2005. Theoretical Neuroscience: Computational And Mathematical Modeling of Neural Systems. Computational Neuroscience, MIT Press, Cambridge, MA. ISBN 9780262541855.

Deisseroth, K., 2011. Optogenetics. Nat. Methods 8 (1), 26–29.

Dempster, A.P., Laird, N.M., Rubin, D.B., 1977. Maximum likelihood from incomplete data via the em algorithm. J. R. Stat. Soc. Ser. B 1–38.

Devor, M., Zalkind, V., 2001. Reversible analgesia, atonia, and loss of consciousness on bilateral intracerebral microinjection of pentobarbital. Pain 94 (1), 101–112.

Dhamala, M., Rangarajan, G., Ding, M., 2008. Analyzing information flow in brain networks with nonparametric granger causality. NeuroImage 41 (2), 354–362.

Diks, C., Wolski, M., 2013. Nonlinear granger causality: guidelines for multivariate analysis. 13, p. 15. University of Amsterdam, CeNDEF Working Paper.

Doherty, J.E., Lebedev, M.A., Ifft, P.J., Zhuang, K.Z., Shokur, S., Bleuler, H., Nicolelis, M.A.L., 2011. Active tactile exploration using a brain-machine-brain interface. Nature 479, 228–231.

Ermentrout, G., Terman, D., 2010. Mathematical Foundations of Neuroscience. Applications of MathematicsSpringer, New York. ISBN 9780387877075.

Fallani, F.D.V., Bassett, D., Jiang, T., 2012. Graph theoretical approaches in brain networks. Comput. Math. Methods Med. 2012. Article ID: 590483.

Fasoula, A., Attal, Y., Schwartz, D., 2013. Comparative performance evaluation of data-driven causality measures applied to brain networks. J. Neurosci. Methods 215 (2), 170–189.

Fenno, L., Yizhar, O., Deisseroth, K., 2011. The development and application of optogenetics. Ann. Rev. Neurosci. 34, 389–412.

Foster, B.L., Bojak, I., Liley, D.T.J., 2008. Population based models of cortical drug response: insights from anaesthesia. Cogn. Neurodyn. 2 (4), 283–296. http://dx.doi.org/10.1007/s11571-008-9063-z.

Foster, B.L., Bojak, I., Liley, D.T.J., 2011. Understanding the effects of anesthetic agents on the EEG through neural field theory. Conf. Proc. IEEE Eng. Med. Biol. Soc. 2011, 4709–4712. http://dx.doi.org/10.1109/IEMBS.2011.6091166.

Gan, Q., 2012. Exponential synchronization of stochastic Cohen-Grossberg neural networks with mixed time-varying delays and reaction-diffusion via periodically intermittent control. Neural Netw. 31, 12–21.

Gandiga, P.C., Hummel, F.C., Cohen, L.G., 2006. Transcranial DC stimulation (TDCS): a tool for double-blind sham-controlled clinical studies in brain stimulation. Clin. Neurophysiol. 117 (4), 845–850.

Ghigliazza, R.M., Holmes, P., 2004. A minimal model of a central pattern generator and motoneurons for insect locomotion. SIAM J. Appl. Dyn. Syst. 3, 671–700.

Gilja, V., Nuyujukian, P., Chestek, C.A., Cunningham, J.P., Yu, B.M., Fan, J.M., Churchland, M.K., Kaufman, M.T., Kao, J.C., Ryu, S.I., Shenoy, K.V., 2012. A high-performance neural prosthesis enabled by control algorithm design. Nat. Neurosci. 15, 1752–1757.

Gilli, M., 1993. Strange attractors in delayed cellular neural networks. IEEE Trans. Circuits Syst. I Fundam. Theory Appl. 40 (11), 849–853.

Grebogi, E.O.C., Yorke, J., 1990. Controlling chaos. Phys. Rev. Lett. 64, 1196–1199.

Gu, H., Pan, B., Xu, J., 2014. Experimental observation of spike, burst and chaos synchronization of calcium concentration oscillations. Europhys. Lett. 106 (5), 50003.

Gu, S., Pasqualetti, F., Cieslak, M., Telesford, Q.K., Yu, A.B., Kahn, A.E., Medaglia, J.D., Vettel, J.M., Miller, M.B., Grafton, S.T., Bassett, D.S., 2015. Controllability of structural brain networks. Nat. Commun. 6, 8414.

Hampson, M., Peterson, B.S., Skudlarski, P., Gatenby, J.C., Gore, J.C., 2002. Detection of functional connectivity using temporal correlations in MR images. Hum. Brain Mapp. 15 (4), 247–262.

Harada, T., Tanaka, H.A., Hankins, M.J., Kiss, I.Z., 2010. Optimal waveform for the entrainment of a weakly forced oscillator. Phys. Rev. Lett. 105, 088301.

Hata, S., Arai, K., Galan, R.F., Nakao, H., 2011. Optimal phase response curves for stochastic synchronization of limit-cycle oscillators by common poisson noise. Phys. Rev. E 84, 016229.

Hauptmann, C., Popovych, O., Tass, P., 2005a. Delayed feedback control of synchronization in locally coupled neuronal networks. Neurocomputing 65–66, 759–767.

Hauptmann, C., Popovych, O., Tass, P., 2005b. Effectively desynchronizing deep brain stimulation based on a coordinated delayed feedback stimulation via several sites: a computational study. Biol. Cybern. 93, 463–470.

Hauptmann, C., Popovych, O., Tass, P., 2005c. Multisite coordinated delayed feedback for an effective desynchronization of neuronal networks. Stochast. Dyn. 5 (2), 307–319.

He, Y., Evans, A., 2010. Graph theoretical modeling of brain connectivity. Curr. Opin. Neurol. 23 (4), 341–350.

He, G., Caoa, Z., Zhua, P., Ogurab, H., 2003. Controlling chaos in a chaotic neural network. Neural Netw. 16, 1195–1200.

Heliot, R., Ganguly, K., Jimenez, J., Carmena, J.M., 2010. Learning in closed-loop brain-machine interfaces: modeling and experimental validation. IEEE Trans. Syst. Man Cybernet. B 40 (5), 1387–1397.

Hodgkin, A.L., Huxley, A.F., 1952. A quantitative description of membrane current and its application to conduction and excitation in nerve. J. Physiol. 117 (4), 500.

Ifft, P.J., Shokur, S., Li, Z., Lebedev, M.A., Nicolelis, M.A.L., 2013. A brain-machine interface enables bimanual arm movements in monkeys. Sci. Transl. Med. 5 (210), 210ra154.

Iolov, A., Ditlevsen, S., Longtin, A., 2014. Stochastic optimal control of single neuron spike trains. J. Neural Eng. 11 (4), 046004.

Izhikevich, E., 2003. Simple model of spiking neurons. IEEE Trans. Neural Netw. 14 (6), 1569–1572.

Izhikevich, E., 2004. Which model to use for cortical spiking neurons? IEEE Trans. Neural Netw. 15 (5), 1063–1070.

Izhikevich, E., 2007. Dynamical Systems in Neuroscience. Computational neuroscience, MIT Press, Cambridge, MA. ISBN 9780262090438.

Jeong, J., Gore, J.C., Peterson, B.S., 2001. Mutual information analysis of the EEG in patients with Alzheimer's disease. Clin. Neurophysiol. 112 (5), 827–835.

Jeong, S.C., Ji, D.H., Park, J.H., Won, S.C., 2013. Adaptive synchronization for uncertain chaotic neural networks with mixed time delays using fuzzy disturbance observer. Appl. Math. Comput. 219 (11), 5984–5995.

Johnston, D., Wu, S., 1995. Foundations of Cellular Neurophysiology. A Bradford bookMIT Press, Cambridge, MA. http://books.google.com/books?id=f8JnQgAACAAJ. ISBN 9780262100533.

Julier, S.J., Uhlmann, J.K., 1997. A new extension of the Kalman filter to nonlinear systems. In: Proceedings of the International Symposium on Aerospace/Defense Sensing, Simulation, and Controls. 3. 26, pp. 3–2.

Kafashan, M., Palanca, B.J., Ching, S., 2014. Bounded-observation Kalman filtering of correlation in multivariate neural recordings. In: Proceedings of the 36th Annual International Conference of the IEEE Engineering in Medicine and Biology.

Kalman, R.E., 1960. A new approach to linear filtering and prediction problems. J. Fluids Eng. 82 (1), 35–45.

Kandel, E., Schwartz, J., 2013. Principles of Neural Science, Fifth Edition. McGraw-Hill Education, New York. ISBN 9780071390118.

Kano, T., Kinoshita, S., 2010. Control of individual phase relationship between coupled oscillators using multilinear feedback. Phys. Rev. E 81, 026206.

Kim, J., Zhu, W., Chang, L., Bentler, P.M., Ernst, T., 2007. Unified structural equation modeling approach for the analysis of multi-subject, multivariate functional MRI data. Hum. Brain Mapp. 28 (2), 85–93.

Kiss, I.Z., Rusin, C.G., Kori, H., Hudson, J.L., 2007. Engineering complex dynamical structures: sequential patterns and desynchronization. Science 316 (5833), 1886–1889.

Kopell, N., Ermentrout, G.B., 1990. Phase transitions and other phenomena in chains of coupled oscillators. SIAM J. Appl. Math. 50, 1014–1052.

Kramer, M.A., Cash, S.S., 2012, Aug. Epilepsy as a disorder of cortical network organization. Neuroscientist 18 (4), 360–372. http://dx.doi.org/10.1177/1073858411422754.

Kramer, M.A., Lopour, B.A., Kirsch, H.E., Szeri, A.J., 2006. Bifurcation control of a seizing human cortex. Phys. Rev. E 73 (4 Pt 1), 041928.

Kramer, M.A., Kolaczyk, E.D., Kirsch, H.E., 2008, May. Emergent network topology at seizure onset in humans. Epilepsy Res. 79 (2–3), 173–186. http://dx.doi.org/10.1016/j.eplepsyres.2008.02.002.

Kumar, G., Kothare, M.V., 2013. On the continuous differentiability of inter-spike intervals of synaptically connected cortical spiking neurons in a neuronal network. Neural Comput. 25 (12), 3183–3206.

Kumar, G., Aggarwal, V., Thakor, N.V., Schieber, M.H., Kothare, M.V., 2011. An optimal control problem in closed-loop neuroprostheses. Proceedings of the 2011 50th IEEE Conference on Decision and Control and European Control Conference, Orlando, FL 53–58.

Kumar, G., Schieber, M.H., Thakor, N.V., Kothare, M.V., 2013. Designing closed-loop brain-machine interfaces using optimal receding horizon control. Proceedings of the 2013 American Control Conference, Washington, DC 5029–5034.

Lagang, M., Srinivasan, L., 2013. Stochastic optimal control as a theory of brain-machine interface operation. Neural Comput. 25 (2), 374–417.

Lajoie, G., Shea-Brown, E., 2011. Shared inputs, entrainment, and desynchrony in elliptic bursters: from slow passage to discontinuous circle maps. SIAM J. Appl. Dyn. Syst. 10 (4), 1232–1271.

Lankarany, M., Zhu, W.P., Swamy, M., 2014. Joint estimation of states and parameters of Hodgkin-Huxley neuronal model using Kalman filtering. Neurocomputing 136, 289–299.

Li, X., Ding, C., Zhu, Q., 2010. Synchronization of stochastic perturbed chaotic neural networks with mixed delays. J. Franklin Inst. 347 (7), 1266–1280.

Li, J.S., Dasanayake, I., Ruths, J., 2013. Control and synchronization of neuron ensembles. IEEE Trans. Autom. Control 58 (8), 1919–1930.

Lian, J., Shuai, J., Durand, D.M., 2004. Control of phase synchronization of neuronal activity in the rat hippocampus. J. Neural Eng. 2, 46–54.

Liley, D.T.J., Cadusch, P.J., Dafilis, M.P., 2002, Feb. A spatially continuous mean field theory of electrocortical activity. Network 13 (1), 67–113.

Marella, S., Ermentrout, G.B., 2008, Apr. Class-II neurons display a higher degree of stochastic synchronization than class-I neurons. Phys. Rev. E 77, 041918.

Marinazzo, D., Liao, W., Chen, H., Stramaglia, S., 2011. Nonlinear connectivity by Granger causality. Neuroimage 58 (2), 330–338.

Marrelec, G., Kim, J., Doyon, J., Horwitz, B., 2009. Large-scale neural model validation of partial correlation analysis for effective connectivity investigation in functional MRI. Hum. Brain Mapp. 30 (3), 941–950.

McIntyre, C.C., Grill, W.M., Sherman, D.L., Thakor, N.V., 2004. Cellular effects of deep brain stimulation: model-based analysis of activation and inhibition. J. Neurophysiol. 91 (4), 1457–1469.

Miranda-Dominguez, O., Gonia, J., Netoff, T., 2010. Firing rate control of a neuron using a linear proportional-integral controller. J. Neural Eng. 7 (6), 066004.

Moehlis, J., Shea-Brown, E., Rabitz, H., 2006. Optimal inputs for phase models of spiking neurons. ASME J. Comp. Nonlin. Dyn. 1, 358–367.

Moritz, C.T., Perlmutter, S.I., Fetz, E.E., 2008. Direct control of paralyzed muscles by cortical neurons. Nature 486, 639–643.

Nabi, A., Moehlis, J., 2011. Single input optimal control for globally coupled neuron networks. J. Neural Eng. 8, 065008. 12pp.

Nabi, A., Mirzadeh, M., Gibou, F., Moehlis, J., 2013a. Minimum energy desynchronizing control for coupled neurons. J. Comput. Neurosci. 34 (2), 259–271.

Nabi, A., Stigen, T., Moehlis, J., Netoff, T., 2013b. Minimum energy control for in vitro neurons. J. Neural Eng. 10 (3), 036005.

Nakayama, T., Hammel, H., Hardy, J., Eisenman, J., 1963. Thermal stimulation of electrical activity of single units of the preoptic region. Am. J. Physiol 204 (1122), 1.

Netoff, T.I., Carroll, T.L., Pecora, L.M., Schiff, S.J., 2006. Detecting coupling in the presence of noise and nonlinearity. In: Handbook of Time Series Analysis: Recent Theoretical Developments and Applications-Wiley Online Library.

Nicholls, J., 2012. From Neuron to Brain. fifth edition, Sinauer Associates, Sunderland. ISBN 9780878936090.

Nicolas-Alonso, L.F., Gomez-Gil, J., 2012. Brain computer interfaces, a review. Sensors (Basel) 12 (2), 1211–1279.

Nicolelis, M.A.L., Lebedev, M.A., 2009. Principles of neural ensemble physiology underlying the operation of brain-machine interfaces. Nat. Rev. 10, 530–540.

Oberlaender, M., de Kock, C.P., Bruno, R.M., Ramirez, A., Meyer, H.S., Dercksen, V.J., Helmstaedter, M., Sakmann, B., 2012. Cell type-specific three-dimensional structure of thalamocortical circuits in a column of rat vibrissal cortex. Cereb. Cortex 22 (10), 2375–2391.

O'Connor, D.H., Hires, S.A., Guo, Z.V., Li, N., Yu, J., Sun, Q.Q., Huber, D., Svoboda, K., 2013. Neural coding during active somatosensation revealed using illusory touch. Nat. Neurosci. 16 (7), 958–965.

Packer, A.M., Roska, B., Häusser, M., 2013. Targeting neurons and photons for optogenetics. Nat. Neurosci. 16 (7), 805–815.

Paninski, L., 2004. Maximum likelihood estimation of cascade point-process neural encoding models. Netw. Comput. Neural Syst. 15 (4), 243–262.

Paninski, L., Pillow, J., Lewi, J., 2007. Statistical models for neural encoding, decoding, and optimal stimulus design. Prog. Brain Res. 165, 493–507.

Parlitz, U., Junge, L., Lauterborn, W., Kocarev, L., 1996. Experimental observation of phase synchronization. Phys. Rev. E 54, 2115–2117.

Pecora, L.M., Carroll, T.L., 1990. Synchronization in chaotic systems. Phys. Rev. Lett.. 92(2).

Penfield, W., Boldrey, E., 1937. Somatic motor and sensory representation in the cerebral cortex of man as studied by electrical stimulation. Brain 9, 389.

Pereda, E., Quiroga, R.Q., Bhattacharya, J., 2005. Nonlinear multivariate analysis of neurophysiological signals. Prog. Neurobiol. 77 (1), 1–37.

Perlmutter, J.S., Mink, J.W., 2006. Deep brain stimulation. Annu. Rev. Neurosci. 29, 229–257.

Piccolino, M., 1997. Luigi Galvani and animal electricity: two centuries after the foundation of electrophysiology. Trends Neurosci. 20 (10), 443–448.

Pikovsky, A.S., Rosenblum, M., Kurths, J., 2001. Synchronization: A Universal Concept in Non-Linear Science. Cambridge University Press, Cambridge, UK.

Plonsey, R., Barr, R.C., 2007. Bioelectricity, A Quantitative Approach. Springer, New York.

Popovych, O., Hauptmann, C., Tass, P., 2005. Effective desynchronization by nonlinear delayed feedback. Phys. Rev. Lett. 94, 164102.

Popovych, O., Hauptmann, C., Tass, P., 2006. Control of neuronal synchrony by nonlinear delayed feedback. Biol. Cybern. 95, 69–85.

Quiroga, R.Q., Kraskov, A., Kreuz, T., Grassberger, P., 2002. Performance of different synchronization measures in real data: a case study on electroencephalographic signals. Phys. Rev. E 65 (4), 041903.

Rauschecker, J., Shannon, R., 2002. Sending sound to the brain. Science 295 (5557), 1025–1029.

Roberts, J.A., Robinson, P.A., 2008, Jul. Modeling absence seizure dynamics: implications for basic mechanisms and measurement of thalamocortical and corticothalamic latencies. J. Theor. Biol. 253 (1), 189–201. http://dx.doi.org/10.1016/j.jtbi.2008. 03.005.

Romo, R., Ruiz, S., Crespo, P., Zainos, A., Merchant, H., 1993. Representation of tactile signals in primate supplementary motor area. J. Neurophysiol. 70 (6), 2690–2694.

Rosenblum, M., Pikovsky, A., 2004a. Controlling synchronization in an ensemble of globally coupled oscillators. Phys. Rev. Lett. 92, 114102.

Rosenblum, M., Pikovsky, A., 2004b. Delayed feedback control of collective synchrony: an approach to suppression of pathological brain rhythms. Phys. Rev. E 70, 041904.

Rossi, U., 2003. The history of electrical stimulation of the nervous system for the control of pain. Pain Res. Clin. Manage. 15, 5–16.

Santaniello, S., Fiengo, G., Glielmo, L., Grill, W.M., 2011. Closed-loop control of deep brain stimulation: a simulation study. IEEE Trans. Neural Syst. Rehabil. Eng. 19 (1), 15–24.

Santaniello, S., Sherman, D.L., Thakor, N.V., Eskandar, E.N., Sarma, S.V., 2012. Optimal control-based Bayesian detection of clinical and behavioral state transitions. IEEE Trans. Neural Syst. Rehabil. Eng. 20 (5), 708–719. http://dx.doi.org/10.1109/TNSRE.2012.2210246.

Sauer, T.D., Schiff, S.J., 2009. Data assimilation for heterogeneous networks: the consensus set. Phys. Rev. E 79 (5), 051909.

Schiff, S., 2012. Neural Control Engineering: The Emerging Intersection Between Control Theory and Neuroscience. Computational neuroscience, MIT Press, Cambridge, MA. ISBN 9780262015370.

Schiff, S.J., Sauer, T., 2008. Kalman filter control of a model of spatiotemporal cortical dynamics. J. Neural Eng. 5 (1), 1.

Shen, B., Wang, Z., Liu, X., 2012. Sampled-data synchronization control of dynamical networks with stochastic sampling. IEEE Trans. Autom. Control 57 (10), 2644–2650.

Shenton, M.E., Dickey, C.C., Frumin, M., McCarley, R.W., 2001. A review of MRI findings in schizophrenia. Sch. Res. 49 (1), 1–52.

Shrimali, M.D., 2009. Pinning control of threshold coupled chaotic neuronal maps. Chaos 19 (3), 033105.

Shrimali, M.D., He, G., Sinha, S., Aihara, K., 2007. Control and synchronization of chaotic neurons under threshold activated coupling. ICANN 954–962.

Sinha, A., Schiff, S.J., Huebel, N., 2013. Estimation of internal variables from Hodgkin-Huxley neuron voltage. In: Proceedings of the 2013 Sixth International IEEE/EMBS Conference on Neural Engineering (NER). pp. 194–197.

Smith, A., Brown, E., 2003. Estimating a state-space model from point process observations. Neural Comput. 15 (5), 965–991.

Smith, S.M., Miller, K.L., Salimi-Khorshidi, G., Webster, M., Beckmann, C.F., Nichols, T.E., Ramsey, J.D., Woolrich, M.W., 2011. Network modelling methods for FMRI. Neuroimage 54 (2), 875–891.

Sporns, O., 2011. The human connectome: a complex network. Ann. NY Acad. Sci. 1224 (1), 109–125.

Sporns, O., Tononi, G., Kötter, R., 2005. The human connectome: a structural description of the human brain. PLoS Comput. Biol. 1 (4), e42.

Stigen, T., Danzl, P., Moehlis, J., Netoff, T., 2011. Controlling spike timing and synchrony in oscillatory neurons. J. Neurophysiol. 105 (5), 2074–2082.

Sun, Y., Cao, J., 2007. Adaptive lag synchronization of unknown chaotic delayed neural networks with noise perturbation. Phys. Lett. A 364 (3–4), 277–285.

Sun, Y., Cao, J., Wang, Z., 2007. Exponential synchronization of stochastic perturbed chaotic delayed neural networks. Neurocomputing 70, 2477–2485.

Sussillo, D., Nuyujukian, P., Fan, J.M., Kao, J.C., Stavisky, S.D., Ryu, S.I., Shenoy, K.V., 2012. A recurrent neural networks for closed-loop intracortical brain-machine interface decoders. J. Neural Eng. 9, 1–10.

Tass, P., 1999. Phase Resetting in Medicine and Biology: Stochastic Modelling and Data Analysis. Springer, Berlin, Heidelberg, New York.

Tass, P., 2000. Effective desynchronization by means of double-pulse phase resetting. Europhys. Lett. 53, 15–21.

Tian, L., Wang, J., Yan, C., He, Y., 2011. Hemisphere-and gender-related differences in small-world brain networks: a resting-state functional MRI study. Neuroimage 54 (1), 191–202.

Tokuda, I., Nagashima, T., Aihara, K., 1997. Global bifurcation structure of chaotic neural networks and its application to traveling salesman problem. Neural Netw. 10, 1673–1690.

Tononi, G., Sporns, O., 2003. Measuring information integration. BMC Neurosci. 4 (1), 31.

Ullah, G., Schiff, S.J., 2009. Tracking and control of neuronal Hodgkin-Huxley dynamics. Phys. Rev. E 79 (4), 040901.

Van Essen, D.C., Ugurbil, K., Auerbach, E., Barch, D., Behrens, T., Bucholz, R., Chang, A., Chen, L., Corbetta, M., Curtiss, S.W., et al., 2012. The human connectome project: a data acquisition perspective. Neuroimage 62 (4), 2222–2231.

Voss, H.U., Timmer, J., Kurths, J., 2004. Nonlinear dynamical system identification from uncertain and indirect measurements. Int. J. Bifurcation Chaos 14 (06), 1905–1933.

Walsh, V., Pascual-Leone, A., Kosslyn, S.M., 2003. Transcranial Magnetic Stimulation: A Neurochronometrics of Mind. MIT Press, Cambridge, MA.

Wang, J., Zhang, T., Che, Y., 2006. Chaos control and synchronization of two neurons exposed to elf external electric field. Chaos Solitons Fract 34 (3), 839–850.

Weber, D.J., London, B.M., Hokanson, J.A., Ayers, C.A., Gaunt, R.A., Torres, R.R., Zaaimi, B., Miller, L.E., 2011. Limb-state information encoded by peripheral and central somatosensory neurons: implications for an afferent interface. IEEE Trans. Neural Syst. Rehabil. Eng. 19 (5), 501–513.

Wei, Y., Ullah, G., Parekh, R., Ziburkus, J., Schiff, S.J., 2011. Kalman filter tracking of intracellular neuronal voltage and current. In: Proceedings of the 2011 50th IEEE Conference on Decision and Control and European Control Conference (CDC-ECC). pp. 5844–5849.

Whalen, A.J., Brennan, S.N., Sauer, T.D., Schiff, S.J., 2015. Observability and controllability of nonlinear networks: the role of symmetry. Phys. Rev. X 5 (1), 011005.

Wilson, H.R., Cowan, J.D., 1972. Excitatory and inhibitory interactions in localized populations of model neurons. Biophys. J. 12, 1–24.

Wilson, H.R., Cowan, J.D., 1973. A mathematical theory of the functional dynamics of cortical and thalamic nervous tissue. Kybernetik 13, 55–80.

Wilson, D., Moehlis, J., 2014. A Hamilton-Jacobi-Bellman approach for termination of seizure-like bursting. J. Comput. Neurosci. 37 (2), 345–355.

Wilson, D., Moehlis, J., 2014. Locally optimal extracellular stimulation for chaotic desynchronization of neural populations. J. Comput. Neurosci. 37, 243–257. http://dx.doi.org/10.1007/ s10827-014-0499-3.

Wu, Z.G., Park, J.H., 2013. Synchronization of discrete-time neural networks with time delays subject to missing data. Neurocomputing 122, 418–424.

Yu, W., Cao, J., 2007. Synchronization control of stochastic delayed neural networks. Phys. A 373, 252–260.

Zhang, C., Guo, Q., Wang, J., 2014. Finite-time synchronizing control for chaotic neural networks. Abstr. Appl. Anal. 2014. Article ID: 938612.

Zhou, J., Chen, T., Xiang, L., 2006. Robust synchronization of delayed neural networks based on adaptive control and parameters identification. Chaos Solitons Fract. 27 (4), 905–913.

Zhou, D., Thompson, W.K., Siegle, G., 2009. Matlab toolbox for functional connectivity. Neuroimage 47 (4), 1590–1607.

Zlotnik, A., Li, J.S., 2011. Optimal asymptotic entrainment of phase-reduced oscillators. In: Proceedings of the ASME Dynamic Systems and Control Conference. Arlington, VA.

Zlotnik, A., Li, J.S., 2012. Optimal entrainment of neural oscillator ensemble. J. Neural Eng. 9 (4), 046015.

Zou, C., Feng, J., 2009. Granger causality vs. dynamic Bayesian network inference: a comparative study. BMC Bioinform. 10 (1), 122.

Zou, F., Nossek, J.A., 1993. Bifurcation and chaos in cellular neural networks. IEEE Trans. Circuits Syst. I Fundam. Theory. Appl. 40 (3), 166–173.

Chapter 4

Testing the Theory of Practopoiesis Using Closed Loops

D. Nikolić

Max Planck Institute for Brain Research, Frankfurt/M, Germany; Frankfurt Institute for Advanced Studies (FIAS), Frankfurt/M, Germany; Ernst Strüngmann Institute (ESI) for Neuroscience in Cooperation with Max Planck Society, Frankfurt/M, Germany; University of Zagreb, Zagreb, Croatia

1 INTRODUCTION

The main goal of systems neuroscience is to eventually build an explanatory bridge between the physiological mechanisms on one side and the mental operations on the other side. We have made a great set of discoveries, a set that continues to grow at a rapid pace. Nevertheless, the answers to high-level questions somehow remain elusive. Despite the progress, we seem to stay equally far from making a satisfactory connection between the processes occurring at the physiological levels and those occurring at the psychological-mental level. The mind-body relationship seems to maintain much of its mystery status no matter how much more we learn about the neural circuitry. It seems we have a shortage of a framework that would guide our research toward faster closing of the mind-body explanatory gap.

Recently, a new theory has been proposed that offers a new approach toward understanding the relationship between the brain and the mind (Nikolić, 2015). In fact, the theory proposes a new approach to understanding how life is organized in general. Hence, the concepts driving this attempt to account for the mental-versus-material relationship presume that the functioning of mental operations is a direct continuation of the very nature by which living organisms function; the same fundamental rules should underlie both thought and digestion.

This new theory is called practopoiesis, meaning "creation of actions." The name reflects the focus of the theory, which is on the creation of novel physical structures (poiesis) and the fact that these novel physical structures are created with the function of generating actions (praxis) that are beneficial for the organism. The theory proposes that the process of practo→poiesis is organized into a hierarchy of operational acting components, whereby the components lower on the hierarchy, through their actions, create components higher on the hierarchy.

The lowest set of acting components on the practopoietic hierarchy of an organism is the gene expression system (genes + ribosomes). On the highest hierarchy lay sensory-motor loops that utilize the pathways of neural activity in order to collect sensory inputs and generate from them overt behavior of that organism (executed by muscles).

Practopoiesis consists of two parts. The first part describes general properties of systems that form this kind of hierarchy, and is founded on the fundamental theorems of cybernetics (Ashby, 1947; Conant and Ashby, 1970). One of the fundamental properties of practopoietic systems is that the interactions across members of the hierarchy have a particular structure: direct internal interactions occur always from the bottom toward the top (ie, low-level structures create higher-level structures), but not the other way around. In contrast, all of the influences that higher structures exert on the lower structures must occur through the mediation of the environment with which the organism interacts; all influences toward lower levels of organization involve feedback from the environment. That way, lower structures receive information from the environment on how a good job they did in creating higher-level structures. In other words, it is the environment that shapes the process of poiesis (Fig. 1). The details of those interactions are provided in the original publication (Nikolić, 2015).

The second part of the theory of practopoiesis describes the hierarchy of a biological mind. The theory predicts that a mind, as we observe it, can be created only if the poietic hierarchy is one level deeper than what brain theories have surmised so far. Classically, brain theories assume that there are two levels at which the brain executes practopoietic processes. These different levels of execution in the practopoietic theory are referred to as *traverses of cybernetic knowledge* or simply *traverses*. The two traverses of the classical brain theory are (1) the lower one:

Closed Loop Neuroscience. http://dx.doi.org/10.1016/B978-0-12-802452-2.00004-4

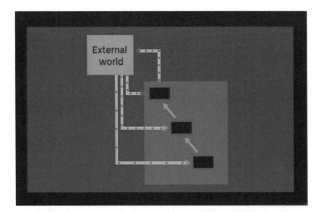

FIG. 1 Interactions across the components of practopoietic systems and with the external world. *Solid arrows*: poiesis. *Dotted arrows*: interactions with the external world (environment).

plasticity mechanisms creating the architecture of the network and driven generally by gene expression and (2) the higher one: neuronal activity driven by electrochemical activity involving hyper- and depolarization of membranes, synaptic transmission and generally described as network computation relying on the mechanisms of excitation and inhibition. Classical theories presume that these two components—the two traverses—are sufficient to explain how the brain works and how it creates the mind.

Practopoiesis permits a more detailed analysis of the adaptive capabilities and limitations of such a two-traversal system. The analysis performed in Nikolić (2015) indicated that these two traverses are not sufficient but, instead, a third traverse is needed in order to achieve a complete account of mental capabilities of the brain. This missing third traverse, according to the arguments, should be located in between the two known traverses. That is, the machinery responsible for executing this novel traverse should be poietically created by the plasticity mechanisms,

while the function of the new traverse should be to adjust—ie, poietically create—the properties of the neural network that employs inhibition and excitation.

Moreover, the argument holds that this third traverse is primarily responsible for the various mental phenomena that so far have escaped satisfactory explanation based solely on a two-traversal system. This middle-level traverse has been argued to be critically responsible for implementing mental processes such as perception, attention, decision-making, and thinking in general. The function of this traverse is primarily not to compute an output based on the sensory input, but instead to adjust the properties of the network such that the network is prepared for more efficient interactions with the incoming sensory inputs. This traverse has a capability to "re-wire" the neural network in a very short period of time (often, less than a second) and has thus the capability to instill a certain type of knowledge into the network and replace it with another knowledge, all of it occurring in real time—while the individual is performing some everyday task. This traverse also has the capacity to reactivate certain knowledge. For example, when one decides to sit on a bike and ride it, the knowledge of how to ride a bike needs to be re-activated. This middle traverse has been proposed to be responsible for that job.

For that reason, the traverse is named *anapoiesis*, from Ancient Greek *ana*, meaning over again. Anapoiesis refers to continual re-creation of knowledge within the neural networks in our brains. Anapoiesis is proposed to be central for generating our cognition and mental operations. See Fig. 2 for the division of responsibilities across the three traverses according to the tri-traversal theory of the mind. Anapoiesis has also been proposed to be the missing link needed to explain consciousness and our capability to understand the world around us (Nikolić, 2015).

FIG. 2 The tri-traversal theory of the mind involving anapoiesis.

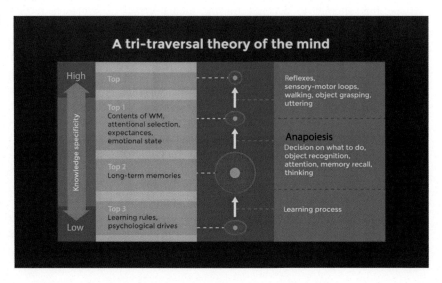

Importantly, practopoiesis generates a number of specific empirical predictions. These predictions are largely based on the presumption that anapoiesis has to operate at the time scales faster than plasticity but a bit slower than the spread of neural activity, based solely on inhibition and excitation. It follows that the physiological mechanisms underlying anapoiesis have to be of that relevant time scale. One (but not the only one) known phenomenon that stands out, that is ubiquitous across the nervous system, is fast neural adaptation (Verhoef et al., 2008; Pearson, 2000; Duncan et al., 1985).

Here we discuss experiments that formulate and test precise predictions on how the mechanisms of fast neural adaptation should behave (here referred to simply as "neural adaptation," and not to be confused with slow adaptive processes that take minutes or longer) if they have the role in our cognition and consciousness as predicted by the theory of practopoiesis. In particular, we explore the tests of the ability of adaptive mechanisms to learn how and when to adapt (pertaining to the concept of *learning to learn* in Nikolić, 2015), and how to integrate these learned capabilities within a global workspace of multimodal conscious interactions across the entire brain (pertaining to the concept of *equi-level interactions* in Nikolić, 2015).

2 NEURAL ADAPTATION

The key physiological mechanism that needs to be explored, according to the theory, and which may need to be assigned the important function of anapoiesis of neural networks, are those related to the *adaptation of neural responses*, but also the opposite ones, *sensitization* of responses. In the phenomenon of adaptation, neurons show a vigorous response only immediately after a stimulus is presented (Verhoef et al., 2008; Pearson, 2000; Duncan et al., 1985). Quickly after the initial response (eg, 100 ms later), neurons typically reduce the firing rates considerably, even though the stimulus remains unchanged and continues to be present. Although adaptation of neural responses is a well-known phenomenon, traditionally it has not been assigned much theoretical significance. Rather, neural adaptation is thought to play a peripheral role. Historically, adaptation of rate responses has been equated with fatigue; it has been commonly presumed that adaptation results simply from lowered activation thresholds of the system. As a consequence, computational models of neuronal activity aimed to account for mental phenomena rarely include adaptation of neural responses. Instead, such models have a tendency to be limited to detailed descriptions of the biophysics of synaptic depolarization or generation of action potentials (eg, Izhikevich and Edelman, 2008).

The situation is similar regarding empirical research. The properties of neural circuitry are investigated with great scrutiny (Toyama et al., 1981; Hellwig, 2000; Stettler et al., 2002) and are integrated into models (eg, Häusler and Maass, 2007; Markram, 2006), but the conditions under which neurons adapt, or sensitize, are left comparably unexplored.

So far, however, enough evidence has been accumulated to reject the fatigue hypothesis and consider adaptation as a much more capable and important component of the nervous system—also contributing more significantly to overall functioning. Existing data show that neural adaptation is a ubiquitous and reliable phenomenon (Verhoef et al., 2008; Pearson, 2000; Duncan et al., 1985) that is much more meaningful than fatigue of a neuron. The most direct evidence indicating that mechanisms underlying adaptation are not trivial comes from the studies of responses to the removal of input drive to neurons (eg, Eriksson et al., 2012; Wacongne et al., 2012). The activity following an abrupt stimulus removal is known as neuronal off-response. A removal of the input drive does not typically result in further silencing of neurons, as would be predicted by the fatigue hypothesis. Instead, off-responses are normally characterized by a sudden increase in neuronal firing rates. The total number of action potentials delivered is often larger than that generated at the onset of the stimulus (ie, during the on-response). This is consistent with the proposals that neural responses signal changes of stimulus features, rather than signaling the presence of those features (Rao and Ballard, 1999; Friston, 2010).

Our theoretically driven ignorance of neural adaptation lead us to ignore much empirical data that otherwise indicated the role and presence of adaptation mechanisms. I will briefly mention several studies made in my laboratory in which data showed not only the presence but possibly also the significance of neural adaptation, and yet, due to the theoretical views at the time, we did not recognize the importance of those findings. One example is the fact that gamma oscillations and neural synchrony in response to grating stimuli appear only 200–300 ms after the stimulus has been presented, ie, after the responses to a stimulus have adapted (see Fig. 3 for a typical trace of local field potentials that exhibit gamma oscillations post-

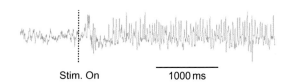

Stim. On   1000 ms

FIG. 3 Gamma oscillations do not begin immediately after a stimulus has been presented but their buildup takes some time, after neurons have adapted. *Vertical dashed line:* The time at which a drifting grating stimulus was presented.

stimulus presentation; also more details can be found in, eg, 8F in Nikolić et al., 2012). To the best of my knowledge, this fact has never been considered for inclusion into theories on how gamma oscillations come about and what their function may be.

In another study, a computer-based readout (ie, classification) was applied to investigate the amount of stimulus-related information in the spiking activity in populations of 100+ simultaneously recorded neurons (Nikolić et al., 2009). We flashed stimuli for 100 ms and found that more information was available later in time, when the neuron had the time to adapt to the initial response, than earlier, when the rate responses were at least as high; off-responses are more stimulus-specific than on-responses and thus carry more information about the stimulus (Nikolić et al., 2006, 2009). However, we seem to have failed to recognize the importance of this finding—which implies that neural adaptation plays a role in providing the system with the needed stimulus-related information (see Fig. 4).

We also found that off-responses were related to a property of the cortex to exhibit fading memory for the recent history of stimulation: Information contained in responses to the most recent stimulus blends with information about the stimulus presented just before that one (eg, 200 ms earlier). As a result, responses to a recent stimulus contained as much information about that stimulus as they contained about the preceding stimulus (Nikolić et al., 2009).

These results on stimulus-specificity and fading memory suggest existence of adaptation mechanisms that operate in the background, covertly, and that affect the overt spiking activity of neurons. The results also suggest that these adaptation mechanisms operate on a time scale slower than hyper-/de-polarization of the membrane but faster than plasticity. Thus, with the lower-bound somewhere at about 100–200 ms and the upper-bound extending up to about a few seconds, these adaptive mechanisms may lay somewhere in between the much faster neuronal activity and the much slower neuronal plasticity. Neural adaptation may represent a sort of a middle-level dynamics within the organizational hierarchy of brain functions—as predicted by the theory of practopoiesis. The results also hint that adaptation mechanism may not be "dull" but may reflect an "intelligent"

and informed form of system adjustment to environmental circumstances.

There is also one more study on adaptation to which I believe we should have paid more attention. In Biederlack et al. (2006), we investigated correlation between neural responses in V1 and subjective phenomenal experiences of stimulus brightness. We found that not only firing rates, but also strength of synchronization between neurons, correlate with the subjectively reported brightness of the stimulus. Thus, we found that synchrony is a neuronal correlate of phenomenal consciousness. However, again, at the time, my coauthors and I did not realize that adaptation may play a role for those phenomena on at least two levels. First, as mentioned earlier, adaptation was needed for synchronous neural activity to appear. Second, we failed in an extensive attempt to build a neural network model of visual cortex that would exhibit the same synchronization behavior as we observed empirically.

Neural adaptation shows up in many other studies. Once one has a theory and is more aware of the phenomenon, it becomes much easier to spot results that could be reinterpreted and data that could be reanalyzed. There are a number of brain phenomena that do not have a good account within the current theories and that may rely heavily on the mechanisms of neural adaptation. Some of those phenomena with possible neural adaptation in disguise may be: trial-to-trial variability (Fox et al., 2005), resting state activity (Deco et al., 2010), mismatch negativity (Näätänen et al., 2007; Wacongne et al., 2012), theta oscillations (Buzsáki, 2002; Klimesch, 1999), noise correlations (Lee et al., 1998), up and down states (Sachdev et al., 2004), and late components of event-related potentials (Donchin and Coles, 1988).

Importantly, neural adaptation does not seem to be a global network effect of the entire nervous system. Rather it is a local phenomenon. This is evidenced by equally vigorous and informative off-responses that occur already in retina (eg, Shapley and Enroth-Cugell, 1984; Nikolić and Gansel, 2012). Neural adaptation likely relies on local mechanisms, which probably employ unique physiological processes largely unknown today.

This all indicates a necessity to learn more about those adaptive operations of the nervous system. Here I propose investigations driven by the following functional

FIG. 4 More stimulus-specific information is typically available later than during the earliest responses, ie, after adaptation, as indicated by the classification performance of the readout.

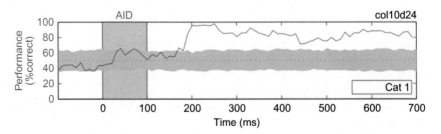

hypothesis about adaptation: *Neural adaptation is not a fixed process determined by genes, but instead it is a flexible mechanism that can be learned and adjusted depending on the past experiences and the degree to which certain adaptation routines satisfy the needs of the organism.* Thus, the system can learn how to adapt.

A key empirical question is whether plasticity mechanisms can change adaptive behavior of neurons and whether such behavior can be learned during the course of an experiment. Because adaptation appears to take place mostly within local circuits, local mechanisms should also be responsible for learning. Hence, experiments may yield results under anesthesia or by using a modality natural to the investigated brain region as feedback (eg, visual, auditory, or tactile stimulation).

Most important for the proposed studies is the design of the experiments. Traditionally, we present single trials in a stimulus-followed-by-response fashion, and we make sure that different stimulus conditions are properly randomized to remove all of the dependencies that may transfer from previous trial to the next. This results in a lot of trail-to-trial variability (eg, Churchland et al., 2010). But what if that variance is not just brain noise, but is generated by mechanisms actually central for our cognition and mental operations? What if by removing this variance we destroy the brain dynamics that enable fast learning and anapoietic restructuring as predicted by the theory of practopoiesis?

A different experiment design may be needed. I propose here that this design should be based on closed loops. Closed loops can allow addressing a key question: *Can the brain learn to learn?* Practopoiesis predicts that an intelligent system should have an adaptive component that lies in between the slow learning mechanisms that provide information for the lifetime and the neural activity that holds information transiently. Neural adaptation offers mechanisms at an intermediate time scale, and may provide a form of a fast "learning" system. The general prediction is the following: If neural adaptation is responsible for adaptive and intelligent functions, then this adaptation should itself be a flexible and adjustable mechanism that learns according to the circumstances and the feedback received from the environment. That way, the brain as a whole should exhibit a capability of learning how to learn. Neural adaptation, in particular, should reflect a mechanism that reactivates our long-term memories not directly stored in synapses, but in the routines by which we adapt neurons to the incoming stimuli.

To test this hypothesis, it will be necessary to conduct a number of studies that will test the capability of the brain to "think through adaptation mechanisms." The key prediction is that the decision on whether or not to adapt the responses of a neuron is made in accordance with how effective similar adaptations were in the past for satisfying the needs of the organism.

Thus, one open possibility is that the brain learns whether and when to adapt responses, and that this learning is driven by the fundamental needs of the organism. For a given type of stimulus properties, the system may adapt or keep active those responses that lead in the past to a "reward." Finally, the decision of whether or not to adapt responses may have direct consequences for the behavior of the organism, and this may be how adaptation ultimately controls what we do.

The direct empirical question then is whether the brain relies on similar adaptive mechanisms. As we interact with the world, perform everyday tasks, etc., do our neurons activate/de-activate largely through their own adaptation mechanisms, and not only through inhibition and excitation received from other neurons? Is it possible that mental operations—perception, attention, decision-making, memory recall, etc.—predominantly rely on learned patterns of neural adaptation?

To test those possibilities, I propose here a general experimental approach using electrophysiological recordings that could be made from various sensory areas in the cortex. These methods can be combined with behavioral tasks and state-of-the-art direct stimulation tools (eg, optogenetics). Also, examples of possible in vitro studies on cortical slices are discussed, as well as the possibilities to make studies on aware human subjects.

All of these experiments would need to have one property in common: They should all rely on an experimental paradigm based on a specific type of a closed-loop experimentation, which would allow the recorded brain signals (in most cases single-unit action potentials) to control the environment, and thus, to control the stimulation delivered back to the nervous system. That way, by experimental manipulations of the properties of the closed-loop, one could create new "virtual" environments to which the brain will experience a pressure to adapt.

The methods also allow one to falsify the above hypothesis. This could be done, for example, by demonstrating that the brain does not have the capability to change its patterns of neural adaptation, despite the fact that such changes in adaptive responses would be beneficial for the organism. Thus, if the responses of neurons were unchanged irrespective of whether we switch the closed-loop on or off, evidence would be provided that the traditional view of the brain computation, which is heavily biased toward explaining cognition based on inhibition and excitation mechanisms, should (for now) be retained. In contrast, if one can demonstrate that the neurons can learn how to adapt their responses, a whole new theoretical perspective would open and one would have to reconsider our understanding of the relationship between the brain and the mind.

A great advantage of closed-loop experimentation is that it fits well with the unified theoretical approach of practopoiesis which entails not only processing of information but also actions of the organism, and the effects that these actions induce back on the organism—ie, on closing the interaction loop with the environment (Nikolić, 2015). Therefore, practopoiesis and closed-loop experiments share in common the fact that they both combine cognition with behavior—a form of embedded cognition (Gibson, 2014; Brooks, 1991; Noë, 2012).

3 USING CLOSED-LOOP EXPERIMENTS

The closed-loop paradigm that should be used to study the predictions of practopoietic theory is illustrated in Fig. 5. Here, neuronal events are detected in real time and consequent changes are swiftly made to the properties of the stimulus. That way, a loop is closed such that it maps direct output of a neuron (or a neural circuit) indirectly to the inputs of the same neuron.

The mapping rules obeyed by the closed-loop machinery are defined by the *rule administrator*, which in turn determine the properties of the "virtual environment" with which the nervous system interacts. Thus, by changing the mappings in the rule administrator, we effectively change the properties of the system's "world." We can make this world more or less complex, more or less stable, or more or less demanding for adjusting to it. As a dependent variable, we can then observe whether, to which degree, and with which speed the system is capable of adjusting.

When we study a *system* with such experiments, we may study not necessarily a network of neurons, but a system that may be implemented by a single neural cell. A single cell may comprise the three different types of adaptive mechanisms described earlier, which may then simultaneously operate and mutually interact: the cell's electrical activity, its adaptation mechanisms, and its plasticity mechanisms. In that case, we would refer to plasticity

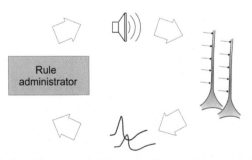

FIG. 5 The closed-loop paradigm used in the presently proposed studies. The action potentials of two simultaneously recorded neurons produce opposing effects on the properties of the stimulus. For example, if the action of one neuron decreases the loudness of the sound, the action of the other neuron increases it.

mechanisms not only responsible for growth of axons and dendrites and for regulating the strength of synapses, but also for learning how to efficiently adapt or sensitize cell's responses.

I will discuss here closed-loop designs for several hypothetical studies based on recording the activity of the sensory systems, some in vivo and others in vitro, some with unnatural and others with natural stimuli, and so on. The discussed designs can be applied to various sensory modalities (tactile, auditory, and visual). I will also discuss possible controls needed in order to minimize the contributions to learning from other brain regions, such as the use of anesthesia and electrical and optogenetic stimulation. All discussed experiments presume multichannel extracellular recordings, as are being routinely performed today (eg, Biederlack et al., 2006; Nikolić et al., 2009; Reyes-Puerta et al., 2014).

Recordings from single units would be advantageous as compared to the use of multiunits, and given the real-time nature of closed-loop experiments, segregation of single-unit activity would need to be performed by template matching, which would need to be performed quickly enough in order to allow for application of response-dependent changes to the stimuli with a total delay no longer than 10 ms (a higher limit of visual stimulation precision determined by a 100-Hz monitor refresh rate). Staying with these small delays in closing the stimulation loop may be essential for inducing adaptive changes to the system under anesthesia. This is because behavioral studies have shown that learning of contingencies that involve delays requires conscious awareness of those delays and thus full functioning of the system. In contrast, contingencies with short delays can be learned without subjects being conscious of those temporal relations (Clark and Squire, 1998). Thus, for studies under anesthesia, short delays may be essential.

One of the properties of the present paradigm is that it will require identifying through trial and error the end states, or the "goals," toward which the system applies changes to the activity of neurons. These goals are not obvious. Therefore, in order to test a general hypothesis that the system is capable of adjusting the firing rates of a neuron, we must also formulate and test in each particular experiment a set of more specific hypotheses: the actual state of the system that it tends to achieve through its adjustment operations.

For sensory systems, those end states or "desired" states are reasonable to postulate only if practopoietic theory is applied. Therefore, if such states exist, they have not been researched yet and hence are currently unknown. While generating hypotheses, one can make educated guesses of what these states may be. From practopoietic theory, it follows that whatever stimulus property we subjectively find aversive or attractive, the same property could (partly) contribute to neurons learning away or toward that stimulus state. Therefore, our own experiences with the stimuli should drive our hypotheses generation of what the neurons in the sensory cortices may "like" or "dislike." One can refer to those goal-related hypotheses as *teleonomic*.

Finding support for teleonomic hypotheses would be an important finding on its merit, and may alone present a sufficient motivation to perform closed-loop experiments. However, this finding would not provide a full support for existence of three traverses as predicted by practopoietic theory. Rather, the tests of whether the system prefers certain states should be used only as tools to test more specific hypotheses. There are two types of such hypotheses: hypothesis I: Whenever a system has the capability to control its own inputs, it is generally able to adjust ie, to learn to produce suitable outputs; hypothesis II: This learning occurs through the mechanisms of neural adaptation, that is, the system learns to learn.

It is only the second set of hypotheses that could provide a proper support (or refusal) of the idea that neural adaptation mechanisms play a central role in our cognition. If one only finds evidence for the teleonomic functions in sensory cortices, but not for learning to learn, there would be no empirical support for tri-traversal organization of the brain.

4 THE BASIC EXPERIMENTAL PARADIGM

The main experimental paradigm is designed to investigate the capability of local circuitry to take over control of its own environment. The hypothesis is that a circuit or a neuron, when given the possibility, is capable of setting up its environment to its own liking or needs.

A simplest imaginable example of such an experimental setup would be a case in which the spiking activity of neurons is connected such to have the power of controlling its own sensory stimulation. For example, each spike of neuron A may make the sound stimulus a bit quieter, a visual stimulus dimmer or a mechanical whisker stimulus less frequent. In the same time, a spike of a neuron B could be wired such to make the stimulus louder, brighter, or more frequent.

If the brain does not have the adaptive capabilities postulated by practopoiesis, the way in which neurons respond to the stimuli would be unaffected by whether they control the stimuli or not, and by whether they control the stimulus up or down. Thus, irrespective of whether a neuron is connected into this closed loop or not, the statistics of responses (and of spontaneous activity) would remain unchanged.

However, if the hypothesis of tri-traversal organization is correct, the brain should exhibit certain preferred properties of the stimulus, and by listening to the feedback it receives for its own actions, the brain should be able to take advantage of the possibility to control its stimulation. The brain should be able to learn to use the granted power to adjust its own inputs, and hence, neurons should adjust their firing rates accordingly. Depending on their preferences, either neuron A would increase its firing rate and neuron B would concurrently reduce its own firing rates, or vice versa. This is an important empirical question that has not been addressed in the past and that can be answered only through closed-loop experimentation.

A full test would also require applying a *reversal condition*. After learning is demonstrated, the rules for controlling the stimuli would need to be reversed across neurons A and B. The prediction is that the neurons should then again change their activity patterns accordingly.

It would be important that these learning capabilities are demonstrated also under anesthesia and therefore without the functioning of the full conscious system. The theory predicts that these learning capabilities operate largely at local levels, ie, each individual living cell possesses all three adaptive traverses. Therefore, it would be expected to find the effects at least at the level of local circuits, if not even at the level of individual neurons.

If such a closed-loop experiment produced positive results, one would show that neurons do not passively receive information, integrate, and pass the result of integration further, but are much more "intelligent," being able to consider the consequences of their own outputs and taking them into account when generating new outputs in the future.

To be able to relate such learning phenomena to the mechanisms of neuronal adaptation, one more important prediction must be tested in such experiments: The neurons should exhibit the capability of *learning to learn*. The three traverses predict not only that the brain exhibits goal-driven learning, but also that it exhibits the goal-driven process of learning how to learn.

This means that neurons should be able to learn to adjust more quickly to familiar experimental setups than to unfamiliar ones. For example, if we repeat the above reversal experiment such that neurons control their environment first according to the original setting, then in a reversed way, then again in the original setting, then again in a reversed way, and so on, we then come to the test of the crucial hypothesis: As we repeat those experiments, the system should become, over time, much more apt in adjusting. It should be able to learn how to adjust such that it detects the reversals of settings and consequently changes its modus operandi. And these changes should be quicker and quicker over time. Thus, after the first ever reversal of the control capabilities, it should take a while until the system "gets it" and finds the way to control stimuli to its teleonomic likings. However, after repeating the same experiment multiple times, the system should be able to

FIG. 6 The proposed research paradigm employing closed loop in which neurons control stimulation (*top*), with the predictions about the times that the system will need to adjust as it accumulates experience about its environment (*bottom*). The progress of the experiment unfolds from left to right.

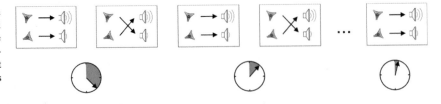

detect the switch in the rules rapidly and thus adjust the firing rates of the neurons rapidly too (see Fig. 6).

If these predictions are confirmed, then what we usually observe as fast neural adaptation of responses after a stimulus is being delivered (in vision already after 100 ms), would have to be interpreted as learned behavior: The neurons must have learned that they are not capable of controlling the stimulus, and hence, they reduce their firing rates. If this interpretation is correct, then this behavior can be also unlearned and replaced by another behavior if the experimental setting is made properly.

5 EXAMPLE EXPERIMENTS

5.1 An Acute Experiment

Animals are to be anesthetized and craniotomy is to be performed to allow access with multiple electrodes to sensory cortices. Any or all of the following sensory modalities could be used: tactile, with recordings from barrel cortex and whisker stimulation; auditory, with recordings from primary auditory areas and sound stimulation; and visual, with recordings from the visual cortex and using grating stimulation. After the electrodes are placed, the quality of signals is determined based on how well the waveforms of extracellularly recorded spikes can be defined for real-time spike sorting.

The steps of the experimental protocol could then unfold as follows:

1. First, a set of standard measurements is performed for characterizing the properties of the receptive fields of neurons, including determination of the corresponding tuning curves (tactile, auditory, and visual).
2. Spontaneous firing rates of neurons are determined by recording (eg, 10 min) of activity without any stimulation.
3. The properties of neural adaptation are characterized for optimal and suboptimal stimuli by presenting stimuli for a prolonged period of time either repeatedly (whisker stimulation) or statically (auditory and visual stimuli). This is an equivalent of an experiment with the closed loop being switched off (also referred to as *open loop*).
4. Finally, the loop is closed to enable the neurons to control their own stimulus. Based on the tuning curves, and the activity in Step 3, the stimuli should be chosen.

In any of the experiments, at least two single units should be involved in control. These units should control stimulus properties in opposing directions (ie, the stimulus is being changed along one-dimension). Examples are: the frequency of tactile stimulation, pitch of the sound, orientation of the grating stimulus on the screen. In addition, two- or n-dimensional stimulus control can be created by increasing the number of neurons recorded simultaneously. An example of a 2D control would be using outputs from four neurons for the control of the location of a visual stimulus on a 2D plane/screen.

When choosing the stimuli to be controlled and the rules by which the stimuli are being controlled, the following *teleonomical* hypotheses about the system may be explored, two of which presume static preferences (a, b) and the other two dynamic preferences of stimulation (c, d):

(a) The system prefers an extreme level of stimulation
(b) The system prefers medium levels of stimulation
(c) The system prefers change
(d) The system prefers stability

To test those hypotheses in a 1D-setting, a single property of a stimulus (intensity, orientation, and frequency) would need to be controlled by two single units, each randomly assigned to either increase or decrease a stimulus parameter by a small step c whenever an action potential has been generated.

5.1.1 Determining c

It seems reasonable to choose the value of c as one-half of the *decay rate* of the stimulus d, with which the stimulus gradually returns to its initial state in case no action potentials have been generated. The decay rates will be determined from the activity of the single units during the recordings in which the same stimulus was presented, but the loop was not closed (Step 3). Thus, the closed loop would operate with the property $d = 2c$.

Therefore the system would have the chance to change the stimulus only if the ratio between the firing rates of the two neurons is adjusted by at least the factor of two. Those eventual changes in the firing rates can be detected for example, by analyzing post-stimulus time-histograms.

Critically, after an appropriate period of time (eg, 20 min), the initial assignments of single units would need to be reversed; the unit whose outputs used to increase the value of stimulus property would now have to decrease it, and vice versa. Most importantly, the reversal of assignments would need to be repeated in multiple cycles (Fig. 6). These cycles will be essential to test whether the system is capable of learning to adjust to new circumstances (new control assignments) more quickly as the reversals progress. In other words, this would test whether the control of the stimulus allows the system to learn how to learn by applying the neural adaptation in a flexible way—only to those units that needed to be silenced in order to reach the goal states of the system. Only if one obtains affirmative results in this test, one can claim to have evidence supporting the idea that the mechanisms of neural adaption play a central role in cognitive processes.

5.1.2 Example Experimentation Protocol

To illustrate this experimental protocol, here I provide a hypothetical example study. After placing electrodes in the visual cortex, a researcher first found two single units with overlapping receptive fields and with differing orientation preferences, eg, 30 and 70 degrees, respectively. The researcher then decided to place a medium bright light bar in the middle position (45 degrees), and to record the activity of the two units in the loop-off condition (open loop). From those responses, the researcher calculated d and c values, and subsequently activated a closed-loop such that each spike of unit A increased the brightness of the bar and each spike of unit B decreased the brightness. The researcher then recorded the activity of the neurons and observed a systematic change in the bar brightness. Then after 20 min, the researcher reversed the rules; now unit B increased brightness and unit A decreased it. The researcher repeated the reversal procedure 10 times in a row: original assignment, reversed assignment, original assignment, and so on.

As a result, the researcher observed changes in the brightness of the bar at the end of each 20-min period, which corresponded to differences in the changes of the firing rates of those neurons. Moreover, if the light bar became consistently dimmer in all loop-on conditions, irrespective of the unit assignment, the researcher may have determined (1) that the system took advantage of the closed loop by taking control over stimuli and (2) that at least one teleonomic function existed in the visual cortex and this was a reduction of stimulus intensity.

The researcher then investigated whether the bar became dim faster in the later 20-min sessions than in the earlier ones. If this was clearly demonstrated, then it could be concluded that the anesthetized brain has learned to detect a change in environmental circumstances and to adjust to them more efficiently. In a way, it has developed adjustment skill. If so, the researcher could conclude that the neural adaptation observed in a traditional step of the experiment (in which the loop was off) also reflected an adaptive act in which the neuron reduced firing rates because it lacked an ability to control the stimuli.

To test this conclusion, the researcher performed a follow-up experiment in which 20-min recording periods were added, such that the closed loop was switched off. The researcher may find that in this case, the firing rates of both neurons are reduced, consistent with the idea that the system has adapted its responses due to their inefficiency to control stimuli. Moreover, if adaption can be learned, this congruent reduction of rate responses during open loop should be faster in later than in early recording periods.

5.2 Control by Disruptive Stimulation

Practopoietic theory predicts that the capability of learning to adapt should be largely a result of local mechanisms, operating even within the very same neuron whose action potentials have the power to control the stimuli. Local circuitry may assist that learning, but these learning effects should be under no circumstances a result of a global network. That is, the commands to adapt responses or to increase the firing rates should not arrive as feedback from higher brain areas. This prediction could only be partly ruled out by performing the earlier study under anesthesia. To address the question of locality of the learning effects more precisely, one may need to disrupt the local learning processes by unspecific (optogenetic) stimulation that would follow shortly after a unit has generated an action potential and a change in a stimulus has been performed. The disrupting stimulation would need to be delivered such that it affects only local circuitry. If the learning process relies mostly on the global network rather than local processes, learning should not be impeded much by local stimulation. In contrast, if learning depends heavily on local processes, such stimulation should have strong disruptive effects.

To ensure that only local circuitry is stimulated optogenetically (by light), one may inhibit selectively only inhibition of pyramidal cells via light by transfecting them with ArchT AAV virus. Also, recently silicone probes have been developed that contain a row of miniature light-emitting diodes (LEDs) on the same shank on which the recording sites are located (eg, recently developed at the University of Freiburg). By activating different numbers of LEDs on these probes, one can parametrically increase the total area stimulated by light. In addition, one can investigate the effect of light stimulation as a function of the distance from the recording site at which the activity of a control unit is being recorded.

To ensure that optogenetic stimulation has maximal disrupting effect with shortest possible light exposure, one can also limit the light exposure to the time period during which the most informative events for learning are likely to take place: when the feedback information arrives about the direction of the change in the stimulus that the action potential has inflicted. An estimate would be that this "critical" period takes place somewhere between 50 and 150 ms after the action potential has been generated. During this period of time, the local circuitry should be prevented from taking into account the feedback information on how spike affected the inputs to the circuits. To ensure that any action potentials generated during the optogenetic stimulation do not affect stimulus properties, the loop would need to be always switched off during delivery of light stimulation.

With such local stimulation, global circuitry should be minimally affected and should still be able to take advantage of feedback information.

As a control, the optogenetic stimulation of the same duration can be delivered during a time period that is not presumed to be critical for detecting the contingencies between the output of a neuron and changes in the stimulus properties (eg, 150–250 ms post-action potential).

5.3 Closed-Loop Experimentation In Vitro

An even more direct approach toward investigating the capability of local circuitry to learn to adapt (or learn how to learn) would be experiments performed on cortical slices. Neural responses to direct electrical and current stimulation adapt, too (Avoli and Olivier, 1989; Sanchez-Vives et al., 2000). Thus, no natural stimulation is necessarily needed to study neural adaptation. In an in vitro preparation, one can also investigate the capabilities of the local circuitry to learn to adapt their responses to electrical or optogenetic stimulation. For stimulation and recording, one would likely need to use at least two electrodes. To preserve the health of a cell during the entire experiment (which may need to last several hours with reversals of assignments) it would probably be a good idea to use cell-attached techniques, rather than using techniques that penetrate the membranes of cells. A loose-seal current-clamp should be good enough for recording and a tight-seal voltage-clamp for stimulation (see, eg, Perkins, 2006).

One should perform more or less the same set of experiments as described earlier for anesthetized preparations. One can close the loop and provide a cell in a dish with the power to control its own (electrical/optogenetic) stimulation. An exception to the earlier experiments could be that, for the sake of simplicity, in an in vitro study, the outputs of only one cell could be investigated at a time. Therefore, by closing the loop between the recording electrode and the stimulation electrode through a rule administrator, one would enable a single cell in a dish to control its own electric stimulation. One can then investigate whether the cell will increase the rates of its action potentials to reduce/increase the stimulation to a degree that counteracts the decay. One can begin by testing whether cells change their output patterns when the loop is switched on as opposed to loop-off condition. Next, one can compare the outputs of the cells after reversing the loop rules, ie, a change in whether an action potential decreases or increases the frequency/intensity of stimulation. Finally, one can investigate whether the cell responds to switching off the rules faster in later than in early parts of the experiment.

The question would be whether local circuitry can exhibit a primitive form of "intelligence" by adapting to the behavior of the environment with which it interacts. If a slice does not have such adaptive capabilities, one should not be able to observe the described learning capabilities. However, if local capacity for learning to adapt plays a crucial role in implementing cognitive functions, as predicted by practopoietic theory, then a slice of a cortex may also possess such learning capabilities. Experiments similar to those described here should be able to answer those questions.

5.4 Studies in Awake Animals (Integration With Behavior)

Yet another type of study can be designed to investigate whether a learned ability to use local outputs from a single neuron to satisfy local goals can be integrated into the global functions of the brain, such that the learned knowledge can be used to the advantage of the global needs of the organism. For example, if an animal has learned to remove aversive stimuli by generating action potentials in a particular sensory neuron, the question may be then whether the animal is now capable of quickly learning to use the same neuron to obtain general rewards such as water or food. Practopoietic theory predicts that the global intelligent adaptive behavior emerges from interactions among numerous local adaptive elements (neurons), each individual element being capable of learning how to adapt and hence, possessing all three traverses. If this is correct, any locally learned capacity to adapt should be efficiently used globally. This transition from local to global could only occur in awake state.

In such experiments, one could use animals with chronically implanted electrodes such that experiments can be performed both under anesthesia and in an awake state. In the first phase, the learning paradigm would need to be performed as in the study described earlier—under anesthesia. This would ensure (bias toward) local learning. For example, one could rely on sound stimulus as depicted in Fig. 6 and by making recordings from the auditory cortex.

In the second phase, the animal would be awakened and put in the following operant conditioning setup: The rule administrator is now re-wired such that each action potential of a recorded neuron accumulates toward (for neuron A) or away from (neuron B) the total amount of reward (eg, orange juice) delivered once every several seconds (with random delays, eg, 1–3 s). As mentioned earlier, in awake states, such longer time delays should be acceptable. The total amount of reward delivered should be proportional to the sum of actions potentials of neuron A, minus the sum of actions potentials of neuron B. The empirical question is whether and how quickly the single units forming a closed loop will detect the contingencies and use them correctly, increasing the rates of the unit that brings more orange juice and reducing (adapting) those of the unit that reduces the amount of orange juice.

A corresponding question is whether this conditioning process will be faster for the single unit that was trained during anesthesia as opposed to a control unit that was not trained during anesthesia. This would address the question of transfer of knowledge acquired during anesthesia; those units already trained under anesthesia to control sound intensity should now, according to the theory, be better suited to learn how to control delivery of reward.

Another prediction of practopoietic theory is the following: a local circuit does not have any other possibility to engage effectively in global processes, but first must be actively engaged in its local operations, ie, in adjustments of sounds through newly granted control powers (in practopoietic theory these interactions are described as equi-level interactions through environmental feedback, see Nikolić, 2015). Thus, a prediction is that learning should be greatly facilitated (both for neurons trained under anesthesia and those not trained under anesthesia) if sounds are presented during operant conditioning of controlling reward. The facilitating function of sounds should be detected even if these sounds are not being controlled by the unit and thus are not contingent with the reward.

To further establish this relationship between local and global processes, one should also perform a third phase of the experiment in which the learning process would continue with an additional protocol for classical conditioning of the responses: Each delivery of reward would need to be preceded by a stimulation at another modality (eg, a whisker or visual stimulation). If the hypotheses of the local-global interactions through adaptation are correct, the firings of the auditory neurons should be conditioned: the firing events should concentrate on the time period between the stimulation at another modality and reward delivery.

Phase four: If the rule administrator is set now such that another pair of auditory neurons (neurons B and C) are given the power of controlling the stimulation from other modality, we would have the following problem to be solved by an awake brain: Auditory neurons B and C need to accordingly fire or not fire to cause external stimulation in another sensory modality, which would then internally cause auditory neurons A and B to fire/not-fire in a way that is necessary to deliver the reward.

This conditioning protocol that combines operant and classical conditioning would also provide a preparation for yet another important phase of the experiment. In this fifth phase, one would anesthetize the animals again and would test which of the learned adaptations can be detected under anesthesia—ie, by the operations of the local mechanisms. It follows from practopoietic theory that only the original sound-driven learned adaptation mechanism should remain operational under anesthesia, while the later acquired associations between another modality and auditory cells should not be detectable under anesthesia. This would be despite the auditory neurons having shown in awake state the capability to control reward and stimulation in other modalities.

These studies on awake animals can also profit from additional controls made with optogenetic stimulation as discussed earlier.

5.5 Studies in Humans

A great advantage of having human subjects in a study is that one can instruct them verbally to perform a certain task so that the subject can create new sensory-motor contingencies practically immediately. Hence, subjects can "mentally re-wire" the flow of information in an instant. This is done by using their vast knowledge and the appropriate attentional mechanisms. This capability of the human brain to use language for comprehension and conscious attentional efforts for immediately creating completely new types of actions is the highest form of adaptive intelligent behavior that we know of. The question is then: Can this flexible attention-based flow of information also be explained by learned patterns of adaptation at the level of individual neurons?

One can design a study to investigate the degree to which the putative learning-to-adapt mechanisms are an integral part of the global conscious workspace of a human intellect. To this end, one can conceive of an experiment that consists of two phases. First, one would need to induce a local learning to adapt. Second, one instructs subjects to perform a certain task that relies on the mechanisms learned in the first phase.

There is a necessity for electrophysiological recordings in such a study. To this end, epileptic patients with subdurally placed electrode grids (Andreas et al., 2002) could be asked to participate in the study while waiting for a seizure to occur, or for a surgery. The patients could be recruited in collaboration with clinical epilepsy diagnostics teams and with an approval from the appropriate ethics committee.

If the electrodes do not make it possible to record neuronal spiking activity directly, the control loops can be closed on the basis of the broad-spectrum signals in the high-gamma range (>80 Hz), as those signals tend to reflect multiunit spiking activity (eg Ray and Maunsell, 2011). These (relatively) local signals picked up by electrodes comprise a much larger number of cells than one single unit. Nevertheless, the relative locality of the signals should be still sufficient for the type of questions that need to be addressed. As epileptic foci are most commonly found in the temporal lobes and this is where the electrodes are typically implanted, it is more likely that a researcher would find an electrode that picks up signals from primary auditory areas than from any other primary sensory area. Thus, a paradigm with human subjects is likely to rely on sound stimuli. In the second phase, however, one can add any other modality, such as a visual task.

In the first phase, it would be necessary to establish learning of adaptive behavior similar to the one described earlier under anesthesia. But this would need to be done without anesthetizing subjects (their bodies are under enough stress already undergoing various surgeries). Thus, learning would need to be performed such that subjects are not aware that they have learned certain "behavior."

This could be achieved such that subjects are required to perform a demanding unrelated conscious task while their neurons in auditory cortex are learning a very simple task. For example, the subjects could be watching a long sequence of movie clips and would need to keep and update three counts in their memory: the number of times a female, a male, or a child actor has appeared in the clips. This demanding distractor task would be a cover for an undergoing unconscious adaptation-learning task about which the subjects would be informed only later, during the debriefing of the experiment. During this task, the signals recorded by electrode grids would be closed in a loop such that they control the loudness of the sound of the movie. Some clips would then begin with uncomfortably loud sounds and others with sounds too quiet to discern the speech of the actors. Subjects' auditory cortex would now have the power to adjust those sounds, although the subjects would most likely not be aware of that capability (see Clark and Squire, 1998 for a related experiment).

In this first phase, one would make investigations similar to those in the studies described earlier, determining whether the subjects can learn to adjust the sound, and whether the switching of the rules (which will take place, eg, on average every 10 movie clips) becomes gradually faster as the experiment progresses.

In the second phase, subjects would be asked explicitly to attempt to perform control "by their mind" of certain stimuli presented on a screen (similar to brain-machine interface experiments). In that case, another modality could be introduced. So, subjects could control a picture on a screen (position of an arrow on a line) but the same modality could also be used to control consciously, eg, the loudness of a sound. One would then test the hypothesis that the subjects are capable of transferring the skill learned during the first phase as compared to a control situation in which other neurons (electrodes) would be connected into the loop that have not had a chance to be trained in the phase one. The prediction is that the subjects will be able to do the task much easier with the previously trained neurons than with those that were not trained.

In addition, a prediction is that unconsciously learned knowledge would be transferred to both modalities, auditory and visual, but the transfer would be faster with control of auditory stimuli. Finally, the transfer to the control over visual information should be facilitated by random presentations of sounds (not connected to the control loop and hence not contingent with the visual information).

Only at the very end, during debriefing, one can check with a post-hoc interview whether subjects were aware during the first phase of their power to control the loudness of the stimulus.

If successful, this study would not only have implications for understanding how global conscious processes emerge from local adaptation mechanisms, but would also have practical implications for better developments of brain-machine interfaces.

6 CONCLUSIONS

It is possible to study the brain and its relation to behavior by approaches different from what has been traditionally done. By using a closed-loop approach, we can create experiments that are not focussed on stimulus-response statistics, but approach the real lives of biological systems; whenever we act on our environment, the environment "acts" back. Identical stimulation inputs rarely repeat and even if they did, the brain is not stationary and cannot respond in the same ways. The classical approach may have driven neuroscience toward a nonoptimal path forcing researchers to throw away much of the relevant information on how the brain creates a mind. This then may have led to severely incomplete theories. We can address these issues after we perform the above-mentioned experiments.

If the hypotheses derived from the theory of practopoiesis happen to receive support, the typical neuroscience experiments could no longer be considered as sufficient to disentangle the functioning of the brain. The trial-to-trial variability could no longer be considered noise of little relevance for brain's information processing. Rather, this variability would have to be considered reflecting key physiological mechanisms for the workings of the brain. More precisely, this would indicate that we should consider the brain as performing anapoiesis and hence, applying a total of three traverses for generation of conscious behavior.

ACKNOWLEDGMENTS

This work was supported by Hertie-Stiftung and Deutsche Forschungsgemeinschaft.

REFERENCES

Andreas, H.-J., et al., 2002. Visualization of subdural strip and grid electrodes using curvilinear reformatting of 3D MR imaging data sets. Am. J. Neuroradiol. 23 (3), 400–403.

Ashby, W.R., 1947. Principles of the self-organizing dynamic system. J. Gen. Psychol. 37, 125–128.

Avoli, M., Olivier, A., 1989. Electrophysiological properties and synaptic responses in the deep layers of the human epileptogenic neocortex in vitro. J. Neurophysiol. 61 (3), 589–606.

Biederlack, J., Castelo-Branco, M., Neuenschwander, S., Wheeler, D.W., Singer, W., Nikolić, D., 2006. Brightness induction: rate enhancement and neuronal synchronization as complementary codes. Neuron 52, 1073–1083.

Brooks, R.A., 1991. Intelligence without representations. Artif. Intell. J. 47, 139–159.

Buzsáki, G., 2002. Theta oscillations in the hippocampus. Neuron 33 (3), 325–340.

Churchland, M.M., Yu, B.M., Cunningham, J.P., Sugrue, L.P., Cohen, M.R., Corrado, G.S., Newsome, W.T., Clark, A.M., Hosseini, P., Scott, B.B., et al., 2010. Stimulus onset quenches neural variability: a widespread cortical phenomenon. Nat. Neurosci. 13, 369–378.

Clark, R.E., Squire, L.R., 1998. Classical conditioning and brain systems: the role of awareness. Science 280 (5360), 77–81.

Conant, R.C., Ashby, W.R., 1970. Every good regulator of a system must be a model of that system. Int. J. Syst. Sci. 1 (2), 89–97.

Deco, G., Jirsa, V.K., McIntosh, A.R., 2010. Emerging concepts for the dynamical organization of resting-state activity in the brain. Nat. Rev. Neurosci. 12 (1), 43–56.

Donchin, E., Coles, M.G.H., 1988. Is the P300 component a manifestation of context updating? Behav. Brain Sci. 11 (03), 357–374.

Duncan, G.E., Paul, I.A., et al., 1985. Rapid down regulation of beta adrenergic receptors by combining antidepressant drugs with forced swim: a model of antidepressant-induced neural adaptation. J. Pharmacol. Exp. Ther. 234 (2), 402–408.

Eriksson, D., Wunderle, T., Schmidt, K., 2012. Visual cortex combines a stimulus and an error-like signal with a proportion that is dependent on time, space, and stimulus contrast. Front. Syst. Neurosci. 6, 26.

Fox, M.D., et al., 2005. Coherent spontaneous activity accounts for trial-to-trial variability in human evoked brain responses. Nat. Neurosci. 9 (1), 23–25.

Friston, K., 2010. The free-energy principle: a unified brain theory? Nat. Rev. Neurosci. 11 (2), 127–138.

Gibson, J.J., 2014. The Ecological Approach to Visual Perception. Psychology Press, New York. ISBN: 0-89859-959-8.

Häusler, S., Maass, W., 2007. A statistical analysis of information-processing properties of lamina-specific cortical microcircuit models. Cereb. Cortex 17 (1), 149–162.

Hellwig, B., 2000. A quantitative analysis of the local connectivity between pyramidal neurons in layers 2/3 of the rat visual cortex. Biol. Cybern. 82 (2), 111–121.

Izhikevich, E.M., Edelman, G.M., 2008. Large-scale model of mammalian thalamocortical systems. Proc. Natl. Acad. Sci. 105 (9), 3593–3598.

Klimesch, W., 1999. EEG alpha and theta oscillations reflect cognitive and memory performance: a review and analysis. Brain Res. Rev. 29 (2), 169–195.

Lee, D., et al., 1998. Variability and correlated noise in the discharge of neurons in motor and parietal areas of the primate cortex. J. Neurosci. 18 (3), 1161–1170.

Markram, H., 2006. The blue brain project. Nat. Rev. Neurosci. 7 (2), 153–160.

Näätänen, R., et al., 2007. The mismatch negativity (MMN) in basic research of central auditory processing: a review. Clin. Neurophysiol. 118 (12), 2544–2590.

Nikolić, D., 2015. Practopoiesis: or how life fosters a mind. J. Theor. Biol. 373, 40–61.

Nikolić, D., Gansel, K., 2012. Der Kontext entscheidet. Hirnforschung: Neurotheorie. Spektrum der Wissenschaft. Gehirn Geist 5, 20.

Nikolić, D., et al., 2006. Temporal dynamics of information content carried by neurons in the primary visual cortex. In: Schlökopf, G., Platt, J., Hofmann, T. (Eds.), Advances in Neural Information Processing Systems, vol. 16. MIT Press, Cambridge, MA, pp. 1041–1048.

Nikolić, D., et al., 2009. Distributed fading memory for stimulus properties in the primary visual cortex. PLoS Biol. 7 (12), e1000260.

Nikolić, D., et al., 2012. Scaled correlation analysis: a better way to compute a cross-correlogram. Eur. J. Neurosci. 35 (5), 742–762.

Noë, A., 2012. Varieties of Presence. Harvard University Press, Cambridge, MA. ISBN: 9780674062146.

Pearson, K.G., 2000. Neural adaptation in the generation of rhythmic behavior. Annu. Rev. Physiol. 62 (1), 723–753.

Perkins, K.L., 2006. Cell-attached voltage-clamp and current-clamp recording and stimulation techniques in brain slices. J. Neurosci. Methods 154 (1), 1–18.

Rao, R.P.N., Ballard, D.H., 1999. Predictive coding in the visual cortex: a functional interpretation of some extra-classical receptive-field effects. Nat. Neurosci. 2 (1), 79–87.

Ray, S., Maunsell, J.H.R., 2011. Different origins of gamma rhythm and high-gamma activity in macaque visual cortex. PLoS Biol. 9 (4), e1000610.

Reyes-Puerta, V., Sun, J.J., Kim, S., Kilb, W., Luhmann, H.J., 2014. Laminar and columnar structure of sensory-evoked multineuronal spike sequences in adult rat barrel cortex in vivo. Cereb. Cortex 25 (8), 2001–2021.

Sachdev, R.N.S., Ebner, F.F., Wilson, C.J., 2004. Effect of subthreshold up and down states on the whisker-evoked response in somatosensory cortex. J. Neurophysiol. 92 (6), 3511–3521.

Sanchez-Vives, M.V., Nowak, L.G., McCormick, D.A., 2000. Cellular mechanisms of long-lasting adaptation in visual cortical neurons in vitro. J. Neurosci. 20 (11), 4286–4299.

Shapley, R., Enroth-Cugell, C., 1984. Visual adaptation and retinal gain controls. Prog. Retin. Res. 3, 263–346.

Stettler, D.D., et al., 2002. Lateral connectivity and contextual interactions in macaque primary visual cortex. Neuron 36 (4), 739–750.

Toyama, K., Kimura, M., Tanaka, K., 1981. Cross-correlation analysis of interneuronal connectivity in cat visual cortex. J. Neurophysiol. 46 (2), 191–201.

Verhoef, B.-E., Kayaert, G., et al., 2008. Stimulus similarity-contingent neural adaptation can be time and cortical area dependent. J. Neurosci. 28 (42), 10631–10640.

Wacongne, C., Changeux, J.-P., Dehaene, S., 2012. A neuronal model of predictive coding accounting for the mismatch negativity. J. Neurosci. 32 (11), 3665–3678.

Chapter 5

Local Field Potential Analysis for Closed-Loop Neuromodulation

N. Maling* and C. McIntyre*,†

*Case Western Reserve University, Cleveland, OH, United States, †Louis Stokes Cleveland Veterans Affairs Medical Center, Cleveland, OH, United States

1 INTRODUCTION

Local field potentials (LFPs) have several qualities that make them well suited for use in closed-loop neurophysiology. Using appropriate analysis techniques is a key component in extracting the rich information contained within an LFP recording. This chapter explores several of the most fundamental ways to analyze LFPs with an emphasis on those useful to clinically relevant closed-loop neuromodulation applications.

1.1 Definition of the LFP

To discuss the analysis of LFPs, it is first necessary to have a clear definition. The LFP is typically described as *the low-frequency component of the extracellular electrical potential* (Destexhe and Goldberg, 2015). Historically, the canonical frequency bands of interest span 1–100 Hz (Buzsaki et al., 2012; Niedermeyer and Silva, 2004), however the last decade has seen a shift of interest toward faster rhythms (Fig. 1). Consequently, researchers have broadened their considerations to accommodate these faster frequencies of interest. The "low frequency" part of the definition is therefore to some degree a mutable constraint. While the range of frequencies considered to be an LFP has varied, the "extracellular" part of the LFP is integral to its identity. A cardinal feature of this signal is that it represents activity from many neurons in the vicinity. The sampling of local neuronal populations captures network dynamics that would otherwise be missed by single-cell recordings. It is these network dynamics that define the LFP, and make it a signal worth recording.

In practice, analyzing LFPs is very similar to analyzing rhythmic brain activity recorded from other modalities such as electroencephalography (EEG) or electrocorticography (ECoG). The primary difference is that the LFP is recorded from a depth electrode while EEG and ECoG are recorded on the surface of the head or brain, respectively. However, the propagation of electrical fields through the skull and brain parenchyma reduces both the signal amplitude and the spatial specificity of surface recordings (Nunez and Srinivasan, 2006; Srinivasan et al., 1998). Consequently, LFPs capture a more spatially localized field that is typically larger in magnitude than an EEG. Perhaps more importantly, LFPs can be recorded from regions deeper in the brain, thus representing activity from neuronal populations less accessible to noninvasive methods. Nonetheless, many insights learned from the decades of EEG research can be translated into LFP analysis.

Although LFPs are recorded at depth, the type of electrode can vary, and the shape and impedance of the electrode also influence the recorded signal (Nelson and Pouget, 2010). Typical LFPs may be recorded from a low-impedance stimulating electrode, a high-impedance glass microelectrode, or a high-density microelectrode array. All of these electrodes implanted into the same brain region will yield a different representation of the underlying activity. The LFP recorded at the electrode is a spatial average of current sources in the surrounding brain area. The size and shape of this sampling field is in part dependent on the size and impedance of the electrode, but primarily based on the synchrony of neuronal firing in the region (Lempka and McIntyre, 2013; Lindén et al., 2011). Perhaps the most important electrode consideration in a discussion of LFP analysis is the presence of multiple contacts. Multiple recording locations can enable analysis techniques that describe not only the activity of a single region, but the relationship between multiple local populations. This kind of information is helpful to describe network-level phenomena. For example, the phase of an oscillation is not so interesting to describe by itself. However, if two distal oscillations are phase-locked, you may conclude that those two regions are transmitting information in one or both directions. These kinds of comparisons are not possible with a single contact recording.

Closed Loop Neuroscience. http://dx.doi.org/10.1016/B978-0-12-802452-2.00005-6

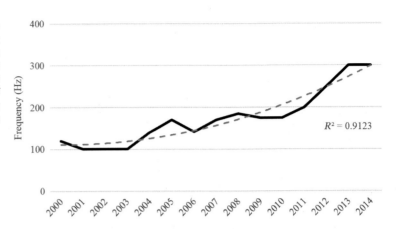

FIG. 1 Highest frequency of interest. The highest frequency analyzed in a subset of publications from each year between 2000 and 2014. Thirty random articles were chosen from Pubmed from the given year that matched the search term "local field potential brain." Values represent either (1) an explicit definition of the LFP, (2) the upper limit of the highest passband analyzed, or (3) the low-pass filter cutoff. Articles were excluded if none of the above values were reported.

1.2 The LFP as a Control Signal

In order to be useful for responsive stimulation applications, LFPs must capture a physiological trigger that reliably represents conditions where stimulation is desired, also known as a biomarker LFP signal. To this end, biomarker LFPs must be relatively time invariant over the lifetime of the stimulation device. Recording from depth electrodes presents a handful of challenges in this regard. These challenges can be biological in origin (such as formation of a glial scar around the recording electrode), or arise from physical changes (like an electrode shift or change in position). For comparison, silicon electrodes used for chronic single-cell recordings have a high failure rate, with up to 72% electrodes failing within a year (Barrese et al., 2013). LFPs are more robust, and in fact when single-cell activity has been lost, LFP recordings can still be salvaged from the same electrodes (Flint et al., 2012). However, in the majority of currently available responsive stimulation devices, LFPs are recorded from macroelectrodes that are used for recording the biomarker LFP signals, as well as delivering the responsive stimulation. Under these conditions, stable LFPs can be reliably recorded for several years (Abosch et al., 2012; Bergey et al., 2015; Maling et al., 2012; Stypulkowski et al., 2013).

When attempting to identify and/or define a biomarker LFP signal for use in a responsive stimulation device, understanding the spatial extent of the recording volume can be useful. Under physiologically realistic conditions it is reasonable for neurons several millimeters away to be included in the recording volume and contributing to the LFP signal (Lempka and McIntyre, 2013; Lindén et al., 2011). This wide spatial sampling contributes to the stability of the LFP in two ways. The first result is that it mitigates the influence of small-scale electrode shift on the signal. Electrode shift should be expected in depth electrode recordings, from resolution of edema, tissue relaxation, or tugging. The average electrode shift for a deep brain stimulation (DBS) electrode in humans is about

2 mm (Khan et al., 2008; Winkler et al., 2005). A relatively large recording volume helps ensure that recordings performed at the time of surgery at least partially overlap with later recordings in a chronic setting. This stability is necessary to implement responsive stimulation detection algorithms that are expected to be effective for several years. The second result is that a wider recording volume reduces sampling error. Sampling error is caused by observing a small sample whose activity is not representative of the population. In the case of neurophysiology, this translates into recording from neuronal populations that do not encode information relevant to either the control signal or the disease state. Because brain activity is heterogeneous and often functionally compartmentalized, a small sampling field can undermine attempts to capture a robust trigger if that trigger arises from a spatially discrete neuronal population. Consequently, the larger recording volume reduces the potential for sampling error. The challenge in designing a responsive stimulation algorithm lies in capturing a signal that can be reliably used to apply stimulation when needed. However, no amount of sophistication in the analysis can compensate for a vapid signal.

Another concern using an LFP as a control signal is the engineering requirements of recording and analysis. In a responsive stimulation device, the LFP must be filtered and sampled using an on-board computer. From there, the signal could conceivably be sent to an external device for analysis, however the burden of sampling will always fall to the implanted device. Here, lower sampling rates and low-frequency content are a boon that results in longer battery life for detection systems, and less bandwidth for wireless systems that depend on off-board analysis. Ideal sampling rates for LFP depend on the frequency content you are interested in recording. As mentioned above, interest in higher frequencies will push sampling rates upward. LFP-based detectors that utilize high-frequency activity can be very powerful in their predictive capability, but in turn require more resources from the implanted device.

2   ANALYSIS

This section introduces fundamental concepts of LFP analysis with an emphasis on closed-loop neuromodulation applications. In the context of responsive stimulation, the purpose of LFP analysis is to extract a physiological biomarker that will act as a control signal for electrical stimulation. Although the nature of this biomarker must be neural in origin, neural signals can manifest in a variety of ways. The techniques described here represent those signal analysis techniques that are sensitive to the energy constraints of on-board processors, as well as useful for detecting changes in neural states. Once the detection event occurs, a stimulation pattern can be delivered. LFP analysis techniques can be divided into time domain and frequency domain methods. Because much of the information content of the LFP is oscillatory in nature, a large portion of the analysis techniques depend on Fourier decomposition or related frequency domain techniques to extract a trigger. The goals of the analysis techniques that are described below are to characterize relationships between the experimental environment and features in the LFP.

We begin by talking about necessary considerations before analysis begins, such as sampling and filtering. Then we discuss simple time domain methods of analysis that are suitable for implementation in closed-loop devices. We finish by discussing frequency domain analysis techniques.

2.1   Sampling and Filtering

2.1.1   Nyquist Theorem and Aliasing

The process of converting an analog signal into a discrete sequence is known as sampling. For a time domain signal, the sampling rate (Fs) is the rate at which the voltage of the analog signal is measured and digitized. In order to fully describe an analog oscillatory signal of frequency F, the Nyquist theorem states that you must use a sampling rate of at least twice the frequency of the oscillation you are trying to record. The two values, $2F$ and Fs/2, are known as the Nyquist rate and Nyquist frequency, respectively. This consideration is particularly important in frequency

domain analysis. Therefore, there is an implicit assumption in all recordings that the highest frequency of interest is no greater than the Nyquist frequency. Assuming the signal recorded is bandlimited such that it fulfills this criterion, then the time domain signal may be perfectly reconstructed. The signal is said to be bandlimited if it contains no power in frequency bands above the Nyquist frequency. In other words, if a time series contains no frequencies higher than the Nyquist frequency (Fs/2), it is completely described by giving ordinates recorded at the Nyquist rate.

When a signal is sampled below the Nyquist rate, frequencies above the Nyquist frequency will become indistinguishable from lower frequencies (Fig. 2). This process is called aliasing. While the Nyquist theorem presents a lower bound for sampling frequency, in practice it is common to sample at faster rates to avoid aliasing. This "oversampling" is not detrimental to the signal, but mitigates aliasing as any real signal likely contains power in frequency bands above the Nyquist frequency. It is for this reason that most digital sampling equipment contains antialiasing filters that attenuate frequency bands above ½ the sampling rate.

2.1.2   Filtering

Filtering is the process of removing unwanted frequency components from a signal. There are many different ways to go about filtering, including electronic filters built into the sampling circuitry, or digital filters that can be applied "in post" while the signal is being analyzed. An ideal filter is one that allows frequencies of interest through unmodified, without "ripple," while stopping any frequency information outside of that passband. Another way of saying that is that an ideal filter is maximally flat in its passband while having perfect attenuation in the stopband (or an infinitely steep roll-off). Any real filter will have some combination of ripple in the passband and/or a gentle roll-off. The most commonly used filters therefore represent tradeoffs between sharp roll-off and flatness in the passband. On one end of the spectrum, a Butterworth is a maximally flat filter with a slow roll-off that will yield a high-fidelity signal, but not perfectly attenuate frequencies outside of the passband. On the other end of the spectrum,

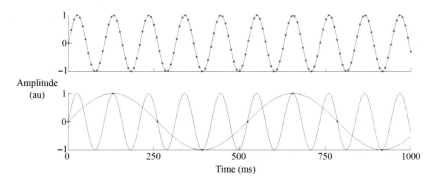

FIG. 2 A 10 Hz wave that is sufficiently sampled (top) can be accurately represented with a sampling rate greater than the Nyquist rate of 20 Hz. Undersampling (bottom) creates spurious lower frequency 2 Hz wave that results from aliasing of frequency information above the Nyquist frequency.

elliptical filters have the sharpest roll-off, but also contain significant ripple in both the passband and the stopband. This results in very low power in the stopband but introduces noise into your passband. Other families of filters can be designed to have a relatively flat passband with a sharp roll-off, but contain ripple in the stopband that can be troublesome in noisy environments.

Filter design is therefore an important consideration in the analysis of LFPs, as the filtering characteristics will have a significant effect on the signal that you are analyzing. For example, if the signal you are recording has a low signal to noise ratio, then you may desire a filter with a sharp roll-off that will reduce the power of frequency bands outside the desired range. On the other hand, if you have a large signal to noise ratio, the quality of the signal could be improved by using a maximally flat signal that may let in some contamination from other frequencies. Ripple in the stopband can be a problem when you have high-amplitude noise that, while might be attenuated, is still higher amplitude than your signal. When doing analysis, it is therefore important to keep in mind how the signal has been filtered, and temper your analysis techniques so that information in the stopband is not considered.

2.2 Time Domain Analysis

2.2.1 Energy

The raw energy in a time domain signal can be a simple and useful measure of neural activity, particularly in seizure detection (Correa et al., 2009). Although more accurate frequency-dependent measures are discussed below, the instantaneous energy can be estimated by simply taking the square of the signal. Therefore the instantaneous energy of a time domain signal X_t can be estimated as X_t^2. Applying a window of length W yields

$$E(t) = \frac{1}{W} \sum_{i=t-W+1}^{t} x(i)^2 \tag{1}$$

This is perhaps the most simple detector that can be used to identify changes in overall signal energy independent of frequency. This kind of detection can be paired with a narrow-passband to create frequency-specific power measure. To create a detection algorithm based on this value, simply include history dependence. Achieve the cumulative sum of this value by sliding a moving average of window length A with overlap O, over the energy function. Doing so yields accumulated energy

$$Ea(t) = \sum_{i=O(t-1)+1}^{O(n-1)+A} E(t) + Ea(t-1) \tag{2}$$

The absolute difference of a short-time windowed average and a long-time historical energy measurements

can be used to make a threshold-based detection. While such a detection algorithm is simple, it can be effective (Litt et al., 2001) and has the benefit of being computationally cheap.

2.2.2 Line-Length

Related to the above estimate of energy, the average line-length of a signal can be used to identify changes in both amplitude and frequency (Esteller et al., 2001). Line-length here is defined by the absolute difference between iterative samples: $D_t = |X_t - X_{t-1}|$. Substituting this difference for the square of the signal above yields a formula for line-length using a window of length W:

$$L(t) = \frac{1}{t} \sum_{i=t-W}^{t} |X_i - X_{i-1}| \tag{3}$$

This method has the advantage that the detector is sensitive to changes in either amplitude or frequency, with increased D_t equating to either increased power or higher frequency. Such economical measures are often preferred in the context of responsive stimulation, even though they are less specific due to their reduced energy cost.

2.2.3 Band-Passed Power

Halfwaves represent another economical measure of energy that can be used to assess band-specific power. A halfwave detector is more sophisticated than simple power measurements and allows for discrimination of changes in frequency-specific power, while still maintaining computationally efficiency. A signal that has been segmented into local minima and maxima is composed of halfwaves. The distance between a local minima and local maxima is a pseudo-wavelength, which can be used to estimate the frequency content of a signal. The amplitude of a halfwave is expressed relative to the background. Background amplitude should be constantly calculated as the average amplitude of the sequences from the previous 5 s. Relative amplitude of a halfwave is then the ratio of the absolute amplitude of the halfwave to the background (Gotman, 1982; Qu and Gotman, 1997). By detecting and counting halfwaves of specific pseudo-wavelength and above, a programmer defined amplitude within a specific time window, frequency-specific power detection can be achieved. For a more exhaustive discussion see the work of Gotman and Gloor (1976).

If it is known a priori that an LFP biomarker exists in a well-defined frequency band, one simple way to determine the power within a particular frequency band is simply to take the magnitude of a time domain signal that has been filtered with a narrowband pass with cutoffs around the frequency of interest. What remains in this signal is only the frequency content of interest, and consequently, the time-varying amplitude of this signal. A simple rectified amplitude detection algorithm, like one described above in Eq. (2), is sufficient to estimate power in the passband. This method is often very economical to assess frequency-specific power, however definition of the frequency band of interest must be known a priori and thus prohibits exploration of other frequency bands as biomarkers.

2.2.4 Evoked Potentials and Triggered Averages

Transient deflections, or potentials, in an LFP can occur with a wide range of frequency components, and can evade detection from spectral analysis. These types of potentials can occur spontaneously, or in response to an external stimulus. This stimulus could be a sensory cue, electrical stimulation, or any discrete time presentation. The average neurophysiological response of a neuron or group of neurons in response to the presentation of a discrete time stimulus is the evoked response (ER). Neurophysiological experiments often use ER measurements to represent how excitable neural tissue is in relation to a given stimulus, as the ER is a surrogate of propagating activity through a given network. Aberrant ERs are hallmarks of several disease states and imply disruption somewhere along the signal transduction pathway.

For closed-loop stimulation paradigms, the response of neural tissue to electrical stimulation could itself be used as a control signal to adjust parameters of future stimulation pulses. Presumably, a prescribed brain area will have a consistent response to a conserved electrical stimuli. Systematically varying the stimulation parameters and recording ERs is one way to sweep through a complex parameter space. This method has been the basis for closed-loop DBS programming protocols that seek to identify which stimulation parameters are optimal by evaluating the physiological response to varying stimulation parameters (Gaynor et al., 2008; Hanajima et al., 2004; Kent et al., 2015). Alternatively, presenting a well-characterized stimulation pulse could be used as a state detector in diseases where symptoms vary in a time-dependent manner. For example, the excitability of neural tissue may vary as a function of the subject's mood. In this case, using an evoked potential can be a surrogate for clinical assessment.

Mathematically, the ER is the average peri-stimulus signal. To compute the ER, the signal in the time window following each event is extracted, and the resulting snippets are averaged. For a given time domain signal S assessed after N events with a window of length W, the ER would be defined as:

$$\text{ER}(N) = \frac{1}{W} \sum_{i=N}^{N+W} S_i \qquad (4)$$

Alternatively, the LFP may be analyzed retrospectively to determine what the average LFP signal looks like preceding a symptom event. For example, one may be interested in determining what kinds of variations occur in the time preceding a seizure. Mathematically, this is essentially the same as an evoked potential but instead of using a user-defined event, you use a spontaneously occurring event. This is known as a triggered average. The triggered average is simply the average signal in a time window W preceding the spontaneously occurring event.

$$\text{ER}(N) = \frac{1}{W} \sum_{i=N-W}^{N} S_i \qquad (5)$$

2.3 Frequency Domain Analysis

2.3.1 Spectral Analysis

A discussion on spectral analysis requires an understanding of the Fourier transform, as it is the basis for most spectral analysis techniques. The Fourier transform decomposes a time domain function into its constituent frequencies, transforming it from a time domain signal to a frequency domain representation. It is a central tenant of spectral analysis that any periodic time domain signal can be described at a discrete time by the summation of all of its constituent frequency components (Fig. 3). The Fourier transform of a signal is a complex with the absolute values representing the amplitude and the complex values representing the phase offset of a sinusoidal wave of that frequency. In addition, an inverse Fourier transform can also be applied, transforming a frequency domain signal back into the time domain. This is known as Fourier synthesis, and combines all the frequency domain signals to recover the input signal in the time domain. Therefore the time and frequency domains of a signal are equivalent and we can transform the signal between them using Fourier analysis.

$\text{FT}(f)$ is the continuous Fourier transform of the bandlimited time domain function $x(t)$:

$$\text{FT}(f) = \int_{-\infty}^{\infty} x(t) e^{-2\pi i t f} dt \qquad (6)$$

Alternatively, the Fourier transform might be desired of a prescribed discrete time interval. This is often the case when analyzing digitized signals such as an LFP. Here the discrete Fourier transform $\text{DFT}(n)$ of discrete time domain signal $f(k)$ is given by:

$$\text{DFT}(n) = \sum_{k=0}^{N-1} f(k) e^{-2\pi i n k / N} \qquad (7)$$

where N is the number of frequencies at which the signal is evaluated.

Lastly, the fast Fourier transform (FFT) is a special case of the discrete Fourier transform that reduces the number of computations required. It is therefore more efficient and

FIG. 3 Assumption for Fourier decomposition. Any time domain signal can be represented as the sum of various periodic inputs. The Fourier decomposition transforms a composite signal into its constituent sin waves.

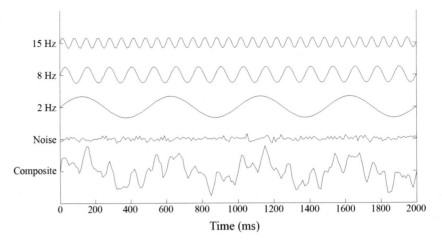

common. The FFT is the same as a DFT, however the number of points N must be a power 2. Such algorithms are typically what are used in modern signal processing software to reduce the computational load.

The results of a Fourier transform can yield the power of a specific frequency band within a time domain signal. A power spectra is the square of the absolute value of the Fourier transform. By averaging the power spectra across a specific frequency band you can obtain the band-specific power of the original signal.

2.3.2 Multi-Taper Methods

Conventional Fourier analysis is limited in that a Fourier transform represents only a single estimate of the spectral power within a given single. While the results of a Fourier transform inform some of the spectral information from the signal, the decomposition is sometimes inaccurate and contains bias when multiple frequency components are blurred together. The blurring of frequencies close together is known as narrow-band bias, while the blurring of distant frequencies is broad-band bias. By multiplying the original data by a set of data tapers, rather than a single taper, you can realize several different iterations of a spectra. The average of these realizations is a more accurate representation of the spectral content that can reduce the bias present in a Fourier transform.

The tapers used in this method are designed to reduce spectral leakage that results from finite data sets. The typical tapers used are discrete prolate spheroidal sequences, or "slepian" sequences. Independent estimates of the power spectrum are then computed by multiplying the data by these tapers. Averaging these estimates yields a better and more stable estimate than do traditional or "single-taper" methods. The tapers themselves are the discrete set of eigenfunctions, which solve the variational problem of minimizing leakage outside of a frequency band

of half bandwidth pfn where $fn = 1/Ndt$ is the Rayleigh frequency, and p is an integer.

2.4 Time Frequency Analysis

Time frequency analysis includes techniques designed to analyze a signal in both the time domain and the frequency domain simultaneously. These kinds of techniques are necessary once one realizes that an assumption of traditional Fourier analysis is that the signal is stationary and infinite in time. Practically recorded signals are neither, containing time-varying frequency components and a prescribed end point. The simplest example of a time frequency technique is the short-time-Fourier transform. This is the Fourier transform of a prescribed time interval and can be computed by windowing the continuous Fourier transform. Typically a Gaussian window is used, although many other types of windows are available. If $w(t)$ is the windowing function with width τ then the short-time-Fourier transform $STFT(f, \tau)$ is:

$$STFT(f, \tau) = \int_{-\infty}^{\infty} x(t)w(t - \tau)e^{-2\pi itf} dt \qquad (8)$$

This transform yields a two-dimensional matrix known as a spectrogram. Spectrograms are visualized to reveal oscillatory activity that varies in amplitude, frequency, or both.

2.5 Windowing

When considering time frequency analysis, choosing an appropriate window size is an important factor. Assuming that the power in a frequency band varies over time, the discriminations must be made in both the frequency and time domain. Unfortunately, the FFT has a fixed resolution, and by altering the window size you cannot exchange resolution in the time domain, for resolution in the frequency domain.

The analysis window has a fixed resolution, which determines whether there is either a good frequency resolution—frequency components close together can be separated—or good time resolution—the time at which frequencies change. A wide window gives better frequency resolution, but poor time resolution. A narrower window gives good time resolution, but poor frequency resolution. These are called narrowband and wideband transforms, respectively. The size of the FFT can improve the frequency definition of the analysis.

Another consideration is that using smaller window sizes can effectively reduce your capacity to evaluate power in slower frequency bands. Each frequency also has an ideal window size to capture. Window size must be $\sim 5 \times$ the period of the signal. In an FFT, temporal resolution comes at the cost of frequency resolution. There is therefore an inherent limitation in doing time frequency analysis that examines multiple frequencies. Because there is a unique "ideal" window size for each frequency, analyzing the broadband spectrum with a single window size will give more accurate representations of certain frequencies and a poorer representation of other frequencies.

A Hamming window is a type of window optimized to reduce the amplitude of the nearest sidelobe, and is common in signal analysis. The hamming window is given by:

$$W(n) = \alpha - \beta \cos \frac{2\pi n}{\tau - 1} \qquad (9)$$

where $\alpha = 25/46$ and $\beta = 21/46$. This window is commonly used in signal processing of LFPs.

2.6 Phase

The Hilbert transform is a powerful tool that can be used to derive the analytic representation of a signal. Transforming a signal that has been filtered in a narrow-passband yields a bare sine wave whose instantaneous phase can be easily estimated. However, if the signal contains frequency information outside of a narrow prescribed range, phase slip can occur, where the estimated phase of the oscillation discontinuously jumps.

The analytic signal is the sum of its real part $A(t)$ and its imaginary part $h(t)$ given by the Hilbert transform

$$u(t) = A(t) + ih(t) \qquad (10)$$

The real portion of the signal is equivalent to the band-pass filtered signal $A(t)$. The imaginary part of the signal is defined using the Cauchy principal value (PV) and is given as:

$$h(t) = \frac{1}{\pi} \text{PV} \int_{-\infty}^{\infty} u(t') / (t - t') dt' \qquad (11)$$

Alternatively, the Hilbert transform can be defined explicitly as:

$$h(u)(t) = -\frac{1}{\pi} \lim_{\varepsilon \to 0} \int_{\varepsilon}^{\infty} \frac{u(t + t') - u(t - t')}{t'} dt' \qquad (12)$$

The analytic phase of signal can then be extracted as:

$$P(t) = a \tan \frac{h(t)}{A(t)} \qquad (13)$$

2.7 Phase-Amplitude Coupling

Phase-amplitude coupling (PAC) is a method for quantifying how oscillations of different frequencies interact. An example of PAC is below (Fig. 4), where the phase of a slow oscillation, for example, theta, controls the amplitude of a higher frequency oscillation, for example, gamma. This process is also known as nesting, or cross frequency coupling. PAC is an emerging technique with

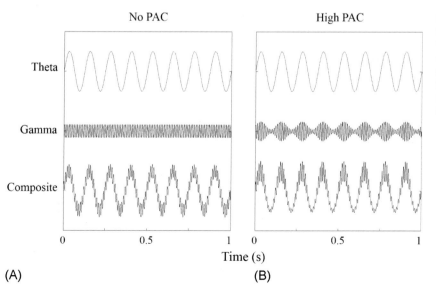

FIG. 4 Construction of a phase-amplitude relationship. (A) A theta wave exhibiting no PAC with a gamma oscillation. (B) A theta wave modulating the amplitude of a gamma wave. The amplitude of the gamma wave is highest when the theta wave is at its peak. This relationship can exist between any phases of the lower frequency oscillation, not just the peak.

various methods to compute a metric of PAC. All methods begin with an estimation of the phase and amplitude of the signal. Both of these measures can be taken from the analytic signal with phase and amplitude, $A(t)$ and $P(t)$, respectively. These measures are related through the Hilbert transform as described above in Eq. (13). Once phase and amplitude vectors are computed, a PAC metric can be determined by computing the mean vector of the composite signal:

$$PAC(t) = A(t)e^{iP(t)} \qquad (14)$$

Computing PAC in real time is complicated by several factors. First, accurate estimation of phase requires several cycles of an oscillation to achieve an accurate measurement (Dvorak and Fenton, 2014). Second, meaningful PAC measures require assessment of phase-amplitude coupling for many different frequency pairs. These multiple comparisons require the signal to be filtered differently for each phase and frequency bin. Lastly, computing a metric for significance of PAC requires nonparametric surrogate techniques that are quite computationally expensive. If the frequency/phase pair of interest was well defined a priori, and some baseline magnitude of PAC was also defined a priori, then computing PAC in real time would be possible. It should be noted that as this is an emerging technique, there is no consensus on the best way to compute a PAC metric (Dvorak and Fenton, 2014).

3 CLOSED-LOOP STIMULATION DESIGN

The development of closed-loop systems is dependent on the both a working understanding of underlying neurophysiology and adequate technological sophistication of the device. Only when both of these conditions are met can electrophysiological recordings be related to disease states and ultimately translated into optimal stimulation parameters. Knowledge of how pathophysiological states, clinical states, and stimulation delivery are related is required to have an effective closed-loop paradigm. The implementation of this paradigm begins with an accurate and robust classifier algorithm that represents the clinical states of interest. These states must then be mapped to stimulation parameters via a control policy that effectively manages the transition between states by delivering electrical stimulation.

3.1 Classifier Algorithm

Classifier algorithms convert the raw LFP signal into an estimate of the state of the nervous system. LFPs are an attractive component of classifier algorithms for their ease of recording and usefulness in representing clinical states. How an LFP represents the clinical state is, of course, of paramount importance (Carlson et al., 2013), and depends on the indication. Detection of Parkinson's disease (PD)

off states represented by increased beta activity will be a classifier algorithm analogous to an energy-dependent seizure detector in epilepsy. The clinical indication will also determine how often a state estimation needs to be made and will depend on the frequency with which clinical symptoms vary. A classifier algorithm that estimates states on the scale of seconds will allow for fine control of stimulation parameters at the expense of energy efficiency. A more course detection regime might be appropriate if the clinical state fluctuates slowly, like a mood disorder, where state estimation might not change much over small time scales. Much of this chapter deals with reducing LFPs to state estimators, but this component remains highly dependent on the clinical needs.

An example of a simple classifier algorithm is one based on a beta power threshold in the LFP. This concept has received great attention in the context of closed-loop DBS of the subthalamic region for the treatment of PD (Modolo et al., 2012). This is because high beta power in the subthalamic nucleus (STN) has been associated with PD symptoms. If the beta power of an LFP is above a certain power threshold, the classifier algorithm defines this state to be suboptimal. Once stimulation has been delivered and beta power is below a certain threshold, the classifier algorithm determines that the clinical state is now optimal. These state estimates are used by the control policy to turn stimulation on or off.

One challenge for a classifier algorithm is to reject artifacts from stimulation delivery (Stanslaski et al., 2009). Because these devices are generally sensing at all times, even during stimulation, the massive artifacts created by current injection must be attenuated. A typical stimulation pulse is measured in volts, while a brain signal is typically on the order of microvolts-millivolts (Stanslaski et al., 2012). One simple way to attenuate this stimulation noise is to restrict yourself to a bipolar recording montage sandwiched around the unipolar stimulating electrode. In this setup the symmetry of the monopolar stimulation field will result in equal representation of the artifact on both recording contacts. This equal representation on both the reference and active contact will result in attenuation of the artifact in the bipolar recording. However, even if this kind of stimulation and recording montage is utilized, the control policy should account for stimulation-related artifacts and parse them out as separate from triggers of biological origin.

Once a classifier algorithm is in place, it is likely to require further refinement (Bergey et al., 2015). Adjusting the classifier algorithm should be done by the clinician until it satisfactorily represents the desired clinical states. In this sense, the refinement of the classifier algorithm is a "backdoor" in closed-loop paradigms for the clinician to play the role of critic. By informing the classifier algorithm of its error the clinician is "opening the loop." A true closed-loop stimulation paradigm might incorporate an actor critic model of machine learning to automatically update the classifier algorithm to represent the true state space. This kind of machine learning algorithm would rely on a secondary detector to determine error in the primary state estimation algorithm. For example, an electromyogram lead to detect tremor could give objective feedback of how well the classifier algorithm is performing. By combining the detection of beta power with tremor levels, a more precise beta threshold could be used for stimulation delivery.

3.2 Control Policy

A control policy is a map of how states in the classifier algorithm map on to different stimulation conditions. A simple control policy might simply turn stimulation on following a detection event with clinician defined, empirically determined stimulation settings. A more complex control policy may integrate multiple detectors to not only make a determination of whether or not to stimulate, but also with what parameters to deliver the stimulation. Various paradigms for control policies exist. A continuously adjusted or adaptive stimulation paradigm might be appropriate for disease with chronic symptom expression, where some element of the classifier algorithm controls the amplitude of the chronically delivered stimulation. For example, rather than using beta power to turn the stimulator on or off, a continuously adjusted stimulation paradigm might stratify beta power into different levels and adjust the amplitude of the continuous stimulation based on the beta power. For paroxysmal detection, a responsive stimulation paradigm that delivers one or more stimulation pulses in response to an event detection might be appropriate. One key assumption of a control policy is static input/output pairs: that is that the stimulation response to a state estimation reliably changes the state of the brain to the desired state. A time variant stimulation response would violate this assumption and require further refinement of the control policy.

In addition to the key considerations of classifier algorithms and control policies, practical and safety concerns must also be met. One potential benefit of closed-loop stimulation is energy savings. If the energy saved on more sparse stimulation is squandered on detection, then a closed-loop stimulation paradigm is of limited value, especially if the clinical benefit is similar. Consideration of sampling rates and detection algorithms must therefore be made when designing a closed-loop paradigm. Furthermore, this must be done within the confines of the implanted device. If a detector requires input from multiple electrodes in different brain regions, then the paradigm will require increased surgical burden. Sensing from the same structure that is the stimulation target therefore has the advantage of decreased surgical risk. Furthermore, closed-loop stimulation brings the possibility of overdelivery of stimulation.

A control policy should therefore be manually bounded to include only stimulation parameters in a safe range. Current density should also be measured and controlled to mitigate the possibility of tissue damage. Malfunction in a detector may also result in overdelivery of stimulation, requiring a shutoff mechanism independent of the classifier algorithm. Lastly, stimulation-induced side effects may also be of interest in the classifier algorithm. For example, repeatedly burdening the patient with stimulation-induced dyskinesia would not be preferable to improved battery life.

4 CURRENT STATE OF CLINICAL RESPONSIVE STIMULATION

This section seeks to evaluate the current state of responsive stimulation detection algorithms and how they pertain to LFP analysis.

4.1 Epilepsy

Seizures potentially represent the perfect substrate for a responsive stimulation paradigm. They are paroxysmal, which means that continuous stimulation is unnecessary. Responsive stimulation therefore has lots of upside in the treatment of seizures and can reduce energy usage and side effects. Secondly, seizures generate tell-tale electrophysiological signals that are amenable to capture with low-energy detectors (Esteller et al., 2001; Litt et al., 2001). Although detecting seizures is not a solved problem, decades of research on seizure detection has reduced the problem to one that can be addressed within the strict constraints of an implanted device. To date, the only closed-loop stimulation capable device that has passed clinical trials is the Neuropace responsive neurostimulator system (RNS) for partial-onset epilepsy (Heck et al., 2014). This device is a programmable neurostimulator that contains sensing, recording, and stimulating circuitry as well as a battery in a hermetically sealed titanium case implanted into the cranium (Sun and Morrell, 2014). The device can support up to two leads, each handling either a depth electrode or ECoG strip, typically placed on or around the focus of the seizures.

A critical component of the RNS device is the ability to communicate with a programmer via short-range wireless telemetry. Although this device is capable of closed-loop stimulation, this two-way communication enables the clinician to observe and adjust detection and stimulations settings into the treatment course in an open loop way. The detection pattern can therefore be optimized as a priori settings are tuned and outcomes are assessed. The ability to record and store electrophysiological activity is of great value in this process, as you can store segments of activity around events of interest to inform future programming decisions. This allows the creation of a library of electrophysiological data built from segments recorded around detection of prespecified electrophysiological abnormalities, delivery of a stimulation pulses, or regularly scheduled recordings.

4.1.1 Sensing

The RNS device continuously monitors activity from its contacts, typically via a bipolar recording. The four recording channels can be spread across contact pairs from either of the connected leads, allowing for simultaneous recording of both surface (ECoG) and depth electrodes. The activity is constantly streamed to the device with a 90 s buffer. When an event that triggers a recording is encountered, the device can then save the previous 45 s and the following 45 s. These parameters can be changed such that the device saves any combination of time before and after the event totaling 90 s or less. Several events can trigger a recording including a "detection" event, time of day, magnet swipe (manual input), amplifier saturation, or noise. These recorded events can then be evaluated by the clinician to make a determination about how accurate the detection algorithm is. For example, if a lot of false positives are detected using a line-length filter, increasing the detection threshold might allow for more specific seizure detection (Sun et al., 2008).

4.1.2 Detection

A detection event occurs when the sensed data matches a preprogrammed pattern. Because the battery used for powering the sensing and detection circuitry also powers the stimulation, economy of detection is a paramount consideration. The algorithms used in this device represent those that are computationally efficient for use in an "always listening" detection paradigm. This device is capable of three different customizable detection algorithms including line-length, area under the curve, and bandpass. All of these detectors are described above in the analysis section. Two separate detectors can be programmed for two separate channels. Having multiple detection algorithms active at once can allow a high-sensitivity low-specificity low-energy detection algorithm (like line-length) to always run, then use a more computationally greedy algorithm with higher specificity, such as bandpass to make the second-tier detection. This approach allows for good energy economy while maintaining high-sensitivity and specificity.

The output of the bandpass, line-length, and area detection tools from the same or different EEG channels may be combined to identify specific electrographic events. In addition, the output of a tool may be assigned a persistence duration to facilitate detection of events in a sequence. For example, a seizure onset consisting of a series of spike-wave elements, followed by low-voltage fast activity may be identified by a line-length detection tool configured to detect a sudden increase in signal amplitude followed by a bandpass detection tool configured to detect high-frequency activity.

4.1.3 Current Delivery

The RNS device delivers current-controlled, charge-balanced biphasic pulses and can be programmed by the physician to deliver bursts of stimulation with pulses of between 40 and 1000 ms, pulse frequencies between 1 and 333 Hz, current amplitudes of between 0.5 and 12 mA, and burst durations between 10 and 5000 ms. The RNS can be configured to deliver current between any combination of electrodes. The titanium neurostimulator canister may also be configured as part of the stimulation path. Up to five individually configured sequential therapies of electrical stimulation may be programmed, where each therapy is composed of two independently configurable bursts. The RNS will attempt to redetect the ECoG pattern after each therapy is delivered. If the pattern is still detected, the next therapy will be delivered. If the pattern is no longer detected, the remaining therapies in the sequence will not be delivered. The therapy sequence will refresh with the detection of each new episode. Typical initial responsive therapy settings are a frequency of 200 Hz, pulse width of 160 ms, and burst duration of 100 ms. Current is initially programmed at a low amplitude (eg, 1.0 mA) and titrated upward as tolerated to a maximum of 12 mA. In general, stimulation is delivered to the leads and electrodes from which electrographic patterns of interest are observed. For example, if electrographic activity of interest is observed on all channels, then the stimulation pathway could be configured to stimulate across all electrodes. However, if electrographic activity is observed on only two channels, then the stimulation pathway could be configured such that current is delivered through only those electrodes with electrographic activity of interest.

4.1.4 Clinical Outcomes

In the Neuropace clinical trial of closed-loop stimulation, 191 epilepsy patients were evaluated over the course of several years after being implanted with the RNS system (Heck et al., 2014). All patients were programmed to use a line-length detector to control a responsive stimulation paradigm, however the exact parameters of the detector were fine-tuned by physicians for each patient. Seizure frequency outcomes were significantly better in the responsive stimulation group compared to the sham group. The treatment group saw an immediate reduction in seizure frequency of 37.9% compared to preimplant, while the sham group saw a decrease of 17.3%. This decrease in the sham group was likely due to the surgery itself, and over the course of the trial the closed-loop stimulation group progressed to a 41.5% seizure reduction while the sham group regressed to 9.4% reduction. Importantly, this seizure reduction also translated into a significant improvement in quality of life measures as well. This treatment remains palliative however, as only 9% of subjects were seizure free.

4.2 Parkinson's Disease

Although the cardinal symptoms of PD are not necessarily paroxysmal, they are time varying, particularly when the kinetics of dopamine replacement therapy are considered. Consequently closed-loop stimulation paradigms have potential to be beneficial in the treatment of PD. The primary finding as it relates to closed-loop stimulation for PD is that beta power is correlated with rigidity and bradykinesia (Little and Brown, 2012). This represents a simple and somewhat reliable substrate on which to base a detector for closed-loop stimulation. However, the relationship between brain states and PD symptoms is slightly more complex. Beta power does not correlate with tremor, or unified Parkinson's disease rating scale (UPDRS) scores, and not all patients with PD have resting beta (Crowell et al., 2012). In addition, the other symptoms of PD such as tremor, freezing of gait, and cognitive deficits may also have their own spectral representation (Oswal et al., 2013; Reck et al., 2009; Tattersall et al., 2014). Furthermore, beta activity is modulated both by regular movement, and the frequency of maximal power changes with dopamine medication.

Given the potential limitations of beta power alone, phase-amplitude coupling between beta and gamma oscillations in the motor cortex may represent a reliable alternative biomarker for use for controlling rigidity in PD (de Hemptinne et al., 2013, 2015). Compared to epilepsy and dystonia patients, PD patients show increased PAC in M1 that was reduced by STN stimulation. Peaks in M1γ-amplitude are coupled to, and precede, the STN beta-trough (Litvak et al., 2011). The results prompt a model of the basal ganglia-cortical circuit in PD, incorporating phase-amplitude interactions and abnormal cortico-subthalamic feedback and suggest that M1 LFPs could be used as a control signal for automated programming of basal ganglia stimulators.

While this section focuses on rigidity and bradykinesia, other PD symptoms may also be well controlled by closed-loop stimulation. For example, gait is usually worsened by traditional high-frequency (100 Hz+) DBS, however low-frequency stimulation (below 50 Hz) of the pedunculopontine nucleus (PPN) may be effective to treatment of gait (Morita et al., 2014; Stefani et al., 2007). PPN stimulation also has possible therapeutic effects for postural instability. LFPs from the PPN often show that alpha and beta oscillations are modulated by dopaminergic medication (Tattersall et al., 2014; Thevathasan et al., 2012). Although these relationships are still poorly understood, future studies may elucidate a relationship that can be captured in closed-loop paradigms.

Additionally, stimulation-induced dyskinesia's may also be fertile ground for closed-loop DBS. Patients who exhibited dyskinesias showed an increase in theta and alpha activity in the contralateral hemisphere of the affected limb (Halje et al., 2012; Silberstein et al., 2005). However dyskinesias are also related to reduced beta power (Little and Brown, 2012). While the idea of aborting a stimulation-induced dyskinesia is within the realm of possibility for a closed-loop DBS system, this finding also suggests that using a single biomarker may be insufficient given the complexity of symptom presentation and biomarker diversity in PD.

4.2.1 Sensing

Current closed-loop approaches in PD take the form of intraoperative cases (Little et al., 2013). This work is encouraging and has led to the development of the first wave of implantable sensing and stimulation devices implanted in PD. However, these devices are not capable of closed-loop stimulation on their own, but require connection to a second device that is programmed by the clinician. Intraoperative recordings are typically done immediately following the DBS implantation surgery, while the leads are implanted, but the wires are still exposed. The leads are internalized when the pacemaker is implanted a few days later. This allows a few days of experimentation to evaluate closed-loop stimulation paradigms. One proof of a principle study demonstrated that a beta power threshold was effective as a controller of stimulation on and off times for reducing PD symptoms (Little et al., 2013). This adaptive stimulation paradigm is a promising result.

4.2.2 Detection

The paradigm that has been reported most exhaustively is the use of a beta threshold to drive an adaptive DBS approach (Little et al., 2013). In this paradigm, the experimenters recorded activity of two pairs of contacts from the implanted DBS electrodes in STN-DBS for PD. Experiments took place the day after surgery while the leads were externalized, and the patients had withdrawn from antiparkinsonian medication. All recordings were filtered to accommodate the beta band with a bandpass between 3 and 37 Hz. Importantly, all subjects in this study showed elevated beta power in their STN at rest. The experimenters determined where the maximal beta power was found across the two contact pairs on the more affected hemisphere. LFPs were rectified and amplitude was computed as the average amplitude in a 400 ms window. Beta power was then assessed in real time and a detection event was defined as the beta power exceeding a user-defined threshold.

4.2.3 Current Delivery

This detection event was used to trigger stimulation with an optically connected stimulus generator. The stimulation parameters were determined beforehand using standard of care programming that consisted of systematically raising the amplitude in 0.5 V increments until clinical benefit was achieved but before paresthesia occurred. Stimulation was monopolar and delivered at 130 Hz regardless of amplitude. The stimulation contact was sandwiched between the two contacts used for recording in the bipolar configuration in order to reduce stimulation artifact. The beta power threshold for delivering stimulation was defined as the beta power that would achieve 50% reduction in "on time" from a test block. Once a detection event occurred, stimulation ramped up over the course of 250 ms and remained on until beta power fell below baseline, when it would ramp down over another 250 ms.

4.2.4 Clinical Outcomes

The eight patients who participated saw reduction of UPDRS scores 30% greater when getting adaptive DBS compared to the continuous DBS while receiving 50% of the current (Little et al., 2013). While the reduction in time stimulated is encouraging, whether or not this resulted in a net power savings is unclear because of difficulties quantifying how much energy was used in the detection. Nonetheless, the efficacy of closed-loop stimulation is an important result. Also of note, although beta power is correlated with rigidity and bradykinesia, the adaptive stimulation paradigm was effective in reducing tremor as well. The main outcome for this study was the total UPDRS, but removal of rigidity and bradykinesia subscores still resulted in a significant reduction. The experiments also compared adaptive stimulation of random stimulation to control for any effects of intermittent stimulation, and discounted intermittent stimulation as a cause for clinical effect which may happen if accommodation were occurring. Encouragingly, the clinical benefit was proportional to beta power decrease, suggesting a tight relationship between the two and further supporting the role of beta power in responsive stimulation paradigms for PD.

4.3 Other Diseases

The potential for closed-loop paradigms is not limited to epilepsy and PD. Neuropsychiatric diseases such as obsessive compulsive disorder, Tourette syndrome, and depression also have great need for stimulation that is tailored to the occurrence of events. Depression, like other mood disorders, is cyclical and time varying. If an appropriate trigger could be found, perhaps from theta oscillations, or alpha asymmetry in frontal lobe for depression (Allen and Cohen, 2010; Broadway et al., 2012; Gotlib et al., 1998), then a tailored stimulation regiment could be implemented. Tourette syndrome is a paroxysmal disease like epilepsy and so also stands to gain a lot from responsive stimulation paradigms. Detection of tics may be particularly difficult given that they are often indistinguishable from ordinary behaviors save for their context, however if differentiation between tics and volitional movement can be achieved, then responsive stimulation could be effective (Almeida et al., 2015). Obsessive compulsive disorder also displays intermittent symptoms with variable expressivity in its compulsions.

5 CONCLUSIONS

Decades of research linking brain oscillations to behavior make LFPs an attractive signal for closed-loop applications. Across species and individuals, reliable relationships between LFP activity and clinical symptoms make them robust inputs to classifier algorithms. The distributed nature of the origin of the signal is also more likely to yield robust representations of clinical states.

Beyond the clinical implications of closed-loop paradigms, there is also the rich opportunity to pursue mechanistic basic science research. If a biomarker is suitable as an effective stimulation trigger, there is likely a biological reason for this relationship. Any device that is capable of responsive stimulation is also capable of carrying out basic science research, and it is incumbent on the clinicians and scientists involved in this work to explore this area to determine how oscillations may result in pathological phenotypes. Understanding the pathophysiology of the disease and determining an effective trigger for closed-loop stimulation are inextricably linked. By keeping a bidirectional and translational approach to closed-loop research, it will propagate its own success, as basic science understanding expands into new avenues to achieve clinical benefits.

ACKNOWLEDGMENT

This work was supported by 5R01MH106173-02.

REFERENCES

Abosch, A., Lanctin, D., Onaran, I., Eberly, L., Spaniol, M., Ince, N.F., 2012. Long-term recordings of local field potentials from implanted deep brain stimulation electrodes. Neurosurgery 71 (4), 804–814. http://dx.doi.org/10.1227/NEU.0b013e3182676b91.

Allen, J.J.B., Cohen, M.X., 2010. Deconstructing the "resting" state: exploring the temporal dynamics of frontal alpha asymmetry as an endophenotype for depression. Front. Hum. Neurosci. 4, 232. http://dx.doi.org/10.3389/fnhum.2010.00232.

Almeida, L., Martinez-Ramirez, D., Rossi, P.J., Peng, Z., Gunduz, A., Okun, M.S., 2015. Chasing tics in the human brain: development of open, scheduled and closed loop responsive approaches to deep brain stimulation for tourette syndrome. J. Clin. Neurol. 11 (2), 122. http://dx.doi.org/10.3988/jcn.2015.11.2.122.

Barrese, J.C., Rao, N., Paroo, K., Triebwasser, C., Vargas-Irwin, C., Franquemont, L., Donoghue, J.P., 2013. Failure mode analysis of silicon-based intracortical microelectrode arrays in non-human primates. J. Neural Eng. 10 (6), 066014. http://dx.doi.org/10.1088/1741-2560/10/6/066014.

Bergey, G.K., Morrell, M.J., Mizrahi, E.M., Goldman, A., King-Stephens, D., Nair, D., Seale, C.G., 2015. Long-term treatment with responsive brain stimulation in adults with refractory partial seizures. Neurology 84 (8), 810–817. http://dx.doi.org/10.1212/WNL.0000000000001280.

Broadway, J.M., Holtzheimer, P.E., Hilimire, M.R., Parks, N.A., DeVylder, J.E., Mayberg, H.S., Corballis, P.M., 2012. Frontal theta cordance predicts 6-month antidepressant response to subcallosal cingulate deep brain stimulation for treatment-resistant depression: a pilot study. Neuropsychopharmacology 37 (7), 1764–1772. http://dx.doi.org/10.1038/npp.2012.23.

Buzsaki, G., Anastassiou, C.A., Koch, C., 2012. The origin of extracellular fields and currents—EEG, ECoG, LFP and spikes. Nat. Rev. Neurosci. 13 (6), 407–420.

Carlson, D., Linde, D., Isaacson, B., Afshar, P., Bourget, D., Stanslaski, S., Denison, T., 2013. A flexible algorithm framework for closed-loop neuromodulation research systems. In: Proceedings of the Annual International Conference of the IEEE Engineering in Medicine and Biology Society, EMBS, pp. 6146–6150. http://dx.doi.org/10.1109/EMBC.2013.6610956.

Correa, A.G., Laciar, E., Orosco, L., Gómez, M.E., Otoya, R., Jané, R., 2009. An energy-based detection algorithm of epileptic seizures in EEG records. In: Conference Proceedings: Annual International Conference of the IEEE Engineering in Medicine and Biology Society. IEEE Engineering in Medicine and Biology Society. Conference, 2009, pp. 1384–1387. http://dx.doi.org/10.1109/IEMBS.2009.5334114.

Crowell, A.L., Ryapolova-Webb, E.S., Ostrem, J.L., Galifianakis, N.B., Shimamoto, S., Lim, D.A., Starr, P.A., 2012. Oscillations in sensorimotor cortex in movement disorders: an electrocorticography study. Brain J. Neurol. 135 (Pt 2), 615–630. http://dx.doi.org/10.1093/brain/awr332.

de Hemptinne, C., Ryapolova-Webb, E.S., Air, E.L., Garcia, P.A., Miller, K.J., Ojemann, J.G., Starr, P.A., 2013. Exaggerated phase-amplitude coupling in the primary motor cortex in Parkinson disease. Proc. Natl. Acad. Sci. U. S. A. 110 (12), 4780–4785. http://dx.doi.org/10.1073/pnas.1214546110.

de Hemptinne, C., Swann, N.C., Ostrem, J.L., Ryapolova-Webb, E.S., San Luciano, M., Galifianakis, N.B., Starr, P.A., 2015. Therapeutic deep brain stimulation reduces cortical phase-amplitude coupling in Parkinson's disease. Nat. Neurosci. 18 (5), 779–786. http://dx.doi.org/10.1038/nn.3997.

Destexhe, A., Goldberg, J.A., 2015. LFP analysis: overview. In: Jaeger, D., Jung, R. (Eds.), Encyclopedia of Computational Neuroscience. Springer, New York, pp. 52–55. http://doi.org/10.1007/978-1-4614-6675-8_782.

Dvorak, D., Fenton, A.a., 2014. Toward a proper estimation of phase-amplitude coupling in neural oscillations. J. Neurosci. Methods 225, 42–56. http://dx.doi.org/10.1016/j.jneumeth.2014.01.002.

Esteller, R., Echauz, J., Tcheng, T., Litt, B., Pless, B., 2001. Line length: an efficient feature for seizure onset detection. In: 2001 Conference Proceedings of the 23rd Annual International Conference of the IEEE Engineering in Medicine and Biology Society, vol. 2(3), pp. 1707–1710. http://dx.doi.org/10.1109/IEMBS.2001.1020545.

Flint, R.D., Lindberg, E.W., Jordan, L.R., Miller, L.E., Slutzky, M.W., 2012. Accurate decoding of reaching movements from field potentials in the absence of spikes. J. Neural Eng. 9 (4), 046006. http://dx.doi.org/10.1088/1741-2560/9/4/046006.

Gaynor, L.M.F.D., Kühn, A.A., Dileone, M., Litvak, V., Eusebio, A., Pogosyan, A., Brown, P., 2008. Suppression of beta oscillations in the subthalamic nucleus following cortical stimulation in humans. Eur. J. Neurosci. 28 (8), 1686–1695. http://dx.doi.org/10.1111/j.1460-9568.2008.06363.x.

Gotlib, I.H., Stanfo, U., Sa, U., 1998. Frontal EEG alpha asymmetry, depression, and cognitive functioning. Cognit. Emot. 12 (3), 449–478. http://dx.doi.org/10.1080/026999398379673.

Gotman, J., 1982. Automatic recognition of epileptic seizures in the EEG. Electroencephalogr. Clin. Neurophysiol. 54 (5), 530–540. http://dx.doi.org/10.1016/0013-4694(82)90038-4.

Gotman, J., Gloor, P., 1976. Automatic recognition and quantification of interictal epileptic activity in the human scalp EEG. Electroencephalogr. Clin. Neurophysiol. 41 (5), 513–529. http://dx.doi.org/10.1016/0013-4694(76)90063-8.

Halje, P., Tamtè, M., Richter, U., Mohammed, M., Cenci, M.A., Petersson, P., 2012. Levodopa-induced dyskinesia is strongly associated with resonant cortical oscillations. J. Neurosci. 32 (47), 16541–16551. http://dx.doi.org/10.1523/JNEUROSCI.3047-12.2012.

Hanajima, R., Dostrovsky, J.O., Lozano, A.M., Hutchison, W.D., Davis, K.D., Chen, R., Ashby, P., 2004. Somatosensory evoked potentials (SEPs) recorded from deep brain stimulation (DBS) electrodes in the thalamus and subthalamic nucleus (STN). Clin. Neurophysiol. 115 (2), 424–434. http://dx.doi.org/10.1016/j.clinph.2003.09.027.

Heck, C.N., King-Stephens, D., Massey, A.D., Nair, D.R., Jobst, B.C., Barkley, G.L., Morrell, M.J., 2014. Two-year seizure reduction in adults with medically intractable partial onset epilepsy treated with responsive neurostimulation: final results of the RNS System Pivotal trial. Epilepsia 55 (3), 432–441. http://dx.doi.org/10.1111/epi.12534.

Kent, A.R., Swan, B.D., Brocker, D.T., Turner, D.A., Gross, R.E., Grill, W.M., 2015. Measurement of evoked potentials during thalamic deep brain stimulation. Brain Stimul. 8 (1), 42–56. http://dx.doi.org/10.1016/j.brs.2014.09.017.

Khan, M.F., Mewes, K., Gross, R.E., Skrinjar, O., 2008. Assessment of brain shift related to deep brain stimulation surgery. Stereotact. Funct. Neurosurg. 86 (1), 44–53. http://dx.doi.org/10.1159/000108588.

Lempka, S.F., McIntyre, C.C., 2013. Theoretical analysis of the local field potential in deep brain stimulation applications. PLoS One 8 (3), e59839. http://dx.doi.org/10.1371/journal.pone.0059839.

Lindén, H., Tetzlaff, T., Potjans, T.C., Pettersen, K.H., Grün, S., Diesmann, M., Einevoll, G.T., 2011. Modeling the spatial reach of the LFP. Neuron 72 (5), 859–872. http://dx.doi.org/10.1016/j.neuron.2011.11.006.

Litt, B., Esteller, R., Echauz, J., D'Alessandro, M., Shor, R., Henry, T., Vachtsevanos, G., 2001. Epileptic seizures may begin hours in advance of clinical onset: a report of five patients. Neuron 30 (1), 51–64. http://dx.doi.org/10.1016/S0896-6273(01)00262-8.

Little, S., Brown, P., 2012. What brain signals are suitable for feedback control of deep brain stimulation in Parkinson's disease? Ann. N. Y. Acad. Sci. 1265, 9–24. http://dx.doi.org/10.1111/j.1749-6632.2012.06650.x.

Little, S., Pogosyan, A., Neal, S., Zavala, B., Zrinzo, L., Hariz, M., Brown, P., 2013. Adaptive deep brain stimulation in advanced Parkinson disease. Ann. Neurol. 74 (3), 449–457. http://dx.doi.org/10.1002/ana.23951.

Litvak, V., Jha, A., Eusebio, A., Oostenveld, R., Foltynie, T., Limousin, P., Brown, P., 2011. Resting oscillatory cortico-subthalamic connectivity in patients with Parkinson's disease. Brain 134 (Pt 2), 359–374. http://dx.doi.org/10.1093/brain/awq332.

Maling, N., Hashemiyoon, R., Foote, K.D., Okun, M.S., Sanchez, J.C., 2012. Increased thalamic gamma band activity correlates with symptom relief following deep brain stimulation in humans with Tourette's syndrome. PLoS One 7 (9), e44215. http://dx.doi.org/10.1371/journal.pone.0044215. pii: PONE-D-12-18619.

Modolo, J., Beuter, A., Thomas, A.W., Legros, A., 2012. Using "smart stimulators" to treat Parkinson's disease: re-engineering neurostimulation devices. Front. Comput. Neurosci. 6, 1–3. http://dx.doi.org/10.3389/fncom.2012.00069.

Morita, H., Hass, C.J., Moro, E., Sudhyadhom, A., Kumar, R., Okun, M.S., 2014. Pedunculopontine nucleus stimulation: where are we now and what needs to be done to move the field forward? Front. Neurol. 5, 243. http://dx.doi.org/10.3389/fneur.2014.00243.

Nelson, M.J., Pouget, P., 2010. Do electrode properties create a problem in interpreting local field potential recordings? J. Neurophysiol. 103 (5), 2315–2317. http://dx.doi.org/10.1152/jn.00157.2010.

Niedermeyer, E., Silva, F.H.L.Da, 2004. Electroencephalography: Basic Principles, Clinical Applications, and Related Fields, vol. 1. Lippincott Williams and Wilkins, Philadelphia. Retrieved from, http://books.google.com/books?id=tndqYGPHQdEC&pgis=1.

Nunez, P.L., Srinivasan, R., 2006. Electric Fields of the Brain: The Neurophysics of EEG. Oxford University Press, Oxford. http://dx.doi.org/10.1093/acprof:oso/9780195050387.001.0001.

Oswal, A., Litvak, V., Brücke, C., Huebl, J., Schneider, G., Kühn, A.A., Brown, P., 2013. Cognitive factors modulate activity within the human subthalamic nucleus during voluntary movement in Parkinson's disease. J. Neurosci. 33 (40), 15815–15826. http://dx.doi.org/10.1523/JNEUROSCI.1790-13.2013.

Qu, H., Gotman, J., 1997. A patient-specific algorithm for the detection of seizure onset in long-term EEG monitoring: possible use as a warning device. IEEE Trans. Biomed. Eng. 44 (2), 115–122. http://dx.doi.org/10.1109/10.552241.

Reck, C., Florin, E., Wojtecki, L., Krause, H., Groiss, S., Voges, J., Timmermann, L., 2009. Characterisation of tremor-associated local field potentials in the subthalamic nucleus in Parkinson's disease. Eur. J. Neurosci. 29 (3), 599–612. http://dx.doi.org/10.1111/j.1460-9568.2008.06597.x.

Silberstein, P., Oliviero, A., Di Lazzaro, V., Insola, A., Mazzone, P., Brown, P., 2005. Oscillatory pallidal local field potential activity inversely correlates with limb dyskinesias in Parkinson's disease. Exp. Neurol. 194 (2), 523–529. http://dx.doi.org/10.1016/j.expneurol.2005.03.014.

Srinivasan, R., Nunez, P.L., Silberstein, R.B., 1998. Spatial filtering and neocortical dynamics: estimates of EEG coherence. IEEE Trans. Biomed. Eng. 45 (7), 814–826. http://dx.doi.org/10.1109/10.686789.

Stanslaski, S., Cong, P., Carlson, D., Santa, W., Jensen, R., Molnar, G., Denison, T., 2009. An implantable bi-directional brain-machine interface system for chronic neuroprosthesis research. In: Proceedings of the 31st Annual International Conference of the IEEE Engineering in Medicine and Biology Society: Engineering the Future of Biomedicine, EMBC 2009. pp. 5494–5497. http://dx.doi.org/10.1109/IEMBS.2009.5334562.

Stanslaski, S., Afshar, P., Cong, P., Giftakis, J., Stypulkowski, P., Carlson, D., Denison, T., 2012. Design and validation of a fully implantable, chronic, closed-loop neuromodulation device with concurrent sensing and stimulation. IEEE Trans. Neural Syst. Rehabil. Eng. 20 (4), 410–421. http://dx.doi.org/10.1109/TNSRE.2012.2183617.

Stefani, A., Lozano, A.M., Peppe, A., Stanzione, P., Galati, S., Tropepi, D., Mazzone, P., 2007. Bilateral deep brain stimulation of the pedunculopontine and subthalamic nuclei in severe Parkinson's disease. Brain 130 (6), 1596–1607. http://dx.doi.org/10.1093/brain/awl346.

Stypulkowski, P.H., Stanslaski, S.R., Denison, T.J., Giftakis, J.E., 2013. Chronic evaluation of a clinical system for deep brain stimulation and recording of neural network activity. Stereotact. Funct. Neurosurg. 91 (4), 220–232. http://dx.doi.org/10.1159/000345493.

Sun, F.T., Morrell, M.J., 2014. The RNS system: responsive cortical stimulation for the treatment of refractory partial epilepsy. Expert Rev. Med. Devices 11 (6), 563–572. http://dx.doi.org/10.1586/17434440.2014.947274.

Sun, F.T., Morrell, M.J., Wharen Jr., R.E., 2008. Responsive cortical stimulation for the treatment of epilepsy. Neurotherapeutics. 5 (1). http://dx.doi.org/10.1016/j.nurt.2007.10.069. pii: 68–74. S1933-7213(07)00257-7.

Tattersall, T.L., Stratton, P.G., Coyne, T.J., Cook, R., Silberstein, P., Silburn, P.A., Sah, P., 2014. Imagined gait modulates neuronal network dynamics in the human pedunculopontine nucleus. Nat. Neurosci. 17 (3), 449–454. http://dx.doi.org/10.1038/nn.3642.

Thevathasan, W., Pogosyan, A., Hyam, J.A., Jenkinson, N., Foltynie, T., Limousin, P., Brown, P., 2012. Alpha oscillations in the pedunculopontine nucleus correlate with gait performance in parkinsonism. Brain J. Neurol. 135 (Pt 1), 148–160. http://dx.doi.org/10.1093/brain/awr315.

Winkler, D., Tittgemeyer, M., Schwarz, J., Preul, C., Strecker, K., Meixensberger, J., 2005. The first evaluation of brain shift during functional neurosurgery by deformation field analysis. J. Neurol. Neurosurg. Psychiatry 76 (8), 1161–1163. http://dx.doi.org/10.1136/jnnp.2004.047373.

Chapter 6

Online Event Detection Requirements in Closed-Loop Neuroscience

P. Varona*, D. Arroyo*, F.B. Rodríguez* and T. Nowotny†

Autonomous University of Madrid, Madrid, Spain, †University of Sussex, Brighton, United Kingdom

1 INTRODUCTION

Neuroscience research essentially relies on experimental work and therefore has to deal with the fundamental problem that the nervous system is, and will always remain, only partially observable. We typically can only access one or a few signals involved in the complex and high-dimensional neural computation mechanisms. We usually consider one type of recording modality at a time: voltage (intracellular or extracellular), calcium imaging, optical readout, blood-oxygen-level-dependent (BOLD) signal, etc. However, neural information processing relies on many different interacting spatial and temporal scales and is not likely to be fully reflected in the time series of one single modality, as recorded in a classical stimulus-response experimental paradigm. In single neuron studies, the most popular, observed variable is membrane voltage recorded at the soma. Calcium imaging is another common alternative. Multielectrode recordings and multielectrode array (MEA) recordings have increasingly gained interest in modern neuroscience, but we still can typically only target a few tens or hundreds of neurons. BOLD signals can cover large areas of the brain, but have low temporal resolution; in some cases these recordings are therefore complemented with electroencephalographic (EEG) recordings at higher update frequencies, but these have poor spatial accuracy. Even in the combined case, it is hard, if not impossible to fully capture all important aspects of neural dynamics.

Spatial and temporal resolution and coverage are however not the only aspects that limit gaining insights about the nervous system. Most experimental protocols in neuroscience research are based on a classical stimulus-response paradigm, ie, the particular nervous system under observation is stimulated and the response of the system is then analyzed offline. Other open-loop variants include recordings of spontaneous activity. Temporal aspects of input signals are often addressed by delivering an a priori prepared stimulus with a given temporal structure. However, neural activity is mostly transient and nonstationary; the associated information

processing is history-dependent and contextual, in many cases involving feedback computations, which adds to the inherent observation intricacy. In this context, closed-loop technologies can allow the design of novel experimental protocols to address the partial observation nature of neuroscience experimental research. In addition, the closed-loop methodology can be applied to deal with the transient aspects of neural activity by exploring neural dynamics through online interaction. Furthermore, these technologies can be used to bridge between disparate levels of analysis, including the study of the interplay between different spatial and temporal scales in neural computation, even when addressing just a single observation modality.

However, closed-loop experimental protocols in neuroscience are not easy to design or implement. A closed-loop interaction requires effective online event detection protocols, fast algorithms to drive identification/control procedures, and effective stimulus exploration/control strategies to achieve the overall closed-loop goal. Although in general, closed-loop experiments require a specific design for each particular preparation/protocol, there are several common issues that can be addressed from a general perspective. In this chapter, we will inspect the common challenge of online event detection and processing with examples that range from in vitro preparations to the realm of brain machine interfaces and behavioral experiments. Identifying challenges at different description levels and from different technical perspectives, and formulating a general framework in which to address them can help conceptualize closed-loop approaches in experimental neuroscience research, make it easier to apply them, and consequently make them more successful.

2 EVENT DETECTION FOR CLOSED-LOOPS AT THE CELLULAR LEVEL

At the level of single neurons, closed-loop interactions have been designed to create artificial ionic or synaptic

Closed Loop Neuroscience. http://dx.doi.org/10.1016/B978-0-12-802452-2.00006-8

conductances. Since the early 1990s, dynamic clamp, a pioneering closed-loop technology in neuroscience, has been used for this purpose in electrophysiological preparations (Destexhe and Bal, 2009; Prinz et al., 2004; LeMasson et al., 2002; Robinson and Kawai, 1993; Sharp et al., 1993). Attracting the interest of many research groups worldwide, several dynamic clamp implementations have been developed and tested on a wide variety of preparations (see, eg, Butera et al., 2001; Dorval et al., 2001; Harsch and Robinson, 2000; Kullmann et al., 2004; Linaro et al., 2015; Muniz et al., 2005, 2009; Nowotny et al., 2006; Pinto et al., 2001; Samu et al., 2012).

The dynamic clamp technique has been a significant step toward biophysically realistic interaction and manipulation of individual neurons and synapses. Beyond the context of single cell experiments, the dynamic clamp has also opened a new line of research with hybrid circuits composed of biological as well as artificial neurons (eg, Elson et al., 1998; Linaro et al., 2015; Nowotny et al., 2003, 2008; Olypher et al., 2006; Pinto et al., 2000; Szucs et al., 2000; Varona et al., 2001). This technology offers biophysically realistic, real-time interaction between living neurons and computer-based neuron models or standalone electronic circuits, and has become a widely used tool for studying neural systems at the cellular and circuit levels.

In most dynamic clamp protocols, there is an instantaneous observation and stimulation task: voltage is measured through an electrode and a current is injected into the target neuron. The injected current is calculated online after solving a model that describes ionic or synaptic conductances or simulates the dynamics of individual neurons or a neural network (Nowotny and Varona, 2015). In the most traditional form where dynamic clamp simulates ion channels and its input is the direct membrane potential measurement, event detection is typically not an issue. However, beginning with synapse models that are triggered by spikes, event detection and processing become important (see top row in Fig. 1). Other examples besides spike detection include burst detection and online rate estimates.

The same principles used in dynamic clamp protocols can be generalized to develop new technology of closed-loop interaction beyond, or in addition to, instantaneous electrophysiological recordings/injections (Chamorro et al., 2009, 2012; Muniz et al., 2008). Expanding immediate interaction as defined by a conductance or a current equation, this generalization can take into account discrete sequential events in the monitored signal to define a continuous or discrete actuation law. Such generalizations can also consider and combine different observation and stimulation modalities (ie, electrical readouts, imaging, drug micro-injections, optical manipulation, etc.) for activity-guided stimulation which can increase the possibilities of bidirectional interaction in an unprecedented manner. For example, online event detection in dual electrophysiology and imaging observations can raise the traditionally restricted time windows in which imaging recordings can be done to deliver the right amount of stimulation at the most effective time.

On the other hand, modern closed-loop technology also permits a tighter relation between models and experiments. Realistic single neuron models are known to be highly sensitive to some of their parameters (eg, maximum conductance values, time constants, and kinetic parameters). This sensitivity severely limits the usefulness of models for expanding neuroscience research. Typically, the reasonable parameter regions in parameter space are also nonconvex such that averaging from different observed neurons leads to spurious solutions (Golowasch et al., 2002). Furthermore, most models cannot reproduce many of the bifurcations observed in real neurons when subjected to different stimuli (Nowotny et al., 2007). Online closed-loop interactions between models and living cells have the potential of overcoming some of these problems. In particular, novel closed-loop techniques enable new methods for determining the biophysical properties of neurons, perform model construction and fitting, and analyzing changes of properties over time, including the effects of drugs. This being the case, online event detection needs more consideration. On a very basic level, one can compare the raw signal of measured membrane potential against model output of membrane potential and no advanced event detection or processing would be required. However, this simple method is vulnerable against small time shifts in the signal, in particular when neurons are spiking. To get more robust estimates of the match of model dynamics and neuron behavior, more advanced metrics are necessary, including convolutional methods like Euclidean distance of spike density functions (SDFs) (Szucs, 1998) and van Rossum distance (van Rossum, 2001), and spike pattern-based techniques, like the Victor-Purpura measure (Victor and Purpura, 2009), or modern spike timing-based distance metrics (Kreuz et al., 2013). Most of them can and have been employed in online contexts, but neuroscience research on the cellular level would benefit from a more systematic view on online event detection and processing techniques.

3 EVENT DETECTION FOR CLOSED-LOOPS AT THE NETWORK LEVEL

Similarly to the strategies described in the previous section, activity-guided stimulation can also be addressed in small or large circuits at the network level. Closed-loops at this level have been traditionally restricted by limited multisite recordings and stimulation capabilities. Modern MEAs in vivo and in vitro, including cultured neuronal networks, have provided a first solution to this problem (Lewis et al., 2015; Newman et al., 2012; Rolston et al., 2010; Spira and

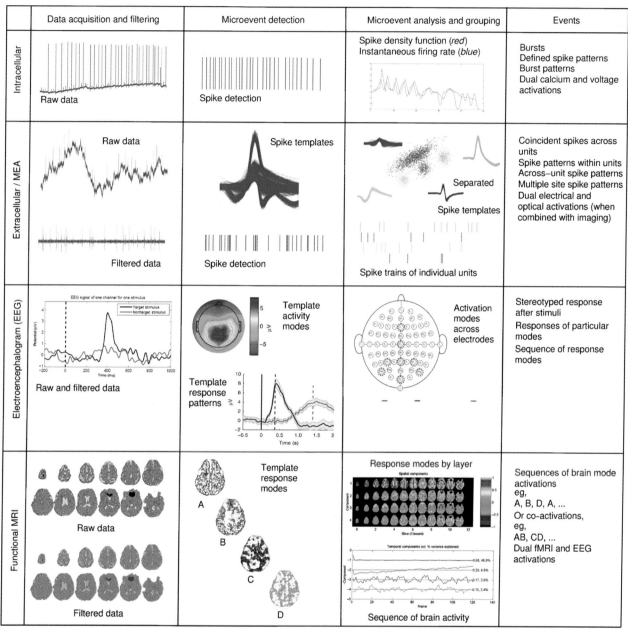

FIG. 1   Examples of online event detection needs at different experimental scales.

Hai, 2013). In particular, cultured neuronal networks are now widespread experimental models in studies on neural circuit dynamics, rhythm generation, and synaptic plasticity (Obien et al., 2015). These biological models, and particularly the ones obtained from mammalian central nervous systems, have been widely used to study joint cellular and network properties underlying information processing and learning (see, eg, Bi and Poo, 1998; Jimbo et al., 1999; Obien et al., 2015; Ruaro et al., 2005). One of the reasons why cultures are extremely suitable for establishing stimulus-response loops is that pre- and postsynaptic neurons are readily accessible and identifiable. Using

multiple intracellular electrodes—or an array of extracellular electrodes—one can identify neuronal circuits and deliver spatially and temporally structured stimuli to targeted neurons.

Hippocampal and cortical cultures are particularly advantageous to reveal long-term changes of neural circuitry because they can be maintained for extended periods of time (Hales et al., 2012; Potter and DeMarse, 2001). Significant progress in the research on long-term memory formation and the discovery of spike timing-dependent synaptic plasticity have been accomplished in cultures (Bi and Poo, 1998; Debanne et al., 1998; Froemke and

Dan, 2002; Markram et al., 1997; Wittenberg and Wang, 2006). One of the important applications of closed-loop technologies in this context is the automatic mapping of functional circuit connectivity in developing cultures to understand the highly complex and often oscillatory or burst-like natural firing patterns (Paulsen and Sejnowski, 2000; Wagenaar et al., 2006). This technology can help understand the relationship between synaptic connectivity and the origin of various modes of network dynamics such as recurrent activity, bursting, or oscillations. Efforts in this direction could greatly benefit from automating the identification and evaluation of spontaneous and evoked episodes. When addressing neural dynamics and plasticity in biological neural networks, experimenters have mainly used relatively simple event detection and stimulus delivery protocols. To further enhance these protocols, there is a need for online detection/sorting of individual spikes, bursts, local field potentials, and combinations of multiple site spike timing relationships associated with any chosen aspect of the dynamics or the spatial localization to drive stimuli at single or multiple sites (Franke et al., 2012), see second row in Fig. 1.

The most reliable and specific spike responses (Mainen and Sejnowski, 1995; Stevens and Zador, 1998; Szucs et al., 2004) and changes in the synaptic transmission (Dobrunz and Stevens, 1999) of biological neural networks can be expected in response to activity-guided biologically realistic stimuli, delivered at the right time in the ongoing neural dynamics, as they occur during sensory processing or motor planning. Sampling the high-dimensional space of possible stimuli in a meaningful closed-loop manner with a real-time monitoring and stimulus-generating system can overcome much of the limitations of open-loop experiments. Also, as for the cellular level, network closed-loop approaches can use bidirectional interaction among neural models and neuronal tissue in hybrid experiments that support communication at the level of individual action potentials.

Recent advances in optogenetics have provided novel ways of optical observation and stimulation at the circuit level (Buzsáki et al., 2015; Grosenick et al., 2015). Optogenetics allows millisecond-scale observation and control. In pioneering closed-loop optogenetics experiments, dynamic clamp tools were used to build the feedback loop (Sohal et al., 2009). These techniques have been rapidly generalized to different recording and stimulation modalities (Grosenick et al., 2015). One of the advantages of optogenetics compared to other methods for stimulation and observation is the opportunity for increased temporal and cellular specificity, and the possibility of doing this in vitro, in vivo, and in behavioral experiments. The development of novel nano- and micro-electrophysiological technologies enabling simultaneous, long-term, multiscale, closed-loop event detection and stimulation is a crucial step toward understanding both local and global processing, and elucidating how recurrent activity is dynamically shaped by intrinsic neural dynamics.

4 EVENT DETECTION FOR CLOSED-LOOPS AT WHOLE BRAIN LEVEL

Closed-loop technology can also be applied at the whole brain level in humans. In fact, all existing brain-computer interfaces (BCIs) in experimental and applied biomedical research work with an intrinsic closed-loop between subject and machine (van Gerven et al., 2009). Using the performance feedback, the subject can try to adapt his/her brain activity to improve BCI efficiency. BCIs typically rely on EEG recordings to enable interaction with a computer or a device (Birbaumer, 2006; Fazel-Rezai et al., 2012; Nicolas-Alonso and Gomez-Gil, 2012; Wolpaw and Wolpaw, 2012), although they have also been proposed for electrocorticography (Schalk and Leuthardt, 2011), magnetoencephalography (Jensen et al., 2011), functional magnetic resonance imaging (fMRI) (Weiskopf et al., 2004), and near-infrared spectroscopy (Naseer and Hong, 2015). The need for adaptive BCIs is increasingly being recognized (see, eg, Faller et al., 2014; Orsborn et al., 2012; Vidaurre et al., 2011; Wander and Rao, 2014). In this context, further progress is needed for a fundamental advance to low effort and personalized BCIs based on real-time interaction. Current BCI adaptation strategies often use offline analysis and training processes that rely mostly on subjects' adaptation efforts.

Traditional studies treat neural activity as a variable, which is dependent on the effect of stimulation or behavior, ie, they assess effects of visual, auditory, olfactory, and other stimuli on brain activity. Neurofeedback techniques have shown that it is possible that healthy subjects or patients learn to control brain activity by actively training themselves with feedback of specific neural events as a reward (Ruiz et al., 2014; Sulzer et al., 2013; van Boxtel and Gruzelier, 2014). Self-learned regulation of brain signals has been used for several decades now for communication with severely paralyzed patients, for suppression of epileptic activity, for treating symptoms of attention-deficit/hyperactivity disorders, and for the control of prostheses, etc. (Birbaumer et al., 1999; Cincotti et al., 2003; Kotchoubey et al., 2001; Kübler et al., 2001). In addition to these tasks and machine control with brain signals, neurofeedback is also a powerful experimental approach in cognitive neuroscience.

Recently, the concept of assisted closed-loops (ACL) in BCIs (Fernandez-Vargas et al., 2013) has been proposed for automatic subject-specific adaptation based on the interactive nature of closed-loop stimulation exploration, while providing online information to the subject on how his/her brain activity is being used in relation to the BCI goal. This kind of closed-loop activity-dependent stimulation technology for observation and control is still largely underdeveloped for BCIs, but the need for it is beginning to be recognized (Fazel-Rezai et al., 2012; Mohseni and Ghovanloo, 2012; Ruiz et al., 2014; Schiff, 2012; Sulzer et al., 2013; van Boxtel and Gruzelier, 2014). ACL technology can provide a new generation of BCIs that take into account subject- and context-specificities (including those related to online event detection), improve efficiency, provide new BCI efficiency predictors, and address the BCI illiteracy problem, which affects many disabled people (Fernandez-Vargas et al., 2013; Guger et al., 2009; Vidaurre and Blankertz, 2010). Moreover, it can bring together the, at times, fractionated efforts in BCIs, increase the synergy among groups, and promote new biomedical applications.

All BCI modalities have severe restrictions for real-time event detection in a closed-loop protocol, even in simple neurofeedback experiments, due to the temporal and spatial preprocessing required (see bottom rows in Fig. 1). While EEG offers only low spatial resolution and ambiguous allocation of neuronal activity, fMRI allows noninvasive recordings of activity across the brain with high spatial resolution in the range of millimeters. Traditionally, fMRI has been limited to offline analysis of brain function, due to the intrinsic delay between the neuronal activation and the BOLD signal changes and its time-consuming data analysis. Recently, faster image analysis and communication procedures have allowed obtaining fMRI results in much shorter time frames. Online analysis is already implemented in some scanners, with analysis times of a few seconds. This allows some quality control of the data and the results of the analysis while the subject is in the scanner. In experimental settings, much faster processing times have been achieved, and recent studies have shown the feasibility of real-time fMRI feedback (Ruiz et al., 2014; Sulzer et al., 2013). Real-time fMRI is a noninvasive method allowing feedback regulation of deep subcortical brain regions. The combination of different sources of stimuli (olfactory, visual, and auditory) with temporal structure can be used to expose brain pathways and connectivity (Plailly et al., 2008). By controlling these stimuli via closed-loop interaction, and modulating them online as a function of specific event detections, sequential activations associated with the encoding of sensorial input including cognitive responses can be unveiled.

Thus, closed-loop technologies can be used to select the best principle of BCI control for each subject, eg, to personalize the events used to drive the closed-loop, to optimize the stimulation parameters, to induce additional feedback by having goal-oriented activity-dependent stimulation, and to map activity in real-time. By sending sensorial cues as a function of the association of event detection with successful control of the BCI (sharing with the subject the learning task and informing him/her of these associations), ACL protocols can contribute to reducing the learning time and to optimizing control. Furthermore, they can adapt the BCI parameters online during the training phase to improve performance. Following the same logic, there are also new possibilities for closed-loop technologies in the context of transcranial and neuromodulation stimulation therapies (Berenyi et al., 2012; Beuter et al., 2014; Hariz et al., 2013).

5 EVENT DETECTION FOR CLOSED-LOOPS AT THE BEHAVIORAL LEVEL

Feedback control protocols are also increasingly used in behavioral experiments (Cowan et al., 2014; Roth et al., 2014), although offline analysis of behavioral data clearly remains the most common choice in neuroethology experimental design. Relating neural activity patterns to behavior has been a long-standing goal in experimental neuroscience. Real-time event detection and processing of neural activity during behavior typically requires robust, fast, and portable solutions. Recordings of online local field potentials, unit bursts or the firing patterns formed by individual spikes, and sequences of these events, can be related online with specific aspects of the monitored behavior (eg, mode of locomotion, spatial location, behavioral modality, etc.). Event detection and associated stimulus delivery in such electrophysiological recordings poses the same challenges described in our previous examples with the burden of dealing with motion and associated artifacts. Recent progress in miniaturized complementary metal-oxide-semiconductor technology has also produced opto-electronic systems capable of real-time awake optical imaging (Yu et al., 2015).

Behavioral event detections require typically real-time video-monitoring or specific sensors tracking hallmarks of the behavioral status (Aguiar et al., 2007). In closed-loop protocols, these behavioral events must be synchronized and combined with the online analysis of neural activity. An opportunity to naturally integrate dual recordings of behavioral and electrophysiological activity in closed-loop paradigms has recently come from optogenetic protocols to drive instantaneous feedback stimulation (Grosenick et al., 2015). In specific experimental models, there are abridged opportunities to build closed-loops protocols. For example, weakly electric fish provide simplified experimental models to implement closed-loop interaction considering behavioral monitoring and remote electrophysiological recordings (Chamorro et al., 2012; Forlim et al., 2013; Forlim et al., 2015). Novel studies employing these methods will allow for determining local and global interactions of neural activity and for addressing how recurrent activity is dynamically modified and shaped by tasks and behavioral contexts.

6 A COMMON NEED: AUTOMATIC EVENT DETECTION AND SYMBOLIC REPRESENTATION

As illustrated at the different levels described in the previous sections, a closed-loop bidirectional interaction with neural systems first requires the automatic detection of relevant events for characterizing the system state and identifying effective stimuli to achieve the experimental goal. From a general point of view, we can distinguish three stages in the event detection process. First, typically preprocessing algorithms are required to process the online signals (eg, filters to deal with artifacts, noise, drift, and adaptation of partially observable dynamics). Secondly, in many cases event detection can be parameterized and adjusted as a function of the closed-loop goal. In other words, event detection and characterization in these cases can use model-based procedures. Finally, signals are conveniently codified to ease fast interpretation and further processing. Signal codification is typically realized by means of classification and detection techniques.

In all observation modalities, neural events in a closed-loop experiment can be defined as a sequence of microevents. The whole process of automatic identification for closed-loop interaction entails online filtering, microevent detection, and the actual event detection. While the specific types of data differ when interacting with single neurons in vitro or the whole brain in vivo, data processing and event detection share many common features. For data from electrical recordings (intra- or extracellular electrophysiology and EEG) the primary features are fast voltage (current) transients that give rise to more complex, higher order events. For example, for intracellular neuron recordings, microevents are individual action potentials or subthreshold microevents, like evoked postsynaptic potentials. Higher order features are, eg, bursts of spikes or specific sequential activations of spikes.

The existence of different interleaved time-scales is a major challenge in neuroscience. Although neural activity is strongly determined by (continuous-time) oscillatory activity, it is also necessary to take into account that discrete-time events can modify the overall behavior of neural systems (Halliday et al., 1995). These concerns must be properly addressed and treated in experimental setups oriented to characterize neural activity in closed-loop technologies. As we have underlined in previous works (Arroyo et al., 2013; Chamorro et al., 2012), we can (and should) properly combine information theory, symbolic dynamics, and time-frequency signal characterization to achieve such a goal. In the vein of (Halliday et al., 1995), it is highly advisable to construct procedures built upon traditional time series analysis and point processes theory (Kass et al., 2014).

In general, there is a need for fast algorithms that can achieve feature extraction and event detection from electrophysiological and optical time series data. Features (microevents) can include individual action potentials extracted from intracellular recordings and action potentials extracted from extracellular recordings, which are deconvolved to recover individual neuronal responses (spikesorting). Higher order features can include burst properties, correlation measures, SDFs (Szucs, 1998) at several time resolutions, integrated SDFs, inter-spike-interval return map densities (Dekhuijzen and Bagust, 1996; Segundo et al., 1998; Szücs et al., 2003), phase response analysis (Schultheiss et al., 2012), etc.

Calcium or voltage imaging in vitro and in vivo or fMRI for the whole brain, require event detection on sequences of spatiotemporal patterns. There is a need for real-time algorithms to automatically identify regions of interest (ROIs). Each ROI is an image region showing correlated dynamical behavior over time in response to known stimulus conditions. The algorithm typically follows two steps: automatic image segmentation to identify candidate subsets of pixels, and analysis of covariance to extract functionally equivalent ROIs from image sequences. In most cases, time series extracted from ROIs can be treated similarly to those obtained from electrophysiological recordings. For fMRI, the raw data consists of the voxel intensity over time. The events can be related to the detection of brain activations modeled as gamma functions (Miezin et al., 2000). After anisotropic filtering of the signal (Solé et al., 2001), one can explore both spatial and spectral (Ngan et al., 2009) detection methods.

For event detection in EEG data, preprocessing techniques are needed to remove movement artifacts, direct current shifts, and power line interferences (see Chapter 7 in Cohen, 2014). Oversampling is used frequently to increase the signal-to-noise ratio (SNR) and perform source derivations. Source derivations are a very effective way to enhance the SNR by averaging or subtracting electrodes from each other (bipolar derivations, common average reference, Laplacian derivations). Parameter estimation can be done with bandpower estimation in reactive frequency bands, Hjorth parameters, adaptive autoregressive (AR) parameters, fast Fourier transform, and other algorithms that give a representation of time and frequency responses. Classification is performed most successfully with linear discriminant analysis and support vector machine classifiers. In a traditional BCI system, the preprocessing, parameter estimation, and classification algorithms and parameters are selected in offline mode.

One of the major challenges here is to categorize the involved data coherently with its transient nature. The characterization of transient dynamics cannot be performed using methods based on asymptotic analysis, and thus the effort must be focussed on developing precise and efficient procedures for classifying nonstationary time series. This can be tackled by reducing the complexity of the involved signals using different transformations in the joint domain of time and frequency, ie, through the proper combination of tools as the Wavelet and Hilbert-Huang transforms (Cohen and Voytek, 2013; Sweeney-Reed and Nasuto, 2007), and considering the applied theory of symbolic dynamics (Porta et al., 2015).

For more general metafeature (event) extraction, tools such as correlation and causality measures between signals (Kleinberg and Hripcsak, 2011; Marinazzo et al., 2008; (Pereda et al., 2005), symbolic dynamics (Daw et al., 2003), recurrence plots (Schinkel et al., 2007), and different measures of entropy and complexity (Bandt and Pompe, 1993; Costa et al., 2005; Martin et al., 2006; Rosso et al., 2001) can be used to generate useful codes.

A further challenge is the robust identification of relevant events in the presence of different types and sources of noise and sensory drift. Classical parametric models such as AR filters can be further extended to grasp and reproduce nonlinear dynamics, drifts, or shifts (Aguirre and Letellier, 2012; Kerschen et al., 2006). Furthermore, in nonstationary environments those models can be improved, for example, by piecewise autoregression with exogenous inputs, switched AR, switched autoregressive exogenous (ARX), switched nonlinear ARX, and PW nonlinear ARX systems (Sayed-Mouchaweh and Lughofer, 2012). Certainly, machine learning, Bayesian inference and information theory can be adequately combined to construct adaptive filtering techniques that can be applied in real-time analysis to address the challenge of nonlinearity and nonstationarity at the different levels of electrophysiology and imaging setups. However, independent of the selected set of tools, we have to be very careful about the underlying assumptions (Amarasingham et al., 2015; Cohen and Kohn, 2011) and the limitations of practical setups (Smirnov and Bezruchko, 2012). Data-driven strategies should be developed under a unified framework for the quantification of causal couplings (Smirnov, 2014), which can be further enriched by the possibility of integrating multiple data sources and processing them in a parallel and fast way (Sejnowski et al., 2014). The so-called *Big Data* approach comes with a set of powerful methods to unveil correlations and to support inference and interpretation. This new technological reality is likely to determine an even stronger relationship between theory and practice in the coming years, but this new regime should be approached cautiously (Fan et al., 2014). In the specific case of neuroscience, this paradigm shift entails a technoscientific ecosystem arising from biology, physics, and computational science. A proper combination of computational advances and experimental and theoretical proposals should be allowed to evolve naturally from previous analysis techniques to modern, data-based procedures.

7 DISCUSSION

Modern technological advances in experimental neuroscience allow a wide variety of effective monitoring and stimulation protocols at different description levels (see, eg, Assad et al., 2014; Lewis et al., 2015). While these developments promote new results using classical stimulus-response paradigms, closed-loop approaches largely complement traditional open-loop protocols to address the partial observation nature of experimental research in neuroscience. Closed-loop approaches can also address the study of transient neural dynamics and history-dependent neural computations, and bridge between disparate levels of analysis. In this chapter we have emphasized some technological needs for online event detection in closed-loop technologies at different description levels. Some of these needs are common throughout all description scales and observation modalities, and could be addressed with joint efforts from different experimental perspectives.

Closed-loop technologies encourage further synergy from the fields of experimental research, data analysis, theoretical descriptions, identification and theoretical control paradigms, and modern hardware and software design in neuroscience. While there is an increasing use of closed-loop protocols all the way from in vitro preparations to behavioral studies, BCIs and biomedical applications, progress seems to be rather independent of advances at each description level and field of interest. Further cooperation and integration of knowledge, analysis, and technological tools could boost the success of closed-loop approaches in neuroscience research and associated biomedical applications.

Recently, several open source software tools have emerged to help the task of designing and implementing closed-loop interactions with neuronal systems. Table 1 lists examples of efforts in this direction, including platforms that consider different description levels. Some of these platforms incorporate effective online event detection tools, but in general the availability of methods and libraries for this task is limited. A few modern algorithms for event detection and dynamical characterization are developed with a perspective for their online application (see, eg, Kreuz et al., 2013; Rey et al., 2015), but this is not a general trend yet.

Online or real-time processing of neural signals is becoming more feasible now through the ever-increasing computational power of modern microprocessors, graphics processing units (GPUs), and field-programmable gate arrays (Wu et al., 2016). For example, modern high-end GPUs have thousands of cores, allowing suitable algorithms to run two to three orders of magnitude faster than on a normal CPU. Online neural event detection often falls into the category of suitable algorithms because it typically involves large processing steps that are fully parallelizable (eg, processing the data from a large MEA or the voxels of fMRI data).

TABLE 1 Examples of Open Source Tools for Closed-Loop Protocol Implementations at Different Description Levels

Closed-Loop Software	Application Level	URL/Reference	Platform	Comments
NeuroRighter	Network electrophysiology, MEAs, optogenetics	https://sites.google.com/site/neurorighter/ (Newman et al., 2012)	Windows	Closed-loop multichannel control Can be integrated with other software and hardware
RTXI	Celullar, network, cardiac, deep brain stimulation	http://rtxi.org/ (Ortega et al., 2014)	RT Linux	Modular format, custom user-made modules
RTBiomanager	Cellular, behavioral studies, EEG-based brain-computer interfaces	http://www.ii.uam.es/~gnb (Chamorro et al., 2012; Muniz et al., 2009)	RT Linux	Activity-dependent drug microinjection Step-motor control for mechanical stimulation. Online video analysis
Real-time neuron-computer interface system	Cellular electrophysiology	https://bitbucket.org/mgiugliano/pc_neuron_simulink (Biró and Giugliano, 2015)	Matlab/Simulink (Windows/Linux)	Visual programming language
Dynamic Clamp StdpC	Cellular electrophysiology	https://sourceforge.net/projects/stdpc/ (Kemenes et al., 2011; Samu et al., 2012)	Windows	Interface for activity pattern generation and data replay in artificial neurons
BCILAB	EEG-based brain-computer interfaces	http://sccn.ucsd.edu/wiki/BCILAB (Delorme et al., 2011; Kothe and Makeig, 2013)	Windows/Linux/Mac	Complementary to other tools such as ERICA and EEGLAB

In conclusion, in this chapter we have briefly reviewed the use and benefits of closed-loop experimentation in neuroscience and have discussed in more detail the need for advanced and systematic online event detection. We believe that formalizing the notion of events and hence the computation steps that are necessary to detect and process them will further facilitate progress in the exciting area of real-time, closed-loop neuroscience.

ACKNOWLEDGMENTS

Authors acknowledge funding from ONRG grant N62909-14-1-N279 (PV), MINECO/FEDER TIN DPI2015-65833-P (PV, DA), TIN2012-30883 (PV, DA), TIN2010-19607 (FBR), TIN2014-54580-R (FBR). TN was supported by a Senior Research Fellowship from the Royal Academy of Engineering and The Leverhulme Trust.

REFERENCES

Aguiar, P., Mendonça, L., Galhardo, V., 2007. OpenControl: a free open-source software for video tracking and automated control of behavioral mazes. J. Neurosci. Methods 166 (1), 66–72.

Aguirre, L.A., Letellier, C., 2012. Nonstationarity signatures in the dynamics of global nonlinear models. Chaos 22 (3), 33136.

Amarasingham, A., Geman, S., Harrison, M.T., 2015. Ambiguity and non-identifiability in the statistical analysis of neural codes. Proc. Natl. Acad. Sci. 112 (20), 6455–6460.

Arroyo, D., Chamorro, P., Amigó, J.M., Rodriguez, F.B., Varona, P., 2013. Event detection, multimodality and non-stationarity: ordinal patterns, a tool to rule them all? Eur. Phys. J. Spec. Top. 222, 457–472.

Assad, J., Berdondini, L., Cancedda, L., De Angelis, F., Francesco Diaspro, A., Dipalo, M., Fellin, T., Maccione, A., Panzeri, S., Sileo, L., 2014. Brain function: novel technologies driving novel understanding. In: Cingolani, R. (Ed.), Bioinspired Approaches for Human-Centric Technologies. Springer International Publishing, Cham, pp. 299–334.

Bandt, C., Pompe, B., 1993. The entropy profile—a function describing statistical dependences. J. Stat. Phys. 70 (3/4), 967–983.

Berenyi, A., Belluscio, M., Mao, D., Buzsaki, G., 2012. Closed-loop control of epilepsy by transcranial electrical stimulation. Science 337, 735–737.

Beuter, A., Lefaucheur, J.P., Modolo, J., 2014. Closed-loop cortical neuro-modulation in Parkinson's disease: an alternative to deep brain stimulation? Clin. Neurophysiol. 125, 874–885.

Bi, G.Q., Poo, M.M., 1998. Synaptic modifications in cultured hippocampal neurons: dependence on spike timing, synaptic strength, and postsynaptic cell type. J. Neurosci. 18 (24), 10464–10472.

Birbaumer, N., 2006. Breaking the silence: brain-computer interfaces (BCI) for communication and motor control. Psychophysiology 43 (6), 517–532.

Birbaumer, N., Ghanayim, N., Hinterberger, T., Iversen, I., Kotchoubey, B., Kübler, A., Perelmouter, J., Taub, E., Flor, H., 1999. A spelling device for the paralysed. Nature 398 (6725), 297–298.

Biró, I., Giugliano, M., 2015. A reconfigurable visual-programming library for real-time closed-loop cellular electrophysiology. Front. Neuroinform. 9, 17.

Butera, R.J., Wilson, C.G., Delnegro, C.A., Smith, J.C., 2001. A methodology for achieving high-speed rates for artificial conductance injection in electrically excitable biological cells. IEEE Trans. Biomed. Eng. 48 (12), 1460–1470.

Buzsáki, G., Stark, E., Berényi, A., Khodagholy, D., Kipke, D.R., Yoon, E., Wise, K.D., 2015. Tools for probing local circuits: high-density silicon probes combined with optogenetics. Neuron 86 (1), 92–105.

Chamorro, P., Levi, R., Rodriguez, F.B., Pinto, R.D., Varona, P., 2009. Real-time activity-dependent drug microinjection. BMC Neurosci. 10, P296.

Chamorro, P., Muñiz, C., Levi, R., Arroyo, D., Rodríguez, F.B., Varona, P., 2012. Generalization of the dynamic clamp concept in neurophysiology and behavior. PLoS ONE 7(7).

Cincotti, F., Mattia, D., Babiloni, C., Carducci, F., Salinari, S., Bianchi, L., Marciani, M.G., Babiloni, F., 2003. The use of EEG modifications due to motor imagery for brain-computer interfaces. IEEE Trans. Neural Syst. Rehabil. Eng. 11 (2), 131–133.

Cohen, M.X., 2014. Analyzing Neural Time Series Data: Theory and Practice. MIT Press, Cambridge, MA.

Cohen, M.R., Kohn, A., 2011. Measuring and interpreting neuronal correlations. Nat. Neurosci. 14 (7), 811–819.

Cohen, M.X., Voytek, B., 2013. Linking nonlinear neural dynamics to single-trial human behavior. In: Multiscale Analysis and Nonlinear Dynamics: From Genes to the Brain. Wiley-VCH Verlag GmbH & Co. KGaA, Weinheim, pp. 217–232.

Costa, M., Goldberger, A., Peng, C.-K., 2005. Multiscale entropy analysis of biological signals. Phys. Rev. E 71 (2), 1–18.

Cowan, N.J., Ankarali, M.M., Dyhr, J.P., Madhav, M.S., Roth, E., Sefati, S., Sponberg, S., Stamper, S.A., Fortune, E.S., Daniel, T.L., 2014. Feedback control as a framework for understanding tradeoffs in biology. Integr. Comp. Biol. 54, 223–237.

Daw, C.S., Finney, C.E.A., Tracy, E.R., 2003. A review of symbolic analysis of experimental data. Rev. Sci. Instrum. 74 (2), 915–930.

Debanne, D., Gähwiler, B.H., Thompson, S.M., 1998. Long-term synaptic plasticity between pairs of individual CA3 pyramidal cells in rat hippocampal slice cultures. J. Physiol. 507 (Pt 1), 237–247.

Dekhuijzen, A.J., Bagust, J., 1996. Analysis of neural bursting: non-rhythmic and rhythmic activity in isolated spinal cord. J. Neurosci. Methods 67, 141–147.

Delorme, A., Mullen, T., Kothe, C., Akalin Acar, Z., Bigdely-Shamlo, N., Vankov, A., Makeig, S., 2011. EEGLAB, SIFT, NFT, BCILAB, and ERICA: new tools for advanced EEG processing. Comput. Intell. Neurosci. 2011, 130714.

Destexhe, A., Bal, T. (Eds.), 2009. Dynamic-Clamp: From Principles to Applications. Springer, New York.

Dobrunz, L.E., Stevens, C.F., 1999. Response of hippocampal synapses to natural stimulation patterns. Neuron 22 (1), 157–166.

Dorval, A.D., Christini, D.J., White, J.A., 2001. Real-time linux dynamic clamp: a fast and flexible way to construct virtual ion channels in living cells. Ann. Biomed. Eng. 29 (10), 897–907.

Elson, R.C., Selverston, A.I., Huerta, R., Rulkov, N.F., Rabinovich, M.I., Abarbanel, H.D.I., 1998. Synchronous behavior of two coupled biological neurons. Phys. Rev. Lett. 81, 5692–5695.

Faller, J., Scherer, R., Costa, U., Opisso, E., Medina, J., Müller-Putz, G.R., 2014. A co-adaptive brain-computer interface for end users with severe motor impairment. PLoS ONE 9(7).

Fan, J., Han, F., Liu, H., 2014. Challenges of big data analysis. Nat. Sci. Rev. 1 (2), 293–314.

Fazel-Rezai, R., Allison, B.Z., Guger, C., Sellers, E.W., Kleih, S.C., Kübler, A., 2012. P300 brain computer interface: current challenges and emerging trends. Front. Neuroeng. 5, 14.

Fernandez-Vargas, J., Pfaff, H.U., Rodriguez, F.B., Varona, P., 2013. Assisted closed-loop optimization of SSVEP-BCI efficiency. Front. Neural Circ. 7 (Article 27).

Forlim, C., Muniz, C., Pinto, R., Rodríguez, F., Varona, P., 2013. Behavioral driving through on line monitoring and activity-dependent stimulation in weakly electric fish. BMC Neurosci. 14, P405.

Forlim, C.G., Pinto, R.D., Varona, P., Rodríguez, F.B., 2015. Delay-dependent response in weakly electric fish under closed-loop pulse stimulation. PLoS One 10, e0141007.

Franke, F., Jäckel, D., Dragas, J., Müller, J., Radivojevic, M., Bakkum, D., Hierlemann, A., 2012. High-density microelectrode array recordings and real-time spike sorting for closed-loop experiments: an emerging technology to study neural plasticity. Front. Neural Circ. 6, 105.

Froemke, R.C., Dan, Y., 2002. Spike-timing-dependent synaptic modification induced by natural spike trains. Nature 416 (6879), 433–438.

Golowasch, J., Goldman, M.S., Abbott, F., Marder, E., 2002. Failure of averaging in the construction of a conductance-based neuron model. J. Neurophysiol. 87, 1129–1131.

Grosenick, L., Marshel, J.H., Deisseroth, K., 2015. Closed-loop and activity-guided optogenetic control. Neuron 86 (1), 106–139.

Guger, C., Daban, S., Sellers, E., Holzner, C., Krausz, G., Carabalona, R., Gramatica, F., Edlinger, G., 2009. How many people are able to control a P300-based brain-computer interface (BCI)? Neurosci. Lett. 462 (1), 94–98.

Hales, C.M., Zeller-Townson, R., Newman, J.P., Shoemaker, J.T., Killian, N.J., Potter, S.M., 2012. Stimulus-evoked high frequency oscillations are present in neuronal networks on microelectrode arrays. Front. Neural Circ. 6, 29.

Halliday, D.M., Rosenberg, J.R., Amjad, A.M., Breeze, P., Conway, B.A., Farmer, S.F., 1995. A framework for the analysis of mixed time series/point process data—theory and application to the study of physiological tremor, single motor unit discharges and electromyograms. Prog. Biophys. Mol. Biol. 64 (2), 237–278.

Hariz, M., Blomstedt, P., Zrinzo, L., 2013. Future of brain stimulation: new targets, new indications, new technology. Mov. Disord. 28, 1784–1792.

Harsch, A., Robinson, H.P., 2000. Postsynaptic variability of firing in rat cortical neurons: the roles of input synchronization and synaptic NMDA receptor conductance. J. Neurosci. 20 (16), 6181–6192.

Jensen, O., Bahramisharif, A., Oostenveld, R., Klanke, S., Hadjipapas, A., Okazaki, Y.O., van Gerven, M.A.J., 2011. Using brain-computer interfaces and brain-state dependent stimulation as tools in cognitive neuroscience. Front. Psychol. 2, 100.

Jimbo, Y., Tateno, T., Robinson, H.P., 1999. Simultaneous induction of pathway-specific potentiation and depression in networks of cortical neurons. Biophys. J. 76 (2), 670–678.

Kass, R.E., Eden, U.T., Brown, E.N., 2014. Analysis of Neural Data. Springer, New York.

Kemenes, I., Marra, V., Crossley, M., Samu, D., Staras, K., Kemenes, G., Nowotny, T., 2011. Dynamic clamp with StdpC software. Nat. Protoc. 6 (3), 405–417.

Kerschen, G., Worden, K., Vakakis, A.F., Golinval, J.-C., 2006. Past, present and future of nonlinear system identification in structural dynamics. Mech. Syst. Signal Process. 20 (3), 505–592.

Kleinberg, S., Hripcsak, G., 2011. A review of causal inference for biomedical informatics. J. Biomed. Inform. 44 (6), 1102–1112.

Kotchoubey, B., Strehl, U., Uhlmann, C., Holzapfel, S., König, M., Fröscher, W., Blankenhorn, V., Birbaumer, N., 2001. Modification of slow cortical potentials in patients with refractory epilepsy: a controlled outcome study. Epilepsia 42 (3), 406–416.

Kothe, C.A., Makeig, S., 2013. BCILAB: a platform for brain-computer interface development. J. Neural Eng. 10 (5), 056014.

Kreuz, T., Chicharro, D., Houghton, C., Andrzejak, R.G., Mormann, F., 2013. Monitoring spike train synchrony. J. Neurophysiol. 109 (5), 1457–1472.

Kübler, A., Kotchoubey, B., Kaiser, J., Wolpaw, J.R., Birbaumer, N., 2001. Brain-computer communication: unlocking the locked in. Psychol. Bull. 127 (3), 358–375.

Kullmann, P.H.M., Wheeler, D.W., Beacom, J., Horn, J.P., 2004. Implementation of a fast 16-bit dynamic clamp using LabVIEW-RT. J. Neurophysiol. 91 (1), 542–554.

LeMasson, G., Masson, S.R.-L., Debay, D., Bal, T., 2002. Feedback inhibition controls spike transfer in hybrid thalamic circuits. Nature 417, 854.

Lewis, C.M., Bosman, C.A., Fries, P., 2015. Recording of brain activity across spatial scales. Curr. Opin. Neurobiol. 32, 68–77.

Linaro, D., Couto, J., Giugliano, M., 2015. Real-time electrophysiology: using closed-loop protocols to probe neuronal dynamics and beyond. J. Vis. Exp. 100.

Mainen, Z., Sejnowski, T., 1995. Reliability of spike timing in neocortical neurons. Science 268, 1503.

Marinazzo, D., Pellicoro, M., Stramaglia, S., 2008. Kernel method for nonlinear granger causality. Phys. Rev. Lett. 100 (14), 144103.

Markram, H., Lübke, J., Frotscher, M., Sakmann, B., 1997. Regulation of synaptic efficacy by coincidence of postsynaptic APs and EPSPs. Science (New York, N.Y.) 275 (5297), 213–215.

Martin, M.T., Plastino, A., Rosso, O.A., 2006. Generalized statistical complexity measures: geometrical and analytical properties. Physica A 369 (2), 439–462.

Miezin, F.M., Maccotta, L., Ollinger, J.M., Petersen, S.E., Buckner, R.L., 2000. Characterizing the hemodynamic response: effects of presentation rate, sampling procedure, and the possibility of ordering brain activity based on relative timing. NeuroImage 11 (6 Pt 1), 735–759.

Mohseni, P., Ghovanloo, M., 2012. Guest editorial closing the loop via advanced neurotechnologies. IEEE Trans. Neural Syst. Rehabil. Eng. 20 (4), 407–409.

Muniz, C., Arganda, S., Rodriguez, F.B., de Polavieja, G.G., Varona, P., 2005. Realistic stimulation through advanced dynamic clamp protocols. Lect. Notes Comput. Sci. 3561, 95–105.

Muniz, C., Levi, R., Benkrid, M., Rodriguez, F.B., Varona, P., 2008. Real-time control of stepper motors for mechano-sensory stimulation. J. Neurosci. Methods 172 (1), 105–111.

Muniz, C., Rodriguez, F.B., Varona, P., 2009. RTBiomanager: a software platform to expand the applications of real-time technology in neuroscience. BMC Neurosci. 10, P49.

Naseer, N., Hong, K.-S., 2015. fNIRS-based brain-computer interfaces: a review. Front. Hum. Neurosci. 9, 3.

Newman, J.P., Zeller-Townson, R., Fong, M.-F., Arcot Desai, S., Gross, R.E., Potter, S.M., 2012. Closed-loop, multichannel experimentation using the open-source NeuroRighter electrophysiology platform. Front. Neural Circ. 6, 98.

Ngan, S.C., Hu, X., Tan, L.H., Khong, P.L., 2009. Improvement of spectral density-based activation detection of event-related fMRI data. Magn. Reson. Imaging 27 (7), 879–894.

Nicolas-Alonso, L.F., Gomez-Gil, J., 2012. Brain computer interfaces, a review. Sensors 12 (2), 1211–1279.

Nowotny, T., Varona, P., 2015. Dynamic clamp technique. In: Jaeger, D., Jung, R. (Eds.), Encyclopedia of Computational Neuroscience. Springer, New York, pp. 1048–1051.

Nowotny, T., Zhigulin, V.P., Selverston, A.I., Abarbanel, H.D.I., Rabinovich, M.I., 2003. Enhancement of synchronization in a hybrid neural circuit by spike timing dependent plasticity. J. Neurosci. 23, 9776–9785.

Nowotny, T., Szucs, A., Pinto, R.D., Selverston, A.I., 2006. StdpC: a modern dynamic clamp. J. Neurosci. Methods 158 (2), 287–299.

Nowotny, T., Szücs, A., Levi, R., Selverston, A.I., 2007. Models wagging the dog: are circuits constructed with disparate parameters? Neural Comput. 19 (8), 1985–2003.

Nowotny, T., Levi, R., Selverston, A.I., 2008. Probing the dynamics of identified neurons with a data-driven modeling approach. PLoS ONE 3(7).

Obien, M.E.J., Deligkaris, K., Bullmann, T., Bakkum, D.J., Frey, U., 2015. Revealing neuronal function through microelectrode array recordings. Front. Neurosci. 8, 423.

Olypher, A., Cymbalyuk, G., Calabrese, R.L., 2006. Hybrid systems analysis of the control of burst duration by low-voltage-activated calcium current in leech heart interneurons. J. Neurophysiol. 96 (6), 2857–2867.

Orsborn, A.L., Dangi, S., Moorman, H.G., Carmena, J.M., 2012. Closed-loop decoder adaptation on intermediate time-scales facilitates rapid BMI performance improvements independent of decoder initialization conditions. IEEE Trans. Neural Syst. Rehabil. Eng. 20 (4), 468–477.

Ortega, F.A., Butera, R.J., Christini, D.J., White, J.A., Dorval, A.D., 2014. Dynamic clamp in cardiac and neuronal systems using RTXI. Methods Mol. Biol. 1183, 327–354.

Paulsen, O., Sejnowski, T.J., 2000. Natural patterns of activity and long-term synaptic plasticity. Curr. Opin. Neurobiol. 10 (2), 172–179.

Pereda, E., Quiroga, R.Q., Bhattacharya, J., 2005. Nonlinear multivariate analysis of neurophysiological signals. Prog. Neurobiol. 77, 1–37.

Pinto, R.D., Varona, P., Volkovskii, A.R., Szücs, A., Abarbanel, H.D.I., Rabinovich, M.I., 2000. Synchronous behavior of two coupled electronic neurons. Phys. Rev. E 62 (2), 2644–2656.

Pinto, R.D., Elson, R.C., Szücs, A., Rabinovich, M.I., Selverston, A.I., Abarbanel, H.D., 2001. Extended dynamic clamp: controlling up to four neurons using a single desktop computer and interface. J. Neurosci. Methods 108 (1), 39–48.

Plailly, J., Howard, J.D., Gitelman, D.R., Gottfried, J.A., 2008. Attention to odor modulates thalamocortical connectivity in the human brain. J. Neurosci. 28 (20), 5257–5267.

Porta, A., Baumert, M., Cysarz, D., Wessel, N., 2015. Enhancing dynamical signatures of complex systems through symbolic computation. Philos. Trans. R. Soc. Lond. A. 373 (2034). 20140099.

Potter, S.M., DeMarse, T.B., 2001. A new approach to neural cell culture for long-term studies. J. Neurosci. Methods 110 (1-2), 17–24.

Prinz, A.A., Abbott, L.F., Marder, E., 2004. The dynamic clamp comes of age. Trends Neurosci. 27, 218–224.

Rey, H.G., Pedreira, C., Quian Quiroga, R., 2015. Past, present and future of spike sorting techniques. Brain Res. Bull. 119, 106–117. http://dx.doi.org/10.1016/j.brainresbull.2015.04.007.

Robinson, H.P., Kawai, N., 1993. Injection of digitally synthesized synaptic conductance transients to measure the integrative properties of neurons. J. Neurosci. Methods 49, 157.

Rolston, J.D., Gross, R.E., Potter, S.M., 2010. Closed-loop, open-source electrophysiology. Front. Neurosci. 4 (31), 1–8.

Rosso, O.A., Blanco, S., Yordanova, J., Kolev, V., Figliola, A., Schürmann, M., Basar, E., 2001. Wavelet Entropy: a new tool for analysis of short duration brain electrical signals. J. Neurosci. Methods 105, 65–75.

Roth, E., Sponberg, S., Cowan, N.J., 2014. A comparative approach to closed-loop computation. Curr. Opin. Neurobiol. 25, 54–62.

Ruaro, M.E., Bonifazi, P., Torre, V., 2005. Toward the neurocomputer: image processing and pattern recognition with neuronal cultures. IEEE Trans. Biomed. Eng. 52 (3), 371–383.

Ruiz, S., Buyukturkoglu, K., Rana, M., Birbaumer, N., Sitaram, R., 2014. Real-time fMRI brain computer interfaces: self-regulation of single brain regions to networks. Biol. Psychol. 95, 4–20.

Samu, D., Marra, V., Kemenes, I., Crossley, M., Kemenes, G., Staras, K., Nowotny, T., 2012. Single electrode dynamic clamp with StdpC. J. Neurosci. Methods 211 (1), 11–21.

Sayed-Mouchaweh, M., Lughofer, E., 2012. Learning in Non-Stationary Environments: Methods and Applications. Springer Science & Business Media, New York.

Schalk, G., Leuthardt, E.C., 2011. Brain-computer interfaces using electrocorticographic signals. IEEE Rev. Biomed. Eng. 4, 140–154.

Schiff, S.J., 2012. Neural Control Engineering. The MIT Press, Cambridge, MA.

Schinkel, S., Marwan, N., Kurths, J., 2007. Order patterns recurrence plots in the analysis of ERP data. Cogn. Neurodyn. 1 (4), 317–325.

Schultheiss, N.W., Prinz, A.A., Butera, R.J. (Eds.), 2012. Phase Response Curves in Neuroscience. Springer, New York.

Segundo, J.P., Sugihara, G., Dixon, P., Stiber, M., Bersier, L.F., 1998. The spike trains of inhibited pacemaker neurons seen through the magnifying glass of non linear analyses. Neuroscience 87, 741–766.

Sejnowski, T.J., Churchland, P.S., Movshon, J.A., 2014. Putting big data to good use in neuroscience. Nat. Neurosci. 17 (11), 1440–1441.

Sharp, A.A., O'Neil, M.B., Abbott, L.F., Marder, E., 1993. Dynamic clamp: computer-generated conductances in real neurons. J. Neurophysiol. 69 (3), 992–995.

Smirnov, D.A., 2014. Quantification of causal couplings via dynamical effects: a unifying perspective. Phys. Rev. E 90 (6), 62921.

Smirnov, D.A., Bezruchko, B.P., 2012. Spurious causalities due to low temporal resolution: towards detection of bidirectional coupling from time series. Europhys. Lett. 100, 10005.

Sohal, V.S., Zhang, F., Yizhar, O., Deisseroth, K., 2009. Parvalbumin neurons and gamma rhythms enhance cortical circuit performance. Nature 459 (7247), 698–702.

Solé, A.F., Ngan, S.C., Sapiro, G., Hu, X., López, A., 2001. Anisotropic 2-D and 3-D averaging of fMRI signals. IEEE Trans. Med. Imaging 20 (2), 86–93.

Spira, M.E., Hai, A., 2013. Multi-electrode array technologies for neuroscience and cardiology. Nat. Nanotechnol. 8 (2), 83–94.

Stevens, C.F., Zador, A.M., 1998. Input synchrony and the irregular firing of cortical neurons. Nat. Neurosci. 1 (3), 210–217.

Sulzer, J., Haller, S., Scharnowski, F., Weiskopf, N., Birbaumer, N., Blefari, M.L., et al., 2013. Real-time fMRI neurofeedback: progress and challenges. NeuroImage 76, 386–399.

Sweeney-Reed, C.M., Nasuto, S.J., 2007. A novel approach to the detection of synchronisation in EEG based on empirical mode decomposition. J. Comput. Neurosci. 23 (1), 79–111.

Szucs, A., 1998. Applications of the spike density function in analysis of neuronal firing patterns. J. Neurosci. Methods 81 (1-2), 159–167.

Szucs, A., Varona, P., Volkovskii, A.R., Abarbanel, H.D.I., Rabinovich, M.I., Selverston, A.I., 2000. Interacting biological and electronic neurons generate realistic oscillatory rhythms. Neuroreport 11 (3), 563–569.

Szücs, A., Pinto, R.D., Rabinovich, M.I., Abarbanel, H.D.I., Selverston, A.I., Szucs, A., 2003. Synaptic modulation of the interspike interval signatures of bursting pyloric neurons. J. Neurophysiol. 89 (3), 1363–1377.

Szucs, A., Vehovszky, Á., Molnár, G., Pinto, R.D., Abarbanel, H.D.I., 2004. Reliability and precision of neural spike timing: simulation of spectrally broadband synaptic inputs. Neuroscience 126 (4), 1063–1073.

Van Boxtel, G.J.M.M., Gruzelier, J.H., 2014. Neurofeedback: introduction to the special issue. Biol. Psychol. 95, 1–3.

Van Gerven, M., Farquhar, J., Schaefer, R., Vlek, R., Geuze, J., Nijholt, A., et al., 2009. The brain-computer interface cycle. J. Neural Eng. 6 (4), 41001.

Van Rossum, M.C., 2001. A novel spike distance. Neural Comput. 13 (4), 751–763.

Varona, P., Torres, J.J., Abarbanel, H.D.I., Rabinovich, M.I., Elson, R.C., 2001. Dynamics of two electrically coupled chaotic neurons: experimental observations and model analysis. Biol. Cybern. 84, 91–101.

Victor, J.D., Purpura, K.P., 2009. Metric-space analysis of spike trains: theory, algorithms and application. Netw. Comput. Neural Syst. 8 (2), 127–164.

Vidaurre, C., Blankertz, B., 2010. Towards a cure for BCI illiteracy. Brain Topogr. 23 (2), 194–198.

Vidaurre, C., Sannelli, C., Müller, K.-R., Blankertz, B., 2011. Co-adaptive calibration to improve BCI efficiency. J. Neural Eng. 8 (2), 025009.

Wagenaar, D.A., Pine, J., Potter, S.M., 2006. An extremely rich repertoire of bursting patterns during the development of cortical cultures. BMC Neurosci. 7, 11.

Wander, J.D., Rao, R.P., 2014. Brain–computer interfaces: a powerful tool for scientific inquiry. Curr. Opin. Neurobiol. 25, 70–75.

Weiskopf, N., Mathiak, K., Bock, S.W., Scharnowski, F., Veit, R., Grodd, W., Goebel, R., Birbaumer, N., 2004. Principles of a brain-computer interface (BCI) based on real-time functional magnetic resonance imaging (fMRI). IEEE Trans. Biomed. Eng. 51 (6), 966–970.

Wittenberg, G.M., Wang, S.S.-H., 2006. Malleability of spike-timing-dependent plasticity at the CA3-CA1 synapse. J. Neurosci. 26 (24), 6610–6617.

Wolpaw, J., Wolpaw, E. (Eds.), 2012. Brain-Computer Interfaces: Principles and Practice. Oxford University Press, New York.

Wu, G., Nowotny, T., Chen, Y., Li, D.D.-U., 2016. GPU acceleration of time-domain fluorescence lifetime imaging. J. Biomed. Opt. 21 (1), 017001.

Yu, H., Senarathna, J., Tyler, B.M., Thakor, N.V., Pathak, A.P., 2015. Miniaturized optical neuroimaging in unrestrained animals. NeuroImage 113, 397–406.

Chapter 7

Closing Dewey's Circuit

A. Wallach*, S. Marom[†] and E. Ahissar[‡]

*University of Ottawa, Ottawa, ON, Canada, [†]Technion—Israel Institute of Technology, Haifa, Israel, [‡]The Weizmann Institute of Science, Rehovot, Israel

1 INTRODUCTION

One hundred and twenty years ago, the American philosopher and psychologist John Dewey (Fig. 1) published his seminal paper *The Reflex Arc Concept in Psychology* in the *Psychological Review* (Dewey, 1896). In this paper, Dewey challenged the reflex arc, a by-then (and since then, as we will soon demonstrate) consensual unifying framework for action and perception. The framework, advanced by European experimental psychologists, pertains to a general interpretation that was referred to, at the time, as "structural psychology." Advocates of this framework, which Dewey so fiercely opposed, viewed psychological processes as sequences of distinct steps that are executed in a timely and precise manner. The structure of the system—the identity and location of the elements executing the different stages of the process—is at the epicenter of interest. Dewey, a key proponent of the philosophy of Pragmatism and one of the founders of the Chicago group of psychologists, sought to advance an alternative, "functional psychology," focussing on system-environment interactions and their value in achieving functionally meaningful goals. Functional psychologists did not ignore or deny the impacts of structures and mechanisms inherent to the individual in the emergence of behavior; likewise, structural psychologists did not ignore or deny the functional attributes of behavior. Rather, the main difference between these two strands of academic psychology pertained to the methods that are most likely to produce understanding of behavior. Structuralists sought to identify program-like processes hidden deep inside the machine, the human brain; functionalists could not see how the emergence of behavior can be understood in isolation from coupled subject-environment dynamics. In this context, Marom (2015) offered to use the more indicative terms *structural-programmatic* and *functional-dynamic* to designate the two stances.[1]

In this chapter, we juxtapose these two stances, discuss the methodological reasons for the dominance of the structural-programmatic stance in neuroscience, and offer closed-loop methodologies as a functional-dynamic alternative for studying action and perception as a complete circuit.

2 THE BABY AND CANDLE SCENARIO

Let us follow Dewey's footsteps by first illustrating the structural-programmatic framework with the "baby and candle" example, originally appearing in William James's *Principles of Psychology* (James, 1890). The example describes how an infant interacts with a novel object (see Fig. 2), and aims to demonstrate the mechanisms underlying existing behaviors and their modification. The scenario begins with the light of a candle stimulating the child's retina. This leads to an internal "idea" of light, which leads to a reflexive motor response of hand reaching toward the candle. A secondary chain of events then commences with the flame's heat stimulating the skin of the child's fingers, leading to an "idea" of pain, which prompts a reflexive motor response of hand retraction. Association by Succession, a well-accepted and studied principle underlying learning, ties the internal ideas of candlelight and pain to inhibit future hand reaching, and so behavior is modified.

The above analysis thus describes two *reflex arcs*, each starting with a peripheral sensory event-designated *stimulus* (flame image, fingertip burning), followed by the generation of an internal *idea* (neural representation of the stimulus) and ends with a motor action-designated *response* (hand reaching or retracting). When the scenario is described in these terms, the reflex arc seems to be the simplest model that can account for the observed behavior and thus the preferred scientific model.

However, several seemingly unrelated questions may come to the reader's mind: what caused the candlelight to appear in the baby's fovea? What was the baby seeing and feeling during either hand motion? And of course,

---

1. For a concise review of the structural versus functional split in behavioral and brain sciences at the turn of the 20th century, see Marom (2015, pp. 138–148).

Closed Loop Neuroscience. http://dx.doi.org/10.1016/B978-0-12-802452-2.00007-X

FIG. 1   John Dewey.

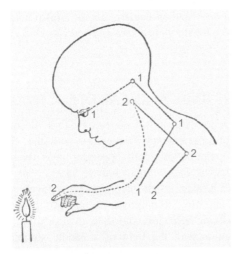

FIG. 2   Baby and candle. Two reflex arcs are depicted: seeing a candle, which leads to hand reaching (labeled 1) and feeling the candle heat, which leads to hand retraction (labeled 2). *(From James, W., 1890. The Principles of Psychology. Holt, Reinhart and Winston, New York.)*

where were the baby's parents, who so irresponsibly left an unattended baby within reach of an open flame? The picture we are left with is of a baby in solitary confinement, staring blankly and motionlessly into the void. The candle appears suddenly and unexpectedly "out of the blue" in his or her field of view. The motor responses are uncontrolled spasmatic jolts. Such overly reduced scenarios had profound impacts on the experimental research of brain and behavior.

Dewey identified the definitions of stimulus and response—the beginning and end of the arc—as the foundation stones upon which the reflex arc is built. The perceptual process is set into motion with a purely sensory event. Thought is an intermediate step and involves neither sensation nor action. And finally, the process comes to its fulfillment with a purely motor event. Each of these components—sensation, thought, and action—is regarded as being distinct in both its timing and nature, running in

turns like workstations in an efficient assembly line or subroutines in a computer program. The breaking up of the arc into these separate, disjoint processes is an inescapable consequence of viewing sensation as the initiation (stimulus) and action as the conclusion (response) of the perceptual process.

3   THE CIRCUIT ALTERNATIVE

The arc concept fails to acknowledge the circular nature of natural perception. "What we have is a circuit, not an arc or broken segment of a circle" says Dewey, referring to the fact that actions are precursors, in fact prerequisites to sensation. Dewey termed the complete circuit *coordination*—a dynamic sensory-motor process that underlies perception. Each of the stages in the baby and candle example (seeing the candle, forming an idea of the candle, reaching the hand toward it, etc.) involves in fact both sensation and action, working at the same time and having equal importance in the emergence of perception. This necessary connection between action and sensation was demonstrated in countless elegant experiments since Dewey published his paper (Ahissar and Arieli, 2001; Held and Hein, 1963; Kleinfeld et al., 2006; Konig and Luksch, 1998; Land, 2006; Lederman and Klatzky, 1987; Prescott et al., 2011; Rucci et al., 2007; Schroeder et al., 2010; Yarbus, 1967).

When all stages of an interaction are acknowledged to be both motor and sensory, the choice of a start- and end-point to the perceptual process becomes completely arbitrary; instead, what we have is a continuous sequence of motor-sensory coordinations, all linked to one another (a "circle," in Dewey's terminology); the occulomotor coordination of seeing the candle is preceded by a different coordination, for instance, by smelling smoke (an olfactory-sniffing coordination), and followed by scanning the candle and hand reaching. These motor-sensory coordinations are in fact constituents of a larger-scale coordination, that of interacting with an object. This coordination is, in turn, part of an even larger one, for instance that of exploring a new room, and so on in a nested, hierarchical manner.[2] Thus the association of a flame with pain is a natural outcome of the image and the pain pertaining to the same coordination, rather than necessitating an additional, external process that mediates between two distinct reflex arcs. Each large-scale coordination also provides context to its constituents, a context which is essential for their interpretation. For instance, the smell of smoke carries a different meaning and entails a different response if one is lighting a candle, baking a cake, or driving a car.

The circuit view removes the need for any cause (either internal or external) for setting the perceptual process in motion; it is always in motion, "one uninterrupted, continuous redistribution of mass," with the different events

---

2. Another type of coordination not addressed by Dewey, that of interactions between subjects, is discussed later in this chapter.

and interactions only reshaping and redirecting it. In other words, rather than viewing perception as a program or an algorithm, it is viewed as a physical dynamical process, similar to the ever-changing wind currents in the atmosphere; winds are entailed by air-pressure gradients, which they in turn equilibrate. Yet neither the wind nor the pressure is regarded as "stimulus" or "response."

4 DYNAMICS OF PERCEPTION

A school of researchers have advanced the characterization of perception as a dynamical process (Ahissar and Vaadia, 1990; Ashby, 1952; Kelso, 1997; O'Regan and Noe, 2001; Port and Van Gelder, 1995; Powers, 1973; Wiener, 1949). First, like all dynamical processes, perception takes time. The first wave of sensory-driven neuronal activity typically reaches most of the relevant cortical areas within ∼100 ms, and with vision quick saccadic reports on the crude category of the perceived item can be generated as fast as 150 ms (Wu et al., 2014). Yet, identification of more delicate categories, or perception of item details, takes typically hundreds of ms from first sensor-object encounter, a period during which perceptual acuity continuously improves (Micheyl et al., 2012; Packer and Williams, 1992; Saig et al., 2012).

Studies performed in rodent and human tactile perception revealed that this gradual improvement in acuity coincides with a behavioral and neuronal process of convergence. For instance, object features, such as location and texture, are perceived via a sequence of several sensor-object interactions whose motor aspects show a pattern of convergence toward asymptotic values (Chen et al., 2015; Horev et al., 2011; Knutsen et al., 2006; McDermott et al., 2013; Saig et al., 2012; Saraf-Sinik et al., 2015). Convergence processes, ie, dynamic processes during which the state of the system gradually approaches a steady state, are hallmarks of closed-loops. When the different components of the system—sensation, motor, and neuronal—all converge together, they are likely to play part in the same closed-loop.

As noted earlier, the closed-circuit view of perception grants equal importance to motor and sensory dynamics (Ahissar and Vaadia, 1990; Diamond et al., 2008; Friston, 2010; Gibson, 1962; Jarvilehto, 1999; O'Regan and Noe, 2001). Sensory signals may often be ambiguous if processed without the motor signals that yielded them. This has been explicitly demonstrated so far only for tactile perception (Gamzu and Ahissar, 2001; Horev et al., 2011; Saig et al., 2012; Saraf-Sinik et al., 2015), but should hold for any scanning sensory modality, as the nature and details of sensory signals depend on both the sampling movement (its direction, speed, and amplitude) and the sampled object (Ahissar and Arieli, 2012; Ahissar and Zacksenhouse, 2001; Bagdasarian et al., 2013). An example is the

curvature signal generated at the base of a whisker upon its contact with an object. The same curvature can be generated when contacting objects at different locations, an ambiguity that is resolved if the angle by which the whisker is rotated is taken into account (Bagdasarian et al., 2013).[3] Such interdependencies can be termed motor-sensory contingencies; O'Regan and Noe (2001) described how these contingencies may serve as the basis for perception and showed that they eliminate the need for an explicit internal representation of external objects.

Perception of the external environment was thus described as a process in which the brain temporarily "grasps" external objects and incorporates them in its motor-sensory loops (Ahissar and Assa, 2016; Assa and Ahissar, 2015). Such a process starts with a perturbation, internal or external, and gradually converges toward a complete inclusion, though in practice, never reaches that state. A laboratory-induced flashed stimulus, whose predominance in neuroscientific research will be discussed shortly, probes only the initiation of the perceptual process.

5 THE PULL OF THE STRUCTURAL-PROGRAMMATIC VIEW

Reflecting on the study of brain and behavior in the 120 years since the publication of Dewey's paper, and particularly over recent decades, it becomes apparent that the structural-programmatic paradigm reigns supreme; it is, by and large, the guiding framework to which almost all neuroscientific endeavors adhere to, including those commanded by proponents of the functional-dynamic alternative, such as the authors of the current chapter. One can easily categorize the studies reported in the literature, especially (but not exclusively) those performed in vivo, into one of the three arc components: sensation (where we find themes such as neural encoding and sensory adaptation) (Szwed et al., 2003; Wallach et al., 2016), thought (internal representation, working memory, decision making) (Shahaf et al., 2008), and action (neural pattern generation and biological motor control) (Simony et al., 2010). What, thus, pulls us so irresistibly toward the structural-programmatic view, notwithstanding the overwhelming evidence for the active dynamic nature of perception?

The structural-programmatic approach is a form of reductionism, the idea that complex "macroscopic" systems and processes can (and should) be explained as the net outcome of their (supposedly) simpler "microscopic" components. In this regard, the disintegration of the reflex arc into a disjoint succession of sensation, thought, and action

---

3. This challenge cannot be alleviated by adding efference copy information to open-loop perceptual processing—efference copies are not accurate enough to account for perceptual accuracy (Ahissar and Assa, 2016; Simony et al., 2010).

is to the reductionist not so much an inescapable evil, but rather a desired good. This reductionist approach places the structural-programmatic view within the Newtonian tradition of the natural sciences, which also involves the isolation of systems from their environment so that they may be studied as *closed systems*. When a system cannot be hermetically "closed," or when its behavior when closed becomes trivial, environmental variables must be added and these are typically represented by generic entities such as field, boundary conditions, or, in our case, stimulus. The underlying assumption in this procedure is that the studied system possesses some intrinsic, context-independent mechanisms that may be uncovered in isolation of its natural environment. Thus, the structural-programmatic view provides an easy-to-follow methodology, with centuries of demonstrated success in the fields of Physics and Chemistry. Before we describe this methodology in detail, we note that the applicability of reductionist approaches to living systems is hindered by two important factors: the overwhelming complexity of the systems' components (Marom, 2015) and the ubiquity of top-down modulating processes, from the macroscopic system to its components (Noble, 2008).

To demonstrate the prevailing methodology, let's consider a prototypic experiment in neuroscience: the perceptual task developed by Vernon B. Mountcastle and broadly utilized by many others (Britten et al., 1992; Fassihi et al., 2014; Mountcastle et al., 1975; Romo and Salinas, 2003) (Fig. 3). The experiment begins with an animal (originally a monkey) being placed in a designated chamber, well isolated from any distracting interactions with fellow subjects or with the experimenter. The relevant sensor (eg, eyes or fingers) is positioned motionlessly (either by physical restrainment or by diligent training) so that it awaits stimulation (Fig. 3A, eg, fixating the eyes on a blank screen or pressing the finger against an idle vibrator). A sensory stimulus (eg, displayed image or tactile vibration) is flashed, ie, briefly presented and then removed, while the animal stays patiently idle (Fig. 3B); sometimes, the stimulation is designed to be so brief that any concomitant motor action (eg, saccades) is precluded. After an additional "waiting" epoch elapses—time for the animal to form ideas and make decisions (Fig. 3C)—a "go" signal is given and the animal provides perceptual report by executing a brief stereotypic motor action (Fig. 3D, eg, pressing a button, moving the eyes left or right). Correlations between

concurrently recorded neural activity and the behavioral signals are used to identify neuronal populations as being related to "stimulus encoding" (sensation), "working memory" and "decision making" (thought), "motor planning and execution" (action), or mixtures thereof.

It should become obvious by now that this experimental setup is designed with the primary objective of reenacting the "baby and candle" scenario in full within the laboratory walls. Sophisticated software and hardware and arduous training are dedicated to cast the animal's behavior into the structural-programmatic mold, to ensure that sensation leads to thought leads to action, with minimal confluence between these components. Action is prevented from affecting sensation by prohibiting the subject from actively exploring the presented object. Furthermore, individual trials are usually designed to be mutually independent, and therefore the perceived object is, in itself, "nonrelational"— it is indifferent to the actions of the subject (an important point to which we will come back toward the end of this section).

Immobilization of the subject, essential for the experimental realization of the reflex arc, is also technically advantageous. Unlike motor responses of free-to-act subjects, sensory objects are easy to manipulate. This controllability of the experimental conditions entails reduction in the variance of the sensory input, which should (so the structural-programmatic view predicts) lead to reduction in the variance of internal processes, motor responses, and perception; such reduction of variance is always sought after, as it entails stronger, statistically significant findings. Moreover, immobilization greatly simplifies measurements of physiological and behavioral processes (eg, neuronal recordings, eye or whisker tracking, etc.); indeed, some measurements are extremely difficult to perform when the animal is awake, let alone moving about. The "reflex-arc" methodology described earlier may be implemented, to a certain degree, even in anesthetized animals, a preparation that offers unparalleled accessibility and controllability. However, since an anesthetized animal does not perceive, think, make decisions, or respond, studies in anesthetized animals focus on systematically analyzing either the ascending (sensory to thought) (Hubel and Wiesel, 1962) or descending (thought to motor) (Brecht et al., 2004) pathways of perception.

Of all the objects in an animal's environment, the most critical for its survival are other subjects—be they parents,

FIG. 3 Popular perceptual task. (A) Sensory organ is positioned motionlessly and awaits stimulation. (B) Sensory object is presented to the subject. (C) Objects is removed; subject continues fixating while internally processing the stimulus. (D) Subject responds with a motor action.

(A)　　　(B)　　　(C)　　　(D)

potential mates, prey items, or predators. These are also the most challenging to perceive, as their relations with the subject, and not just their attributes, need to be discerned. Such intersubject relations are, by definition, of closed-loop nature; it is of no wonder therefore that parents are absent from the "baby and candle" reflex-arc scenario. Consequently, the reflex-arc experimental method described earlier tends almost exclusively to focus on inanimate, non-relational sensory objects. This preference stems, in part, from the wish to minimize trial-to-trial variance as much as possible. But it also originates from the reductionist assumption that "complex" objects are just the sum of their "simple" components, as discussed earlier. Thus, to the structuralists, the face of the subject's mother is a collection of lines and surfaces; first, they think, we uncover how lines and surfaces are perceived, and later we can see how these are combined into a face "downstream" (Marr, 1982; Tanaka et al., 2014). The functional-dynamic view, however, claims that perception of mother has nothing to do with lines and surfaces. It is yet another type of high-level coordination: the on-going, closed-loop coordination between the two subjects, the child and the mother. This *relational* process, deeply rooted in phylogenetic and ontogenetic histories, brings about a functional, adaptive perception of the situation (Marom, 2015). Indeed, recent evidence suggests that relational and nonrelational objects lead to distinct neural dynamics, even in brain regions taken to be strictly sensory (Bobrov et al., 2014). In order to allow the assessment of the relevance of such processes to perception, methodologies that aim at studying perception should better allow for the inclusion of meaningful, dynamical, and responsive objects.

In this section, we described the prevailing reflex-arc like experimental setup. One cannot deny that this open-loop methodology yielded exciting results and advanced our understanding of possible workings of components of the brain. In rare cases, a complete sensory motor circuit was analyzed this way to provide a satisfactory explanation of ethologically relevant behaviors (Heiligenberg, 1991). However, given the points we raised earlier, such rare cases cannot justify the application of the structural-programmatic approach in more complex cases. Importantly, the extent of reduction imposed upon the studied system by the design of the experimental setup, choice of controlling and measured variables (Rosen, 1991), and constraints on the temporal scale of the experiment (Marom, 2010) are often overlooked. Once the open-loop constraints have been imposed on the animal and it is compelled to reenact the reflex arc, it is no wonder we see no evidence for closed-loop perceptual processes. The fact that animals can perform limited perceptual tasks within these open-loop constraints perhaps reveals the lower bounds for perception, and how far they may be extended via training, but certainly not the processes dominating natural perception.

6 REDEFINING STIMULUS AND RESPONSE

We have seen that the foundations of the reflex arc are the definitions of stimulus and response. Dewey suggests replacing these definitions with ones befitting the functional-dynamic view. Instead of a purely sensory event, the stimulus is "something to be discovered; to be made out," namely a state in the motor-sensory coordination space in which a "conflict [arises] within the coordination." In the language of dynamical systems, we would call such a state a *separatrix*—a region in state-space from which the dynamics may progress toward one of several basins of attraction. Likewise, the response is not a motor action, but "whatever will serve to complete the disintegrating coordination"; it is the dynamics in the coordination which settle the above-mentioned conflict. In other words, it is the aforementioned convergence of the motor-sensory dynamics into a stable attractor (Ahissar and Assa, 2016; Assa and Ahissar, 2015). Such convergence need not be complete (in fact it never is, as this would require infinite time); it suffices that stability of the coordination is reconstituted (Horev et al., 2011; Saig et al., 2012). Because of the nested, hierarchical organization of different levels of closed-loop perceptual processes, once a low-level coordination converges sufficiently, the high-level one in which it is embedded may stimulate it once more, throwing it into a new separatrix (eg, executing a saccade to another, yet-to-be-perceived object). With these new definitions we can now offer alternative methodologies to advance the functional, close-circuit approach.

7 USING CLOSED-LOOP METHODOLOGY

As we have demonstrated, the root cause for the dominance of the structural-programmatic view in neuroscience is the open-loop methodology which reenacts the reflex arc. Proponents of the functional-dynamic view fell short of providing a satisfactory alternative methodology. It is perhaps ironic that the functional-dynamic view, which pertains to the Pragmatic philosophy, failed at the acid-test for ideas in that philosophy: "… to develop a thought's meaning, we need only determine what conduct it is fitted to produce."[4] What is the experimental conduct that is derived from the functional-dynamic view? What simple and applicable methodology allows for the study of action and perception in a holistic manner?

One option is to experiment with animals that are free to move and act. This approach is gaining popularity as the technology needed for such experiments is rapidly advancing: miniaturization allows recordings of neural

---

4. James, citing Pierce, in Pragmatism, Lecture II.

activity from freely walking (O'Keefe and Dostrovsky, 1971), swimming (Canfield and Mizumori, 2004), or even flying (Yartsev and Ulanovsky, 2013) subjects, while advanced video processing and data analysis enable high-resolution tracking of free behavior (Jun et al., 2014; Knutsen et al., 2005). Such experiments yielded important findings, particularly in the study of memory and navigation, and appear to be highly promising for studying unconstrained perception. Some studies demonstrated how neuronal activity changes in the freely moving versus immobilized contexts (Oram et al., 2015). This approach faces two major challenges. First, it is highly demanding in terms of recording techniques and experimental design. Secondly, it requires forsaking much of the control over the perceptual process, as the subject chooses how to interact with the environment; each subject will do so differently, and even the same subject will never repeat the same trajectory twice. This lack of control is further aggravated if one wishes to study relational dynamics between subjects, as this involves several (at least two) acting agents.

The alternative we wish to advance here is of using closed-loop methodology as a way to perform systematic studies of perception while allowing natural interactions between the subject and the environment. Like the classic perceptual task described earlier, the sensory input in the relevant modality is controlled by the experimental setup; this may involve partial or complete immobilization and even anesthesia. The difference, however, is that this sensory flow is not predetermined and fixed, but responds on-line to the actions of the subject that are detected in real-time. What the experimenter controls and manipulates is the set of rules governing these environmental responses. In other words, the closed-loop experimental setup completes the reflex arc into a circuit; how the circuit is closed (ie, what sensations are generated in response to any detected action) defines the nature of the controlled environment in which the subject is situated.

Several recent studies followed this direction of using closed-loop methodology to create a well-controlled responsive environment. An emerging technique is to construct a "virtual reality" environment to provide the illusion of free exploration in immobilized animals (Dombeck et al., 2010; Tammero and Dickinson, 2002). The strongest drive for the development of such setups is technical, as it allows using advanced techniques (eg, calcium imaging or fMRI) to study subject-environment interactions. Most of these studies, while providing fascinating new results, still base their analyses on the structural-programmatic stimulus-response definitions described earlier.

A number of studies went the extra mile of using closed-loop methodology to study how different environmental contexts affect neural and behavioral dynamics. Ahrens et al. (2012) manipulated the rules of environmental feedback to study the convergence processes that govern closed-loop locomotion in zebra fish larva. Keller et al. (2012) recorded neuronal activity in the primary visual cortex, a region that (according to the structural-programmatic view) is regarded as an early stage in the "algorithm" that extracts basic features (eg, line orientation and color) from the visual scene (Marr, 1982). By comparing the neuronal activity under closed-loop virtual reality settings and when the loop is opened, Keller and colleagues demonstrated that this region integrates both sensory and motor information. Marom and Wallach (2011) asked human subjects to report on the perception of ambiguous objects displayed in a very similar method to Mountcastle's perceptual task described earlier. The circuit was closed by allowing the level of ambiguity to respond to the subject's perceptual reports, and the dynamics under this closed-loop relational context were compared with those in the static "reflex-arc" case. Contrary to the prediction of the structural-programmatic view, the added trial-to-trial variability of the sensory input in the closed-loop setting reduced (rather than enhanced) the variance in perception, indicating that perceptual variability is not simply reflecting internal "noise." Rather, it suggests an innate exploration for relational dynamics with the perceived object.

Closed-loop methodology can be used to study the functioning of individual components within a circuit. For example, as sensory pathways evolved to operate in conjunction with the motor pathways that actuate their sensory organs, studying these pathways in open-loop conditions might miss their major functional characteristics. In animal models, one can close the loop between a given sensory station and its relevant sensory organ via artificial actuators and motor control algorithms. Such hybrid systems (Wallach et al., 2016) allow both the study of the sensory pathway in a closed-loop condition, and the investigation of the plausibility of candidate motor control algorithms.

Closed-loop methodology also enables the introduction of relational objects that mimic the behavior of ethologically significant subjects. The rules of feedback can be altered to give rise to different intersubject dynamics. For instance, in the case of conspecific rivalry, one can study what aspects of these relational dynamics determine whether a subject converges into a dominant or a subordinate role.

Finally, analysis of perceptual dynamics could be aided by complementing the closed-loop methodology with the functional-dynamic definitions of stimulus and response discussed in the previous section. This involves confronting the subject with perceptual ambiguities (stimuli) and analyzing the processes of convergence (responses) both in behavior (of the subject and of the responsive environment) and in physiology. The closed-loop setup can be used to investigate, in the laboratory, the dynamics around naturally occurring states. For example, subjects can be actively directed toward the aforementioned "separatrices" and maintained there; in the above-mentioned study by Marom and Wallach, for instance, the feedback was used to keep the object just at the threshold of detection. Similar approaches can be used to keep bi-stable objects with dual-meanings (eg, Necker cubes) just at the borderline between two basins of attractions. Adhering to Dewey's alternative definitions of stimulus and response may thus help us anchor our attention to the functional-dynamic interpretations, an antidote or counterbalance to the ever-present temptation of structural-programmatic simplicity.

8 CONCLUSIONS

In this chapter we aim to show how the closed-circuit, functional-dynamic framework so eloquently described by Dewey, failed to supersede the reflex arc view of perception, despite the persuasive evidence accumulated over the past 120 years. This dominance of the structural-programmatic framework is largely owing to the simple experimental and analytical reduction that supports it, illustrated here by the baby and candle scenario and its laboratorial reenactments. Closed-loop methodology, complemented by Dewey's functional-dynamic definitions of stimulus and response, is offered as a way to complete the arc into a full circuit.

REFERENCES

Ahissar, E., Arieli, A., 2001. Figuring space by time. Neuron 32, 185–201.

Ahissar, E., Arieli, A., 2012. Seeing via miniature eye movements: a dynamic hypothesis for vision. Front. Comput. Neurosci. 6, 89.

Ahissar, E., Assa, E., 2016. Perception as a closed-loop convergence process. eLife 12830.

Ahissar, E., Vaadia, E., 1990. Oscillatory activity of single units in a somatosensory cortex of an awake monkey and their possible role in texture analysis. Proc. Natl. Acad. Sci. U. S. A. 87, 8935–8939.

Ahissar, E., Zacksenhouse, M., 2001. Temporal and spatial coding in the rat vibrissal system. Prog. Brain Res. 130, 75–88.

Ahrens, M.B., Li, J.M., Orger, M.B., Robson, D.N., Schier, A.F., Engert, F., et al., 2012. Brain-wide neuronal dynamics during motor adaptation in zebrafish. Nature 485 (7399), 471–477.

Ashby, W.R., 1952. Design for a Brain. Chapman & Hall, London.

Assa, E., Ahissar, E., 2015. Motor-sensory closed-loop perception (CLP) models and robotic implementation. In: Paper Presented at the IROS 2015 Workshop on Sensorimotor Contingencies for Robotics.

Bagdasarian, K., Szwed, M., Knutsen, P.M., Deutsch, D., Derdikman, D., Pietr, M., et al., 2013. Pre-neuronal morphological processing of object location by individual whiskers. Nat. Neurosci. 16 (5), 622–631.

Bobrov, E., Wolfe, J., Rao, R.P., Brecht, M., 2014. The representation of social facial touch in rat barrel cortex. Curr. Biol. 24 (1), 109–115.

Brecht, M., Schneider, M., Sakmann, B., Margrie, T.W., 2004. Whisker movements evoked by stimulation of single pyramidal cells in rat motor cortex. Nature 427, 704–710.

Britten, K.H., Shadlen, M.N., Newsome, W.T., Movshon, J.A., 1992. The analysis of visual motion: a comparison of neuronal and psychophysical performance. J. Neurosci. 12 (12), 4745–4765.

Canfield, J.G., Mizumori, S.J., 2004. Methods for chronic neural recording in the telencephalon of freely behaving fish. J. Neurosci. Methods 133 (1), 127–134.

Chen, J.L., Margolis, D.J., Stankov, A., Sumanovski, L.T., Schneider, B.L., Helmchen, F., 2015. Pathway-specific reorganization of projection neurons in somatosensory cortex during learning. Nat. Neurosci. 18, 1101–1108.

Dewey, J., 1896. The reflex arc concept in psychology. Psychol. Rev. 3 (4), 357.

Diamond, M.E., von Heimendahl, M., Knutsen, P.M., Kleinfeld, D., Ahissar, E., 2008. 'Where' and 'what' in the whisker sensorimotor system. Nat. Rev. Neurosci. 9 (8), 601–612.

Dombeck, D.A., Harvey, C.D., Tian, L., Looger, L.L., Tank, D.W., 2010. Functional imaging of hippocampal place cells at cellular resolution during virtual navigation. Nat. Neurosci. 13 (11), 1433–1440.

Fassihi, A., Akrami, A., Esmaeili, V., Diamond, M.E., 2014. Tactile perception and working memory in rats and humans. Proc. Natl. Acad. Sci. 111 (6), 2331–2336.

Friston, K., 2010. The free-energy principle: a unified brain theory? Nat. Rev. Neurosci. 11 (2), 127–138.

Gamzu, E., Ahissar, E., 2001. Importance of temporal cues for tactile spatial-frequency discrimination. J. Neurosci. 21 (18), 7416–7427.

Gibson, J.J., 1962. Observations on active touch. Psychol. Rev. 69, 477–491.

Heiligenberg, W., 1991. Neural Nets in Electric Fish. MIT Press, Cambridge, MA.

Held, R., Hein, A., 1963. Movement-produced stimulation in the development of visually guided behavior. J. Comp. Physiol. Psychol. 56 (5), 872.

Horev, G., Saig, A., Knutsen, P.M., Pietr, M., Yu, C., Ahissar, E., 2011. Motor-sensory convergence in object localization: a comparative study in rats and humans. Philos. Trans. R. Soc. Lond. B Biol. Sci. 366 (1581), 3070–3076.

Hubel, D.H., Wiesel, T.N., 1962. Receptive fields, binocular interaction and functional architecture in the cat visual cortex. J. Physiol. 160, 106–154.

James, W., 1890. The Principles of Psychology. Holt, Reinhart and Winston, New York.

Jarvilehto, T., 1999. The theory of the organism-environment system: III. Role of efferent influences on receptors in the formation of knowledge. Integr. Physiol. Behav. Sci. 34 (2), 90–100.

Jun, J.J., Longtin, A., Maler, L., 2014. Long-term behavioral tracking of freely swimming weakly electric fish. J. Vis. Exp. 85, e50962.

Keller, G.B., Bonhoeffer, T., Hübener, M., 2012. Sensorimotor mismatch signals in primary visual cortex of the behaving mouse. Neuron 74 (5), 809–815.

Kelso, J.S., 1997. Dynamic Patterns: The Self-Organization of Brain and Behavior. MIT Press, Cambridge, MA.

Kleinfeld, D., Ahissar, E., Diamond, M.E., 2006. Active sensation: insights from the rodent vibrissa sensorimotor system. Curr. Opin. Neurobiol. 16 (4), 435–444.

Knutsen, P.M., Derdikman, D., Ahissar, E., 2005. Tracking whisker and head movements in unrestrained behaving rodents. J. Neurophysiol. 93 (4), 2294–2301.

Knutsen, P.M., Pietr, M., Ahissar, E., 2006. Haptic object localization in the vibrissal system: behavior and performance. J. Neurosci. 26 (33), 8451–8464.

Konig, P., Luksch, H., 1998. Active sensing—closing multiple loops. Z. Naturforsch. C 53 (7–8), 542–549.

Land, M.F., 2006. Eye movements and the control of actions in everyday life. Prog. Retin. Eye Res. 25 (3), 296–324.

Lederman, S.J., Klatzky, R.L., 1987. Hand movements: a window into haptic object recognition. Cogn. Psychol. 19 (3), 342–368.

Marom, S., 2010. Neural timescales or lack thereof. Prog. Neurobiol. 90 (1), 16–28.

Marom, S., 2015. Science, Psychoanalysis, and the Brain. Cambridge University Press, Cambridge.

Marom, S., Wallach, A., 2011. Relational dynamics in perception: impacts on trial-to-trial variation. Front. Comput. Neurosci. 5, 16.

Marr, D., 1982. Vision. W. H. Freeman, San Francisco, CA.

McDermott, J.H., Schemitsch, M., Simoncelli, E.P., 2013. Summary statistics in auditory perception. Nat. Neurosci. 16 (4), 493–498.

Micheyl, C., Xiao, L., Oxenham, A.J., 2012. Characterizing the dependence of pure-tone frequency difference limens on frequency, duration, and level. Hear. Res. 292 (1), 1–13.

Mountcastle, V.B., Lynch, J., Georgopoulos, A., Sakata, H., Acuna, C., 1975. Posterior parietal association cortex of the monkey: command functions for operations within extrapersonal space. J. Neurophysiol. 38 (4), 871–908.

Noble, D., 2008. The Music of Life: Biology Beyond Genes. Oxford University Press, Oxford.

O'Keefe, J., Dostrovsky, J., 1971. The hippocampus as a spatial map. Preliminary evidence from unit activity in the freely-moving rat. Brain Res. 34 (1), 171–175.

Oram, T., Ahissar, E., Yizhar, O., 2015. Head-motion modulation of the activity of optogenetically tagged neurons in the vibrissal thalamus. In: Paper Presented at the SfN Annual Meeting, Chicago.

O'Regan, J.K., Noe, A., 2001. A sensorimotor account of vision and visual consciousness. Behav. Brain Sci. 24 (5), 939–973. discussion 973–1031.

Packer, O., Williams, D.R., 1992. Blurring by fixational eye movements. Vis. Res. 32, 1931–1939.

Port, R.F., Van Gelder, T., 1995. Mind as Motion: Explorations in the Dynamics of Cognition. MIT Press, Cambridge, MA.

Powers, W.T., 1973. Feedback: beyond behaviorism. Science 179 (71), 351–356.

Prescott, T.J., Diamond, M.E., Wing, A.M., 2011. Active touch sensing. Philos. Trans. R. Soc. Lond. B Biol. Sci. 366 (1581), 2989–2995.

Romo, R., Salinas, E., 2003. Flutter discrimination: neural codes, perception, memory and decision making. Nat. Rev. Neurosci. 4 (3), 203–218.

Rosen, R., 1991. Life Itself: A Comprehensive Inquiry into the Nature, Origin, and Fabrication of Life. Columbia University Press, New York.

Rucci, M., Iovin, R., Poletti, M., Santini, F., 2007. Miniature eye movements enhance fine spatial detail. Nature 447 (7146), 851–854.

Saig, A., Gordon, G., Assa, E., Arieli, A., Ahissar, E., 2012. Motor-sensory confluence in tactile perception. J. Neurosci. 32 (40), 14022–14032.

Saraf-Sinik, I., Assa, E., Ahissar, E., 2015. Motion makes sense: an adaptive motor-sensory strategy underlies the perception of object location in rats. J. Neurosci. 35 (23), 8777–8789.

Schroeder, C.E., Wilson, D.A., Radman, T., Scharfman, H., Lakatos, P., 2010. Dynamics of active sensing and perceptual selection. Curr. Opin. Neurobiol. 20 (2), 172–176.

Shahaf, G., Eytan, D., Gal, A., Kermany, E., Lyakhov, V., Zrenner, C., et al., 2008. Order-based representation in random networks of cortical neurons. PLoS Comput. Biol. 4 (11), e1000228.

Simony, E., Bagdasarian, K., Herfst, L., Brecht, M., Ahissar, E., Golomb, D., 2010. Temporal and spatial characteristics of vibrissa responses to motor commands. J. Neurosci. 30 (26), 8935–8952.

Szwed, M., Bagdasarian, K., Ahissar, E., 2003. Encoding of vibrissal active touch. Neuron 40 (3), 621–630.

Tammero, L.F., Dickinson, M.H., 2002. Collision-avoidance and landing responses are mediated by separate pathways in the fruit fly, *Drosophila melanogaster*. J. Exp. Biol. 205 (18), 2785–2798.

Tanaka, Y., Tiest, W.M.B., Kappers, A.M., Sano, A., 2014. Contact force and scanning velocity during active roughness perception. PLoS ONE 9 (3), e93363.

Wallach, A., Bagdasarian, K., Ahissar, E., 2016. On-going computation of whisking phase by mechanoreceptors. Nat. Neurosci. 19 (3), 487–493.

Wiener, N., 1949. Cybernetics. John Wiley & Sons, New York.

Wu, C.-T., Crouzet, S.M., Thorpe, S.J., Fabre-Thorpe, M., 2014. At 120 msec you can spot the animal but you don't yet know it's a dog. J. Cogn. Neurosci. 27 (1), 141–149.

Yarbus, A.L., 1967. Eye Movements and Vision. Plenum, New York.

Yartsev, M.M., Ulanovsky, N., 2013. Representation of three-dimensional space in the hippocampus of flying bats. Science 340 (6130), 367–372.

Chapter 8

Stochastic Optimal Control of Spike Times in Single Neurons

A. Iolov*,†, S. Ditlevsen† and A. Longtin*

*University of Ottawa, Ottawa, Canada; †University of Copenhagen, Copenhagen, Denmark

1 INTRODUCTION

Controlling or changing the spiking activity in single neurons through electrical stimulation has applications in brain-machine interfaces or neuroprosthetics. The goals may be many. Maybe it is desired that a neuron follow a specific fixed firing pattern (Feng and Tuckwell, 2003; Moehlis et al., 2006; Ahmadian et al., 2011; Dasanayake and Li, 2011; Nabi et al., 2013a,b; Iolov et al., 2014b). Alternately, one may wish that interspike intervals attain either minimum or maximum lengths (Danzl et al., 2009; Nabi and Moehlis, 2012; Wu et al., 2008). A pathological synchronous firing pattern in clusters of neurons may need to be broken (Nabi et al., 2013a,b; Nabi and Moehlis, 2011; Feng et al., 2007), which is relevant for neurological disorders such as epilepsy and Parkinson's Disease. The goal may be simpler, namely, to reveal important information about the neuronal subthreshold dynamics leading to spiking. The variable to be controlled may be a firing rate, or a precise firing time. The ability to do one or the other will depend on the underlying mathematical model of the neural system of interest. For example, in Parkinson's and epilepsy research, some studies have focussed on controlling models for the evolution of the firing rate such as Wilson-Cowan (see Schiff, 2012 and references therein). Another field of potential application is the accumulation of evidence prior to a response or decision, where drift-diffusion processes to a threshold are of interest (Bogacz et al., 2006).

In this chapter, we will explain how methods from stochastic optimal control can be implemented in closed-loop, ie, when the controller has running access to measurements of the membrane potential. A major obstacle in these types of problems is that the electrical activity of single neurons is noisy, and the presence of nonnegligible noise makes the control problem much harder. In particular, the accuracy of the controlled spike times degrades with increasing intensity of the noise. Here, we will only consider targeting fixed spike times using a cost function proportional to the square of the deviation of the spike time from the desired one (see the following section); but the methodology can be applied more generally by redesigning the cost function for a specific problem.

The model for the membrane potential is given by a current balance equation of the form

$$C\frac{dV}{dt} = \text{sum of currents} + \text{noise}$$

where C is the cell membrane capacitance, and V is the membrane potential evolution. Sometimes the constant C is not specifically stated and absorbed into other parameters. The currents are ions, such as sodium, potassium, calcium, and chloride, flowing in and out of the cell through ion channels in the cell membrane, as well as input currents received from other neurons in the surrounding network, or injected current controlled by the experimentalist. The noise models the inherent stochasticity of neural activity. While all ionic currents have an associated noise, eg, caused by thermal and channel conductance fluctuations, for simplicity here we model the noise as an additive term on the current balance equation.

When trying to control the evolution of V a simple model is needed. Firstly, calculations have to be done online as measurements of V become available, in order to update the control on the fly. This is only feasible for relatively simple models. Secondly, and more importantly, more realistic biophysical models are highly nonlinear, which makes the system much more sensitive to model misspecification, jeopardizing the success of the control. A simple model will be more robust, and the online control will correct for possible errors, so they do not accumulate. We therefore only consider the class of integrate-and-fire (LIF) models, where the dynamics of the membrane potential between two consecutive neuronal firings are represented by a scalar drift-diffusion process $X = \{X_t\}_{t\geq 0}$ indexed by the time t, and given by the Itô-type stochastic differential equation (SDE)

$$dX_t = F(X_t, t; \alpha)dt + G(X_t, t)dW_t, \quad X_0 = x_0 \quad (1a)$$

where $F(\cdot)$ is the drift coefficient function, parameterized by the control $\alpha(t)$, and $G(\cdot)$ is the diffusion coefficient

Closed Loop Neuroscience. http://dx.doi.org/10.1016/B978-0-12-802452-2.00008-1

function. It is also possible to consider a control in the diffusion coefficient G, if this is more relevant for a given application; we will confine our attention to a control in the drift term. Here, W_t is a standard Wiener process, a continuous random process with independent and zero-mean normally distributed increments. For the standard leaky LIF model, we set $F(X_t, t; \alpha) = -\frac{X_t}{\tau} + \mu + \alpha(t)$ and $G(X_t, t) = \beta$. This is simply an Ornstein-Uhlenbeck process for the voltage, where μ is a bias current acting on the cell, τ is the decay time, and β is the strength of the stochastic fluctuations.

We consider the possibility of dynamically changing the external control current $\alpha(t)$ impinging on the neuron based on feedback from online measurements of the membrane potential. A simplified picture of the effect of such a control is illustrated in Fig. 1 for the case of an Ornstein-Uhlenbeck model for the membrane voltage X. There the control successively takes on four different constant values. One observes that each control value serves to bias the drift-diffusion process, making it head upwards or downwards, depending on the current control value.

Then the problem is to design the control $\alpha(t)$ such that the spike times are as close as possible to a target spike train. Further, constraints need to be imposed on the control to ensure that the cell is kept in a healthy biophysical environment. The most common is a simple box constraint, where $\alpha(t) \in [\alpha_{\min}, \alpha_{\max}]$ for all t, typically with $\alpha_{\min} = -\alpha_{\max}$. To control a system under higher noise comes at the price of allowing a more powerful control, ie, by allowing a larger $\alpha_{\max}$. Below we will sometimes write $\alpha_t = \alpha(t)$.

We will write $f(x, t | y, s)$ for the transition density function, ie, the probability density function for the process X to be at x at time t given that the process was initially at y at time $s < t$. To ease notation we will often just write $f(x, t) = f(x, t | y = x_0, s = 0)$ if the initial conditions are clear from the context. Then f satisfies the Fokker-Planck equation,

$$\partial_t f(x, t) = -\partial_x (F(x, t; \alpha) f(x, t)) + \partial_x^2 \left(\frac{G^2(x, t)}{2} f(x, t) \right) \quad (1b)$$

Spike events or action potentials are produced when X exceeds an upper voltage threshold x_{th} for the first time. Formally, the spike time T is then identified with the first-passage time of the threshold,

$$T = \inf \{ t > 0 | X_t \geq x_{\text{th}} \} \quad (1c)$$

where after X is reset to x_0. It follows that if X is time-homogenous, the interspike intervals form a renewal process, and the initial time following a spike can always be identified with 0. The spike threshold manifests itself as an absorbing boundary condition in Eq. (1b) by setting $f(x_{\text{th}}, t) = 0$.

2 STOCHASTIC OPTIMAL CONTROL

Optimal Control Theory has three main components—a state, x, a control, α, and an objective J, which is a functional of (x, α), and which we try to either minimize or maximize. In *stochastic* optimal control, x follows a stochastic process, and the objective J is most often expressed in terms of some average or expectation over the random realizations of x. The general theory of optimal control, and in particular the special case dealing with random systems, has two main analytical techniques—the *maximum principle* and *dynamic programming*. A classic reference is Fleming and Rishel (1975). The maximum principle uses a variational approach to characterize the optimal pair, $(x_{\text{opt}}, \alpha_{\text{opt}})$, optimizing J, while dynamic programming recursively builds up the optimal solution with a backward induction from the terminal conditions. Both techniques have their advantages and disadvantages. It turns out that there is a close relation between the two approaches, which is well known in the deterministic finite-dimensional case (Fleming and Rishel, 1975; Evans, 1983).

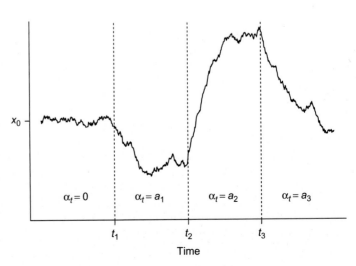

FIG. 1 Illustration of externally controlled dynamics with piecewise constant controls. The voltage evolves according to the Ornstein-Uhlenbeck process, Eq. (1a), with $F(X_t, t; \alpha) = -\frac{X_t}{\tau} + \mu + \alpha(t)$ and $G(X_t, t) = \beta$, where $\tau = 10$, $\mu = 1$, and $\beta = 0.3$. Up to time t_1 the dynamics are evolving freely with $\alpha(t) = 0$ from initial condition $X_0 = 10$. Between time t_1 and t_2 a constant control of size $\alpha(t) = a_1 = -0.5$ is applied. At time t_2 the control changes to $a_2 = -1.0$, and again at time t_3 the control changes to $a_3 = -0.2$. The integration time step is $\Delta = 0.1$, and 1000 points have been generated here drawn from the exact distribution, which is a Gaussian distribution with conditional mean $E(X_{t_{i+1}} | X_{t_i} = x) = xe^{-\Delta/\tau} + (\mu + \alpha(t_i))\tau \left(1 - e^{-\Delta/\tau} \right)$ and conditional variance $V(X_{t_{i+1}} | X_{t_i} = x) = \beta^2 \tau / 2 \left(1 - e^{-2\Delta/\tau} \right)$. The goal below is to link the values of the control, here chosen arbitrarily, to the time evolution of the process to achieve a specific spike time target.

Consider the following functional

$$J(\alpha) = E\left(\int_0^{t_f} L(X_s, \alpha_s) ds + M\left(X_{t_f}\right) \right) \qquad (2)$$

where the expectation is taken over realizations of X_t governed by an Itô SDE as in Eq. (1a). Here L is a running cost function, which depends on the trajectory and the applied control, while M is the terminal cost function, which only depends on the value of the trajectory at the terminal time, t_f. We assume that t_f is fixed a priori, although this is not necessary in general. We seek α^*, which minimizes J in Eq. (2),

$$\alpha^* = \arg \min_{\alpha \in \mathcal{U}} J(\alpha) \qquad (3)$$

where the optimization in Eq. (3) is done over the space $\mathcal{U}$ of stochastic processes that are measurable with respect to the filtration generated by the underlying Wiener process, W_t. Further constraints on $\mathcal{U}$ may be imposed by a specific problem. The optimization could of course also be a maximization, depending on the specific problem and formulation of $J(\alpha)$.

The simple-looking Eq. (2) can give rise to different variations. For example, the final time t_f may be variable, or we can impose path or terminal constraints for X_t. In the stochastic context such constraints are only enforceable in a probabilistic sense and they can easily make the problem much more difficult. However, we will not deal with either of these complications and in the sequel, we assume that t_f is fixed and that there are no constraints on X_t other than that it satisfies an SDE.

Given the objective and the optimization equations (2) and (3), the maximum principle relies on the forward Kolmogorov (Fokker-Planck) equation and the expectation of the SDE, while dynamic programming relies on the relation between an SDE expectation and the backward Kolmogorov equation.

3 THE MAXIMUM PRINCIPLE FOR TRANSITION PROBABILITY DENSITY PARTIAL DIFFERENTIAL EQUATIONS

The maximum principle uses a variational principle to characterize the optimal control, α, and the corresponding transition density. The expectation in the objective, Eq. (2), can be written in terms of the transition probability density $f(\cdot)$ of the state X_t,

$$J(\alpha) = \int_0^{t_f} \int_x L(x, \alpha_s) f(x, s | x_0, 0) dx ds + \int_x M(x) f\left(x, t_f | x_0, 0\right) dx \qquad (4)$$

The stochastic problem is therefore reduced to a deterministic optimization problem, but for partial differential equations (PDEs), given that f satisfies the forward PDE in Eq. (1b). Thus, the maximum principle for stochastic optimal

control is in this context really a maximum principle for PDEs. Its variational argument is similar in spirit to the Euler-Lagrange equations from the calculus of variations; one can think of it as a generalization of the zero-tangent rule (Fermat's Rule) for finding optima in single-variable calculus. In the same way as both the Euler-Lagrange equations and Fermat's Rule, the maximum principle provides necessary, but not sufficient, conditions for an optimum.

Originally, the maximum principle was developed for finite-dimensional deterministic systems, ie, systems described by ordinary differential equations (ODEs). In that context it is known as the *Pontryagin* maximum principle and for ODEs, the theoretical results of existence and uniqueness of optimal controls are strongest. In the infinite dimensional case, for PDEs, generalization of the maximum principle is discussed in many books (see Fattorini, 1999; Borzì and Schulz, 2012; Lenhart and Workman, 2007). Controls of the Fokker-Planck PDE is a special case of the general theory and recently there has been a series of publications on PDE control applied to the Fokker-Planck equation (see Annunziato and Borzì, 2010, 2013; Annunziato et al., 2014). In particular, Annunziato et al. (2014) discusses the relation between the dynamic programming approach to stochastic control and the approach based on PDE optimization of the Fokker-Planck equation.

The basic idea of the maximum principle is to adjoin the PDE of the dynamics to the objective and introduce a Lagrange multiplier, which in this case is called the *adjoint state*. We illustrate with a one-dimensional SDE, as in Eq. (1a). Its forward density is governed by the Fokker-Planck equation, Eq. (1b). We assume that for all $\alpha(t)$, both the SDE and the PDE have unique solutions. For notational convenience, we will write Eq. (1b) as

$$\partial_t f = \mathcal{L}_\alpha f \qquad (5)$$

where $\mathcal{L}_\alpha$ is the differential operator corresponding to the right-hand side of Eq. (1b). For now we will assume that there are no boundary conditions, and the domain of X is all of $\mathbb{R}$.

The key concept in the maximum principle for PDEs is to adjoin the dynamics, Eq. (1b), multiplied by the adjoint state, p, to the objective, Eq. (4),

$$J(\alpha) = \int_0^{t_f} \int_x L(x, \alpha_s) f(x, s | x_0, 0) dx ds + \int_x M(x) f\left(x, t_f | x_0, 0\right) dx$$
$$- \int_0^{t_f} \int_x p(x, s)(\partial_s f(x, s) - \mathcal{L}_\alpha f(x, s)) dx ds$$

Since f satisfies Eq. (5), the last term equals zero. However, this allows us to transfer the time and space derivatives from f to p. We will then be able to form the "variation" of J with respect to the control α and either set it to zero or use this as a gradient in an optimization procedure.

To illustrate, let us perform this transfer explicitly. It is simply an application of integration-by-parts; in larger dimension this is also commonly called *Green's identities*. We have that

$$\int_0^{t_f} p\partial_s f ds = pf|_0^{t_f} - \int_0^{t_f} \partial_s pf ds$$

which transfers the time-derivative to p. Moreover, assuming that the order of integration can be interchanged, and that f and all its partial derivatives go to zero uniformly for $|x|$ large enough, we obtain

$$\int_x p\mathcal{L}_\alpha f dx = \int_x \left(F\partial_x p + \frac{G^2}{2}\partial_x^2 p\right) f dx = \int_x (\mathcal{L}_\alpha^* p) f dx$$

which transfers the space-derivative to p. Boundary conditions on f would appear in the spatial terms. The differential operator $\mathcal{L}^*$ is the adjoint, in a Banach-space sense, to $\mathcal{L}$ in Eq. (5). Actually, $\mathcal{L}^*$ is just the generator of the SDE in Eq. (1a). Using the above integration-by-parts, the objective can be written as

$$J(\alpha) = \int_0^{t_f} \int_x L(x, \alpha_s) f(x, s|x_0, 0) dx ds + \int_x M(x) f\left(x, t_f|x_0, 0\right) dx \\ - \int_x \left(pf|_0^{t_f} - \int_0^{t_f} (\partial_s p + \mathcal{L}_\alpha^* p) f ds\right) dx \qquad (6)$$

Now assume a small variation around a given control α, also affecting the transition density f,

$$\alpha + \epsilon\delta\alpha \quad \text{and} \quad f + \epsilon\delta f$$

Then the variation of the objective J, with respect to the control variation of α, can be calculated by

$$\frac{dJ}{d\epsilon}\Big|_{\epsilon=0} = \int_0^{t_f} \int_x (\nabla_\alpha L(x, \alpha_s)\delta\alpha f(x, s) + L(x, \alpha_s)\delta f(x, s)) dx ds \\ + \int_x M(x)\delta f\left(x, t_f\right) dx - \int_x p\left(x, t_f\right)\delta f\left(x, t_f\right) dx \qquad (7) \\ + \int_x \int_0^{t_f} [(\partial_s p(x, s) + \mathcal{L}_\alpha^* p)\delta f(x, s) + \nabla_\alpha F(x, s; \alpha)\delta\alpha\partial_x p(x, s) f(x, s)] ds dx$$

A few notes are required to explain Eq. (7). The initial conditions are fixed and therefore $\delta f|_{t=0} \equiv 0$, that is, the variation in the control does not change $f(x, 0)$. Thus, only the term $pf|_{t=t_f}$ is retained from Eq. (6).

Heuristically, we would like to infer from Eq. (7) a gradient with respect to the control. However, this is impossible since there are also variations with respect to f. Recall that p is the free variable—the Lagrange multiplier. Thus, choosing p appropriately can eliminate δf from the expression. Let p evolve (backward) according to:

$$\begin{cases} -\partial_t p = \mathcal{L}^* p + L \\ p\left(x, t_f\right) = M(x) \end{cases} \qquad (8)$$

Then Eq. (7) simplifies to

$$\frac{dJ}{d\epsilon}\Big|_{\epsilon=0} = \int_0^{t_f} \int_x (\nabla_\alpha L(x, \alpha_s) f + \nabla_\alpha F(x, s; \alpha)\partial_x pf)\delta\alpha dx ds \qquad (9)$$

From Eq. (9) it follows that

$$\nabla_\alpha J(x, t) = \nabla_\alpha L(x, \alpha_t) f(x, t) + \nabla_\alpha F(x, t; \alpha_t)\partial_x p(x, t) f(x, t) \qquad (10)$$

can be considered a pointwise gradient of the objective with respect to changes in the control, given the current control. It is then natural to expect that setting Eq. (10) equal to zero will provide a necessary condition for an optimal α. If equating Eq. (10) to zero and solving for α is possible, we would then have an expression for α in terms of the state, f, and the adjoint p; the equations for these two latter functions, with α given in terms of f and p, could then be solved. This explicit representation of α is not always possible for general cost/rewards, L, or drift fields, F. Even if it is possible, the pair of now nonlinear forward/backward equations for f and p still have to be solved. It should be clear by now why the maximum principle theory for PDEs faces many practical challenges.

There are two iterative approaches to obtain numerical results given Eq. (10). If α can be expressed in terms of f and p, then a nonlinear forward-backward system can be obtained, by replacing α in the forward and backward equations for f and p, respectively. Unfortunately, due to the coupled nature of the equations, this is usually not possible to solve explicitly. Thus, iterative methods are required. These come in two main flavors; the gradient descent methods, and fixed point iterations. The fixed point approach is advocated in Lenhart and Workman (2007), but only for simple pedagogic examples. Alternatively, a gradient-based optimization method can be applied. Since Eq. (10) provides a gradient, the current control α can be incremented in the direction of $\nabla_\alpha J$. However, usually the objective is not convex with respect to the control, and as with all gradient-based optimization methods, the standard disclaimer about local minima and optimization initialization applies. As a minimum requirement, the procedure should be run from different initializations to explore if a given minimum is local or whether it might be global. More rigorous treatments can for example be found in Fattorini (1999), Borzì and Schulz (2012), and Ahmed and Teo (1981).

4 DYNAMIC PROGRAMMING FOR STOCHASTIC OPTIMAL CONTROL

Dynamic programming uses backward recursion to tabulate the optimal control starting from the terminal time. The basic object in dynamic programming is the value function, v. First the definition of the objective J is extended to consider a later starting time $t > 0$, and allowing the state X_t to have different initial conditions. The running cost-to-go corresponding to Eq. (2) becomes

$$J(\alpha; x, t) = E\left(\int_t^{t_f} L(X_s, \alpha_s) ds + M\left(X_{t_f}\right)\right), X_t = x \qquad (11a)$$

Thus, $J(\alpha; x_0, 0) = J(\alpha)$ is the original objective from Eq. (2).

Now introduce the *value function, v*, also called the optimal cost-to-go, defined by:

$$v(x,t) = \inf_{\alpha \in \mathcal{U}} J(\alpha; x, t) \qquad (11b)$$

The terminal conditions on v are

$$v(x, t_f) = M(x)$$

since if $X_{t_f} = x$, there is no more time for a control to be applied and no more running cost, and we just incur the terminal cost corresponding to wherever X is now, ie, $M(x)$. It turns out that the value function, v, can be characterized as the solution to the following nonlinear PDE:

Theorem 1 *The value function v is solution to the Hamilton-Jacobi-Bellman equation*

$$\begin{cases} -\partial_t v(x,t) = \min_{\alpha \in \mathcal{U}} (L(x,\alpha) + F(x,t;\alpha)\partial_x v) + \dfrac{G^2(x,t)}{2}\partial_x^2 v \\ v(x, t_f) = M(x) \end{cases}$$

$$(12)$$

A heuristic derivation of Theorem 1, adapted from Evans (1983), is as follows. Suppose at time t that $X_t = x$. Consider a time increment $(t, t+h)$ and assume that during this time a constant control α is applied, and subsequently the optimal control α^* is applied. The running cost then decomposes as

$$J(\alpha; x, t) = E\left(\int_t^{t+h} L(X_s, \alpha_s) ds + v(X_{t+h}, t+h) \right)$$

Since $v(x,t) = \inf_\alpha J(\alpha; x, t)$, ie, ν is an infimum over J, then $v(x,t) \leq J(\alpha; x, t)$ for all J.

Rearranging, we obtain

$$0 \leq E\left(\int_t^{t+h} L(X_s, \alpha_s) ds \right) + E(v(X_{t+h}, t+h) - v(x,t))$$

Using Dynkin's Formula, the last term on the right-hand side can be expressed as

$$E(v(X_{t+h}, t+h) - v(x,t)) = \int_t^{t+h} \left(\partial_t v + F \partial_x v + \frac{G^2}{2}\partial_x^2 v \right) ds$$

and we get

$$0 \leq E\left(\int_t^{t+h} \left(L(X_s, \alpha_s) + \partial_t v(x,s) + F(x,s;\alpha_s)\partial_x v(x,s) + \frac{G^2(x,s)}{2}\partial_x^2 v(x,s) \right) ds \right)$$

Letting $h \to 0$ we obtain

$$0 \leq L(x,\alpha) + \partial_t v(x,t) + F(x,t;\alpha)\partial_x v(x,t) + \frac{G^2(x,t)}{2}\partial_x^2 v(x,t)$$

We conjecture that for the optimal control, the inequality becomes an equality:

$$\begin{cases} 0 = L(x,\alpha^*) + \partial_t v(x,t) + F(x,t;\alpha^*)\partial_x v(x,t) + \dfrac{G^2(x,t)}{2}\partial_x^2 v(x,t) \\ \alpha^*(t) = \arg \min_{\alpha \in [\alpha_{\min}, \alpha_{\max}]} (L(x,\alpha) + F(x,t;\alpha)\partial_x v(x,t)) \end{cases} \quad (13)$$

A rigorous proof of Theorem 1 can, for example, be found in Krylov (2008) and Fleming and Soner (2006). We will assume that solving Eq. (12) numerically is sufficient.

5 NUMERICAL SOLUTIONS TO PDEs—FINITE DIFFERENCE METHODS

Both the variational approach of the maximum principle and the backward induction of dynamic programming result in having to solve PDEs in order to obtain the optimal control. For practical purposes, solving these PDEs requires numerical discretization. In all cases the PDEs are *parabolic* (see Press et al., 1992), which in general can be written as

$$\partial_t f = \Gamma f(x,t) \qquad (14)$$

for some differential operator, Γ. In one spatial dimension, a finite difference discretization of Eq. (14) is to select a set of space- and time-nodes $\{x_i\}, \{t_k\}$ and approximate f by solving for $f(x_i, t_k)$, given the initial conditions $f(x_i, t_0)$ and possible boundary conditions.

A classic technique for parabolic PDEs is the *Crank-Nicholson* scheme, which time discretizes Eq. (14) as

$$\frac{f(x_i, t_{k+1}) - f(x_i, t_k)}{t_{k+1} - t_k} = \frac{\Gamma f(x_i, t_{k+1}) + \Gamma f(x_i, t_k)}{2}$$

and then solves for the resulting linear system before stepping iteratively forward in time. See Chapter 19 in Press et al. (1992) for a brief discussion on the theoretical properties of the Crank-Nicholson method.

This solution approach requires a finite spatial domain. If the spatial domain is infinite, it has to be truncated and some reasonable boundary conditions applied at the artificial boundary which approximates the solution of the infinite space.

6 APPLICATION TO THE ORNSTEIN-UHLENBECK LEAKY INTEGRATE-AND-FIRE MODEL

To illustrate the theory above, we will now use dynamic programming on the Ornstein-Uhlenbeck model to target predefined spike times. Let

$$F(x,t;\alpha) = -\frac{x}{\tau} + \mu + \alpha(t) \quad \text{and} \quad G(x,t) = \beta$$

as above in the model given by Eqs. (1a), (1c). The goal is to control the spike time given in Eq. (1c), ie, to choose a control $\alpha(t)$ such that $T \approx t^*$ for some target time t^*. We therefore have to choose some cost function that measures the distance between the realized and the desired spike time. A natural optimal control objective would be the least squares solution

$$\alpha(\cdot) = \arg \min_{\alpha(\cdot)} E(T - t^*)^2 \qquad (15)$$

where the expectation is taken with respect to the distribution of the trajectories of X. Furthermore, we impose symmetric box constraints, $\alpha(t) \in [-\alpha_{max}, \alpha_{max}]$. We will also add to our objective a running energy cost based on the control. This regularizes the problem by eliminating the subtleties of singular-control situations; it serves to avoid excessive control as well as to avoid excessive charge building up on the cell (see Ahmadian et al., 2011). As the objective given in Eq. (2) we take the function

$$J(\alpha(\,\cdot\,)) = E\left(\varepsilon \int_0^T \alpha^2(s)\,ds + (T - t^*)^2 \right)$$

where ε measures how much weight we put on minimizing the energy cost. If $\varepsilon = 0$ it reduces to Eq. (15). The optimal control α^* is then given by Eq. (3). In the closed-loop setting, it is assumed that the controller has online access to the value of the stochastic process X, and it is natural to take the dynamic programming approach. Given $X_t = x$ at time t, let $(T - t)$ be the remaining unknown time before the spike occurs. Then the remaining cost objective in Eq. (11a) becomes

$$J(\alpha(\,\cdot\,); x, t) = E\left(\varepsilon \int_t^T \alpha^2(s)\,ds + ((T - t) - (t^* - t))^2 \,|\, X_t = x \right)$$

That is, if time t has elapsed without a spike occurring, we now want to minimize the difference between $(T - t)$ and $(t^* - t)$, given the current state x. A similar problem is discussed analytically at length in the book on optimal control by Whittle (1996), but with no discussion of the numerics required to solve it.

The Hamilton-Jacobi-Bellman equation, Eq. (12), is obtained as follows. The value function $v(x, t)$ from Eq. (11b) is given by

$$v(x, t) = \min_{\alpha(\,\cdot\,) \in \mathcal{U}_t} J(\alpha(\,\cdot\,); x, t)$$

where $\mathcal{U}_t$ is the set of admissible control functions running from time t. Then v satisfies Eq. (13), which takes the form

$$\partial_t v(x, t) + \frac{\beta^2}{2}\partial_x^2 v(x, t) + \min_{\alpha(t) \in \mathcal{U}_0} \left(\varepsilon\alpha^2(t) + \left(\mu - \frac{x}{\tau} + \alpha(t) \right)\partial_x v(x, t) \right) = 0$$

The special feature of this equation, in contrast to a generic parabolic PDE, is that it contains an embedded optimization that depends on the solution, v. The optimal control, $\alpha^*(x, t)$, can be found analytically as

$$\alpha^*(x, t) = \arg\min_{\alpha \in \mathcal{U}_t} \left(\varepsilon\alpha^2 + \left(\mu - \frac{x}{\tau} + \alpha \right)\partial_x v \right)$$
$$= \min\left(\alpha_{max}, \max\left(\alpha_{min}, -\frac{\partial_x v}{2\varepsilon} \right) \right)$$

We need boundary conditions for v. If $X_t = x_{th}$ then the neuron spikes and $T = t$. Thus, at the threshold, the value function equals the squared difference between the desired spike time and the realized one,

$$v(x_{th}, t) = (t - t^*)^2$$

For large negative values of x, v should not be significantly affected by the change in x, ie, for some lower boundary x_- is $\partial_x v(x_-, t) = 0$. Such a boundary condition will be justified if we choose x_- such that the probability for the process to take values smaller than x_- is small.

It is important to realize that dynamic programming and the Hamilton-Jacobi-Bellman equation work backward. The evolution of the value function requires some terminal condition at some point in the future from which to start incrementing v using the dynamics and the boundary conditions. The idea is simple (Iolov et al., 2014b): if we reach t^* before spiking, we apply maximum control in the positive direction, ie, $\alpha(t) = \alpha_{max}$ for $t \geq t^*$. This yields

$$v(x, t^*) = E\left((T - t^*)^2 \,|\, X_{t_f} = x, \alpha(t) = \alpha_{max} \right) \quad (16)$$

This ignores the energy term, $\varepsilon\alpha^2$, in the objective for $t \geq t^*$, which is less crucial for $\varepsilon \ll 1$. Alternatively, this approximate terminal condition could be imposed at some $t_f > t^*$.

Eq. (16) is the second moment of the remaining time to reach the threshold starting at $X_{t^*} = x$ and applying α_{max} throughout. This quantity can be found easily for all x in the domain by solving a stationary backward Kolmogorov equation. Finally, the Hamilton-Jacobi-Bellman equation in its fully specified form is

$$\partial_t v(x, t) + \frac{\beta^2}{2}\partial_x^2 v(x, t) + \varepsilon\alpha^2(x, t) + \left(\mu - \frac{x}{\tau} + \alpha(x, t) \right)\partial_x v(x, t) = 0 \quad (17)$$

$$\alpha^*(x, t) = \min\left(\alpha_{max}, \max\left(\alpha_{min}, -\frac{\partial_x v(x, t)}{2\varepsilon} \right) \right)$$

$$\begin{cases} v(x_{th}, t) = (t - t^*)^2 & \text{upper boundary condition} \\ \partial_x v(x_-, t) = 0 & \text{lower boundary condition} \\ v(x, t^*) = \text{Eq. (16)} & \text{terminal condition} \end{cases}$$

We are solving $v(x, t)$ over the domain $[x_-, x_{th}] \times [0, t_f]$.

6.1 Numerical Method for Solving the Hamilton-Jacobi-Bellman Equation

Since the equation is one-dimensional in space, it is straightforward to apply the standard centered finite difference using the Crank-Nicholson scheme to step in time. The nonlinear term in the PDE, namely $(\partial_x v)^2$ appearing in α^2, can be treated as a mixed implicit-explicit term (Iolov et al., 2014b)

$$(\partial_x v(x, t_k))^2 \approx \partial_x v(x, t_k)\,\partial_x v(x, t_{k+1})$$

The implicit term is in the previous time t_k instead of, as is conventional, the next time t_{k+1}, because we are solving for v backward in time from t_{k+1} to t_k and down to $t_0 = 0$. The discretization in time and space should be chosen in relation to the parameter setting. The time discretization

Δx should be chosen such that numerically we are in the diffusion-dominated regime, by setting it less than β divided by the largest possible absolute value of the drift $\left(\mu - \dfrac{x}{\tau} + \alpha\right)$ for $\alpha \in [\alpha_{\min}, \alpha_{\max}]$, $x \in [x_-, x_{\text{th}}]$. We set Δt to be twice the value of Δx divided by the largest possible absolute value of $\left(\mu - \dfrac{x}{\tau} + \alpha\right)$.

While there are a number of parameter settings that can be investigated for controllability, it is useful to classify parameter space into the usual sub- vs suprathreshold regimes. In the former case, for zero control, the system needs noise in order to spike, ie, we have noise-activated spiking. This occurs for $\mu \le 1/\tau$. In the suprathreshold case $\mu > 1/\tau$, the system spikes periodically on its own, and the noise jitters the spike times around the deterministic period. Within this sub/supra division, one can also consider the high- and low-noise cases.

In Fig. 2 we show an example of the solution of Eq. (17), giving the value function v and the corresponding optimal control α for parameters corresponding to the supra-threshold high-noise regime. Other regimes (suprathreshold low noise and subthreshold low and high noise) can be found in Iolov et al. (2014b). Recall that lower values for the value function v are preferred, because we are minimizing. At the end of the interval, $v(x, t^*)$ is monotonically decreasing in x, which is to be expected given its terminal condition in Eq. (16). That is, the lower the value of X_{t^*}, the longer it takes for a spike to occur. On the other hand, as we go back in time v inverts, with a clear peak near x_{th}. That is, for intermediate values of t there is a risk of spiking too early, represented by the high value of v near the upper boundary, or the risk of spiking too late, represented by the high values of v at the lower boundary. As we progress even further back in time to the beginning of the interval, the peak near the threshold rises further, since we are spiking earlier, while the peak at the lower end flattens, because now there is enough time to reach threshold despite

starting far from it. The full surface plots for $v(x, t)$ in Fig. 2 can be thought of as the interpolation between these three basic phases.

Now let us focus on the control in Fig. 2. Naturally, at the end of the interval, the control takes its maximum value, $\alpha(x, t^*) = \alpha_{\max}$ for all x, ie, it gives the maximum available push for the neuron to spike. This is consistent with Eq. (16). However, going back in time the control α decreases for $x \ll x_{\text{th}}$, and it becomes negative, ie, inhibitory for $x \lesssim x_{\text{th}}$. That is intuitively consistent with the problem objective. For $t < t^*$, we want to bring X_t closer to the threshold, but not too close, to avoid early spiking.

The behavior of the closed-loop control is illustrated in Fig. 3 for two values of the energy penalty. In both top panels, we see that the control current has a net increasing trend, and becomes close to maximal near the desired spike time; this forces the deterministic dynamics + control to override the randomizing effect of the noise. The factor ε multiplies the integral of the square of the control over time, and thus determines the predominance in the cost function of this "energy" relative to the expected squared deviation between actual and desired spike time. The higher energy penalty yields an overall smoother control with lower energy; the lower penalty promotes control values that approach (and may alternate fairly wildly between) the allowed extrema of the control, which is reminiscent of *two-state bang-bang control*, defined as a control that switches abruptly between the two most extreme states.

Results on control of Eq. (1a) in Iolov et al. (2014b) reveal that it is easiest to control the system for low noise, as expected. It is easiest in the sense that, in this regime, the average (over realizations of the model) of the squared deviation between the desired and actual spike time is the lowest. The higher noise leads to much uncertainty that the control has a hard time overcoming as the system heads strongly toward its threshold. Furthermore, if as we discussed, below the target interspike interval is relatively

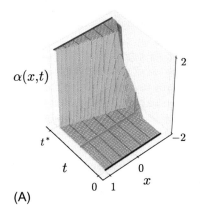
(A)

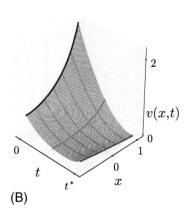
(B)

FIG. 2 (A) Plot of the control as a function of state and time. The target spike time is $t^* = 1.5$. The control is applied in the suprathreshold high-noise regime for which $\beta = 1.5$, $\tau = 0.5$, and $\mu = 3.0$. The threshold is set at $x_{\text{th}} = 1$. In the absence of noise and control, the model simply spikes periodically. (B) Numerically determined solution of the value function $v(x, t)$ in the Hamilton-Jacobi-Bellman partial differential equation corresponding to the control obtained in (A). The control bounds are $\alpha \in [-2, 2]$. In both (A) and (B), the colored lines on the surface correspond to, respectively, the state-dependent control and value function at three different times: $t = 0$ (*blue*), $t = t^*/2$ (*green*), and $t = t^*$ (*red*). The energy penalty is chosen as $\varepsilon = 0.001$.

FIG. 3 An example of closed-loop stochastic optimal control for Eq. (1a) with absorbing ("spiking") threshold at $x_{th} = 1$. The goal is for the integrate-and-fire model to fire at the target spike time $t^* = 1.5$. The chosen regime is again the suprathreshold high-noise regime with the same parameters as in Fig. 2. *Top panels* illustrate the time course of the voltage, while the bottom plots show the corresponding control. The energy penalty is chosen as $\varepsilon = 0.001$ for the left plots (A) and $\varepsilon = 0.1$ for the right plots (B).

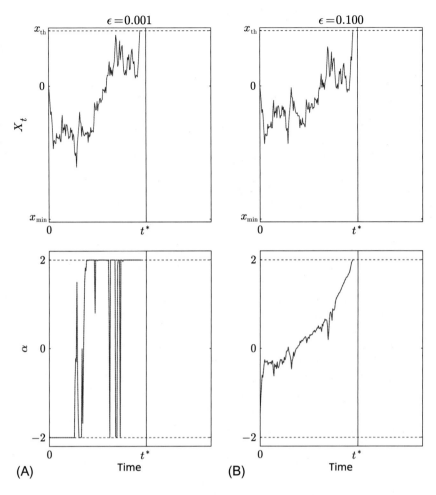

short compared to the mean period of the LIF without control, suprathreshold dynamics are beneficial, whereas for a relatively long target interspike intervals, the subthreshold regime is most easily controlled. Also, one might think that a useful targeting strategy in the subthreshold case would be to apply a control that brings the system above threshold. This may be possible, but is constrained by the energy penalty; deep in the subthreshold regime, even providing the maximal control may not be sufficient.

In Figs. 4 and 5 we illustrate the application of the closed-loop control method to a more realistic neural model based on Hodgkin-Huxley-type dynamics, namely the two-dimensional Morris-Lecar model (see Iolov et al., 2014b for details). Here we present new results that contrast the control of this model in the sub- vs suprathreshold regimes. The idea is to estimate an approximating Ornstein-Uhlenbeck process with threshold (ie, LIF model with additive Gaussian white noise) using maximum likelihood estimators of the bias μ, membrane time constant τ and noise level β on data generated from the Morris-Lecar model. With this in hand, the idea is then to control the

approximating LIF with the dynamic programming method, with boundary conditions as described earlier. One now has to allow for the finite width of the spikes by stopping the control during the spike and incorporating a fixed refractory period in the LIF description. One finds that the control is possible using the simple LIF model, even for this biophysically more realistic case. For the target train in Fig. 4, the control happens to be easier for the subthreshold than for the suprathreshold regime. The control for the suprathreshold regime deteriorates mostly for the long target interspike intervals (the second and third target spike times in the figures). That is to be expected since, in the suprathreshold regime, the model neuron has a high natural frequency of firing (ie, mean firing rate without control); this is related to the steep increase in firing frequency above threshold for the deterministic control-free model (the chosen parameters are not right near the threshold for firing). Hence, it is difficult to inhibit firing for a long period, unless strong inhibitory controls are allowed. In this case the control reduces to maximal inhibition until the targeted spike time followed by maximal excitation. In the subthreshold

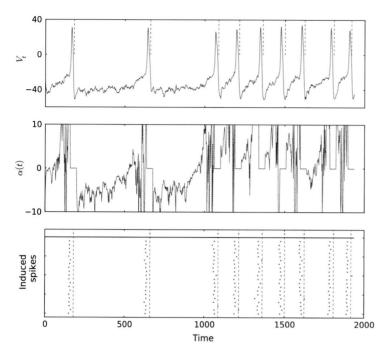

FIG. 4 *Top panel*: A realization of the Morris-Lecar spiking model with additive Gaussian white noise on the current balance V, along with targeted spike times represented by *red, dashed vertical lines*. Parameters are as in Iolov et al. (2014b), except that here the bias parameter is $I = 38$, which places these type 1 dynamics in the subthreshold regime. The noise is considered moderate here. The approximating Ornstein-Uhlenbeck process has been estimated to have $\mu = 4.02$, $\tau = 0.14$, and $\beta = 0.80$. These parameters satisfy the $\mu < 1/\tau$ condition for subthreshold dynamics in the Ornstein-Uhlenbeck model, ie, both the Morris-Lecar model and its approximating Ornstein-Uhlenbeck model are subthreshold, highlighting the consistency of our approach. *Middle panel*: External additive current control derived from application of the closed-loop dynamic programming technique to the estimated Ornstein-Uhlenbeck process. Note that the control is set to zero during the observed stereotypical spike portion of the V trajectory. *Bottom panel*: Multiple realizations of the control process using different sequences of random number sequences for the model evolution. Again the *dashed red lines* are the targeted spike times. The time interval to a spiking target is re-evaluated once the system has spiked.

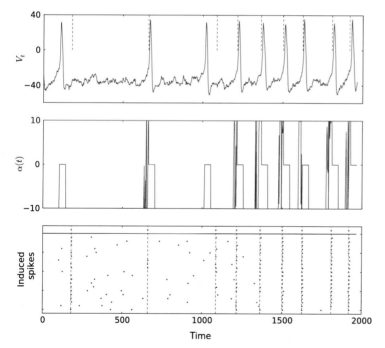

FIG. 5 Control of Morris-Lecar model in the type 1 regime as in Fig. 4, this time in the suprathreshold regime with bias parameter $I = 48$. Panels and Morris-Lecar model parameters are as in Fig. 4. The approximating Ornstein-Uhlenbeck process has been estimated to have $\mu = 1.28$, $\tau = 3.62$, and $\beta = 0.34$. These parameters satisfy the $\mu > 1/\tau$ condition for suprathreshold dynamics in the Ornstein-Uhlenbeck model, ie, both the Morris-Lecar model and its approximating Ornstein-Uhlenbeck model are suprathreshold.

regime, the control gradually builds up from inhibitory to excitatory, becoming more jittery to correct for noise-induced perturbations in the last interim before the target spike. The relative ease with which one can control the targeting for this model is dependent upon both the regime and the target train, and is thus highly problem/context specific.

7 CONTROLLING SPIKE TIMES WITHOUT CONTINUOUS OBSERVATIONS

It is also possible to control to firing times without the controller having access to continuous (or frequently sampled) voltage measurements in the LIF. Given the relative lack of information in this case, it is not surprising that this yields higher averaged squared deviations between desired and actual firing times. Nevertheless, the results are interesting in that the method works in principle and for numerical examples (Iolov et al., 2014b). For a single trial where the system and controller are initialized and a single spike time is targeted, the open-loop method of choice is based on the transition density for the stochastic dynamics. Specifically, since this transition density is based on a deterministic PDE, namely the Fokker-Planck equation, Eq. (1b), the idea is to apply a maximum principle as described above to this Fokker-Planck PDE to obtain the optimal control (Annunziato and Borzì, 2010). In particular, the cost function for our spiking problem has to be reformulated in terms of the transition density, and a gradient descent method has to be used for optimization.

One can also consider the case where, again, the controller has access only to the spike times, and nothing in between these times, but where the goal is to hit multiple target spike times in a sequence. This multispike open-loop case, treated in Iolov et al. (2014b) for a single cell, can be considered a higher form of "closed-loop" control. This is because after a spike time is obtained, the new time to the next target spike is computed, and the associated desired control is determined as a consequence of this new piece of information.

8 OUTLOOK

This chapter has focussed on closed-loop stochastic optimal control of spike times in a one-dimensional scalar drift-diffusion process. It was illustrated in the LIF model as well as in a biophysically more realistic two-dimensional spiking model from which a one-dimensional LIF approximation was computed. The optimality refers to the fact that the control minimizes a certain cost functional in the face of stochasticity. The procedure can in principle be applied to even more biophysically realistic models, such as those incorporating synaptic input that acts via a conductance. Notably, a channel conductance has the added constraint of being greater or equal to zero, so the actual result of the optimization procedure of a model with more realistic conductances is perhaps not trivial, and it is certainly worthy of future investigation. Relatedly, the elaboration of optimal controls for neural models with multiplicative noise (ie, noise on synaptic conductance) would be another interesting avenue to pursue. One can envisage also strategies where the control is meant not so much to target spike times, but perhaps a certain mean spiking rate over a characteristic time interval, or perhaps even a certain distribution of rates.

The technique described here suffers from the curse of dimensionality in that it becomes rapidly unwieldy to solve many PDEs in parallel in an online manner to target the spiking of multiple cells. This is in contrast to models based

on controlling the probability of spiking in nondiffusion-type neural models, such as spike response models (Ahmadian et al., 2011), a technique however constrained so far to low-noise activity. Approximation procedures are thus needed for such higher dimensional diffusion cases. This is especially so if the control is in the aforementioned "open-loop" control case. Problems can then also occur where an actual spike occurs much beyond a desired spike, eg, beyond even the next spike in the sequence. It remains to be seen what constitutes reasonable expectations for an optimal control of a stochastic neuron, which must fire close to a specified sequence. It may be that certain parts of the brain oversee the firings of other parts in a manner inspired by the concepts put forward in this chapter.

Finally, it is important to consider how much information needs to be gathered about the neuron prior to the application of the control. We have assumed for the LIF model that the membrane time constant, drift, and noise strength are known parameters. In practice one may have to resort to estimation procedures (Lansky and Ditlevsen, 2008), especially if one has only access to spike times in the open-loop case (Iolov et al., 2014a), or even require that the estimation and the control happen simultaneously, which might be termed "adaptive control."

Another use of optimal control might be to design ideal probes to estimate parameters of the system itself or to optimize its response to specific time-varying inputs. Ultimately, good knowledge of the parameters helps alleviate potential problems associated with any optimization problem, ie, that of clever initialization of eg, the gradient descent algorithms to overcome the presence of multiple local minima.

REFERENCES

Ahmadian, Y., Packer, A.M., Yuste, R., Paninski, L., 2011. Designing optimal stimuli to control neuronal spike timing. J. Neurophysiol. 106 (2), 1038–1053.

Ahmed, N.U., Teo, K., 1981. Optimal Control of Distributed Parameter Systems. Elsevier Science, New York.

Annunziato, M., Borzì, A., 2010. Optimal control of probability density functions of stochastic processes. Math. Model. Anal. 15 (4), 393–407.

Annunziato, M., Borzì, A., 2013. A Fokker-Planck control framework for multidimensional stochastic processes. J. Comput. Appl. Math. 237, 487–507.

Annunziato, M., Borzì, A., Nobile, F., Tempone, R., 2014. On the connection between the Hamilton-Jacobi-Bellman and the Fokker-Planck control frameworks. Appl. Math. 5 (16), 2476–2484.

Bogacz, R., Brown, E., Moehlis, J., Holmes, P., Cohen, J.D., 2006. The physics of optimal decision making: a formal analysis of models of performance in two-alternative forced-choice tasks. Psychol. Rev. 113, 700–765.

Borzì, A., Schulz, V., 2012. Computational Optimization of Systems Governed by Partial Differential Equations. Society for Industrial and Applied Mathematics, Philadelphia, PA.

Danzl, P., Hespanha, J., Moehlis, J., 2009. Event-based minimum-time control of oscillatory neuron models: phase randomization, maximal spike rate increase, and desynchronization. Biol. Cybern. 101 (5–6), 387–399.

Dasanayake, I., Li, J.S., 2011. Constrained minimum-power control of spiking neuron oscillators. In: IEEE Conference on Decision and Control and European Control Conference, pp. 3694–3699.

Evans, L.C., 1983. An Introduction to Mathematical Optimal Control Theory: Lecture notes, version 0.1. University of California. https://math.berkeley.edu/~evans/control.course.pdf.

Fattorini, H.O., 1999. Infinite Dimensional Optimization and Control Theory. Encyclopedia of Mathematics and Its Applications. Cambridge University Press, Cambridge. p. 816.

Feng, J., Tuckwell, H.C., 2003. Optimal control of neuronal activity. Phys. Rev. Lett. 91, 1–4.

Feng, X.J., Greenwald, B., Rabitz, H., Shea-Brown, E., Kosut, R., 2007. Toward closed-loop optimization of deep brain stimulation for Parkinson's disease: concepts and lessons from a computational model. J. Neural Eng. 4 (2), L14–L21.

Fleming, W.O., Rishel, R.W., 1975. Deterministic and Stochastic Optimal Control. Springer-Verlag, Berlin.

Fleming, W.H., Soner, H., 2006. Controlled Markov processes and viscosity solutions. Stochastic Modelling and Applied Probability, vol. 25. Springer-Verlag, New York.

Iolov, A., Ditlevsen, S., Longtin, A., 2014a. Fokker-Planck and Fortet equation-based parameter estimation for a leaky integrate-and-fire model with sinusoidal and stochastic forcing. J. Math. Neurosci. 4, 4.

Iolov, A., Ditlevsen, S., Longtin, A., 2014b. Stochastic optimal control of single neuron spike trains. J. Neural Eng. 11 (4), 046004.

Krylov, N.V., 2008. Controlled Diffusion Processes. Springer, New York.

Lansky, P., Ditlevsen, S., 2008. A review of the methods for signal estimation in stochastic diffusion leaky integrate-and-fire neuronal models. Biol. Cybern. 99, 253–262.

Lenhart, S., Workman, J.T., 2007. Optimal Control Applied to Biological Models. Mathematical & Computational Biology, Chapman and Hall/CRC, Boca Raton.

Moehlis, J., Shea-Brown, E., Rabitz, H., 2006. Optimal inputs for phase models of spiking neurons. J. Comput. Nonlinear Dyn. 1 (4), 358–367.

Nabi, A., Moehlis, J., 2011. Single input optimal control for globally coupled neuron networks. J. Neural Eng. 8 (6), 065008.

Nabi, A., Moehlis, J., 2012. Time optimal control of spiking neurons. J. Math. Biol. 64 (6), 981–1004.

Nabi, A., Mirzadeh, M., Gibou, F., Moehlis, J., 2013a. Minimum energy desynchronizing control for coupled neurons. J. Comput. Neurosci. 34 (2), 259–271.

Nabi, A., Stigen, T., Moehlis, J., Netoff, T., 2013b. Minimum energy control for in vitro neurons. J. Neural Eng. 10 (3), 036005.

Press, W.H., Flannery, B.P., Teukolsky, S.A., Vetterling, W.T., 1992. Numerical Recipes in C: The Art of Scientific Computing, second ed. Cambridge University Press, Cambridge (Chapter 19).

Schiff, S.J., 2012. Neural Control Engineering. MIT Press, Cambridge, MA.

Whittle, P., 1996. Optimal Control: Basics and Beyond. Wiley, New York.

Wu, Y., Peng, J., Luo, M., 2008. First spiking dynamics of stochastic neuronal model with optimal control. In: Advances in Neuro-Information Processing. 15th International Conference, ICONIP 2008, Auckland, New Zealand, November 25–28, Revised Selected Papers, Springer, Berlin, Heidelberg, 2009, pp. 129–136.

Chapter 9

Hybrid Systems Neuroscience

E.M. Navarro-López*, U. Çelikok[†] and N.S. Şengör[‡]

*The University of Manchester, Manchester, United Kingdom, [†]Boğaziçi University, Istanbul, Turkey, [‡]Istanbul Technical University, Istanbul, Turkey

1 WHAT THE STUDY OF THE BRAIN ENTAILS

Can we simulate a human brain? Not yet; the goal eludes our most determined efforts. The main sticking point is our limited understanding of how brain adaptation works and the overwhelming complexity of how connections of ideas arise from the combination of dynamic chemical, electrical, and control mechanisms that operate simultaneously at multiple levels and time scales. This is an extraordinarily complex problem: we each have about 100 billion neurons, each of which may be connected to some other 10,000 neurons.

Neuroplasticity is the name given to the brain's ability to change due to experience or damage, and is key to understanding the adaptive dynamical processes involved in the formation and consolidation of memory. Neuroplasticity is typically associated with synaptic plasticity, which defines the modification of the biochemical behavior of synaptic connections between neurons, thought to be key in memory formation and learning. But recent experiments have revealed that in memory and learning, not only is synaptic behavior modified, but synapses may also be rewired. This is known as structural plasticity (Lamprecht and LeDoux, 2004). The concept of structural plasticity can be extended to explore changes in the network topology, which is also termed wiring plasticity (Chklovskii, 2004). Our interpretation of neuroplasticity goes beyond synaptic and structural or wiring plasticity, and is used as a generic name for the adaptive and evolving dynamical mechanisms in the human brain.

Different theories from different fields have been used and partial explanations for specific phenomena have been found. Control theory—the art of making the necessary corrections in the inputs of a system by means of a feedback mechanism so that the outputs have a desired value— dynamical analysis tools, network science, and engineering methods have uncovered many of the biophysical processes in neurons and populations of neurons, especially in the study of neurodegenerative diseases (Alamir et al., 2011;

Liu et al., 2011a; McCarthy et al., 2011; Savage, 2013; Schiff, 2010; Tanaka, 2006). Most of these techniques were originally designed for systems different in nature from the brain processes under study, and tend to interpret the brain as an electrical system, a mechanical device, or a complex network as in physics. They fail to understand that the brain is much more than a set of interconnected neuronal complex networks, or a distributed control system. Many of the behavioral patterns of the brain cannot be modeled efficiently with most of the present mathematical and computational tools. Our work fills the gap by proposing novel adaptive dynamical models that combine elements of hybrid automata and complex networks, and introduces the new field of "hybrid systems neuroscience" (Navarro-López, 2014).

We will apply our framework to better understand how working memory operates. Working memory is an abstract model used in cognition to explain the processes involved in the temporary manipulation and storage of information, and is crucial in reasoning and learning, because it links perception, long-term memory, and action (Baddeley, 2003). During working memory computations, modulation of different frequency-band oscillations have been observed. A full explanation of this phenomenon is still needed. Here, we shall consider the interdependency of the basal ganglia, posterior cortex, prefrontal cortex, and thalamus and their role in the generation of subcortical background oscillations, key for working memory operations, and reinterpret the basal ganglia-thalamo-cortical circuits as structures giving rise to multiple adaptive and self-organizing dynamical processes that include closed-loop control and discontinuous mechanisms. In particular, we focus on discontinuous transitions provoking changing dynamical patterns in the basal ganglia functions connected to the working memory. Expanding on the results of Çelikok et al. (2015a)—demonstrating, with a computational framework for the cerebral cortex, how the flexible control of subcortical background oscillations can drive working memory into different computations (eg, loading, maintenance, clearance)—we propose basal ganglia-thalamus' feedback loops as crucial mechanisms for the modulation of these dynamic background oscillations.

Closed Loop Neuroscience. http://dx.doi.org/10.1016/B978-0-12-802452-2.00009-3

This chapter is organized as follows. A brief introduction to the main concepts of hybrid systems will be given in Section 2. The philosophy of "hybrid systems neuroscience" is given in Section 3. This section will elaborate on the use of control engineering and hybrid systems to transform neuroscience, and vice versa: the use of neuroscience to transform control engineering and hybrid systems. Section 3 will also identify some of the switching behaviors and site discontinuous control in the brain, phenomena which are at the heart of hybrid systems. In Section 4, we illustrate these ideas in a novel and biophysically realistic working memory network model, which unifies the influence of dopamine, basal ganglia-thalamocortical circuits, and subcortical background oscillations. We certify the model's performance with the simulation of the firing activity of the main populations of neurons during loading/updating of information in working memory. Conclusions are given in the last section.

The work presented in this chapter does not claim to be comprehensive. It sketches future possibilities and opens up new avenues for the unification of different ideas and theories in neuroscience. It is not our intention in this work to establish final results. It is a starting point for future explorations that may be expected to yield fruitful discoveries.

2 HYBRID SYSTEMS: A NEW KIND OF CONTROL OF COMPLEX SYSTEMS

Hybrid dynamical systems have been studied over the past 20 years as the 2-C science—Computation and Control. Fifteen years ago they were predicted to be the control theory of tomorrow (Matveev and Savkin, 2000), and represent a very active research theme with many unsolved problems in theory and practice. Recently, hybrid dynamical systems have evolved to *cyber-physical systems* and can be regarded as the 3-C science (Computation, Control, and Communication) as defined in Navarro-López (2013). These systems have recently been called cyber-physical systems of systems or cyber-physical networked embedded systems.

Hybrid systems, from the theoretical viewpoint, provide alternative ways of interpreting the abstraction, specification, verification, and design of complex control systems. The term "hybrid system" has been used to label a wide variety of engineering problems, such as: heterogeneous systems, multimodal systems, multicontroller systems, logic-based switching control systems, discontinuous/switched, piecewise smooth and nonsmooth systems, discrete-event systems, reset, jump or impulsive systems, and variable structure systems.

Broadly speaking, a hybrid system is a system that continuously changes and has unpredictable behavior. They are dynamical systems combining continuous dynamics and discontinuous events or transitions, smooth, and abrupt changes. They merge formal computational tools, dynamical systems theory, and control engineering methodologies, and entail a better formulation of complex systems.

There are several representations for hybrid dynamical systems or modeling frameworks. Each representation is oriented to specific types of problems and, indeed, reflects the background of the researchers behind it, be it a computer scientist, a control engineer, or an applied mathematician. The most popular representation, which will be used in our work, is the hybrid automaton framework. Hybrid automata merge continuous dynamics and finite automata theories (Navarro-López, 2009a,b; Navarro-López and Carter, 2011; Carter and Navarro-López, 2011, 2012). An automaton is a computational abstraction of the transitions of a system between discrete states or locations (on and off, for instance). A hybrid automaton, additionally, considers dynamical evolution in each location. For example, the abstraction of the model of the adaptive exponential integrate-and-fire neuron model (Brette and Gerstner, 2005) as a hybrid automaton is shown in Fig. 1. This individual spiking point-like neuron of the threshold-firing type—that is, including an auxiliary reset condition for the generation of the action potentials or spikes—is one of the simplest examples of a reset control system. Reset control systems—similar to a bouncing ball—are systems where the states are abruptly updated at selected instants of time to achieve a desired response.

Modeling a single neuron with a hybrid automaton does not add to the analysis of neural dynamical behaviors. The potential of hybrid automata lies in the modeling and analysis of more complex neuronal structures, as we will point out in the next section. This has already been studied for complex engineering systems with the computational dynamical framework DYVERSE. DYVERSE was proposed with the goal of generating a framework for the modeling, analysis, formal verification, and control of systems with switching dynamics and complex behaviors (Navarro-López, 2013). It stands for the DYnamically driven VERification of Systems with Energy considerations. Here, energy refers to the abstract energy of

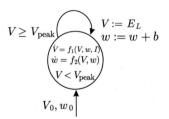

FIG. 1 Hybrid automaton of the adaptive exponential integrate-and-fire neuron, where V is the membrane potential, ω is the adaptation current, and I is the total input current to the neuron. The dot represents the derivative with respect to time. When $V(t)$ reaches V_{peak} at $t = t_{\text{peak}}$, a spike is generated and $V(t_{\text{peak}})$ is reset to a new value V_r, with $V_r = E_L$ and $\omega(t_{\text{peak}})$ is increased.

dynamical systems. It might not have a physical interpretation, and is studied by means of the dissipativity theory, which analyzes dynamical systems behavior by means of the exchange of energy with the environment (Navarro-López, 2002; Navarro-López and Laila, 2013).

The process of checking in an automated way that a system behaves correctly is called "formal verification" or "automated verification" in the theory of computer science; the elimination of negative behaviors falls into the field of control engineering. DYVERSE has provided new insights into the modeling, automated verification—mainly, model checking techniques—and control of nonlinear hybrid systems with real-world applications, and holds the promise of transforming computational neuroscience with the expansion to neuronal networks through the framework Neuro-DYVERSE (Navarro-López, 2014). Automated verification is out of the scope of this chapter, however, it is one strand in the field of hybrid systems neuroscience.

3 HYBRID SYSTEMS IN NEUROSCIENCE

Although computational neuroscience, especially in the last 10 years, has uncovered many of the biophysical processes within a single neuron (Gerstner and Kistler, 2002; Izhikevich, 2007), the collective dynamical behavior of networks of billions of neurons is still poorly understood, as is the relationship to the emergence of learning and memory. Classical nonlinear dynamical models and analysis techniques cannot capture this, and we are compelled to use new modeling and analysis approaches. We propose the combination of hybrid control systems—where continuous and discrete parts interact—and complex network models. However, these models are not adequate for our purposes in their present form. In this section, we will explain how to use control and hybrid systems to influence and better understand brain dynamical processes, and further, how to learn from the brain to propose new dynamical models and mathematical and control systems definitions.

We will start by identifying some of the switching behaviors in the brain and explain what discontinuous control means in this context. These phenomena are key in hybrid systems.

3.1 Switches, Abrupt Changes, and Discontinuous Control in Neural Systems

In engineering terms, the brain is a complex control system consisting of many interdependent closed-loop feedback units that operate simultaneously at different levels and are subject to different types of discontinuous or abrupt transitions that provoke changing activity patterns. Indeed, these discontinuities, switches, or discrete events ensure brain performance and its ability to adapt to damage or sudden unexpected changes. Think about the feeling when you have an illuminating idea: a cascading behavior and a transient synchronization of neurons trigger a snowball of sensations; or the spikes traveling between neurons, causing changes in some of the associated synapses. Microchanges produced in the last few milliseconds or next few minutes might be sufficient to allow recall of an event for life. Another example of a discontinuous process in the brain is the discontinuous control for movement generation (Gross et al., 2002).

The identification of spontaneous emergent behaviors related to these discontinuous changes in activity patterns is crucial to brain dynamics. Cascading dynamics, different types of synchrony and self-organized criticality (Droste et al., 2013; Hesse and Gross, 2014; Plenz and Niebur, 2014) epitomise the tendency of large systems to engender a "critical" state by minimal internal changes. The consequences are unpredictable. The human brain appears to work optimally on the cusp of stability, which allows quick switching between coherent states; this has recently been associated with what is called transient brain dynamics (Rabinovich and Varona, 2012), in contrast to attractor dynamics (eg, limit cycles, chaotic attractors, and equilibrium points).

Phase transitions, a term inherited from physics, or switches between network brain states are other types of discontinuities present in the brain, and are especially relevant to memory processes (Torres et al., 2014; Schmidt et al., 2013), and particularly to working memory. It is likely that the brain uses different control modes for different processes of working memory (Tanaka, 2006). The transitions and switches between network states or operation modes are associated with specific oscillatory activity patterns and are determined by externally driven background oscillations: these may be considered as a control mechanism to regulate neuronal activity, especially in working memory operations (Dippopa and Gutkin, 2013). The oscillatory frequency bands define distinct dynamical regimes (or gating modes) for the working memory network. By modulating the frequency of externally driven oscillations, the working memory switches between dynamical regimes enabling the necessary computations. We suggest that the working memory system selects these gating modes by controlling input oscillations projected to the network via the cortical-basal ganglia-thalamus circuits. This is where the hybrid nature of our system resides. More details will be given in the rest of this section and Section 4.

In this work, we use the interdependency of the basal ganglia nuclei—a collection of subcortical structures, relatively large in primates, particularly in humans—thalamus and cortex as a good example of:

- Interdependent feedback loops entailing discontinuous control mechanisms. Particularly, basal ganglia-thalamo-cortical circuits that activate three different pathways: direct, indirect, and hyperdirect (Frank, 2006).

- Switching behaviors. The basal ganglia implement a selective gating mechanism for working memory (O'Reilly and Frank, 2006). We claim, in this chapter, that this gating function is facilitated by the generation of subcortical background oscillations in the thalamus. Basal ganglia areas inhibit and disinhibit regions of the thalamus, which in turn activate regions of the cortex. Switching is at the heart of the action selection performed by the basal ganglia (switching as the transition of control that goes from one selection to another) (Redgrave et al., 1999). The dopaminergic regulation is key for this switching. Indeed, there are multiple switchings. We claim that the interaction between the basal ganglia and dopamine release is the key to switch on/off background oscillations. The gating function of dopamine may be mediated by a thalamo-cortical circuit, or more widely a cortico-striato-pallido-thalamo-prefrontal network (Tanaka, 2006; Cools and D'Esposito, 2011).
- Cooperation and competition are the chief requirements for self-organization—that is, the emergence of behavioral patterns with unexpected properties and regularities. This will be expanded in the rest of this section and Section 4.

3.2 Hybrid Systems and Closed-Loop Approaches in Neuroscience

A revolution is happening right now in life sciences. There is growing interest in the control community to apply control theory and control engineering methodologies to transform the course of biological and natural processes (Cosentino and Bates, 2012; Cowan et al., 2014; Iglesias and Ingalls, 2010; Samad and Annaswamy, 2011).

It is also significant that electrical and control engineers are currently studying neuronal systems and have contributed to the growth of *neural control engineering* as defined in Schiff (2012) and what was recently called the area of closed-loop approaches in neuroscience. By closed-loop approach, we refer to the use of control techniques and the study of closed-loop/feedback dynamics in the brain. In classical control theory, a control system is based on measuring an output, comparing it to a desired output, and generating an error that is employed to make the necessary corrections in the input so that the system output reaches a desired value. This is what "closing the loop" means. In the brain, the definition of a desired output is not as straightforward as in classical control systems, and most of the time, it will not be feasible.

Closed-loop neuroscience is a nascent field of research that overcomes some of the limitations of the classical stimulus-response (open-loop) paradigm to study neural systems. The growing interest in studying existing feedback loops or building new closed loops in neuronal systems is reflected in different recent issues of prestigious journals such as van Boxtel and Gruzelier (2014), Mohseni and

Ghovanloo (2012), and Potter et al. (2014), in books like Haykin (2012) and Schiff (2012) or at a special session organized at the last *AIMS Conference on Dynamical Systems, Differential Equations and Applications* (Arroyo and Varona, 2014). There are many more examples of attempts to apply classical control analysis tools to neuroscience (Alamir et al., 2011; Danzl et al., 2009; Schiff, 2010; Liu et al., 2011a; Girard et al., 2008; Mao, 2005).

Although hybrid systems have been increasingly used to model and analyze systems in biology for the past 10 years, this has not happened in neuroscience. The use of hybrid system models, particularly hybrid automata, is practically nonexistent in neuronal processes. We can find only two works that propose a hybrid automaton to model some features of a single neuron: the work of Bey and Leue (2011) models the spike-timing dependent plasticity (STDP) and the work of Makin and Abate (2007) proposes an unrealistic one-discrete-location hybrid automaton of a single neuron integrated in the basal ganglia. As we pointed out in Section 2, and as the *motto* of our work, hybrid automata are ineffective for the study of single neurons and best used to probe the dynamical interactions of networks of neurons. However, hybrid automata cannot be used for this purpose in their present form, and new models, such as the one introduced in Section 3.3, are needed.

3.3 Hybrid Automata Model What Complex Networks Cannot Model

The brain is an example of a "complex network": a large number of interdependent systems (neurons and other cells) that interact in nontrivial and nonregular ways (topology) with each other and their environment in order to organize in specific emergent structures. In the brain, there is a chain of interacting processes that are dynamic and evolve with time, including neuroplasticity. There are three main levels of interaction to consider:

1. "Local" dynamical neurons and relationships of neurons to other neurons—connected by synapses—and other cells.
2. "Global" neuronal network topology evolution and structural plasticity.
3. Interdependencies of networks in the brain, and how neuron dynamics influence the evolving topology of the brain networks (structural plasticity), and how neurons' interconnections affect the dynamics of each neuron.

All attempts to model any process of the human brain need to integrate in some manner these mechanisms; some of them are shown in Fig. 2.

Most of these mechanisms are related to discontinuous transitions in brain activity patterns. Indeed, the brain represents information in many regions by generating global activity patterns over populations of neurons, and switches

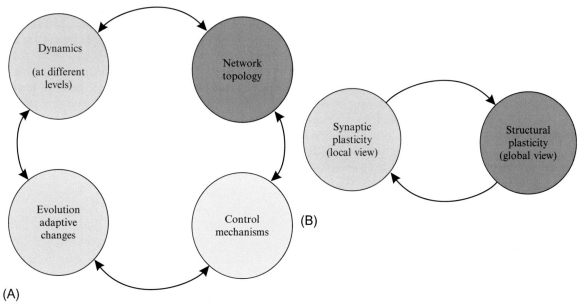

FIG. 2 Elements we need to integrate in our models: (A) The brain's chain. (B) The neuroplasticity's chain.

among these patterns account for the brain's ability to adapt to unexpected changes or to carry out action or decision control. This is well understood in the basal ganglia (Redgrave et al., 1999) or in working memory (Dippopa and Gutkin, 2013; O'Reilly and Frank, 2006).

Great progress has been made in the understanding of large-scale complex networks. Different models have been proposed in order to study the complex topology of real-world networks, namely: small-world networks, random networks, or scale-free networks (Barrat et al., 2010; Dorogovtsev, 2010; Newman, 2010) and, recently, evolving or adaptive networks (Dorogovtsev and Mendes, 2003; Gorochowski et al., 2012; Gross and Sayama, 2009). In addition, crucial collective dynamical processes have been uncovered: mainly, synchronization, polychronization, and other self-organizing behavioral patterns (cascading or spreading dynamics and polychronization, among others). New advances are being made in what is now referred to as networks of networks, interacting networks, and interdependent dynamical networks (Gao et al., 2012). However, in the 15 years since the establishment of network science—the theory of complex networks—as a field of study, the tendency has been to use mathematical measurements to describe the topological properties of the network (mainly size, density, and connectivity). This approach is more statistical than behavioral, and consequently has significant limitations in the analysis of emergent dynamical properties, and network evolution and adaptation. In the case of the brain, discontinuous phenomena and dynamical transitions cannot be completely understood within existing models of neuronal networks (Sporns, 2011), nor with new evolving and adaptive models of complex networks (Dorogovtsev and Mendes, 2003; Gorochowski et al., 2012; Gross and Sayama, 2009), whereas

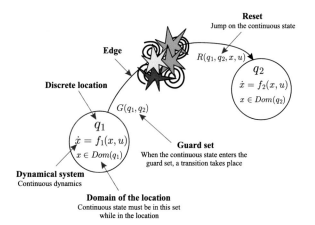

FIG. 3 Breaking the fixed structure of hybrid automata: "plastic" or reconfigurable hybrid automata. Edges, resets, and guards changing with time following adaptive rules: depending on the evolution of the continuous state x of the subsystems, the discrete locations q, and the topology of the connections between the discrete locations.

hybrid system models, especially hybrid automata, are equipped for just such discontinuous processes.

But preexisting hybrid automaton models are still inadequate to describe network switching patterns: we need adaptive and evolving transitions between discrete locations, that is: edges, guards, and resets changing with time that can follow adaptive rules based on the system's dynamics (Fig. 3). These rules depend on the evolution of the continuous state of the subsystems (single neurons or populations of neurons), the discrete locations (operation modes), and the topology of the connections between the discrete locations. Each discrete location may not correspond to a single neuron, but to populations of neurons with similar goals or activity patterns—which may vary over time.

The result is a "plastic" hybrid automaton such as the reconfigurable hybrid automaton (Navarro-López, 2013, 2014), which provides a unique framework to incorporate control laws to drive the network to a desired goal state or sequence of states. Here, networks are thought of as self-organizing structures, and the topology itself can be used as an integral part of the controller, and, in addition, be dependent on the dynamics of the nodes (associated with single neurons or populations of neurons). With this, a new family of adaptive networked dynamical systems is introduced as biologically inspired real-time self-organizing systems—a term originally proposed in Navarro-López (2013).

For our working memory network model, we also need to incorporate computation nodes into the reconfigurable hybrid automaton. The idea of computation nodes has recently been proposed as part of a new hybrid automaton modeling framework to describe multirigid-body systems with multiple impacts and friction (O'Toole and Navarro-López, 2012, 2016). Computation nodes consist of nondynamical (without dynamics) discrete locations, transitions, and reset operations. In the computation node, the executions of the hybrid automaton are assumed to operate over zero time as the discrete states are nondynamical. The inclusion of the computation nodes gives us the flexibility to encode more complex and even iterative algorithms into the hybrid-automaton framework, which has the benefit that we can model more nontrivial neuronal interactions.

3.4 New Interpretation of Control and Hybrid Systems Paradigms for Neuronal Networks

In addition to applying control systems thinking to transform neuroscience, neuroscience can be used to generate a new dynamical systems theory for complex networked systems that can impact in many other areas of research. The thread of thought is: how can we learn from natural processes to improve engineering systems and technological designs? We say emphatically that it is possible to mimic biological systems' mechanisms to better understand human behavior and design engineering and computing systems that were inconceivable not so long ago. In this section, we propose new concepts that are inspired by the neuronal dynamical processes present in the complex cortico-basal ganglia-thalamo networks. All these concepts will be further explained in our working memory network model in Section 4.

3.4.1 Switching Dominance: It Is Not Only About Switching On and Off

The classical interpretation of switching in control is that when one mode or discrete location (in a hybrid automaton) is activated ("on"), the rest are deactivated ("off").

By observing how the populations of neurons in the different nuclei of the basal ganglia behave and how the activation of the three different pathways—direct, indirect, and hyperdirect—work, we know that one pathway is activated. Yet this does not mean that the others are inactive. On the contrary, the competition or balance between them results in a dominant operation mode or pathway: this is what we call *switching dominance.*

This idea entails different levels of activation in neuron populations: one population has a dominant behavior or activity pattern, and the others still have activity with a different behavioral pattern.

3.4.2 The Network as the Controller: Self-Organizing Neuronal Interdependent Control

Reproducing the different brain control mechanisms requires not only the use of new control theory, but a better understanding of collective and cooperative behavior.

There are papers that affirm dopamine as the key element in working memory (Durstewitz and Seamans, 2002; Versace and Zorzi, 2010; Tanaka, 2006) and in the selection of the different pathways in the basal ganglia (Cohen and Frank, 2009; Cools and D'Esposito, 2011; Gruber et al., 2006). However, dopamine in the basal ganglia and its effect in working memory, particularly the dopaminergic neurons in the substantia nigra pars compacta (SNc), should not be interpreted as a controller as in the classical control model: where a control input (dopamine release) to a system (striatum) produces a desired behavior (working memory operation). This would be to oversimplify the problem.

There might be other control mechanisms within the cortico-basal ganglia-thalamo networks that modify the behavior of working memory. We claim that the connections between neurons within a population of neurons and between the populations of different types of neurons may create assemblies or clusters. These assemblies may act as a control or regulation mechanism by influencing the behavior of the whole working memory network. This idea is related to the cluster reverberation of Johnson et al. (2013). This self-organizing network control could integrate a cascading control: a behavior triggered by the dopaminergic neurons and transmitted and transformed along the different networks. Here, the synaptic plasticity and the topological characteristics of the different networks of neurons involved might play a key role. We term this collective phenomena distributed across the network of networks of the basal ganglia, the thalamus, and the cortex as self-organizing neuronal interdependent control (SONIC). The main elements of SONIC are in Fig. 4.

Here, the mathematical and computational challenge is vast and wide-ranging, being beyond the theory of distributed and decentralized control or the most recent

FIG. 4 SONIC architecture: simplified feedback control systems in our working memory network model. Multiscale control system with: feedback loops related to neurons or populations of neurons, the whole network as a stabilizing feedback mechanism, and driver control cells.

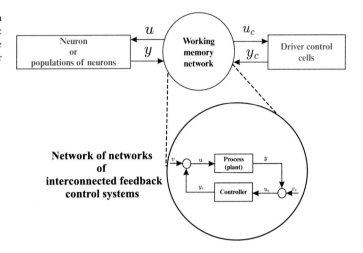

networked embedded control systems. The different networks of neurons of our model are conceived as self-organizing and adaptive structures, with the topology itself as an integral part of the control system. This may be considered a type of homeostatic mechanism, which is a regulatory mechanism that prevents the excitatory activity from growing boundlessly and is key to guaranteeing the stability of the network dynamics in learning processes (Giudice et al., 2003). It is also related in a certain way to homeostatic plasticity (Lazarevic et al., 2013; Turrigiano, 2012).

3.4.3 Driver Control Neurons

The idea of driver nodes (the terminology is inherited from biology), proposed by Liu et al. (2011b) and Gao et al. (2014), can be successfully used within this framework. The dopaminergic cells in the SNc might be interpreted as the driver nodes or *driver control neurons*, since they are the agents that provoke the balance between the main basal ganglia pathways (direct and indirect). The influence of dopamine in the hyperdirect pathway is not so clear. We will not include the hyperdirect pathway in this chapter.

In the case of the reconfigurable hybrid automaton, our basic modeling framework, the consideration of driver control neurons means that the control is not applied to all the discrete locations of the hybrid automaton; it is selectively applied, and not all the dynamical systems in the hybrid automaton are influenced by control inputs.

4 TOWARD A HYBRID AUTOMATON FOR THE SWITCHING CONTROL OF WORKING MEMORY

We propose a multilevel hybrid automaton to describe the switching control of working memory through basal ganglia and subcortical background oscillations produced in the thalamus. We will consider the dynamics of the interdependent activity among key basal ganglia regions and the

dopaminergic cells in the SNc—typically considered within the basal ganglia network.

We propose a multiple-regime mathematical model of the basal ganglia, following the conceptual two-phase model proposed in Lisman (2014). Every regime can be considered as a phase, operation mode or oscillatory state. Consequently, we assume that the decision about the selection of working memory tasks are sent to working memory in a discontinuous manner. This is not problematic if the working memory also operates discontinuously, and there are biological (Frank et al., 2001; O'Reilly and Frank, 2006) and psychological (Baddeley, 1986) evidences for this. The other assumption that we make is that the operation of the basal ganglia with respect to working memory is analogous to how the basal ganglia operate for motion control (Frank, 2006).

The neural populations considered in our model correspond to units of the posterior cortex, prefrontal cortex, striatum, internal, and external segments of globus pallidus (GPi and GPe), subthalamic nucleus (STN), and thalamus. In the GPi's population, we integrate the activity of the substantia nigra pars reticulata (SNr). Each individual neuron will be modeled as a spiking point-like neuron of the threshold-firing type; that is: they include an auxiliary reset condition for the generation of the action potentials or spikes. Models used will be variations of the adaptive exponential integrate-and-fire model (Gerstner and Kistler, 2002) and Izhikevich's model (Izhikevich, 2007). All the neurons in the model are connected according to a random graph and all the probabilities both in local connections within a population and interconnections between neurons of different populations are chosen in accordance with brain imaging studies. We make inhibitory synaptic couplings stronger than excitatory ones to allow inhibitory neurons to adequately regulate excitatory population behavior. The complete model for all neurons and all interconnections is available in the document of Çelikok et al. (2015b).

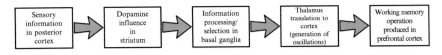

FIG. 5 *Rationale* behind our working memory network model. By generation of oscillations in the thalamus we refer to the generation of background oscillations with different frequencies, mainly: in the beta-gamma, theta, and alpha bands.

4.1 Working Memory Driven by the Basal Ganglia and Subcortical Background Oscillations

The *rationale* behind our working memory network model is illustrated in Fig. 5. It is well known that the basal ganglia circuits are activated/deactivated by the release/shortage of dopamine. We take this idea one step further by investigating the indirect influence of dopamine in the thalamus and the interactions with the prefrontal cortex (where working memory operations are processed). The thalamus can be thought of as a translator which receives a message from the basal ganglia; and this message is translated and projected back to the cortex. The thalamus makes this translation by activating oscillations with different frequencies. Depending on the frequency of this background thalamus activity (mainly, beta-gamma, theta, and alpha frequency bands), information is updated, maintained, or cleared in the working memory. This reinforces the idea that working memory is a dynamical system that evolves with time. A complete analysis of this idea is given in our recent work (Çelikok et al., 2015a,b).

Our work unifies the following threads of thought:

- **Role of background oscillations in working memory phase switching**. It is well known that different bandwidths of oscillatory activity patterns are associated with different aspects of working memory functions (Howard et al., 2003; Klimesch et al., 1999; Spitzer et al., 2010; Tesche and Karhu, 2000). In other words, working memory completes tasks by choosing suitable gating modes with the background oscillations aiding the process. These oscillations are typically subcortical background oscillations (Dippopa and Gutkin, 2013; Çelikok et al., 2015a,b). Background oscillations are a patterned brain stimulation to modulate neuronal activity. That is, they are rhythmic firing patterns which are different to intrinsic oscillations of single neurons. The importance of oscillatory activity patterns to control brain states and switching in memory phases has been already established (Schmidt et al., 2013; Torres et al., 2014). We associate different behaviors in the prefrontal cortex (working memory) with specific incoming oscillatory patterns from the thalamus that are associated with specific pathways in the basal ganglia. The direct pathway allows working memory to be updated/loaded and coincides with beta-gamma frequency band activity within the thalamus (13–120 Hz); theta frequency band activity (4–8 Hz) is directly linked to the maintenance of information while ignoring irrelevant stimulation, which is related to the activation of the indirect pathway. We will not consider the hyperdirect pathway in this chapter.

- **Role of the basal ganglia in working memory**. It has been already established that the brain uses different control modes for different processes of the working memory (via modulation of dopamine) consistent with different basal ganglia pathways (O'Reilly and Frank, 2006; Cools and D'Esposito, 2011; Tanaka, 2006).

- **Dopamine influence in the selection of pathways in basal ganglia**. Dopamine is key for the activation and inhibition of activity in the basal ganglia (Humphries et al., 2009; Redgrave et al., 1999). The balance between direct and indirect pathways is regulated by the differential actions of dopamine on striatal neurons from neurons in the SNc. A big enough release of dopamine by the SNc neurons makes the direct pathway dominant (acting on D_1 receptors in striatal neurons) and reduces activity along the indirect pathway (acting on D_2 receptors). Together, these actions result in reduction of network activity in the GPi and the SNr, and disinhibition of the thalamus activity. Conversely, a decrease in striatal dopamine release would result in the dominant activity of the indirect pathway, and consequently, an increase in the GPi and SNr activity (and thus decreased activity in thalamo-cortical neurons) (DeLong and Wichmann, 2007).

- **Dopamine influence in the selection of working memory operations/phases**. Dopamine has been established as a switching/gating mechanism for working memory, which is mediated by the basal ganglia (Gruber et al., 2006; Tanaka, 2006; Tomkins et al., 2014; Versace and Zorzi, 2010). We formalize how the change of dopamine level and consequent activation of different pathways leads toward different working memory operations by the production of different background oscillations in the thalamus—each of which is uniquely associated with each basal ganglia's pathway and each working memory process. Despite the importance of dopamine in the switch of working memory tasks, dopamine is not the only control mechanism for this (Frank et al., 2001).

- **Relationship between dopamine and subcortical background oscillations**. This has not been widely studied in the literature and the link is still unclear. There is a relationship established in the work of

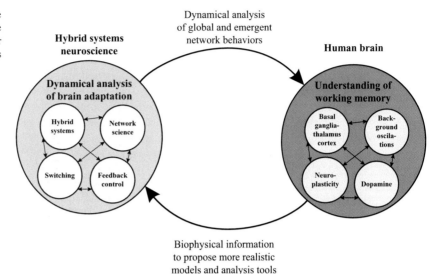

FIG. 6 Building hybrid systems neuroscience for understanding of working memory. Rationale behind the application of hybrid systems to our problem, with its main coupled components and tools.

Dippopa and Gutkin (2013) with reference to theta frequency band activity.

- **Feedback control, discontinuous control, and switching/hybrid behavior in memory and basal ganglia processes.** It is widely accepted the state switching in working memory (Cools and D'Esposito, 2011; Compte et al., 2000; Tanaka, 2006; Versace and Zorzi, 2010) is directly linked to the switching mechanisms present in the basal ganglia.

The unifying factor is that the brain uses different control operation modes or states for different processes of working memory (via dopamine modulation), consistent with the different basal ganglia pathways (direct and indirect) that can be associated with different oscillatory activity patterns in the cortex. Our model for describing the main processes of working memory will combine all these elements, something that has never been attempted before. The different levels of abstraction considered in our analysis are described by the nested "hybrid automaton" given in Fig. 6.

4.2 Basal Ganglia Network Model and Switched Control System for Working Memory

In order to generate the hybrid automaton model of the basal ganglia-thalamo-cortical network, and taking into account the complexity of the dynamical systems considered, we need to design this hybrid automaton for describing specific behaviors and activity patterns. That is, it has to be an goal-oriented model; in our case for working memory operations. We focus on the loading/update and maintenance of information. Here, we will not present a formal definition of the main elements of a hybrid automaton. For the classical

definition, the reader is referred to Navarro-López and Carter (2011).

We follow a hierarchical and multiscale modeling approach, and our goal-oriented hybrid automaton will be a multilevel model with three levels of abstraction: (1) local neuron dynamical behavior, (2) neuron population dynamical behavior, and (3) interdependencies of networks of populations to describe working memory computations. The three levels of abstraction of our modeling framework are represented in Fig. 7.

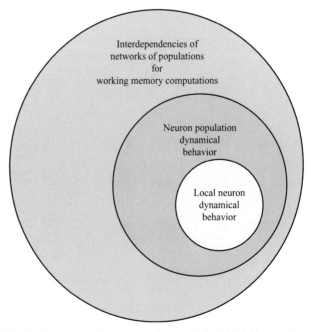

FIG. 7 Three levels of abstraction of our goal-oriented hybrid automaton-based modeling framework for describing the basal ganglia-thalamo-cortical network and the switched control for working memory.

The first level of abstraction corresponds to a "modified" hybrid automaton for a neuron including the relationships with its immediate neighbors (presynaptic and postsynaptic neurons). We will only present here the first-level hybrid automaton for a medium spiny projection neuron (MSN) with D_1-type receptors in the striatum. To understand the connections in this type of neuron, in Fig. 8, we present the populations of the striatum considered in our model. The complete model for all neurons and all their interconnections is available in our recent work (Çelikok et al., 2015b).

Inspired by Humphries et al. (2009) and Izhikevich (2007), the dynamics for the membrane potential (v_{D_1}) and the recovery current (u_{D_1}) for a striatal MSN with D_1-type receptors are:

$$C\frac{dv_{D_1}(t)}{dt} = k[v_{D_1}(t) - v_r][v_{D_1}(t) - v_t] - u_{D_1}(t) + I_{D_1}(t)$$
$$+ \phi_1 g_{DA}[v_{D_1}(t) - E_{DA}],$$
$$\frac{du_{D_1}(t)}{dt} = a\{b[v_{D_1}(t) - v_r] - u_{D_1}(t)\}, \qquad (1)$$

with the following spike-generation conditions and reset of v_{D_1} and u_{D_1} at $t_{\text{peak}_{D_1}}$:

$$\text{if } v_{D_1} \geq v_{\text{peak}_{D_1}} \text{ then} \begin{cases} v_{D_1} \leftarrow c_{D_1}, \\ u_{D_1} \leftarrow u_{D_1} + d_{D_1}, \\ d_{D_1} \leftarrow d_{D_1}(1 - L\phi_1), \end{cases} \qquad (2)$$

where C is the membrane capacitance, $v_{\text{peak}_{D_1}}$ spike cut-off value, v_r the resting membrane potential, v_t the instantaneous threshold potential, $I_{D_1}(t)$ is the total synaptic current flowing into the neuron at time t, and c_{D_1} is the voltage reset value—that is, the value of the membrane potential immediately after the neuron fires. Parameter a is the recovery time constant; k and b are derived from the I–V curve of the neuron and d_{D_1} is tuned to achieve the desired rate of spiking output. Constant $\phi_1 \in [0,1]$ expresses the proportion of active dopamine D_1 receptors and is important in our model for the dominance of the direct and indirect pathways of the basal ganglia. Values of ϕ_1 close to 1 result in over activation of D_1 receptors by modulating both cortical input (see equation for $I^*_{\text{pCtx}_{\text{NMDA}_{D_1}}}(t)$ below) and membrane potential dynamics. For D_2 receptors, we use the constant $\phi_2 \in [0,1]$, which appears in the membrane potential dynamics for MSNs with D_2-type receptors. The higher the value of ϕ_2 is, the more inhibition in D_2 receptors is produced. We model dopamine's effect by expanding the Izhikevich's neuron model as proposed in Humphries et al. (2009). ϕ_1 and ϕ_2 can be interpreted as input control parameters. In this work, they are considered as constants that we vary to show different dominant states in the network. E_{DA} and g_{DA} are the reversal potential and the conductance of dopamine regulation, respectively. Finally, parameter $L \in [0,1]$ is a scaling coefficient for the Ca^{2+} current effect.

For the synaptic currents, we have:

$$I_{D1}(t) = I^*_{\text{pCtx}_{\text{NMDA}_{D_1}}}(t) - I_{\text{Local}}(t) - (1 - \beta_1\phi_1)I_{\text{FSIs}_{D_1}}(t)$$
$$- I_{\text{GPe}_{D_1}}(t) + I_{\text{background}}(t),$$
$$I^*_{\text{pCtx}_{\text{NMDA}_{D_1}}}(t) = (1 + \beta_1\phi_1)I_{\text{pCtx}_{\text{NMDA}_{D_1}}}(t). \qquad (3)$$

The current $I_{\text{pCtx}_{\text{NMDA}_{D_1}}}$ models the excitatory contribution of pyramidal neurons with NMDA-receptors from the

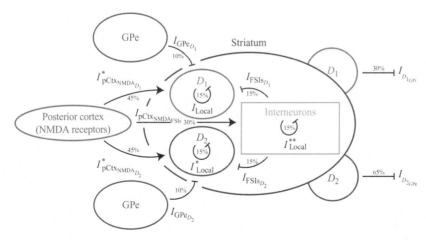

FIG. 8 The striatum consists of three different populations: D_1-type receptor neurons, D_2-type receptor neurons, and fast-spiking interneurons (FSIs). D1 and D2-type receptor neurons are medium spiny neurons (MSNs). D_1 and D_2-type receptor neurons receive cortical input from excitatory neurons of posterior cortex. Other inputs are the inhibitory contributions from the GPe, the FSIs, and local recurrent connections. FSIs have direct input from excitatory neurons of the posterior cortex and from local recurrent connections. MSNs have two main targets: output from D_1-MSNs to the GPi (as part of the direct pathway) and output from D_2-MSNs to the GPe (as part of the indirect pathway). The arrow → corresponds to excitatory connections, and ⊣ corresponds to inhibitory connections. The percentage values indicate connection probabilities of considered connections. (*Reprinted with permission from Çelikok, U., Navarro-López, E.M., Şengör, N.S., 2015b. A computational model describing the interplay of basal ganglia and subcortical background oscillations during working memory processes. Available at arXiv.org. Paper identifier: arXiv:1601.07740 [q-bio.NC], http://arxiv.org/abs/1601.07740.*)

posterior cortex, $I_{\text{FSIs}_{D_1}}$ is the inhibitory contribution from fast-spiking interneurons (FSIs) in the striatum, $I_{\text{GPe}_{D_1}}$ the inhibitory contribution from GPe neurons, and $I_{\text{background}}$ the random background activity in the striatum. $I_{\text{background}}$ is obtained from random values generated with a Poisson probability distribution; in this manner, we can induce random neuronal spike trains in our model. I_{Local} models the recurrent activity of MSNs with D_1-type receptors; recurrent activity of a neural population is crucial, especially for robust maintenance of self-sustained persistent neural activity. $\beta_1 \in (0,1]$ is a scaling coefficient of the dopamine's effect. Each synaptic input current z (where z is one of $\text{pCtx}_{\text{NMDA}_{D_1}}$, FSIs_{D_1}, GPe_{D_1}, Local) is modeled by:

$$I_z(t) = \sum_{t_n} J_{s_z} s_{D_1} \delta(t - t_n), \qquad (4)$$

with the constant J_{s_z} the synaptic strength, $\delta(t - t_n)$ a Dirac delta function that results in the incremental increase of $I_z(t)$ at each spike time t_n of a presynaptic neuron and s_{D_1} the synaptic weight between the neuron D_1-MSN and the presynaptic neuron that has fired at time t_n. We define dynamical connections between neurons, with s_{D_1} obtained in the following manner for every time a presynaptic neuron fires at time t_n:

- For every n, if $t_n \neq t_{\text{peak}_{D_1}}$:

$$\frac{ds_{D_1}(t)}{dt} = -\frac{s_{D_1}(t)}{\tau_{D_1}}. \qquad (5)$$

- For every n, if $t_n = t_{\text{peak}_{D_1}}$:

$$\frac{ds_{D_1}(t)}{dt} = -\frac{s_{D_1}(t)}{\tau_{D_1}} + J_{\text{inc}_{D_1}} \omega \delta(t - t_n), \qquad (6)$$

with τ_{D_1} a decaying-effect constant, $J_{\text{inc}_{D_1}} \in (0,1]$ a parameter, which is the same for each neuron within the

same population, and ω a parameter with a value between 0 and 1 randomly assigned every time the increment of s_{D_1} is done. That is: in our dynamical connections, the synaptic weight between two connected neurons decays over time if their firing times do not coincide, but it increases if their firing times coincide.

The "hybrid automaton" that describes the dynamics of a D_1-MSN neuron in the striatum is shown in Fig. 9. This is not a classical hybrid automaton, since includes computation nodes as they have been defined in O'Toole and Navarro-López (2012, 2016) for mechanical systems with impacts. Here, we use computation nodes (c_1, c_2, c_3, c_4, c_5, and c_6) to calculate the synaptic weights and the synaptic currents. We think that it is important to separate these computations from the membrane dynamics (discrete location q_{D_1}). The computation nodes c_4 and c_6 to calculate the synaptic weights for each postsynaptic neuron consist of two computation nodes of the type of c_1 and c_2, and they have not been included for the sake of clarity.

The network of neurons may be built connecting hybrid automata of the type presented in Fig. 9, changing the dynamics for each type of neuron. However, this can be unfeasible due to the high number of neurons needed within each population to have realistic simulation results. A solution for this problem would be to generate a hybrid automaton describing a second level of abstraction for each population of neurons. We could generate a hybrid automaton analogous to the one presented in Fig. 9, but now, the dynamical discrete location q_{D_1} would represent a whole population of D_1-MSNs. This would require the obtainment of average dynamics and variables describing the dynamics of the population, including the "mean synaptic weights" to connect with other populations and the 1; "mean input currents." This will not be done in this chapter.

The third level of abstraction is in Fig. 10. It is a switching dominance state diagram, which corresponds to the third-level "hybrid automaton" model (with an abuse

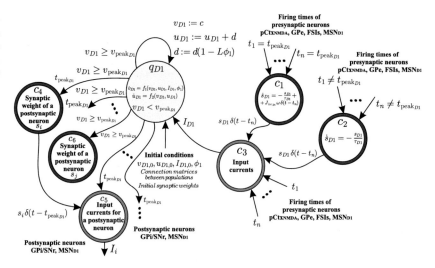

FIG. 9 Nonclassical hybrid automaton including computations nodes for a medium spiny neuron with D_1-type receptors in the striatum. The dot represents the derivative with respect to time. The double circled discrete locations are computation nodes.

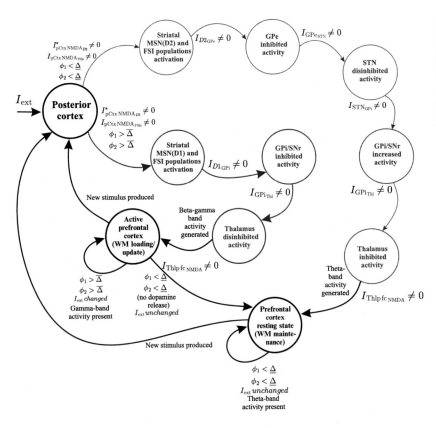

FIG. 10 Higher level of abstraction for the switching dominance states of the basal ganglia explaining the selection of working memory operations. In blue (with *bold line*), we have highlighted the basal ganglia direct pathway, and in red (with a *thinner line*), we have highlighted the basal ganglia indirect pathway. The description of the currents that appear in this figure is given in Table 1.

TABLE 1 Description of Input Currents Appearing in Fig. 10

Input Current	Description
I_{ext}	Excitatory external sensory input
$I^*_{\text{pCtx}_{\text{NMDA}_{D_1}}}$	$\begin{cases} \text{Modified excitatory contribution to striatal } D_1 - \text{MSNs from} \\ \text{NMDA} - \text{pyramidal neurons of the posterior cortex (Eq. 3)} \end{cases}$
$I^*_{\text{pCtx}_{\text{NMDA}_{D_2}}}$	$\begin{cases} \text{Modified excitatory contribution to striatal } D_2 - \text{MSNs} \\ \text{from NMDA} - \text{pyramidal neurons of the posterior cortex} \end{cases}$
$I_{\text{pCtx}_{\text{NMDAFSIs}}}$	$\begin{cases} \text{Excitatory contribution to striatal FSIs} \\ \text{from NMDA} - \text{pyramidal neurons of the posterior cortex} \end{cases}$
$I_{D_1 \text{GPi}}$	Inhibitory contribution to GPi neurons from striatal D_1-MSNs
$I_{\text{GPi}_{\text{Thl}}}$	Inhibitory contribution to thalamic neurons from neurons of the GPi
$I_{\text{Thl}_{\text{pfcNMDA}}}$	$\begin{cases} \text{Excitatory contribution to NMDA} - \text{pyramidal neurons of} \\ \text{prefrontal cortex from thalamic neurons} \end{cases}$
$I_{D_2 \text{GPe}}$	Inhibitory contribution to GPe neurons from striatal D_2-MSNs
$I_{\text{GPe}_{\text{STN}}}$	Inhibitory contribution to STN neurons from neurons of the GPe
$I_{\text{STN}_{\text{GPi}}}$	Excitatory contribution to GPi neurons from neurons of the STN

of notation and nomenclature). It includes the direct and indirect pathways' operation modes of the basal ganglia explaining the selection of the loading/update and maintenance of information in working memory. We emphasize that every discrete location (circles) in this switching

dominance does not represent a population of neurons, but a mode of operation/regime or activity pattern. The conditions for changing among these modes are given.

The exit conditions for each "discrete location" are given at the outgoing arrow from them, and all the written conditions have to be met for a change of operation mode to occur, forming what it is typically termed as guard sets or guard conditions. The main condition to quit a discrete location is if specific input currents are different from zero. This produces a large firing of populations of neurons to disinhibit or inhibit another population of neurons and a switch to the next discrete location.

In our model, the main condition to select the direct or indirect pathway is the type of dopamine release, which is expressed by the control parameters ϕ_1 and ϕ_2. For the direct pathway's dominance, condition $\phi_1 > \overline{\Delta}$ has to be met, which means that the dopamine level is high enough to increase the activation of D_1-receptor neurons, and consequently, to activate the direct pathway, with $\overline{\Delta}$ a constant such that $0 < \overline{\Delta} < 1$ and $\overline{\Delta} >>> 0$. Moreover ϕ_2 has to be big enough to inhibit the population of D_2-receptor neurons. This is also expressed by $\phi_2 > \overline{\Delta}$. The dominance of the indirect pathway is given by conditions $\phi_1 < \underline{\Delta}$ and $\phi_2 < \underline{\Delta}$, with $\underline{\Delta}$ a constant such that $0 < \underline{\Delta} < 1$ and $\underline{\Delta} <<< \overline{\Delta}$. In other words, the dopamine release is small enough and the activity of D_2-receptor neurons is dominant.

Each pathway is seen as a regime of the basal ganglia network (Chakravarthy, 2013; Frank et al., 2001). This agrees with the idea that the activation of the different pathways—which are associated with the different working memory processes—implies that different neurons have dominant activity over others. Here, the idea of switching dominance introduced in Section 3.4.1 is important, because we are considering the dominance of one pathway over another, and if an operation mode is active then the associated population of neurons has dominant activity, but other populations still present some activity.

4.3 Some Simulation Results

In this section, we will illustrate the "working memory's chain" of Fig. 5 with simulation results obtained from our model. The simulations presented here have been obtained with MATLAB and all the details on how to obtain them are given in Çelikok et al. (2015b). We will only analyze the direct pathway of the basal ganglia and the generation of beta-gamma frequency band activity in the thalamus. For this purpose, the values of ϕ_1 and ϕ_2, which are crucial for modeling the effect of dopamine, are chosen as $\phi_1 = \phi_2 = 0.8$. We will confirm the network state transitions shown in the switching dominance state diagram of Fig. 10 with the spiking activity of the populations of neurons of our model.

The neural populations considered in our model correspond to units of the posterior cortex (where sensory information arrives), striatum, GPi, GPe, STN, thalamus, and prefrontal cortex (representing working memory). All the neurons in the model are connected according to a random

graph and all the probabilities both in local connections within a population and interconnections between neurons of different populations are chosen in accordance with brain imaging studies. The complete model for all neurons, interconnections, and parameters used are given in detail in the document of Çelikok et al. (2015b).

In the posterior cortex, we consider two populations of neurons: excitatory pyramidal neurons with NMDA-receptors (80 neurons) and inhibitory GABAergic neurons (20 neurons). We choose the ratio of excitatory to inhibitory neurons to be 4:1. This ratio is determined by the anatomy of a mammalian cortex (Noback et al., 2005). The same arrangement is considered for the prefrontal cortex. In the striatum, as given in Fig. 8, three different populations are considered: D_1-type receptor neurons (100 neurons), D_2-type receptor neurons (100 neurons), and FSIs (100 neurons). The GPe has only inhibitory neurons, and we consider one population of neurons (100 neurons). The GPi, which is the output nucleus of the basal ganglia, regulates the activity of the thalamus; we consider only one population of GPi neurons (100 neurons). For the STN, we consider only one population of neurons (100 neurons). Finally, in the thalamus, for the sake of simplicity, we consider only two populations of neurons: excitatory thalamic neurons (80 neurons) and inhibitory reticular nucleus (RTN) neurons (20 neurons). While the thalamic neurons excites the prefrontal cortex, the RTN neurons provide local inhibition for the thalamus. The thalamus is the source of background oscillations associated with different working memory processes (in the prefrontal cortex).

An external stimulus is applied to NMDA-type neurons within the posterior cortex between 25 and 75 ms. This stimulus is modeled by I_{ext} and represents a random external sensory input which will be loaded, maintained, or ignored in the working memory according to a subcortical decision. In our model, it is obtained from random values generated with a Poisson probability distribution. The stimulus excites the NMDA-receptor neuron population and produces a persistent spiking activity in the posterior cortex (after about $t = 55$ ms). The GABAergic neurons of the posterior cortex and the prefrontal cortex have high firing activity independently of the external stimulation, they regulate the activity of NMDA neurons and do not affect other populations of neurons out of the cortex. Striatal D_1-MSNs and D_2-MSNs have high dopamine receptor occupancy to allow the direct pathway to be dominant over the indirect pathway ($\phi_1 = \phi_2 = 0.8$). Dopamine has different effects for the incoming cortical input: the cortical input is scaled up according to the dopamine level for D_1 receptors (see Eq. 3) and scaled down for D_2 receptors. The firing activity in the posterior cortex excites D_1-MSN population, whose activity is relatively higher than that of D_2-MSNs. The D_1-MSN neurons exhibit low-frequency activity before the activity of NMDA neurons starts, and, after about $t =$

75 ms, they present a persistent activity. The incoming inhibitory signal from striatal D_1-MSNs inhibits the neurons in the GPi and, consequently, the thalamus is disinhibited. The activity of the neurons in the GPe is high, but indeed, this reduces the activity in the indirect pathway. Now, the "gates" from the thalamus to the prefrontal cortex are open and the thalamus will deliver the basal ganglia's message to the prefrontal cortex. Significantly, the thalamic neurons achieve this by using high-frequency bursts in spikes. These bursts may be interpreted as "wake-up calls" to the prefrontal cortex, indicating that a change has occurred in the environment (Sherman, 2001). With regard to RTN neurons, they regulate the activity of thalamic cells and their activity does not depend on the activity of the populations considered here. We will further investigate the firing activity of the thalamus below. Once the prefrontal cortex is activated, new information can be loaded/updated in the working memory. The NMDA neurons in the prefrontal cortex present a persistent activity after 75–100 ms. The cascading dynamics of the firing activity of these populations of neurons may be seen as an example of the SONIC mechanism of Section 3.4.2, and is shown in Fig. 11.

Studies on human electrophysiology have shown that the loading of new information in working memory correlates positively with oscillations in the beta frequency band (13–30 Hz) and the gamma frequency band (30–120 Hz) as established in Howard et al. (2003) and Dippopa and Gutkin (2013). In our model, we have reached the same conclusion: beta-gamma band activity is appreciated from the spectrogram shown in Fig. 12A, corresponding to the activity in the thalamus. From this figure, we can appreciate that the dominant frequency fluctuates with time, within

the limits of the beta and gamma bands. This changing frequency behavior is due to the characteristic bursting activity—with "up" and "down" states—in the thalamus. In the "up" states, the thalamic cells generate spikes with high frequency, which is represented by the gamma-band activity in the spectrogram (30–120 Hz). Every "up" state is followed by a "down" state. In the "down" states, the probability for a neuron to generate a spike is lower, which is represented by the beta-band activity in the spectrogram (13–30 Hz). The power spectrum of the thalamic activity for different frequencies given in Fig. 12B clarifies the dominant frequencies of the oscillations generated in the thalamus. From the figure, it is appreciated that the most dominant frequency is 40 Hz, within the gamma band. Gamma-band bursts allow the prefrontal cortex to get activated, allowing the loading of the external input in working memory. In this case, gamma-band oscillations enable the neurons of the prefrontal cortex to present a bistable regime: a state of low activity before the external stimulus is applied, and a state of persistent activity after the stimulus is applied.

Thalamic cells have two distinctive firing patterns which allow the neurons to respond to inputs in two different ways: bursting activity or tonic spiking (Sherman, 2001). The bursting mode is known to be related to stimulus detection. Significantly, as well as increasing firing activity, thalamic cells switch from tonic spiking to bursting activity. The firing rate presents "up" and "down" states. This is the bursting-like activity in the thalamus: high-frequency spikes are generated, followed by a period of hyperpolarization. When excitable, these bursts are more effective than fast-spiking regimes at producing persistent activity in the prefrontal cortex.

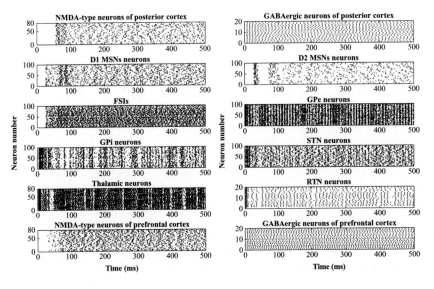

FIG. 11 Raster plots of the spiking activity of the populations of neurons considered in our model for $\phi_1 = \phi_2 = 0.8$, for which the direct pathway of the basal ganglia is dominant. Each black dot corresponds to a spike of the corresponding neuron at a given time.

FIG. 12 Analysis of the firing activity in the thalamus from simulation results where the direct pathway of the basal ganglia is dominant. (A) Spectrogram obtained from the average firing rates of the thalamus (30 trials), where the dominant frequency band is in red. There is a clear differentiation of beta-gamma band activity. The dominant frequency fluctuates with time, however, it is within the expected frequency band. (B) Power spectrum of the thalamic activity for different frequency values. The most dominant frequency is 40 Hz (within the gamma frequency band).

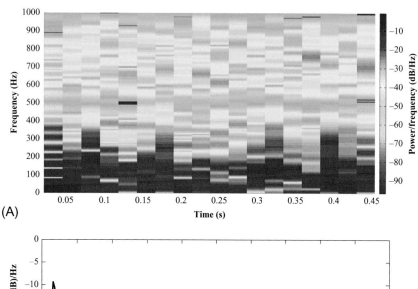

(A)

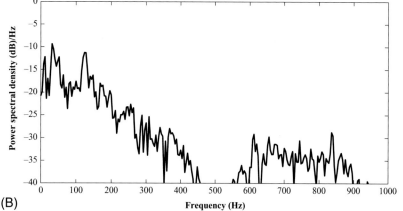

(B)

5 CONCLUSIONS

We have defined the field of hybrid systems neuroscience as the reformulation of hybrid system models, analysis tools, and control schemes for neuronal systems. We have challenged the existing orthodoxy in control systems theory and computer science and offered a new viewpoint in modeling and analyzing the complex and unique behaviors of brain networks. Under the hybrid systems neuroscience framework, we have proposed new concepts like switching dominance, SONIC or driver control neurons, and a new interpretation of hybrid automata. Our ideas have been illustrated in a novel working memory network model, which unifies the influence of dopamine, basal ganglia-thalamo-cortical circuits, and the generation of subcortical background oscillations.

REFERENCES

Alamir, M., Welsh, J., Goodwin, G., 2011. Synaptic plasticity based model for epileptic seizures. Automatica 147, 1183–1192.

Arroyo, D., Varona, P., 2014. Special session theoretical, technical, and experimental challenges in closed-loop approaches in biology. In: Proceedings of the 10th AIMS Conference on Dynamical Systems, Differential Equations and Applications, Madrid, Spain, July 7–11.

Baddeley, A., 1986. Working Memory. Oxford University Press, Oxford, UK.

Baddeley, A., 2003. Working memory: looking back and looking forward. Nat. Rev. Neurosci. 10, 829–839.

Barrat, A., Barthélemy, M., Vespignani, A., 2010. Dynamical Processes on Complex Networks. Cambridge University Press, Cambridge, UK.

Bey, A., Leue, S., 2011. Modeling and analyzing spike timing dependent plasticity with linear hybrid automata. Technical Report, University Konstanz, Germany.

Brette, R., Gerstner, W., 2005. Adaptive exponential integrate-and-fire model as an effective description of neuronal activity. J. Neurophysiol. 94, 3637–3642.

Carter, R., Navarro-López, E.M., 2011. Abstractions of hybrid systems: formal languages to describe dynamical behaviour. In: Proceedings of the 18th IFAC Triennial World Congress, August 31–September 2, pp. 4552–4557.

Carter, R., Navarro-López, E.M., 2012. Dynamically-driven timed automaton abstractions for proving liveness of continuous systems. In: Proceedings of the International Conference Formal Modeling and Analysis of Timed Systems (FORMATS 2012), Lecture Notes in Computer Science, vol. 7595. Springer, pp. 59–74. http://dx.doi.org/10.1007/978-3-642-33365-1_6.

Çelikok, U., Şengör, N.S., Navarro-López, E.M., 2015a. The interplay of working memory and subcortical background oscillations. In: Integrated Systems Neuroscience Workshop. vol. 1.

Çelikok, U., Navarro-López, E.M., Şengör, N.S., 2015b. A computational model describing the interplay of basal ganglia and subcortical background oscillations during working memory processes. Available at arXiv.org. Paper identifier: arXiv:1601.07740 [q-bio.NC], http://arxiv.org/abs/1601.07740.

Chakravarthy, V., 2013. Do basal ganglia amplify willed action by stochastic resonance? A model. PLoS ONE 8 (11), e75657.

Chklovskii, D., 2004. Cortical rewiring and information storage. Nature 431, 782–788.

Cohen, M., Frank, M., 2009. Neurocomputational models of basal ganglia function in learning, memory and choice. Behav. Brain Res. 199, 141–156.

Compte, A., Brunel, N., Goldman-Rakic, P., Wang, X.J., 2000. Synaptic mechanisms and network dynamics underlying spatial working memory in a cortical network model. Cereb. Cortex 10, 910–923.

Cools, R., D'Esposito, M., 2011. Inverted-u-shaped dopamine actions on human working memory and cognitive control. Biol. Psychiatry 69, e113–e125.

Cosentino, C., Bates, D., 2012. Feedback Control in Systems Biology. CRC Press, Taylor & Francis Group, Boca Raton, FL.

Cowan, N., Ankarali, M., Dyhr, J., Madhav, M., Roth, E., Sefati, S., Sponberg, S., Stamper, S., Fortune, E., Daniel, T., 2014. Feedback control as a framework for understanding tradeoffs in biology. Integr. Comp. Biol. 54 (2), 223–237.

Danzl, P., Hespanha, J., Moehlis, J., 2009. Event-based minimum-time control of oscillatory neuron models: phase randomization, maximal spike rate increase, and desynchronization. Biol. Cybernet. 101, 387–399.

DeLong, M., Wichmann, T., 2007. Circuits and circuit disorders of the basal ganglia. Arch. Neurol. 64, 20–24.

Dippopa, M., Gutkin, B., 2013. Flexible frequency control of cortical oscillations enables computations required for working memory. PNAS 110 (31), 12828–12833.

Dorogovtsev, S., 2010. Lectures on Complex Networks. Oxford University Press, Oxford, UK.

Dorogovtsev, S., Mendes, J., 2003. Evolution of Networks. Oxford University Press, Oxford, UK.

Droste, F., Do, A.L., Gross, T., 2013. Analytical investigation of self-organized criticality in neural networks. J. R. Soc. Interface 10 (78), 20120558.

Durstewitz, D., Seamans, J., 2002. The computational role of dopamine D1 receptors in working memory. Neural Netw. 15, 561–572.

Frank, M., 2006. Hold your horses: a dynamic computational role for the subthalamic nucleus in decision making. Neural Netw. 19, 1120–1136.

Frank, M., Loughry, B., O'Reilly, R., 2001. Interactions between frontal cortex and basal ganglia in working memory: a computational model. Cogn. Affect. Behav. Neurosci. 1 (2), 137–160.

Gao, J., Buldyrev, S., Stanley, H., Havlin, S., 2012. Networks formed from interdependent networks. Nat. Phys. 8, 40–48.

Gao, J., Liu, Y.Y., D'Souza, R.M., Barabási, A.L., 2014. Target control of complex networks. Nat. Commun. 5. Article No. 5415.

Gerstner, W., Kistler, W., 2002. Spiking Neuron Models. Single Neurons, Populations, Plasticity. Cambridge University Press, Cambridge, UK.

Girard, B., Tabareau, N., Pham, Q., Berthoz, A., Slotine, J.J., 2008. Where neuroscience and dynamic system theory meet autonomous robotics: a contracting basal ganglia model for action selection. Neural Netw. 21, 628–641.

Giudice, P.D., Fusi, S., Mattia, M., 2003. Modelling the formation of working memory with networks of integrate-and-fire neurons connected by plastic synapses. J. Physiol. 97, 659–681.

Gorochowski, T., Bernardo, M.D., Grierson, C., 2012. Evolving dynamical networks: a formalism for describing complex systems. Complexity 17 (3), 18–25.

Gross, J., Timmermann, L., Kujala, J., Dirks, M., Schmitz, F., Salmelin, R., Schnitzler, A., 2002. The neural basis of intermittent motor control in humans. PNAS 99 (4), 2299–2302.

Gross, T., Sayama, H., 2009. Adaptive Networks: Theory, Models and Applications. Springer-Verlag, Berlin.

Gruber, A., Dayan, P., Gutkin, B., Solla, S., 2006. Dopamine modulation in the basal ganglia locks the gate to working memory. J. Comput. Neurosci. 20, 153–166.

Haykin, S., 2012. Cognitive Dynamic Systems: Perception-Action Cycle, Radar and Radio. Cambridge University Press, Cambridge, UK.

Howard, M., Rizzuto, D., Caplan, J., Madsen, J., Lisman, J., Aschenbrenner-Scheibe, R., Schulze-Bonhage, A., Kahana, M., 2003. Gamma oscillations correlate with working memory load in humans. Cereb. Cortex 13 (12), 1369–1374.

Humphries, M., Lepora, N., Wood, R., Gurney, K., 2009. Capturing dopaminergic modulation and bimodal membrane behaviour of striatal medium spiny neurons in accurate, reduced models. Front. Comput. Neurosci. 3 (26), 1–16.

Iglesias, P., Ingalls, B., 2010. Control Theory and Systems Biology. MIT Press, Cambridge, MA.

Izhikevich, E., 2007. Dynamical Systems in Neuroscience: The Geometry of Excitability and Bursting. MIT Press, Cambridge, MA.

Hesse, J., Gross, T., 2014. Self-organized criticality as a fundamental property of neural systems. Front. Syst. Neurosci. 8, 166.

Johnson, S., Marro, J., Torres, J., 2013. Robust short-term memory without synaptic learning. PLoS ONE 8 (1), e50276.

Klimesch, W., Doppelmayr, M., Schwaiger, J., Auinger, P., Winkler, T., 1999. Paradoxical alpha synchronization in a memory task. Cogn. Brain Res. 7 (4), 493–501.

Lamprecht, R., LeDoux, J., 2004. Structural plasticity and memory. Nat. Rev. Neurosci. 5, 45–54.

Lazarevic, V., Pothula, S., Andrés-Alonso, M., Fejtova, A., 2013. Molecular mechanisms driving homeostatic plasticity of neurotransmitter release. Front. Cell. Neurosci. 7 (244), 1–10.

Lisman, J., 2014. Two-phase model of the basal ganglia: implications for discontinuous control of the motor system. Philos. Trans. R. Soc. B 369, 20130489.

Liu, J., Khalil, H., Oweiss, K., 2011. Model-based analysis and control of a network of basal ganglia spiking neurons in the normal and Parkinsonian states. J. Neural Eng. 8 (4), 045002.

Liu, Y.Y., Slotine, J.J., Barabási, A.L., 2011. Controllability of complex networks. Nature 473, 167–173.

Makin, J., Abate, A., 2007. A neural hybrid-system model of the basal ganglia. Technical Report, University of California at Berkeley.

Mao, Z.H., 2005. Modeling the role of the basal ganglia in motor control and motor programming. PhD Thesis, Massachusetts Institute of Technology.

Matveev, A., Savkin, A., 2000. Qualitative Theory of Hybrid Dynamical Systems (Control and Engineering). Birkhäuser, Boston.

McCarthy, M., Moore-Kochlacs, C., Gu, X., Boyden, E., Han, X., Kopell, N., 2011. Striatal origin of the pathologic beta oscillations in Parkinson's disease. PNAS 108 (28), 11620–11625.

Mohseni, P., Ghovanloo, M., 2012. Closing the loop via advanced neurotechnologies. IEEE Trans. Neural Syst. Rehab. Eng. 20 (4), C1.

Navarro-López, E.M., 2002. Dissipativity and passivity-related properties in nonlinear discrete-time systems. PhD Thesis, Universitat Politècnica de Catalunya, Barcelona, Spain.

Navarro-López, E.M., 2009. Hybrid-automaton models for simulating systems with sliding motion: still a challenge. In: Proceedings of the Third IFAC Conference on Analysis and Design of Hybrid Systems, pp. 322–327.

Navarro-López, E.M., 2009. Hybrid modelling of a discontinuous dynamical system including switching control. In: Proceedings of the Second IFAC Conference on Analysis and Control of Chaotic Systems, pp. 1–6.

Navarro-López, E.M., 2013. DYVERSE: from formal verification to biologically-inspired real-time self-organising systems. In: Zander, J., Mosterman, P.J. (Eds.), Computation for Humanity—Information Technology to Advance Society. CRC Press/Taylor & Francis, Boca Raton, FL, pp. 303–348.

Navarro-López, E.M., 2014. Neuro-DYVERSE: building hybrid systems neuroscience. In: Special Session Theoretical, Technical, and Experimental Challenges in Closed-loop Approaches in Biology, Proceedings of the 10th AIMS Conference on Dynamical Systems, Differential Equations and Applications, Madrid, Spain.

Navarro-López, E.M., Carter, R., 2011. Hybrid automata: an insight into the discrete abstraction of discontinuous systems. Int. J. Syst. Sci. 42 (11), 1883–1898.

Navarro-López, E.M., Laila, D., 2013. Group and total dissipativity and stability of multi-equilibria hybrid automata. IEEE Trans. Autom. Control 58 (12), 3196–3202.

Newman, M., 2010. Networks: An Introduction. Oxford University Press, Oxford, UK.

Noback, C., Ruggiero, D., Demarest, R., Strominger, N., 2005. The Human Nervous System: Structure and Function. Humana Press, New York.

O'Reilly, R., Frank, M., 2006. Making working memory work: a computational model of learning in the prefrontal cortex and basal ganglia. Neural Comput. 18, 283–328.

O'Toole, M.D., Navarro-López, E.M., 2012. A hybrid automaton for a class of multi-contact rigid-body systems with friction and impacts. In: Proceedings of the Fourth IFAC Conference on Analysis and Design of Hybrid Systems, ADHS 2012, Eindhoven, The Netherlands, pp. 299–306.

O'Toole, M., Navarro-López, E.M., 2016. Falsification of safety properties in multi-rigid-body mechanical systems with hybrid automata and constraint satisfaction. Available at http://staff.cs.manchester.ac.uk/~navarroe/papers/otoole_navarro2016.pdf.

Plenz, D., Niebur, E., 2014. Criticality in Neural Systems. Wiley & Sons, New York.

Potter, S., El-Hady, A., Fetz, E., 2014. Closed-loop neuroscience and neuroengineering. Front. Neural Circuits 8, 115.

Rabinovich, M., Varona, P., 2012. Transient brain dynamics. In: Rabinovich, M., Friston, K., Varona, P. (Eds.), Principles of Brain Dynamics: Global State Interactions. MIT Press, Cambridge, MA, pp. 71–92.

Redgrave, P., Prescott, T., Gurney, K., 1999. The basal ganglia: a vertebrate solution to the selection problem? Neuroscience 89 (4), 1009–1023.

Samad, T., Annaswamy, A., 2011. The impact of control technology: overview, success stories, and research challenges. Report published by the IEEE Control Systems Society, available at www.ieeecss.org/main/IoCT-report.

Savage, N., 2013. Decoding dementia. Commun. ACM 56 (3), 13–15.

Schiff, S., 2010. Towards model-based control of Parkinson's disease. Philos. Trans. R. Soc. A 368, 2269–2308.

Schiff, S., 2012. Neural Control Engineering. The Emerging Intersection Between Control Theory and Neuroscience. MIT Press, Cambridge, MA.

Schmidt, S., Scholz, M., Obermayer, K., Brandt, S., 2013. Patterned brain stimulation, what a framework with rhythmic and noisy components might tell us about recovery maximization. Front. Hum. Neurosci. 7 (325), 1–10.

Sherman, S., 2001. A wake-up call from the thalamus. Nat. Neurosci. 4, 344–346.

Spitzer, B., Wacker, E., Blankenburg, F., 2010. Oscillatory correlates of vibrotactile frequency processing in human working memory. J. Neurosci. 30 (12), 4496–4502.

Sporns, O., 2011. Networks of the Brain. MIT Press, Cambridge, MA.

Tanaka, S., 2006. Dopaminergic control of working memory and its relevance to schizophrenia: a circuit dynamics perspective. Neuroscience 139, 153–171.

Tesche, C., Karhu, J., 2000. Theta oscillations index human hippocampal activation during a working memory task. PNAS 97 (2), 919–924.

Tomkins, A., Vasilaki, E., Beste, C., Gurney, K., Humphries, M., 2014. Transient and steady-state selection in the striatal microcircuit. Front. Comput. Neurosci. 7 (192), 1–16.

Torres, J., Elices, I., Marro, J., 2014. Efficient transmission of subthreshold signals in complex networks of spiking neurons. arXiv:1410.3992 [physics.bio-ph].

Turrigiano, G., 2012. Homeostatic synaptic plasticity: local and global mechanisms for stabilising neuronal function. Cold Spring Harb. Perspect. Biol. 4 (a005736), 1–17.

van Boxtel, G., Gruzelier, J., 2014. Neurofeedback. Biol. Psychol 95, 1–134.

Versace, M., Zorzi, M., 2010. The role of dopamine in the maintenance of working memory in prefrontal cortex neurons: input-driven versus internally-driven networks. Int. J. Neural Syst. 20 (4), 249–265.

Chapter 10

Computational Complexity and the Function-Structure-Environment Loop of the Brain

B. Juba*

Washington University, St. Louis, MO, United States

1 INTRODUCTION

The brain is generally understood to be a kind of computer. As a first, gross simplification, it is generally believed that the brain takes sensory inputs and produces motor commands as outputs, for the purpose of prolonging the survival of its host. Although this view can help orient the study of the brain, we would naturally like to understand its function more precisely. We possess a pretty thorough picture of the behavior of the key cells comprising the brain, in particular neurons (Koch, 1998; Koch and Segev, 2000; London and Häusser, 2005; Herz et al., 2006); but, while clarity at the cellular level is necessary to understand the human brain, it is insufficient to understand it. This is because, as we will review in Section 2.1, unlike most other organs, the function of the brain seems particularly sensitive to the arrangement and behavior of even a handful of cells. The cellular characterization of the brain's behavior is therefore inadequate to provide an understanding of its function for two reasons that we will develop in Section 3. The first widely acknowledged reason is that the human brain at the cellular level is therefore too large and complex a system for us to comprehend in full detail. The second reason is less acknowledged, but is in a sense more fundamental: the human brain is therefore crucially sensitive to the idiosyncrasies of individual development and physiology, which we review in Sections 2.2–2.3. It may therefore be incorrect to speak of "the" human brain as a single system, in the same way one might discuss "the" human eye or "the" human respiratory system. Instead, our object of study is the range of functional behaviors of the family of brains that develop in a closed-loop interaction with natural environments. We can seek to address both difficulties through the appropriate use of *abstraction*.

1.1 The Theory of Algorithms Informs the Study of the Brain

The first difficulty stems from our inability to understand the behavior of more than a few cells at a time. Addressing this difficulty thus involves some abstraction of the behavior of such groups of cells as performing some simpler computational function. The use of such abstraction in studying the human brain was advocated in a classic work by Marr (1982). Marr advocated the characterization of algorithms and representations as a bridge between the behavior of individual neurons and a precise characterization of the brain's behavior at a functional level. A description of the brain's function in terms of algorithms and representations gives us a narrative in which we can place the actions of the individual neurons as providing such representations, or as executing such algorithms.

The study of the human brain at this level of algorithms and representations is the focus of the field of cognitive neurosciences. But, such algorithms and representations are studied more generally by computer science. Indeed, the analogous separation of function from implementation in silicon is recognized as crucial in enabling humans to engineer the highly complex computer hardware and software we use today. In computer science, the study of algorithms and representations is divorced from the requirement in cognitive neuroscience that these algorithms capture the particular, natural systems of the brain. This freedom has enabled computer science to develop a reasonably successful suite of techniques for the analysis of such systems, in order to facilitate their design. I will argue that some of these techniques should also be useful in understanding the brain at this abstract level.

Closed Loop Neuroscience. http://dx.doi.org/10.1016/B978-0-12-802452-2.00010-X

131

1.2 The Role of Computational Complexity

In particular, I will argue that the framework of *computational complexity* provides a conceptual orientation and some basic mathematical techniques to help overcome the second difficulty; that the study of "the" human brain is actually the study of the many possible brains that may result from development and experience. We must analyze something more general than a particular brain, or, perhaps, even more general than a particular algorithm. One such kind of analysis, that I will describe in Section 4, draws on the observation at the cellular level that human brains, and structures within these brains, tend to have roughly similar sizes, densities, signaling speeds, and so on. These measurable, quantitative constraints on a typical brain constrain the algorithms that the brain can implement, especially on short time scales. A constraint on the algorithms in turn constrains the possible functions that the brain could be computing. Such relationships are the objects of study in computational complexity—the relationships between the *quantitative "complexity" constraints* of an implementation level in some computational model and the *possible computational behaviors* realized at the abstract, functional level. Specifically, computational complexity addresses questions about whether or not a function can be computed at all within some given quantitative constraints, and about the relative power, that is, relative breadth of functions computable under different constraints and models.

Actually, a surprising outcome of these studies has been that many different computational models, when allowed a reasonable amount of resources (most notably computation time), have essentially the *same* amount of power, that is, compute exactly the same functions. In particular, we do not know of any currently feasible model of computation more powerful than the Turing machine,[1] for example, but we know of many models of equivalent power. This state of affairs motivates the *Strong Church-Turing Thesis:* roughly, that the notion of feasible computation is largely insensitive to the choice of model. Or, in other words, the Turing machine model in particular, when subject to a reasonable constraint on its running time, expresses precisely the family of functions that can be feasibly computed. This hypothesis enables a mathematically precise study of efficient computation *in general.*

In turn, such a study of efficient computation in general enables a kind of "quantitative philosophy" that provides a

second application for computational complexity to the study of the brain. In short, the language of computational complexity allows us to sensibly formulate questions about the high-level abilities of the brain. For example, in Section 5 I will sketch an example explication of "predictive ability." This is enabled by a precise complexity-theoretic formulation of what constitutes a "reasonable" encoding of semantic information content. Note that such questions are philosophically thorny otherwise: in the absence of a known, fixed formal representation scheme, who is to say what was computed and what was not? In particular, there is no single, fixed function (at the highest of Marr's three levels) that captures such abilities in general. Nevertheless, the notions of computational complexity provide a way of capturing such things.

The value of such a definition is that by abstracting away the details of the encoding used by a particular system, it is then possible to ask precisely whether or not such capabilities are possessed by a given region of the brain, the brain of another species, or a computer program. Perhaps most crucially of all, when a system does *not* have such abilities, the mathematical techniques of computational complexity may provide a means to demonstrate their absence.

2 FUNCTIONAL AND STRUCTURAL VARIABILITY IN THE BRAIN

It is hard to understate the functional significance of the structural variation across brains. Even so, it is somewhat surprising how much of the brain's structure, even at larger scales, seems to be determined by the environment in which it develops, as opposed to having been genetically determined. The consequences of this extreme sensitivity and variability motivate turning to a complexity-theoretic model of the brain, so we will review the evidence for this view of the brain.

2.1 Learning and the Sensitivity of Neural Representations

It is widely understood that the brain has the capacity to learn. This learning, in turn, must be realized by some kind of mechanism in the brain. One mechanism for learning was first proposed by Hebb (1949). Recall that much of the brain consists of a network of cells called *neurons*, joined at *synapses*. Neurons are capable of producing an electric *spike*, that can be detected across these synapses by other neurons, sometimes causing them to generate spikes as well, if the overall level of activity is high enough. In Hebb's mechanism(s), the relative sensitivity of a neuron to spiking activity across its various synapses is modified when the

---

1. A Turing machine is a simple, idealized computer, given by a finite control unit with an unbounded length memory tape. The details will not be significant to us, but it behaves as follows. On a single step of the machine, it reads and updates the contents of a single symbol on the tape, updates the state of the control unit, and moves to an adjacent cell of the tape. The "time" taken by a Turing machine corresponds to the number of such steps.

neurons spike simultaneously, or in sequence. Cellular mechanisms capable of supporting Hebb's proposal were subsequently observed, starting with work by Bliss and Lømo (1973). Although it was plausible that learning in the brain was the result of Hebb's mechanism for learning by altering these synaptic weights, Yang et al. (2009) recently confirmed the existence of another mechanism for learning: they observed the creation of new dendritic spines in mice, leading to the creation of *new synapses*, as well as the elimination of existing spines, eliminating synaptic connections. In short, both the network of connections among neurons which happen to have their axons and dendrites near one another, as well as the strength of these connections, appear to be subject to continual change. In simple models of neural networks, as for example, proposed by McCulloch and Pitts (1943), then, no part of the network can be assumed to remain fixed.[2]

This *plasticity* of the brain is perhaps surprising when one considers that it appears that the firing of just a handful of neurons may represent specific content. Quian Quiroga et al. (2005) describe their observation of a *"Halle Berry"* neuron during recordings of the spiking activity of neurons in a neurological patient: the particular neuron in question spiked when the patient was shown a picture of the actress Halle Berry, or when the patient was shown the words "Halle Berry," but not apparently in response to any other stimulus the patient was shown. It thus appears that the firing of one or a few neurons really does encode the concept of "Halle Berry" in this particular patient's brain. Quian Quiroga et al. were able to find a variety of other cells that responded selectively to various images, and Quian Quiroga et al. (2007) argue that this suggests that these concepts are represented by the firing of a population of a small number of neurons. Others (Bowers, 2009) argue more strongly that these findings support the "grandmother cell" hypothesis, essentially that information is represented by the firing of *single* neurons in the brain.

Either way, it does appear that the kind of local changes in the brain's structure that have been observed can radically alter the functional behavior of these networks. Hence, the brain seems particularly sensitive to variation at the cellular level among individuals, and even within the same individual over time during development and learning. It therefore seems necessary to study the function and capabilities of the brain at a more abstract level than the actual cellular network.

2.2 The Role of Environmental Factors in Shaping Architectural Features of the Brain

A classical view of the brain holds that the various skills and functions of the brain are localized in small regions. Beyond the evident features of the brainstem, cerebellum, hippocampus, etc., that can be identified with various functions, a vast array of sensory and motor tasks involved in vision, language, planning, etc., are believed to be carried out by various small patches of the cortex. Much of this view of *localization* was first put forth by Brodmann (1909). These classical views were largely based on the experience of loss of function of various kinds by patients who suffered damage to one part of their brain or another. Most famously, the first identified case was Broca's patient Tan Tan. Tan Tan, apparently due to a lesion in the front of the left hemisphere of his cortex, could only speak one syllable, "tan." His gestures led Broca and others to conclude that he could understand the language of others well enough, though. Thus, the natural hypothesis was that the damaged area was linked to language production. Modern in vivo imaging studies of healthy, functioning brains lend support to this view: for example, Moro et al. (2001) observed that activity in "Broca's area," an anatomical feature approximately located where Tan Tan suffered his injury, can be used to *detect* grammatical errors. More strikingly, a subsequent study by Musso et al. (2003) suggests that activity in this area may be tied to the grammar of natural language in particular, not simply rule learning in general. Concretely, subjects were taught constructed languages containing rules that violate the model of Universal Grammar that captures all known human languages; activity in Broca's area then predicted correct performance for rules that are consistent with Universal Grammar, and predicted performance errors for the inconsistent rules.

In view of the observed phenomenon of localization, a natural first hypothesis would be that the various functions of the cortex develop in portions of the cortical tissue in a genetically determined way. Such a course of development according to a genetically determined schedule would be much like how the other organs of the body differentiate from stem cells during development. Indeed, the observed pattern of connectivity of the cortex, which aided Brodmann's division of the cortex into regions, lends some weak support to such a view.

Strikingly, genetics per se appears *not* to be the primary factor determining the function of the cortex. Sur et al. (1988) determined that the auditory regions of ferret pups could be induced to develop features typical of the visual regions by surgically moving the retinal projections from the region of cortex usually associated with vision to the auditory regions. Sur et al. were thus led to posit that it is the inputs to a region of cortex during development that determines its function. That is, the visual areas develop

---

2. Valiant (2015) suggests a more liberal view of a *potential synapse* as occurring wherever axons and dendrites pass near enough to one another that a synapse could potentially be created. It is still plausible that the network given by these potential synapses does remain fixed.

in response to being connected to optic nerves, the motor regions develop in response to the thalamus and spinal cord, and so on. This view is further supported by the observed recruitment of the unused cortical regions by the remaining sensory modalities in congenitally blind (Cohen et al., 1997) and deaf (Bavelier and Neville, 2002) humans.

This is roughly in line with Lashley's doctrine of *equipotentiality* (Lashley, 1929): that any portion of the cortex has the capacity to carry out the function of any other. Equipotentiality, like localizationism, was originally supported by observations of patients with damaged or missing portions of their brain; but in this case, damage early in development, which leads to no apparent lasting deficiency (in contrast to damage to the developed brain).[3] A beneficial upshot of such hypotheses is that studies of the algorithm controlling development in one region of the cortex should transfer to others. A problematic upshot is that the results of such a developmental process are likely to vary across individuals; and, on account of the apparently localized representation of information in the cortex (cf. Section 2.1), these variations likely have significant impact on the way these fundamental computations are performed across the brains of various individuals.

2.3 Variability in Interconnections

So far, we have observed that it appears that any significant structural feature of the cortex *can* change—be it at the level of weights, synapses, or even functional regions. Moreover now, beyond the variability of the network of synapses in the cortex due to this plasticity, we observe that it appears that even the basic scaffolding of interconnections formed during development is not fixed. Braitenberg and Schüz (1998) describe *Peter's rule*, that this network of interconnections is formed during development by the idiosyncratic pattern of contact between cells. The pattern of interconnectivity in this network is therefore essentially random. Indeed, even the density of interconnections along specific pathways has been found to vary widely across individuals (Hilgetag and Grant, 2000). So, in summary, not only are the weights and synapses of the brain formed by the idiosyncratic processes of learning and development, but moreover, the *initial* configurations and larger scale aggregate structure of these networks are also highly diverse.

---

3. Along similar lines, motivated by observations of the structure and behavior of different regions of cortex, Mountcastle (1978) proposed that the various, hierarchically arranged regions of the cortex all employed the same algorithm for learning. The mechanisms by which cortical regions develop their unique character during development seem fundamentally distinct from the mechanisms underlying other kinds of learning, however (Greenough et al., 1987).

3 SUFFICIENTLY RICH COMPUTATIONAL MODELS ARE UNDERDETERMINED

Everyday experience tells us that the human brain is capable of learning to address a wide variety of tasks. Since the brain's function is apparently determined by its current, conditioned structure, it is perhaps unsurprising that the structure of the brain varies at all levels across individuals, and that the (apparently) localized representation of information in the brain renders the functions computed by portions of the brain highly sensitive to even small variations in this structure.

This poses a major barrier to the accurate modeling the brain of a single individual, let alone a population of individuals. Although it is surely possible to estimate various statistics of the distribution of such variability, we should expect that this diverse range of variation across individuals will be reflected by a functionally significant variation in the lower-level architecture of the brain, specifically, in the architecture of regions of the cortex. We should not expect the corresponding cortical regions of two individuals to compute the same function. And, to the extent that they compute similar functions, we should not expect them to be computed in the same way. So, we should not expect to be able to determine the functional architecture of the brain (across individuals) on the basis of such statistics alone.

The development of theoretical models capturing the workings of a brain is further hindered by the inherent weakness of the link between functional behavior and the underlying model whenever this model is computationally rich. In the first place, the underlying model itself is not determined by the functional behavior. In cognitive psychology, this is known as *nonidentifiability*.

I will review these issues and examine the prospects for addressing them, eg, by measurable "performance characteristics." Ultimately, in response to these difficulties, I will suggest in Section 4 that we should model "the" brain (or, precisely, regions of the brain) as belonging to a *family* of possible *functions* rather than a precise computational architecture.

3.1 Invariance and Nonidentifiability

The algorithms and representations used by the brain, ie, at Marr's intermediate level (Marr, 1982), are studied in cognitive psychology. In a classic work, Anderson (1978) argued that the functional, stimulus-response behavior of subjects is inherently inadequate to determine key aspects of algorithmic architecture. That is, the underlying algorithms and representations used by the brain are not *identifiable* on the basis of the observed competence (or noncompetence) of subjects at various tasks. Anderson's observation is closely related to the various forms of the Church-Turing thesis: essentially, his claim follows from the ability of one algorithmic architecture to simulate the functionality of another. The Church-Turing thesis is based on the observation that such simulations yield many models capturing precisely the same family of functions. The strong version of the Church-Turing thesis is based on the observation that whenever a natural model of time complexity exists, these simulations incur an overhead that is limited by some polynomial function of the running time and input size; so, as long as we allow for some polynomial "slack" in the amount of time, again we find that many natural models of feasible computations are also equivalent. (As we will review later, however, we *can* exhibit differences between the models if we take a more refined view, and don't allow arbitrary polynomial slack.)

A simple example of such nonidentifiability occurs in circuits. Suppose we have a circuit of gates computing the Boolean AND, OR, and NOT functions, where the AND and OR have arbitrary numbers of inputs: recall that the AND function outputs "true" if *all* of its inputs carry "true," and outputs "false" otherwise, whereas the OR function outputs "false" if *all* of its inputs carry "false," and outputs "true" otherwise. So the OR and NOT gates can *simulate* the AND gate: we can exchange the role of "true" and "false" by negating all of the inputs to the OR gate using NOT gates, and then negating its output (this is known as *de Morgan's law*). So, we cannot identify the difference between a network consisting of AND, OR, and NOT gates, and a network consisting only of OR and NOT gates (or AND and NOT gates). Notice, this simulation only uses three layers of gates to simulate a single AND gate, so if the circuit consisted of a constant number of layers ℓ (supposing for simplicity that there are no outputs feeding back to a higher layer), the simulation can be carried out in 3ℓ layers. More radically, consider the NAND function which outputs "false" if all of its inputs are "true" and outputs "true" otherwise. This gate computes the NOT function on a single input, and so can actually be used to simulate both the AND function using two layers of gates, as well as the OR function using two layers of gates; thus, an AND/OR/NOT circuit that is ℓ layers deep can be simulated by a NAND-only circuit that is 2ℓ layers deep and vice versa. Likewise, the total number of gates in this NAND-only circuit only needs to be twice that of the original AND/OR/NOT circuit (plus possibly the size of the input) since we only need to either compute a negation of the output of a gate, or compute the negation of the gate feeding its input.

Moreover, now, suppose we consider the family of *linear threshold* gates: on n inputs $x_1,\ldots,x_n$, these are parameterized by real-valued weights $w_1,\ldots,w_n$, and a threshold θ. Supposing "true" corresponds to 1 and "false" corresponds to 0, the linear threshold gate outputs "true" when $\sum_{i=1}^{n} w_i x_i \geq \theta$ and outputs "false" otherwise. Such gates are rich enough to express AND, OR, and NAND (and

many other functions besides). Therefore, there may be many different threshold circuits that ultimately express precisely the same Boolean computation in different ways. Such threshold gates correspond to the McCulloch-Pitts model of a neuron (McCulloch and Pitts, 1943). Therefore, at least in the McCulloch-Pitts model, we see that we will encounter difficulties if we wish to determine the actual architecture of a circuit based on its input-output behavior alone. Edelman (2004) discusses such systems, in which in particular there are multiple, functionally equivalent structures expressible by the same model. He calls such systems *"degenerate."* Since, as we reviewed in the previous section, the development of the connectivity and weights of the neural circuitry of the cortex appears to be highly idiosyncratic, we would certainly expect to encounter a wide variety of the possible circuits in this particular degenerate system.

The models of computation by our different kinds of gates are particularly similar; it is possible for far more diverse models to simulate one another, with correspondingly more serious consequences. Work on modeling the feedforward path of the ventral stream of the visual cortex, carried out independently by Serre et al. (2005), and Thorpe (2002) and Thorpe et al. (2004), serves as a mild cautionary tale. These two teams of researchers gave quantitative models of the function of the visual cortex constructed from biophysically plausible circuits, and demonstrated via computer simulations that the models so constructed performed well on standard vision tasks. This is a significant achievement, as it is a substantial and necessary test of any would-be theory of the workings of the brain. Consequently, both teams seemed to feel confident that they had proposed a model that would likely become the starting point for developing "the" model of the visual areas, and ultimately the entire brain. Imagine, then, the surprise of the researchers to discover that another team had achieved similar results utilizing quite dissimilar underlying mechanisms! The major difference between the two approaches was how data was encoded: Serre et al. had used a more traditional spike rate-based encoding scheme, whereas Thorpe et al. used a temporal scheme, ie, one in which earlier spikes have higher weight. Thorpe (2002) initially argued for a temporal coding based on the difficulties that a rate-based scheme would encounter in accounting for the fast response times observed in practice, and Guyonneau et al. (2005) have performed a more extensive theoretical study suggesting that the timing of spikes is what shapes a neuron's response. On the other hand, Serre et al. (2005) have tested the performance of their model against the performance of human subjects extensively, and found that it does fairly well at predicting the performance of the human subjects on visual tasks.

The lesson we should take away from this incident is that Marr's hierarchy (Marr, 1982) is not merely a taxonomy. The tasks of identifying the physiology and identifying the functionality are quite distinct, and should not be confused. The fact that two rather different underlying models were able to produce similar results should not surprise us so much. It follows from our prior discussion of equivalences across models of computation that a given class of problems will often have many equivalent formulations. The difference from Edelman's notion of degeneracy is that he is considering equivalent structures in a fixed family of models (eg, the family of McCulloch-Pitts circuits), whereas here we are concerned with the possibility that two different families of models—eg, temporal versus rate-based models here—of the underlying physiology could yield equivalent behavior. So, while the underlying model is critical from the standpoint of neurophysiology and often important from the standpoint of experimental design, studying the functional capabilities will not help us identify which of a set of computationally equivalent models best describes the underlying physiology. So, we must be careful not to claim too much about what such studies say about the physiology. Likewise, if we manage to successfully identify the computational power of the cortex with a class of circuits, further work on identifying the physiology that "truly" yields the computational model (at Marr's lowest level) will generally not contribute to our understanding of its function (at Marr's highest level). The functionality essentially never identifies a unique model, to say nothing of the underlying physiology that (in a sense) implements the model.

3.2 Limitations of Model Fitting

So, in general, one cannot identify "the" computation underlying a given behavior. But surely, one might think, it should be possible to at least capture the behavior itself, irrespective of whether or not we understand how it is computed. This is essentially the problem of *(supervised) machine learning*: given a collection of example inputs and outputs, can we find the function? Or more precisely, can we at least find a function that is close to the actual behavior, close enough that it can predict behavior under novel circumstances? Of course, we expect the actual functional behavior of individuals' cortical regions to vary, so the function we learned would be unique to that individual, somewhat analogous to sequencing their genome.

The work of Sur et al. (1988) suggests that the brain itself is learning these functions, so perhaps we as observers ought to be able to learn them, too. Indeed, one might be optimistic, given not only the simulations of Thorpe (2002) and Thorpe et al. (2004) and Serre et al. (2005), but also related, recent work involving *"deep neural nets."* These models are more loosely inspired by the observed architecture of the cortex, but have moreover sometimes matched human-level performance in tasks such as labeling images (Krizhevsky et al., 2012), speech recognition

(Dahl et al., 2012), and face detection (Taigman et al., 2014). In light of this, one might be optimistic that the functions being trained by these "deep nets" might actually be that learned by the cortex—that because we are getting similar input-output behavior out of computationally feasible model-fitting processes for the same family of models, perhaps they really do capture functionally equivalent representations. Or at least, that they produce models that are representative of "typical" brains.

These hopes are thoroughly dashed by the work of Nguyen et al. (2015). They show that it is easy to construct, for example, images that appear to a human observer to be a slightly "noisy" picture of one type of animal, but that are confidently classified by the deep net as an entirely different type of animal. Note that these counterexamples are consistent with the claim that the trained deep net matching human level performance, as these images are not *natural*: they feature "noise" that has been carefully chosen with knowledge of the deep net's weights. Learning algorithms, for example, in Vapnik's statistical learning model (Vapnik, 1998) and Valiant's computationally bounded model (Valiant, 1984), only guarantee that a good *approximation* to the target is learned, as measured by the distribution of the data. That is, learning algorithms only promise correct labels on most of the images that are likely in the natural distribution. They do not and cannot promise to identify the actual function. While we can improve the approximation by training on more data, it is infeasible to collect a training set large enough to guarantee good predictions on the entire range of inputs of interest. So, not only do the representations produced fail to enable any deeper analysis of the actual representations used or other features of the cortical computation, they also cannot be used to make predictions beyond the labels on the training distribution. Thus, as they cannot be used to answer our natural questions, they are poor models of the cortical region.

3.3 Avenues for Model Identification and Their Challenges

Modeling the function of regions of the cortex in actual subjects is still the most straightforward approach to studying the brain. We will briefly review its prospects. As these approaches face formidable challenges, this review will motivate a more abstract study of the brain in Section 4.

To begin, we note that the function of such circuits must be considered in the context of natural data: The work of Sur et al. (1988) suggests that neural circuits are developed in response to the kind of inputs they receive. As we noted above, the functions produced by such learning systems are only meaningfully fixed in the context of the distribution of their training data. In particular, this suggests that the context of the signaling inputs received from the rest of the cortex will be essential to studying the function of regions that are not immediately connected to sensory inputs; in other words, in vivo studies are therefore essential.

At a sufficiently small granularity, identifiability is not at issue. For example, we can begin to make meaningful sense of the behavior of a single neuron. The studies of Quian Quiroga et al. (2005) have used in vivo recordings of single neurons to great effect here, giving strong evidence that concepts are represented by the firing of a few neurons. The trouble, of course, is that it is hard to record from more than a relatively small number of neurons at a time, on the order of thousands, whereas cortical regions may contain hundreds of millions of neurons. Such techniques therefore cannot say much about more abstract computations in the brain, which depend on an understanding of the patterns of activity in neighboring regions. Indeed, note that the successes of this technique were at identifying activity with images of faces, names, and buildings, which are relatively concrete sensory inputs.

While the prospects are dim for obtaining useful information from direct recording of neurons in these more abstract regions of cortex, functional magnetic resonance imaging (fMRI) has been used profitably to obtain information at a smaller granularity than the totality of human behavior. Although it cannot resolve the activity of individual neurons, it can at least give a rough picture of the activity in cortical regions in vivo. For example, this technique was used by Moro et al. (2001) and Musso et al. (2003) to investigate the hypotheses that the Broca-Wernicke areas use a Chomskian Universal Grammar to decode and generate linguistic structure, and moreover, that this grammar computation is distinct from the general rule-learning capacity of the brain. Donoso et al. (2014) likewise used fMRI to produce evidence for activity in the prefrontal cortex capturing the effectiveness of abstract hypotheses being considered by subjects, and predicting changes between such hypotheses. In particular, they proposed an approximate-Bayesian reinforcement learning model that generates similar activity. While such fMRI studies provide useful tests for such well-developed models, due to the coarseness of the signal they provide, however, they are less useful in the development of models. In particular, because fMRI cannot resolve the actual representation of information by the cortex, it cannot actually determine the functions being computed. We can only check the correlation of the strength of activity with rather specific quantities, such as the probability that a subject makes an error of some kind.

A related approach is to use the pattern of such performance errors and computation time to inform the study of the brain at the algorithmic level (Newell, 1990; Pylyshyn, 1984). Although we know that one computational model can generally simulate another using similar resources, this equivalence breaks down under sufficient precision. For example, Maass (1984) has shown that the basic, single-tape Turing machine cannot determine whether or not a string of letters is a palindrome without taking time that grows quadratically in the length of the string. By contrast, most of our usual architectures (such as the random-access memory model underlying desktop digital computers) can check whether or not a string is a palindrome in time merely proportional to the length of the string. Thus, in principle, the single-tape Turing machine might be ruled out as a computational model for the brain *if* the amount of time taken by subjects to determine whether or not a string was a palindrome appeared to be sub-quadratically related to the length. The primary issue that arises is that if subjects did appear to be taking quadratic time, then we essentially learn nothing—it is possible for stronger architectures to utilize a suboptimal algorithm. Moreover, limits will remain on the extent to which we can nail down the actual architecture; although such separations will allow us to rule out some architectures, not all pairs of models can be separated this way. Some simulations are indeed quite efficient, such as the circuit simulations we described earlier. The actual power of the different models is then essentially the same, even with respect to the resources available. We then cannot use such a strategy for model identification.

4 COMPLEXITY-THEORETIC MODELING OF THE HUMAN BRAIN

So far, we have seen that the actual functions computed within cortical regions are likely to vary widely across subjects, and that the prospects for resolving examples of these functions in detail are dim. I now argue that the brain—or, more precisely, regions of the cortex—is meaningfully studied as *families* or *"classes"* of functions, constrained by the resources available to compute them—eg, number of neurons, energy, depth, etc.—and their learnability.[4] Computational complexity is precisely the study of the relationship between such resource constraints and the functions that can be computed within such limited resources; computational learning theory (Valiant, 1984) is, in turn, the study of the learnability of functions as informed by computational complexity. Such families of circuits thus provide a model for regions of the cortex capturing the range of *potential* functional variation of a typical brain. A key benefit of taking this relaxed stance, that we wish to identify a *class of functions* as opposed to a computational model of the brain, is that model-invariance now

---

4. A similar argument has been made before, see for example, Parberry (1994). But, much more has been discovered in the study of both circuits and learning since that work, and the details of my conclusions will differ.

works in our favor: as the classes of functions are relatively insensitive to the details of the computational models, we expect the mathematical theory to tolerate revisions to the picture of the brain at the cellular level.

4.1 Complexity-Theoretic Models for Short Time Scales

Our starting point for the modeling of the computation performed by the brain is the classical McCulloch-Pitts model of the neuron (McCulloch and Pitts, 1943). In this model, a neuron takes as inputs the presence or absence of spikes along the synapses of its dendrites, encoded by Boolean values, and computes a linearly weighted vote to decide whether or not to itself produce a spike. That is, if the presence or absence of input spikes is encoded by the Boolean variables $x_1,\ldots,x_n$, there are weights $w_1,\ldots,w_n$ indicating the strength of these synapses, as well as whether they are excitatory or inhibitory; the neuron then decides to fire if $\sum_{i=1}^{n} x_i w_i \geq \theta$ for some real-valued threshold θ. In essence, the McCulloch-Pitts neuron thus computes some *linear threshold* gate.

The brain is then modeled as a circuit comprised by such linear threshold gates. More narrowly, we may model the behavior of a region of the cortex on short time scales (eg, 100–200 ms) by an acyclic circuit of highly limited depth, since on such short time scales information appears to flow in only one direction, and the cascade of spikes can only reach neurons a few synapses away (Thorpe et al., 1996). In the coarse-grained framework of computational complexity, we translate such concrete parameters into relative, asymptotic quantities, taking the actual brain to be a typical member of an infinite *family* of such circuits of different sizes. For example, we may naturally choose to cast these parameters as varying relative to the size of the input (number of neurons or other cells with projections to the region of interest), "n," that provides an index for the family. We may then take the "very small" depth to be bounded by some (arbitrary) constant independent of n; by contrast, parameters of size similar to the input, such as the number of neurons involved in the computation, would be regarded as having an arbitrary polynomial relationship to n.

This is, of course not entirely correct—we know gross limits on the sizes of brains and cortical regions, and so such an asymptotic statement is therefore nonsense, strictly speaking. But, the number of layers through with spikes may pass is quite small relative to the hundreds of millions of neurons that may comprise the cortical regions through which such spikes pass. The asymptotic relationship captures the algorithmically significant relationship that one quantity is very small relative to another, without requiring us to specify precisely how much so.

Moreover, modeling the depth of layering of such circuits as bounded by a constant, independent of the input size, is mathematically convenient. In particular, a major benefit of a model formulated in these terms is that it is *invariant* or robust to many variations in its precise formulation; this is the positive side of nonidentifiability. For example, more sophisticated models of the neuron that include computation along the dendritic branches (Koch, 1998; Koch and Segev, 2000; London and Häusser, 2005; Herz et al., 2006) will not alter the computational model we use substantially, although they do generally suggest that a single "layer" of neurons is more aptly modeled as two or possibly three layers of threshold circuits.[5] Since our model is given by threshold circuits with small but otherwise arbitrary numbers of layers, such variation in the exact number of layers will be irrelevant to us.[6] We note that the analog properties of the real neurons comprising these threshold circuits are likewise essentially irrelevant from a computational perspective: Maass and Orponen (1998) have shown that the presence of even small amounts of noise renders such circuits no more powerful than Boolean threshold circuits.

Such circuits are indeed highly parallel; this parallelism allows significant computations to be performed under such severe timing constraints. But, the need for parallelism is *just another constraint on the functions computed.* We note that the variety of functions that can be computed in parallel under such severe constraints is quite limited relative to a desktop computer. In particular, the aforementioned works of Serre et al. and Thorpe et al. demonstrate that cortical regions can be meaningfully simulated on a computer (Serre et al., 2005; Thorpe, 2002; Thorpe et al., 2004).

We now will seek to identify a *sub*class of such circuits that is *learnable* by feasible algorithms (at least, polynomial-time bounded). Learnability is actually quite a severe constraint. A result by Klivans and Sherstov (2009) shows that even very simple two-layer circuits of linear threshold functions may not be learnable: even if the output layer threshold gate always computes an AND, any strategy that could learn even a nontrivial approximation to the function computed by such circuits could be used to break encryption schemes that are believed to be secure. A corollary of such results is that although we do not know what representations can be learned by deep nets in practice, we certainly *don't* believe that they can even be trained to fit arbitrary two-layer neural

---

5. These layers are determined by the connectivity of the neurons—that is, the first "layer" consists of neurons that can be reached directly from the "inputs," the second layer consists of neurons with synapses taking input from neurons in the first layer, and so on. This layering is *not* meant to correspond to the layered structure of the cortical tissue itself.

6. This stands in contrast to the implication of such developments on the variety of learning strategies available at the neuronal level, where the ability to jointly update two layers is significantly different.

networks, even though their architecture may be capable of representing precisely such networks.

More generally, there is a distinction in learning theory between the functions that a learning algorithm can learn, and the kinds of representations that the learning algorithm produces. Both are generally captured by some class of representations, and the expressive power of the class of representations that the algorithm produces necessarily constrains the power of the functions it can learn. So, for example, if we take our constant-layer linear-threshold gate (McCulloch-Pitts neuron) networks as our representation, then we can only learn functions that can be expressed by such circuits. But, the class of functions that are efficiently learnable is often strictly weaker than the representations produced by their efficient learning algorithms, as appears to be the case here.

So, the class of *all* constant-depth, polynomial-size threshold circuits is likely too rich for our purposes. One biologically motivated further restriction on the class of circuits was proposed by Uchizawa et al. (2006): they proposed to consider such circuits with the further restriction that on any input, a relatively small number (logarithmically many) of the model neurons "fires," ie, relatively few of the threshold gates output "true." This restriction is motivated by the observations of Margrie et al. (2002) that the spiking activity in the brain is relatively sparse, and the related *energy conservation principle* proposed by Lennie (2003) on theoretical grounds: namely, a firing rate of just a few percent in each local region of the cortex would be metabolically unsustainable.

Now, the learnability of functions represented by a class of circuits appears to be closely related to the tractability of the mathematical analysis of such circuits, for example, to our ability to establish that some functions are *not* computed by such circuits. An informal connection has long been observed between learnability and such circuit complexity lower bounds, and moreover, work by Fortnow and Klivans (2009) has provided some formal connection. Of course, the tractability of the mathematical analysis of such circuits is of fundamental importance to this program of work in its own right. On this count also, the family of all constant-depth, polynomial-size threshold circuits was known to be too rich, again essentially due to the fact that secure cryptography is believed to be feasible within such circuits (Razborov and Rudich, 1997). Fortunately, however, once the class is further restricted to have limited energy complexity as proposed by Uchizawa et al., such analysis becomes feasible: in particular, Uchizawa and Takimoto (2008) were able to show strong lower bounds for computing a simple function (the Boolean inner product) in this model. We still do not know if this class of circuits is efficiently learnable, but given the observed close connection between learnability and such lower bounds, it appears likely to be learnable. This class is therefore a natural candidate for the functions computed by regions of cortex on short time scales, ie, along feed-forward paths.

4.2 Modeling for Longer Time Scales

On time scales longer than a few hundred milliseconds, there is adequate time for the recurrent (feedback) connections to play a significant role in the functionality. Of course, the backward connections may already play a role in shaping the response to stimuli on the feed-forward path (a view we will revisit in Section 5), but there they cannot respond to the stimulus, which substantially increases the computational power of the circuits: Even in a "constant-depth" model, without the energy restriction of Uchizawa et al. (2006), cyclic circuits can simulate general time-and/or memory-bounded computation (depending on whether the number of iterations is bounded). And, even with a logarithmic energy complexity bound, such circuits can (at least) simulate finite automata with a number of states quadratically larger than the number of gates of the circuit (Šíma, 2014). Finite automata are strongly believed not to be learnable (cf., Kearns and Valiant (1994)), so this fact may not be of much relevance to the computational power of the learned behavior of the individual regions.

On such longer time scales, however, we may be more interested in the architecture of connections between the features and regions of the brain, in the spirit of Brodmann's map. In particular, a modern update of such a map is the objective of the connectome project. We might model the individual regions as computing constant-depth, polynomial-size, logarithmic-energy circuits, imposing the constraint of learnability at the level of the individual cortical regions alone. If we now consider the power of the resulting overall architecture, Šíma's result then gives us some idea of its potential extent.

An alternative, related line of work initiated by Valiant (1994, 2000), has aimed to develop a quantitative computational model of the brain that closely models the quantitative aspects of the underlying physiology while approaching the kind of algorithmic model sought in cognitive science. Of course, in light of the challenges posed by identification at Marr's algorithmic level, Valiant is careful to stress that he is *not* asserting that the algorithms he describes in the neuroidal model are actually used by the brain. Rather, he is only seeking to demonstrate that the model, set with realistic parameters, is adequate to capture the range of behavior exhibited by the brain. A current review of the program and its guiding philosophy is given by Valiant (2014). Work by Valiant (2005) and Feldman and Valiant (2009) has demonstrated that this model can support multiple basic tasks and a plausible amount of storage while using a plausible degree of connectivity, etc. For example, an analysis in Valiant (2006) observed that these ranges of parameters were consistent with the presence of the kind of sparse concept representations conjectured to underlie the observation of the "Halle Berry" neuron by Quian Quiroga et al. (2005).

As the neuroidal model models the brain as performing computations in a way constrained by the quantitative parameters of the observed physiology, it is in particular a computational complexity-theoretic model. One significant distinction between the program underlying the neuroidal model and the program I am suggesting, however, is that I am proposing somewhat looser quantitative constraints than considered thus far by Valiant. The reason is that I would like to understand the relationships between (our models of) the brain and the models considered in the larger complexity-theoretic study of computation. For example, it is natural to ask how the power of a network of neuroids compares to other computational models, such as the threshold circuits of bounded energy complexity discussed earlier. This is not immediate, as although neuroids have a threshold-firing behavior like threshold gates, they also operate according to a nontrivial state-transition model (eg, resetting after a refractory period), unlike these gates. The neuroidal model with a given range of parameters naturally has greater fidelity to the actual brain than the model I wish to consider; but, the kinds of simulations used to establish such comparisons rely on a little bit of slack or overhead in the computational constraints they obey. It is thus theoretically convenient to regard models as having "the same" or "similar" power in spite of having to lose some resources in converting from one to the other. We can also typically give these arguments in such a way that concrete bounds can still be extracted for the purposes of testing a theory.

A complexity model such as the finite automata model used by Šíma (2014) that counts the number of states, as opposed to simply bounding the complexity of the algorithm, is likely crucial here. Primarily this is because it allows us to meaningfully discuss computation on finite inputs, by "charging" the finite automaton for the complexity of the algorithm it uses, which is kept separate from the space and time used in the Turing machine model. That is, such a state-size versus gate-size bound makes concrete, testable predictions about the number of neurons, level of activity, etc. required in order to support a given computation on a given input size. It also respects the fact that the brain's computation only needs to be performed on a limited size input. Although we are currently only aware of this feature of the model giving it additional power under somewhat contrived circumstances, such circumstances do exist.

5 CHARACTERIZING FUNCTIONALITY USING COMPLEXITY-THEORETIC CONCEPTS

A second role for the language and models of computational complexity in the study of the brain is that it allows us to formulate claims and questions about the kinds of computations performed by systems like brains, while abstracting away the particular details of *how* this computation is performed. In this way, it becomes meaningful to pose such questions across different systems that were *not* apparently designed with any particular semantics in mind, eg, in comparative biology. I will give an example of such "quantitative philosophy" by sketching a definition of "predictive ability," as proposed to capture "intelligence" (Hawkins and Blakeslee, 2004; Clark, 2013).

5.1 The Necessity of Abstract and Generalized Theories

In the study of the function of the brain, following Blum et al. (2005), I claim that it is desirable to define properties such as consciousness or intelligence in an abstract or generalized manner. Such a theoretical description of these concepts would of course be complementary to the actual study of the brain's physiology: in the physical study, certain processes should be observed which could be verified to have the necessary properties to be called "conscious" or "intelligent" in the abstract theory. These theories are desirable because they are *necessary* to decide the presence or absence of such high-level features in other systems. For example, a natural question that often arises is whether members of other species—dogs, octopi, etc.—are conscious or intelligent (and *how* intelligent). A more loaded family of questions concern when a human fetus or infant, or a brain-damaged patient should be considered conscious, which may have serious legal consequences.

Specifically, if we wished to decide whether or not a member of another species, or an entire ant colony, or a robot is conscious or intelligent, it would obviously be unsatisfying to conclude that since none of these have precisely the kind of primate brains in which the various studies of neural representation have been carried out, none could be conscious or intelligent. Turing's test (Turing, 1950), which was proposed as a means to identify "intelligence" without having to specify precisely what "intelligence" is, also has some related, well-known shortcomings. In Turing's test, a judge converses with a human and a machine (or some other subject), and attempts to tell which is which; if the judge cannot tell the difference with probability adequately greater than chance, then the test subject is declared intelligent. The troubles with the test include its failing of subjects who refuse to participate or, for whatever reason, cannot communicate in the judge's language. Clearly, we need to develop theories that describe what it is about the function of the brain that gives rise to these properties without being tied to the particular realization in the systems we observe. When a subject is intelligent or conscious, these theories should permit a simple demonstration of this property by describing how the required systems are implemented. The theory should also make it possible in principle to demonstrate conclusively that a system is *not* intelligent or conscious, although demonstrating the absence of these properties may not necessarily be so simple. Establishing such a claim might require demonstrating that the system in question is incapable of computing some necessary functions using some lower-bound argument, or some kind of clever experiment. Nevertheless, possession of such abstract definitions allows us to meaningfully consider such questions, which is the first necessary step to resolving them.

5.2 Outline of a Complexity-Theoretic Study of High-Level Functionality

In our study we wish to isolate the tractable portions of the problem—in this case, the functional or mechanical aspects of our high-level attributes—from the philosophically contentious questions of what an entity really "thinks" or "feels." Koch (2004) gives a nice, analogous discussion of the distinction between the neural correlates of consciousness that he wishes to study and a philosopher's notion of consciousness. We begin by viewing the brain as a formal system, as circuits from a particular class. We continue by asking what properties of those circuits characterize our concepts like intelligence or consciousness: we are seeking the necessary and sufficient conditions that permit our formal system to exhibit the behavioral or mechanical aspects of these high-level properties. We then may define the high-level properties in terms of these conditions.

For example, I will sketch a first attempt at a complexity-theoretic definition of "predictive ability"; Hawkins and Blakeslee (2004) proposed that such an ability to make predictions about one's environment captures our informal notion of "intelligence." (See Blum et al. (2005) for a similar initial attempt at defining "consciousness.") Their proposal was motivated by Rao and Ballard (1999) and Lee and Mumford (2003), who suggested that the feedback connections in the cortex serve to carry predictions about future inputs. (A similar proposal was fleshed out in more detail recently by Clark (2013).) I will attempt to show how this idea might be translated into a formal definition, illustrating where we expect computational complexity theory to play a role. I stress that this definition is agnostic with respect to the underlying mechanisms and encodings used; in particular, it is *not* necessarily tied to the hierarchical Bayesian frameworks in which the ideas were first developed. For example, the "predictive join" operation in the neuroidal model, proposed by Papadimitriou and Vempala (2015) should satisfy our

framework, when applied to the inputs to the model. As we will see, this definition abstracts a level above even Marr's functional level, identifying a property that such *functions* may or may not possess.

We will model time as proceeding in discrete steps. Sequences of one input string for each time step, $x_1, x_2, \ldots, x_t, \ldots$, will then represent an environment; we will later have cause to consider *classes* of environments, which are merely sets of possible input string sequences. We will consider our candidate predicting machine M to have some internal state s_t, which in the language of Turing machines would be the contents of its work tapes or in the language of modern computers would be a dump of the contents of its memory; in a more natural setting, we could think of this as a "snapshot" of neural activity around one moment in time. Regardless of our terminology, if M is in state s_t and sees input x_t, it may perform some action and updates its internal state to s_{t+1}. We assume that the action and s_{t+1} are determined uniquely by s_t and x_t.

In this setting, we will take "the ability to make predictions" to mean that at time t, M frequently has access to x_{t+1}. We remark that although one could consider richer notions of "predictive ability," this simple notion is still nontrivial, and is sufficient to illustrate how we proceed in providing such definitions. Now, in order for M to be making predictions about x_{t+1}, it must be updating its internal state so that eventually s_t, together with x_t will contain sufficient information about x_{t+1} that M can do better than merely guess its contents, but we will *avoid* requiring it to store this information in any particular format. Instead, we will say that M is intelligent with respect to some class of environments $\mathcal{C}$ if there is some efficiently (ie, polynomial-time) computable *deciphering function D_M* such that for any environment $\{x_t\} \in \mathcal{C}$, eventually, $D_M(x_t, s_t) = x_{t+1}$ with frequency "nonnegligibly" better than chance—formally, better than chance by an amount that is at least $1/\text{poly}(|x_t|, |s_t|)$, that is, some inverse polynomial in the encoding length (in bits) of the inputs and states. We may regard such deciphering functions as generalizing the "read-out" classifiers of Hung et al. (2005). By allowing a computation that extracts (or "translates") the relevant information packed into s_t, the use of a custom deciphering function grants our theory the desired robustness to the details of the machine model, information encoding, and so on.[7] A machine M for which such a deciphering function exists has sufficient information about x_{t+1} in such a format that we may consider it to possess such a prediction, whereas any realistic machine for which no such deciphering function can be implemented could not be doing much better than blindly guessing about the next input, so satisfying this property is necessary and sufficient for

---

7. The strong Church-Turing thesis asserts that the definition of "efficiently computable" is similarly robust.

the machine to make predictions about the next input from environments in $\mathcal{C}$.

Something like "nonnegligibility" is generally necessary for us to even observe a statistical difference in the behavior from random guessing. The requirement of a "nonnegligible" advantage also helps guard the definition against declaring M able to predict on the basis of a deciphering functions with a few special rules that obtain a small advantage by looking at x_t directly. On account of the limitation that D_M be fixed, this generally only offers a negligible advantage (in the technical sense) against rich families of environments. Moreover, we note that in the related case of learning, a result of Freund and Schapire (Schapire, 1990; Freund and Schapire, 1997) establishes that several such "weak predictions" can be used to make "strong predictions" in which the predictions are correct with probability that are arbitrarily close to certainty. In other words, the ability to make such weak predictions (on examples weighted by the performance of the other weak predictors) addresses the algorithmic essence of the learning task.

We remark that one feature of this definition is that it is possible to classify the *degree* of intelligence of a machine M by the richness of environment classes that it can successfully predict; it is easy to see that on one end of the spectrum, any machine is able to predict a constant environment, but no machine is able to predict (the class of) all environments. We are particularly interested in the behavior of the machine with respect to "natural" environments, by which we mean informally the class of environments corresponding to the entity's natural environment, encoded appropriately as Boolean strings. We may speculate that such environments, if captured formally, would be produced by a process featuring an infinite "unobserved state" and each symbol in its state at a given time-step would be computed from portions of its state in the previous time-step that were at most some bounded distance away—a locality constraint in its update rules, but more work would need to be done before we would be satisfied with our definition. Part of the Hawkins-Blakeslee notion of "intelligence" would involve characterizing these environments more precisely; machines capable of predicting such "natural" environments would be considered "intelligent."

We should require that any model of the low-level mechanisms in the brain (eg, formal models of neural circuits) be sufficiently powerful to exhibit intelligence, defined in this way. Note that we assume that M's storage is bounded, and hence that M cannot simply offload the task to D_M by storing everything. So, the problem of preparing to make such predictions is nontrivial in general, and this definition does require the models that satisfy it to have some computational power.

Returning to our broader goals, we would like to separate which characteristics of the brain's function should be considered necessary, such as making predictions in the proposal above, and which merely serve to implement those functions. Notice that this definition makes no mention of any details of how the brain carries out the task of making predictions. Rather, it abstractly characterizes or specifies what makes a machine "capable of making predictions." This is analogous to separating the specifications of the neural correlates of consciousness (Koch, 2004), etc., from their implementations. Once this has been accomplished, then we can say that an entity having implementations of the specified functions has the high-level attribute described—intelligence, consciousness, etc.

We expect such definitions to be relevant since this functional description of our high-level properties is precisely what we need for the sorts of abstract and generalized theories discussed earlier. By separating the specifications of the functions from their implementations, we permit implementations in different settings to be constructed or discovered, in particular in other species. In addition, if such theories have been developed and the special case of brains have not yet been fully understood, then we would hope that our definitions would help explain how the function of the brain could give rise to at least the observable aspects of consciousness, intelligence, etc., by clarifying precisely what sort of functionality we are looking for and establishing what kind of underlying models would be necessary or sufficient for exhibiting such functionality.

ACKNOWLEDGMENTS

I thank Sashank Varma and Frank Jäkel for a series of conversations that shaped the section on identifiability and invariance, and their comments on a previous draft. I thank Erick Chastain for much helpful feedback on an early version of this work that appeared in the Carnegie Mellon University Undergraduate Research Journal, *Thought* (Juba, 2006). I also thank Leslie Valiant for an informative discussion on neural modeling, and Manuel Blum, Ryan Williams, and Matt Humphrey for the many influential discussions while working on the CONSCS project. Parts of this work were supported by the National Science Foundation as part of the Aladdin Center (www.aladdin.cmu.edu) under grant CCR-0122581, and associated REU funding. The preparation of the present manuscript was supported by an AFOSR Young Investigator award.

REFERENCES

Anderson, J.R., 1978. Arguments concerning representations for mental imagery. Psychol. Rev. 85 (4), 249–277.

Bavelier, D., Neville, H.J., 2002. Cross-modal plasticity: where and how? Nat. Rev. Neurosci. 3 (6), 443–452.

Bliss, T.V., Lømo, T., 1973. Long-lasting potentiation of synaptic transmission in the dentate area of the anaesthetized rabbit following stimulation of the perforant path. J. Physiol. (Lond.) 232, 331–356.

Blum, M., Williams, R., Juba, B., Humphrey, M., 2005. Toward a high-level definition of consciousness. In: Proceedings of the 20th IEEE

Computational Complexity Conference, San Jose, CA, Invited Talk, http://www.cs.cmu.edu/mblum/research/pdf/CONSCS8.ppt.

Bowers, J.S., 2009. On the psychological plausibility of grandmother cells: Implications for neural network theories in psychology and neuroscience. Psychol. Rev. 116, 220–251.

Braitenberg, V., Schüz, A., 1998. Cortex: Statistics and Geometry of Neuronal Connectivity, second Springer, Berlin.

Brodmann, K., 1909. Vergleichende Lokalisationslehre der Großhirnrinde. Barth-Verlag, Lepzig.

Clark, A., 2013. Whatever next? Predictive brains, situated agents, and the future of cognitive science. Behav. Brain Sci. 36, 181–204.

Cohen, L., Celnik, P., Pascual-Leone, A., Corwell, B., Faiz, L., Dambrosia, J., Honda, M., Sasato, N., Gerloff, C., Catala, M.D., Hallett, M., 1997. Functional relevance of cross-modal plasticity in blind humans. Nature 389, 180–183.

Dahl, G.E., Yu, D., Deng, L., Acero, A., 2012. Context-dependent pretrained deep neural networks for large-vocabulary speech recognition. IEEE Trans. Audio Speech Language Process. 20 (1), 30–42.

Donoso, M., Collins, A.G.E., Koechlin, E., 2014. Foundations of human reasoning in the prefrontal cortex. Science 344, 1481–1486.

Edelman, G.M., 2004. Wider Than the Sky: The Phenomenal Gift of Consciousness. Yale University Press, New Haven, CT.

Feldman, V., Valiant, L.G., 2009. Experience-induced neural circuits that achieve high capacity. Neural Comput. 21, 2715–2754.

Fortnow, L., Klivans, A.R., 2009. Efficient learning algorithms yield circuit lower bounds. J. Comput. Syst. Sci. 75, 27–36.

Freund, Y., Schapire, R.E., 1997. A decision-theoretic generalization of on-line learning and an application to boosting. J. Comput. Syst. Sci. 55 (1), 119–139.

Greenough, W.T., Black, J.E., Wallace, C.S., 1987. Experience and brain development. Child Dev. 58 (3), 539–559.

Guyonneau, R., VanRullen, R., Thorpe, S.J., 2005. Neurons tune to the earliest spikes through STDP. Neural Comput. 17, 859–879.

Hawkins, J., Blakeslee, S., 2004. On Intelligence. Times Books, New York, NY.

Hebb, D.O., 1949. The Organization of Behavior: A Neuropsychological Theory. Wiley, New York, NY.

Herz, A.V.M., Gollisch, T., Machens, C.K., Jaeger, D., 2006. Modeling single-neuron dynamics and computations: a balance of detail and abstraction. Science 314 (5796), 80–85.

Hilgetag, C.C., Grant, S., 2000. Uniformity, specificity and variability of corticocortical connectivity. Philos. Trans. R. Soc. B 355 (1393), 7–20.

Hung, C.P., Kreiman, G., Poggio, T., DiCarlo, J.J., 2005. Fast readout of object identity from macaque inferior temporal cortex. Science 310 (5749), 863–866.

Juba, B., 2006. On the role of computational complexity theory in the study of brain function. Thought (Carnegie Mellon Univ. Undergraduate Res. J.) 1, 32–45.

Kearns, M., Valiant, L., 1994. Cryptographic limitations on learning Boolean formulae and finite automata. J. ACM 41, 67–95.

Klivans, A.R., Sherstov, A.A., 2009. Cryptographic hardness for learning intersections of halfspaces. J. Comput. Syst. Sci. 75 (1), 2–12.

Koch, C., 1998. Biophysics of Computation—Information Processing in Single Neurons. Oxford University Press, New York, NY.

Koch, C., 2004. The Quest for Consciousness: A Neurobiological Approach. Roberts & Company Publishers, Englewood, CO.

Koch, C., Segev, I., 2000. The role of single neurons in information processing. Nat. Neurosci. 3, 1171–1177.

Krizhevsky, A., Sutskever, I., Hinton, G.E., 2012. Imagenet classification with deep convolutional neural networks. In: Advances in Neural Information Processing Systems 25 (NIPS 2012), pp. 1097–1105.

Lashley, K., 1929. Brain Mechanisms and Intelligence: A Quantitative Study of Injuries to the Brain. Chicago University Press, Chicago, IL.

Lee, T.S., Mumford, D., 2003. Hierarchical Bayesian inference in the visual cortex. J. Opt. Soc. Am. 20 (7), 1434–1448.

Lennie, P., 2003. The cost of cortical computation. Curr. Biol. 13, 493–497.

London, M., Häusser, M., 2005. Dendritic computation. Ann. Rev. Neurosci. 28, 503–532.

Maass, W., 1984. Quadratic lower bounds for deterministic and nondeterministic one-tape turing machines. In: Proceedings of the 16th Annual ACM Symposium on Theory of Computing (STOC), pp. 401–408.

Maass, W., Orponen, P., 1998. On the effect of analog noise in discrete-time analog computations. Neural Comput. 10, 1071–1095.

Margrie, T.W., Brecht, M., Sakmann, B., 2002. In-vivo, low-resistance, whole-cell recordings from neurons in anaesthtized and awake mammalian brain. Pflugers Arch. 444 (4), 491–498.

Marr, D., 1982. Vision: A Computational Investigation into the Human Representation and Processing of Visual Information. Freeman, New York.

McCulloch, W.S., Pitts, W., 1943. A logical calculus of ideas immanent in neural activity. Bull. Math. Biophys. 5, 115–133.

Moro, A., Tettamanti, M., Perani, D., Donati, C., Cappa, S.F., Fazio, F., 2001. Syntax and the brain: disentangling grammar by selective anomalies. Neuroimage 13, 110–118.

Mountcastle, V.B., 1978. An organizing principle for cerebral function: the unit module and the distributed system. In: The Mindful Brain. MIT Press, Cambridge, MA, pp. 7–50.

Musso, M., Moro, A., Glauche, V., Rijntjes, M., Reichenbach, J., Büchel, C., Weiller, C., 2003. Broca's area and the language instinct. Nat. Neurosci. 6, 774–781.

Newell, A., 1990. Unified Theories of Cognition. Harvard University Press, Cambridge, MA.

Nguyen, A., Yosinski, J., Clune, J., 2015. Deep neural networks are easily fooled: high confidence predictions for unrecognizable images. In: Proceedings of the 2015 IEEE Conference on Computer Vision and Pattern Recognition (CVPR 2015).

Papadimitriou, C., Vempala, S., 2015. Cortical learning via prediction. In: Proceedings of the Conference on Learning Theory (COLT 2015).

Parberry, I., 1994. Circuit Complexity and Neural Networks. MIT Press, Cambridge, MA.

Pylyshyn, Z., 1984. Computation and Cognition. MIT Press, Cambridge, MA.

Quian Quiroga, R., Reddy, L., Kreiman, G., Koch, C., Fried, I., 2005. Invariant visual representation by single neurons in the human brain. Nature 435, 1102–1107.

Quian Quiroga, R., Kreiman, G., Koch, C., Fried, I., 2007. Sparse but not 'grandmother-cell' coding in the medial temporal lobe. Trends Cogn. Sci. 12 (3), 87–91.

Rao, R.P.N., Ballard, D.H., 1999. Predictive coding in the visual cortex: a functional interpretation of some extra-classical receptive-field effects. Nat. Neurosci. 2, 79–87.

Razborov, A.A., Rudich, S., 1997. Natural proofs. J. Comput. Syst. Sci. 55, 24–35.

Schapire, R.E., 1990. The strength of weak learnability. Mach. Learn. 5 (2), 197–227.

Serre, T., Kouh, M., Cadieu, C., Knoblich, U., Kreiman, G., Poggio, T., 2005. A theory of object recognition: computations and circuits in the feedforward path of the ventral stream in primate visual cortex.

Institute of Technology, Cambridge, MA. Technical Report CBCL Paper #259/AI Memo #2005-036.

Sur, M., Garraghty, P.E., Roe, A.W., 1988. Experimentally induced visual projections into auditory thalamus and cortex. Science 242 (4884), 1437–1441.

Taigman, Y., Yang, M., Ranzato, M., Wolf, L., 2014. Deepface: closing the gap to human-level performance in face verification. In: Proceedings of the 2014 IEEE Conference on Computer Vision and Pattern Recognition (CVPR 2014), pp. 1701–1708.

Thorpe, S.J., 2002. Ultra-rapid scene categorization with a wave of spikes. In: Biologically Motivated Computer Vision. pp. 1–15.

Thorpe, S., Fize, D., Marlot, C., 1996. Speed of processing in the human visual system. Nature 381, 520–522.

Thorpe, S.J., Guyonneau, R., Guilbaud, N., Allegraud, J.M., VanRullen, R., 2004. Spikenet: real-time visual processing with one spike per neuron. Neurocomputing 58–60, 857–864.

Turing, A.M., 1950. Computing machinery and intelligence. Mind 49, 433–460.

Uchizawa, K., Takimoto, E., 2008. Exponential lower bound on the size of constant-depth threshold circuits with small energy complexity. Theor. Comput. Sci. 407, 474–487.

Uchizawa, K., Douglas, R., Maass, W., 2006. On the computational power of threshold circuits with sparse activity. Neural Comput. 18, 2994–3008.

Valiant, L.G., 1984. A theory of the learnable. Commun. ACM 18 (11), 1134–1142.

Valiant, L.G., 1994. Circuits of the Mind. Oxford University Press, New York, NY.

Valiant, L.G., 2000. A neuroidal architecture for cognitive computation. J. ACM 47 (5), 854–882.

Valiant, L.G., 2005. Memorization and association on a realistic neural model. Neural Comput. 17 (3), 527–555.

Valiant, L.G., 2006. A quantitative theory of neural computation. Biol. Cybern. 95 (3), 205–211.

Valiant, L.G., 2014. What must a global theory of cortex explain? Curr. Opin. Neurobiol. 25, 15–19.

Valiant, L.G., 2015. Personal communication. .

Vapnik, V.N., 1998. Statistical Learning Theory. Wiley-Interscience, New York.

Šíma, J., 2014. Energy complexity of recurrent neural networks. Neural Comput. 26 (5), 953–973.

Yang, G., Pan, F., Gan, W.B., 2009. Stably maintained dendritic spines are associated with lifelong memories. Nature 462, 920–924.

Chapter 11

Subjective Physics

R. Brette

Sorbonne Universités, Paris, France; INSERM, Paris, France; CNRS, Paris, France

If we are capable of knowing what is where in the world, our brains must somehow be capable of representing this information.

Marr, 1982

David Marr, one of the most influential figures in computational neuroscience, proposed that perceptual systems can be analyzed at three levels:

1. The computational level: what does the system do?
2. The algorithmic/representational level: how does it do it?
3. The physical level: how is it physically realized?

For example, consider the task of localizing a sound source for an animal with two ears (Fig. 1). The computational level would be: to localize the direction of a sound source, relative to the animal's head (Fig. 1A). When the source is on the left, the sound arrives first at the left ear, then slightly later at the right ear. So the algorithmic level could be: to calculate the delay applied to the left signal that makes it maximally similar to the right signal [the interaural time difference (ITD)], and then calculate the direction that is compatible with that delay (Fig. 1B). In practice, this delay can be calculated by computing the cross-correlation function of the two signals and finding the time lag at which the function peaks. There are intermediate representations, the cross-correlation function and the ITD. Then the physical level could be the Jeffress model (Jeffress, 1948; Joris et al., 1998): monaural neurons from the two ears project to an array of binaural neurons with various conduction delays, and each binaural neuron detects coincidences between spikes arriving from the two sides; the binaural neuron spikes when the ITD matches the difference in conduction delays from the two sides (Fig. 1C). It can be seen that, under some conditions, the array of binaural neurons implements the calculation of the cross-correlation function.

In this example, the three levels are essentially independent of each other, and this is indeed what Marr claimed when he described this methodological subdivision. But this view is not universally shared. For example, in some spike-based theories, physical instantiation is constitutive of both representations and algorithms (Deneve, 2008; Brette, 2012), and therefore levels 2 and 3 are not independent. More importantly for the present essay, as the quote above suggests, Marr considered that the computational level (level 1) can be defined independently of any other level, a view that Thompson et al. called "computational objectivism" (Thompson et al., 1992): the function of a perceptual system is to extract objective properties of the world ("what is where in the world"), as one could describe with the laws of physics.

But on close examination of the example above, it appears that it is not so clear what is meant exactly by "what is where in the world." I will leave aside the question of "what," and focus on "where." First of all, it is obvious that the animal cannot know the absolute location of the sound source (ie, its geographic coordinates), but only the source's direction relative to the animal. Thus, the information that an observer can obtain about the world cannot be entirely independent of itself. This is a trivial point: of course, the direction of the sound source is to be defined in ego-centric coordinates. By coordinates, we mean the angle of the sound source relative to the frontal axis. But do we mean that the animal actually calculates a number of degrees, or radians? Certainly this is absurd: these physical units are arbitrary, and in any case inaccessible to the animal. So what we mean by "angle" is not the absolute value of an angle, which has no meaning in itself, but its value relative to other angles. But then what does it mean for the observer that the angle of a sound source is twice the angle of another sound source? Quite simply: it means that if the animal turns so that it faces the first sound source, then another identical turn will make it face the second sound source (Fig. 1D). So it turns out that the very definition of spatial location is implicitly related to the potential movements of the observer. This remark was made by Poincaré at the beginning of the 20th century, discussing the relativity of space (Poincaré, 1968). In psychology, the idea that perception is tightly interlinked with the movements of the perceiver has been developed in particular by Gibson (1986) and by O'Regan (O'Regan and Noë, 2001).

Closed Loop Neuroscience. http://dx.doi.org/10.1016/B978-0-12-802452-2.00011-1

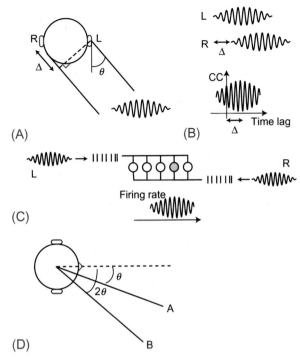

FIG. 1 Three levels of analysis: the sound localization example. (A) Computational level: an animal with two ears receives acoustical waves from a sound source at direction θ, which it must estimate. (B) Algorithmic/representational level: the wave arrives at the right ear with a delay Δ, which can be extracted from the peak of the cross-correlation function. (C) Physical level (Jeffress model): signals are transduced to spike trains, then transmitted with various delays to binaural neurons responding to coincidences. The firing rate represents the cross-correlation. (D) Source B is at an angle twice that of the angle of source A relative to the animal.

It is also related to embodied theories of cognition, in philosophy of perception [in particular (Merleau-Ponty, 2002)], and in robotics (eg, Brooks, 1991).

Thus the claim that the function of a perceptual system is to extract objective properties of the world cannot be taken literally: in fact, the computational level and the physical level are not entirely independent. But if the computational level (what is to be perceived) cannot be described in terms of objective properties of the world, then how can it be described? The aim of this essay is to define the computational level of perceptual systems, from the perspective of a perceiver embedded in its environment. In the computational-objectivist view, what is to be perceived are the properties of the world, as described by physics. Since I consider the computational level from the perspective of the perceiver, I propose that what is to be perceived is the "subjective physics" of the world—I will clarify the notion in the remainder of this text.

The following discussion of subjective physics is largely inspired by the theories of Gibson (1986) (in particular the notion of ecological optics) and O'Regan (O'Regan and Noë, 2001). It is relevant for computational neuroscience, but also for robotics, psychology, and philosophy of perception.

1 PERCEPTUAL KNOWLEDGE

Ever since Descartes, psychology has been held back by the doctrine that what we have to perceive is the 'physical world' that is described by physics. I am suggesting that what we have to perceive and cope with is the world considered as the 'environment'.

Gibson, 1972

1.1 What Is Knowledge?

The goal of this section is to clarify what might be meant by "knowing what is where in the world" for a perceiver (eg, an animal or a robot). As I mentioned above, the meaning of the term "where" is already not so obvious. The term "world" also needs clarification, as shown by the above quote from Gibson. But probably the most problematic term in this phrase is "knowing." What is meant exactly by knowledge? Let us take again the example of sound localization. From any two signals, it is possible to apply a set of operations to extract the time lag at which the cross-correlation function peaks, and then possibly to map this lag to an angle, according to a formula obtained from physics (eg, the Woodworth formula). It could be said that by this process, some information about the source is extracted from the signals. But in fact, strictly speaking, no information is produced by this process, at least in terms of Shannon's information theory: any operation applied to a set of signals can only reduce the amount of information. At the end of the process, we simply have another number, directly derived from the initial numbers. Yet we feel that this process creates some knowledge about the world ("where" the source is). If creating knowledge is not creating information, then what is it?

Following Karl Popper in philosophy of science (Popper, 1959), I suggest that what qualifies as "knowledge" is a universal statement about future observations, a *law*. Unlike a number, a law may be confirmed or falsified by future observations. Imagine that the perceiver in Fig. 1 is a robot that cannot move (no motors). The kind of statements it can make upon capturing the sound waves produced by a source is: the sound wave in the right ear is the same as the sound wave in the left ear, delayed by a specific amount Δ. This is a law that seems to be obeyed by the two sound waves, and it could be invalidated by future observations, for example if the source moves. It is knowledge in the sense that it has a predictive value about future sensory inputs. However, there is nothing intrinsically spatial in such knowledge. If the animal can move, there can be additional knowledge, such as: if I make such movement, I will observe that the two sound waves are identical (corresponding to when the source in the front). In this case, such knowledge is about the expected outcome of actions, given the sensory observations.

Thus knowledge can be defined as universal statements or laws. I suggest calling *weak knowledge* the kind of knowledge that has a predictive value about future sensory inputs, and *strong knowledge* the kind that applies to the effect of actions. If perception is for action, then certainly the most important kind of knowledge is strong knowledge.

Note that I am not making any particular claim here about how knowledge manifests itself in the perceiver, in particular about the controversial notion of "mental representations." I simply wish to give an operational definition of knowledge, independent of the nature of the cognitive system that creates and handles that knowledge, in the same way as physics can be defined independently of the scientists that produce it.

1.2 The Nature of Knowledge for a Perceiver

Physics qualifies as the kind of knowledge defined above. It describes the laws of nature, some of which may be relevant for perceivers, such as mechanics (walking and grasping things), optics (seeing), and acoustics (hearing). For robots and animals to act in the world requires an implicit knowledge of these laws. In physics, these laws of nature are described in external terms, for example mass, tension, waves, and atoms, concepts that cannot be directly grasped by organisms. Therefore organisms cannot literally understand these laws given as such. Rather, the laws that are available to them are those that govern the sensory signals they capture. For example: what happens when a limb is moved in a particular way, or how the visual field changes when the eye moves. I propose the terms "subjective physics" to describe the laws of nature from the perspective of an organism.

To contrast subjective physics with physics, I will again appeal to Popper. To distinguish science from metaphysics, Popper proposed that a scientific statement is one that can potentially be falsified by an observation, whereas a metaphysical statement is a statement that cannot be falsified. For example, the statement "all penguins are black" is scientific, because I could imagine that one day I see a white penguin. On the other hand, the statement "there is a God" is metaphysical, because there is no way I can check. Closer to the matter of this essay, the statement "the world is actually five-dimensional but we live in a three-dimensional subspace" is also metaphysical because independently of whether it is true or not, we have no way to confirm it or falsify it with our senses.

In the same way, knowledge about the world that qualifies as non-metaphysical for a given perceiver depends on the senses it possesses and the actions it can make. For example, imagine that the perceiver in Fig. 1 is a fixed robot, with no ability to move. In this case, all it can grasp is the delay between the two sound waves. From this delay, the angle could be inferred using the Woodworth formula for example, so let us consider the following statement: "the sound source is at angle x." From the point of view of an external observer, this is a scientific statement, because she can measure that angle with a tool. However for the perceiver, this is a metaphysical statement because the statement cannot be falsified. Now consider that the robot can turn its head. Then the statement "the sound source is at angle x" is still a metaphysical statement. A non-metaphysical statement would be "the sound source is at such a location that if I make movement x, the two sound waves will be identical." That is, the location is defined only in terms that are accessible to the perceiver.

Thus, I define subjective physics for a perceiver as the laws that govern the sensory inputs and the effect of actions on them, as they are implied by physics, but only including those laws that are non-metaphysical for that specific perceiver.

1.3 Subjective Physics and Gibson's Ecological Optics

What I just described as subjective physics is very close to what Gibson described as "ecological optics" for vision. The word "ecological" refers to the relation between a specific organism and its environment, and therefore is fully relevant to this matter. But there are a few reasons why I favored the terms "subjective physics" over "ecological physics." The first reason is that I wanted to avoid the confusion with environmental physics or ecology. The second reason is that I want to make clear that subjective physics is not a psychological theory. It is relevant to psychological theories, but it does not rely on specific assumptions about the cognitive abilities of the perceiver, or about what perception is. It is only meant to describe what is intrinsically available to a perceiver, given a specific set of sensors and actuators. Thus, the term "subjective" should only be understood as "from the perceiver's perspective." Finally, a third reason is that Gibson described the laws of ecological optics in terms of the structure of the visual field (the "optic array"), independently of the fact that light is received by sensors. But this fact is highly significant, because the activity of these sensors is only indirectly related to the optic array. Notably, in the eye there is a blind spot, there are inhomogeneities in spatial sampling but also in color sampling. These facts and their significance are taken into account by O'Regan's sensorimotor theory of perception (O'Regan and Noë, 2001; O'Regan, 2011). This extensive quote explains the nature of the problem very well with an experiment of thought:

Imagine a team of engineers operating a remote-controlled underwater vessel exploring the remains of the Titanic, and imagine a villainous aquatic monster that has interfered with the control cable by mixing up the connections to and from the underwater cameras, sonar equipment, robot arms, actuators, and sensors. What appears on the many screens, lights, and dials, no longer makes any sense, and the actuators no longer have their usual functions. What can the engineers do to save the situation? By observing the structure of the changes on the control panel that occur when they press various buttons and levers, the engineers should be able to deduce which buttons control which kind of motion of the vehicle, and which lights correspond to information deriving from the sensors mounted outside the vessel, which indicators correspond to sensors on the vessel's tentacles, and so on.

To rephrase it, in addition to Gibson's ecological optics, subjective physics acknowledges that the specific relation between physical inputs (light) and sensor activity (input signal to the perceptual system) qualifies as metaphysical knowledge for the perceiver—that is, it cannot be known from the sensory inputs alone.

Thus subjective physics has been used before as a component of psychological and philosophical theories of perception, but it has not been developed for itself. The goal of this text is to provide definitions and relevant concepts for subjective physics. I suggest that subjective physics is particularly relevant to computational neuroscience, but it is also relevant to psychology, neuroscience, robotics, and philosophy of perception. It is an attempt to redefine what Marr called the "computational level" of perceptual systems, without recourse to metaphysical knowledge about the world. It provides an alternative to the more traditional approach, in which the computational level is presented as an inverse problem to be solved, that is, recovering objective properties of the world using metaphysical knowledge about the relation between objects and sensors.

But what can be known without metaphysical knowledge about the world, just from a set of sensors and actuators?

2 A DETAILED EXAMPLE

2.1 Subjective Physics of a Hearing Robot

To clarify the problem, I will discuss again the simple example of the hearing robot, but now in more detail (Fig. 2). On top of the robot's head, there are two antennae, with one microphone mounted on each one, at the same height. The two microphones are close to each other. In the world, there are sound sources, which produce sounds repeatedly, and they lie on the floor. There is only one source present in the world at a time, and it is present for a long time. What can the robot know about the world, without metaphysical knowledge?

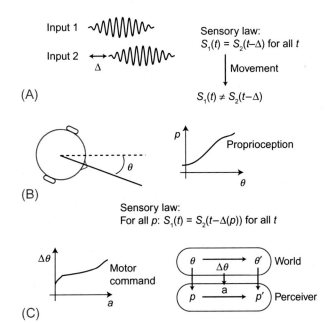

FIG. 2 Subjective physics of a binaural robot. (A) When a source produces a sound, the robot captures two sensory signals S_1 and S_2. The robot can notice that these signals follow a particular law: $S_1(t) = S_2(t - \Delta)$ for all t. The law is spatial because it is falsified when a movement of the head is produced. (B) The robot receives a proprioceptive signal p related to the head's angle (possibly in a nonlinear way). There is a relationship between p and a sensory law followed by the auditory signals, which defines the sound location (note that the function $\Delta(p)$ depends on source location). (C) A motor command a produces a rotation of the head. Each action a produces a change in the proprioceptive signal p. The relationship to physical quantities is unknown (world) but the structure is the same (group action of motor commands on proprioceptive signals).

First of all, when a source produces a sound, two sound waves are captured by the two mics, and these two sound waves have a very special property, which is that they are delayed versions of each other (Fig. 2A). Strictly speaking, the robot cannot examine the sound waves themselves but rather the electrical signals produced by the mics. However, for now, I will assume that this conversion is identical for the two mics, so that it will not affect the following discussion.

To notice this law, or "invariant structure" in Gibsonian terminology, may require quite advanced cognitive abilities, but the present discussion focuses on what the laws of subjective physics are (the computational level), not on how they are extracted (the algorithmic level). The important point to notice here is that this law can be phrased independently of any metaphysical knowledge about sound, or even about the fact the signals are acoustical. It is simply that two signals are delayed copies of each other.

When the same sound source produces another sound, the same property is true, with the same delay. But when another sound source produces a sound, the property holds but with another delay. Although we (external observers) know that the value of this delay is related to the direction of the sound source (angle θ), at this stage the robot has no way to know it—it would require metaphysical knowledge. To see this clearly, consider again the thought experiment proposed by O'Regan and Noë (2001): imagine the wires from the two mics have been mixed up and you do not know which one belongs to the left or to the right mic. Then given the delay between the two signals, it is not possible to tell whether the sound source was on the left hemifield or on the right hemifield. Thus, there cannot be any spatial knowledge at this stage.

We now consider the fact that the robot can rotate its head. When the head rotates, the observed delay changes. Thus the sensory property that was picked up, the fact that the two sound waves are delayed copies of each other with a specific delay, is falsified when a movement is produced. In this sense it can be said that the property is spatial. In addition, when the head rotates, the two sound waves are still delayed versions of each other, but the delay changes in a systematic way with the movement: there is a lawful relation between the head angle and the delay (Fig. 2B). At this point we must be careful: if there is no metaphysical knowledge, then there is, in fact, no such thing as "head angle." Rather, the robot has access to either proprioceptive signals related to that angle, and/or to motor commands. For now, I will assume the former: the robot has access to another sensory input that is related to head angle by a possibly nonlinear relation. Thus, from the perspective of the robot, there is a lawful relation between proprioceptive signal and interaural delay. There is a one-to-one correspondence between this relation and the angle of the sound source, and so in this sense, it may be said that the direction of the sound source can be picked up by the robot, in the form of the higher-order sensory relationship between head proprioception and interaural delay (a lower-order sensory relationship). Note that the villainous monster is not a problem at all here: the same structure exists if the mics are inverted.

Several remarks are in order. The relation that defines the direction of the sound source is not defined directly on the sensory signals. It is defined as a relation between one sensory signal (proprioception) and a set of relations on sensory signals (acoustical inputs). It is a relation between a value and a relation, and therefore it can be described as a higher-order relation. The second point is that the notion of space is very restricted here. In particular, the direction can be known, but not the distance. But even then, the notion of direction is very weak. There is a notion of topology, that is, that directions can be arranged on a circle and not on an infinite line. But there is no metric structure: it cannot be known that an angle is twice another angle, for example. Mathematically, the group structure of angles is not part of the subjective physics of the robot's world.

Consider now that, in addition to proprioceptive signals, the robot also has access to its motor commands—what is called the "efferent copy" in neuroscience (Fig. 2C). I assume that these commands take the form of rotation commands. A rotation command is an action that specifies a rotation with a value that relates to the angle of the rotation, but again the relation between the value and the actual physical angle is unknown. Now when the robot performs an action a, in the form of a specific rotation, the proprioceptive signal changes in a specific way, from p to p'. Thus, action can be considered literally in the mathematical sense as a mapping from (a, p) to p', which one would write $a.p = p'$. It can be seen that this action has the algebraic structure of a group action. Indeed, for every action, there is another action that brings the structure back to its original state (corresponding to the inverse rotation), which is then called the inverse action; the combination of two actions is another action. The group is also commutative because the order of actions has no effect on the end result. Note how this group structure arises even though rotations are not specified as angles or even as quantities linearly related to angles. The subjective physics of the robot now include a much richer notion of space, where actions are isomorphic to the group of rotations. In particular, it now makes sense to say that an angle is twice another angle.

2.2 Inference Versus Pick-up

A valid objection to the above remarks is that observing the sensorimotor relation that defines source direction requires making movements, and so it cannot be picked up on the first presentation of the sound. Wouldn't that mean that a single short sound cannot produce spatial perception? This is where inference becomes important and I must depart from Gibson (Fig. 3).

When a continuous sound is presented at a given location, there is a lawful relation between head position p and interaural delay Δ that can be "picked up" by producing movements. The terms "picked up," based on Gibson terminology, mean that no memory is required to produce this knowledge, it simply derives from the present observations. When only a single observation (p, Δ) is available, such pick-up is not possible (Fig. 3A). However, based on previously observed laws, one may notice that the observation is consistent with one of these laws, in which case it is *inferred* that the future observations (p, Δ) should follow the identified law—for example if the source produces sound again. Since the observation may be consistent with several laws, inference may come with some degree of ambiguity (Fig. 3B). For example, $\Delta = 0$ for sources in the front and in the back, but the delay changes in opposite directions with head angle (sources A and B in Fig. 3B). This is inference and not pick-up, because it relies on previously acquired knowledge (Fig. 3C). But note that unlike

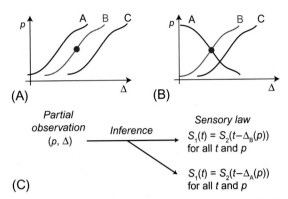

FIG. 3 Inference and pick-up. (A) For each source location (A, B, C), there is a relationship between delay Δ and proprioceptive signal p, which defines that location. This relationship can be picked up by producing movements if the sound is continuous *(red curve)*, but if the sound is transient then only a partial observation (p, Δ) is available *(red dot)*. In this case source location (the full curve) is *inferred* if the partial observation is consistent with some previously observed relationship. (B) Inference can be ambiguous if several relationships are consistent with the observation (for example a sound coming from the front or from the back). (C) In this framework, inference means hypothesizing a particular sensory law from partial observations, which can be an ambiguous process.

in traditional approaches, inference does not rely on metaphysical knowledge (a priori assumptions about what should be inferred). Again, how this inference is made by the perceptual system is an important and difficult question, but one that concerns the algorithmic level, not the computational level. Here I simply point out that there are cases that require inference, and that what is to be inferred is related to previously picked-up laws.

Let us summarize our findings so far. There are two properties that can be picked. There is a low-order sensory property, involving no action, the property that the two monaural signals are delayed copies of each other, with a specific delay. It is spatial in the sense that it is affected by movements. There is then a higher-order property, the relation between head position and the low-order interaural property. Both types of properties are sensory, but additionally, rotation commands form a group action on head position. All this structure can be "picked-up," that is, it can be discovered without relying on preexisting knowledge—this is presumably what Gibson meant when he insisted that perception is "direct." Then through learning, relations can be inferred on the basis of partial observations, using the previously picked-up knowledge—this is arguably not direct perception. The example of the transient sound (eg, a hand clap) is particularly interesting: in this case, one can pick-up the property that the two signals differ by a particular delay, but one must infer the relation between future actions and the change in delay.

We can say more about this example. In particular, we note that the inferential process is indeed ambiguous: the

same interaural delay can correspond to two different directions that are symmetrical with respect to the interaural axis, a "front-back confusion" (Fig. 3B). This is really an ambiguity for the inferential process, not an ambiguity in the laws of subjective physics, as is the case for distance. Indeed there is no front-back confusion in the direction as defined by the sensorimotor relationship. The confusion only exists in the mapping from interaural delay to direction. Thus, the spatial property to be perceived in the robot's world is direction defined on the entire circle, and the inferential process has a front-back ambiguity, which can be resolved by movements.

A final remark regarding inference and pick-up: strictly speaking, given that there can only be a finite number of (possibly unreliable) observations, all knowledge is inferential. For the matter of describing subjective physics, this is not a critical issue. It should be understood that subjective physics describes the underlying laws that can potentially be discovered by the perceiver, in the same way as physics describes the laws of the world even though human observation is limited. The term "pick-up" refers to the ideal limit of unrestrained observation.

2.3 Cognitive Abilities

In this very simple example, it appeared that the information available to the robot without metaphysical knowledge is surprisingly rich. Indeed, the robot can have a sense of direction, isomorphic to a circle, along with its topological structure. By describing this information in a systematic way, we have also uncovered that to obtain and use that information, the robot must display a number of cognitive abilities, such as:

- the "pick-up" of low- and high-order sensory relations
- producing movements
- long-term memory (for inference)
- inference

Thus the aims of subjective physics are twofold: (1) to describe "what is out there," that is, what information is available to the perceiver within an environment, given a set of sensors and actuators, (2) to uncover the cognitive abilities that are necessary for an organism to discover this information. Thus subjective physics hints at the algorithmic level, although it does not describe it. I want to stress again that the notion of "information" about the world must not be understood in the sense of Shannon, because Shannon's information is unstructured and its interpretation requires metaphysical knowledge (the "code"). The information or knowledge we are talking about takes the form of laws, which can be arbitrarily complex and structured—exactly like the laws of physics.

3 DEFINITIONS AND BASIC PRINCIPLES

The goal of this section is to define the core concepts of subjective physics. First of all, I define "subjective physics" as the field of study that analyzes the structure of the sensory and sensorimotor relationships that are available to an organism embedded in an environment, given a set of sensors and actuators, without "metaphysical knowledge" about the world (which I explain in more detail below). I shall call this structure the "subjective structure of the world." It is directly related to what Gibson and followers described as "ecological optics" (Gibson, 1986) and "ecological acoustics" (Gaver, 1993). It is not at all meant to be a psychological field, even if there are relationships with psychological theories. No assumptions are made about the cognitive abilities of the organism. The aims are only to describe what information is *available* to the organism, not what perception is for that organism. The mode of enquiry is agnostic about the psychological question, and in fact it is meant to apply to living organisms as well as to robots. The organism (animal or robot) will be referred to as the "perceptual system."

The aim of subjective physics is to analyze in the highest possible detail the subjective structure of the world, and in particular, under what form the information is available. As we have seen in the example, it is also expected that this descriptive process uncovers some of the cognitive abilities that are necessary for the perceptual system to discover this information.

The analysis is specific to a set of sensors and possible actions on the world. I will call this set the "interface" of the perceptual system. What is meant exactly by "sensors" and "actions?" A sensor is simply something that is modified by the world, and whose modification can be picked up by the perceptual system. For example, acoustical waves make the membrane of a microphone vibrate, and this vibration can be picked up by the robot. An action is something that the perceptual system can do, which produces modifications in the world. In turn, these modifications can affect sensors.

There are a few subtleties in these two concepts that I will discuss in Section 3.2. But to uncover these subtleties, I first need to introduce a methodology that I shall call "ecological reduction" in reference to the concept of "phenomenological reduction" in philosophy.

3.1 Ecological Reduction

Subjective physics aims at describing the subjective structure of the world without recourse to metaphysical knowledge of the world. I introduced the terms "metaphysical knowledge" in reference to Popper's demarcation criterion in philosophy of science. It refers to statements about the world that cannot be falsified or corroborated given a specific interface with the world (sensors and actions). Thus what is considered as metaphysical is always with respect to a specific interface.

There is an interesting parallel to be made with phenomenology in philosophy. The key point of subjective physics is to get rid of any preconception about the nature of the external world. This is not to say that such preconception, such as the Euclidian structure of space, is not real, but we wish to suspend this belief while describing the subjective structure of the world, so as to focus on what is really intrinsic to that structure. This attitude is, in fact, strikingly analogous to the aims of phenomenology, a field of philosophy initially introduced by Edmund Husserl at the beginning of the 20th century, and later developed by a number of philosophers, such as Heidegger, Merleau-Ponty and Sartre. Phenomenology is the philosophical study of the structures of subjective experience and consciousness. This study relies on Husserl's methodology of "phenomenological reduction" (also called "bracketing" or "epoché," which means suspension in Greek), which is the suspension of any a priori judgment about the world. This is not to say that the object of consciousness is denied any objective existence, or to refute that it has objective properties, but simply to refrain from using this knowledge in the analysis of the phenomenon. For example, a phenomenological analysis of color would not allow physicalist statements such as: "red is the conscious experience of light with wavelength 650 nm."

Similarly, subjective physics relies on the methodology of "ecological reduction," that is, the suspension of any prior knowledge about the world on behalf of the perceptual system, in the analysis of the subjective structure of the world. This suspension may extend to the nature or properties of the sensors and the specification of actions (see Section 3.2).

An associated concept in phenomenology is "eidetic variation" or "eidetic reduction" ("eidos" means shape or form in Greek, as in Plato's "ideal Forms"). It consists in varying the perspective on the phenomenon so as to clearly understand what constitutes its essence, what is critical to the phenomenon and what the phenomenon is invariant to. Applied to subjective physics, this means slightly varying the nature of the external world (eg, the nature of the sound sources), the interface with the world (nature of sensors and actions that can be performed, eg, whether the robot has proprioceptive information), or possibly the constraints on the perceptual system (eg, whether it has memory). This technique allows a deeper analysis of the structure of the world, as it reveals how the structure depends on the specification of the world and of the perceptual system.

To be more concrete, O'Regan's "villainous monster" (O'Regan and Noë, 2001) is an application of this reduction technique to the interface with the world: if sensors are mixed up in an unknown way, what knowledge about the world remains? By this technique, one reveals and discards metaphysical knowledge about sensors.

3.2 Sensors and Actions: The Interface of the Perceptual System

3.2.1 Sensors

A sensor is something that is modified by the world, and whose modification can be picked up by the perceptual system. I gave the example of two microphones. This may give the impression that the perceptual system has access to the value of the acoustical pressure at any given moment in time, at both ears. I implicitly assumed it when I described the example. But there are in fact two presuppositions in this statement:

1. The two sensors have the same properties. I will call this assumption the "sensor homogeneity" assumption.
2. The values provided by the sensors can be interpreted as acoustical pressure, which qualifies as metaphysical knowledge.

I will start with the second point. In fact, a microphone does not exactly provide the acoustical pressure, it provides an electrical signal (Fig. 4A). The signal is a function of the acoustical pressure, but to deduce the acoustical pressure from the electrical signal requires the knowledge of this function. We may want to free ourselves from this assumption in the analysis of the subjective structure of

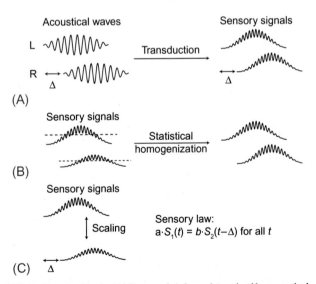

FIG. 4 Sensory signals. (A) Sensory signals are determined by acoustical inputs through an unknown and possibly nonlinear transformation, but here the structure (delay) is preserved. (B) If sensors are not homogeneous, the structure may be changed. One way to solve this problem is statistical homogenization: each signal is normalized with respect to some statistics (eg, average, *dashed line*). (C) Another way is to allow more general sensory laws: here the two signals are related by a temporal shift, which depends on the source, and a scaling transformation, which is universal (depending on the sensors only).

the world. In the robot example, we can easily remove this assumption, because what matters is only that the two sensor signals are delayed copies of each other. Making the sensors nonlinear would also leave the subjective structure of the world unchanged (note how we used eidetic variation to analyze the determinants of this structure).

The sensor homogeneity assumption is more difficult to remove. In the example, it is not a problem at all that the sensor signal is an indeterminate function of acoustical pressure. But if the two sensor signals are differently related to acoustical pressure in the two microphones, then they would not be delayed copies of each other anymore. It is easy to imagine such a case: there is some tolerance in the properties of electrical components, and so the two signals could be similar and proportional to each other but no exactly identical. It is even easier to imagine in an animal. Let alone the variability in receptor properties, we may simply consider the eye: given the blind spot due to the optic nerve, the blood vessels, the optical aberrations and the fact that light must go through several layers of cells before reaching the receptors, it is clear that two photoreceptors do not exactly see the same thing up to a spatial shift—this is a critical observation in the sensorimotor theory of perception (O'Regan and Noë, 2001; O'Regan, 2011).

I will describe two procedures to remove the sensor homogeneity assumption. The first one is "statistical homogenization" (Fig. 4B). Suppose the two microphones produce signals proportional to acoustical pressure, but with different and unknown proportionality factors. One could decide to scale these signals by two different factors, chosen so that the two resulting signals have the same average power (on a long time scale). This procedure solves the problem of sensor inhomogeneity. But we note that it does not provide a way to recover the true acoustical pressure, it only equalizes the signals. Indeed it is impossible to know such an objective value, because only relationships can be observed. More generally, statistical homogenization consists in choosing a particular statistics and transforming the sensor signals so that they have identical long-term values for this statistics. We note that the procedure relies on long-term learning, on behalf of the perceptual system.

Another procedure is to extend the types of relationships that are observed on the sensory signals (Fig. 4C). Suppose the two acoustical signals are given by $X(t)$ and $Y(t)$. The presence of a sound source at a given direction is attested by the relationship $X(t) = Y(t - d)$ for all t, for a specific delay d. But the perceptual system only has access to $S_1(t)$ and $S_2(t)$, which are scaled version of X and Y, with unknown scaling factors. Then the presence of a sound source is attested by the following relationship: there exist two numbers a and b such that $a \cdot S_1(t) = b \cdot S_2(t - d)$. The numbers a and b are universal in the sense that they are identical for all source directions. Thus, the procedure is to allow the discovery of more complex relationships in the sensory signals. Again, this does not provide a way to recover the original acoustical pressure. On behalf of the perceptual system, this means that it must be possible to discover complex relationships in the signals.

What we have just done illustrates the concept of "ecological reduction": we progressively removed implicit metaphysical knowledge of the world, and analyzed what structure remains. It appears that the spatial structure of the world is robust: it remains even when sensors are inhomogeneous and related to acoustics in an unknown way. On the other hand, the objective acoustical pressure or the energy of the signals cannot be known from such sensors. We also observe that dealing with sensor inhomogeneities puts an additional load on the perceptual system, in terms of cognitive abilities. Note that we could go further in this reduction, for example by not assuming that the two sensors have the same spectral response.

The previous observations suggest that the subjective structure of the world can be analyzed at different depths of ecological reduction: at the first level, we consider that sensor data is specified in external terms (acoustical pressure); at the second level, we consider that sensor data is specified in internal terms (transduced quantity); at the third level, we consider that sensors are inhomogeneous.

3.2.2 Actions

An action is something that the perceptual system can do, which produces modifications in the world. In turn, these modifications can affect sensors. For example, the robot can rotate its head. This definition needs to be made more precise (Fig. 5A). Indeed: do we mean that the perceptual system issues a command that makes its head rotate *to* a

particular angle (relative to the fixed body), or *by* a particular angle (relative to the current head angle)? In the first case, the command is issued in an absolute spatial frame. In the second case, the command is issued in a relative spatial frame. This has important consequences for the subjective structure of the world and for the cognitive load on the perceptual system. Indeed, the interaural delay is relative to the head, not to the fixed body. Therefore, if the reference frame for commands is the head, then there is no way that the perceptual system can capture the absolute direction of the sound source, unless there is proprioceptive information. If the reference frame is the fixed body, then inferring the direction from a single sound presentation requires either proprioceptive information or memory of the last command. The sensorimotor structure is also different in these two cases.

Secondly, we can make the same remark about actions as about sensors. Considering that the command is issued as an angle implicitly assumes metaphysical knowledge about the world, that is, about the relationship between the command itself (an electrical signal) and the result of the command in externally defined terms (rotation angle). If we want to remove this assumption, then actions should be thought of as "levers" or switches: some command that can be triggered with an associated value, which has an unknown effect on the world—apart from what is picked up by the sensors.

Actions generate two types of signals for the perceptual system: sensory signals (including proprioception) and "efferent copy." Sensory signals result from the effect of actions on the world. The efferent copy is the copy of the command values issued by the perceptual system, considered as input signals. The sensorimotor structure is the relationship between these two types of signals. The efferent copy depends only on the commands, but the sensory signals depend on both the commands and the world, which is why the sensorimotor structure is informative about the world.

A special type of sensory input is proprioception: these are signals about the body (eg, muscle tension, position of the head relative to the body) rather than about the external world. As far as sensorimotor structure is concerned, these can be considered as sensory signals. However, their sensorimotor structure has distinctive properties, which I will discuss later. In fact, many sensors are both proprioceptive and sensory: for example, in the case of the robot, the acoustical inputs depend both on the direction of the sound source (world) and on the position of the head (body).

A special type of action is when the action influences the sensors themselves rather than the external world (although the distinction is somewhat arbitrary). For example, the pupil in the eye can contract and dilate, which changes the amount of light coming into the retina. In the cochlea, the medial efferent system changes the way incoming sounds put the basilar membrane in motion (Guinan, 2006).

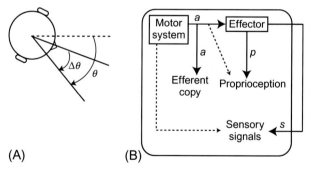

FIG. 5 Actions and their effects on sensorimotor signals. (A) A motor command for the robot may be specified as a target angle θ, or as the angle increment $\Delta\theta$. (B) A motor command a is sent to an actuator (muscle or motor), which results in different types of signals available to the perceptual system: the efferent copy of the command a, the proprioceptive signal p, and the sensory signals s due to the interaction of the actuator with the world. In addition, motor commands may affect proprioceptive or sensory signals without interaction with the world *(dashed lines)*, for example the efferent system in the cochlea (Guinan, 2006).

3.2.3 Relationship Between Sensors and Actions

A critical concept in the study of subjective physics is that action is not considered as caused by sensory input. In this sense, it is not a subfield of psychology, because the object of study is not what determines an organism to act in the way it does, but rather how potential actions modify sensory inputs. It is the opposite perspective of many studies in neuroscience, in which ones observes the effect of stimuli on the nervous system or on behavior. In subjective physics, it is in fact considered that actions are *voluntary*, in the sense that any action can potentially be taken—it is not constrained by the sensory inputs. The term "voluntary" should not be taken in a psychological sense, but rather as meaning free from any determinism.

This notion of "voluntary action" is important because it makes action very different in nature from sensory inputs. One can observe relationships between sensory inputs, but it is not possible to see one input as being caused by another input: only correlates can be observed. In contrast, causality exists in the sensorimotor structure precisely because action can be taken or not taken, independently of sensory inputs. Action solves the problem raised by David Hume, that correlation does not imply causation.

It should be clear that the subjective structure of the world depends not only on the sensors but also on the possible actions. For example, if the robot could not rotate its head, the acoustical structure would remain, but it would not have a spatial character (no movement can disrupt it), and it would not be possible to define a source direction (only an interaural delay).

3.3 On the Notion of Information

To avoid confusion, I have tried to avoid the term "information," and replaced it with the term "structure." The term "information" may indeed be confusing when speaking of knowledge about the world, because in neuroscience it is often meant in the sense of Shannon. In this section, I want to make the distinction as explicit as possible to avoid misunderstandings.

Shannon's information comes from communication theory. There is an emitter that wants to transmit some message to a receiver (Fig. 6). The message is transmitted in an altered form called "code," for example, Morse code, which contains "information" insofar as it can be "decoded" by the observer into the original message (Fig. 6A). The metaphor is generally carried to neuroscience in the following form: there are things in the external world that are described in some way by the experimenter, for example bars with a variable orientation, and the activity of the nervous system is seen as a "code" for this description (Fig. 6B). It may carry "information" about the orientation

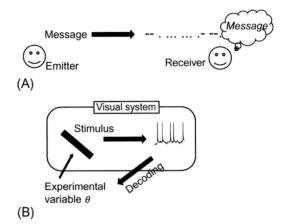

FIG. 6 Information in the sense of Shannon. (A) A communication channel consists of an emitter who wants to transmit some message to a receiver, in an altered form named "code" (here Morse code). The receiver knows the correspondence and can reconstruct the original message. (B) In a neuroscientific context, the emitter is the experimenter, who presents a stimulus (oriented bar) characterized by some objective values (orientation θ). The brain receives the message in the form of neural activity, from which it infers information about the stimulus (θ). However the decoding process, that is, the relationship between neural activity and the experimental variable, is metaphysical knowledge for this perceptual system.

of the bar insofar as one can reconstruct the orientation from the neural activity.

It is important to realize how weak and specific this notion of information is. In a communication channel, the two ends agree upon a code, for example on the correspondence between letters and Morse code. For the receiving end, the fact that the message is information in the common sense of the word, that is, knowledge about the world, relies on two things: (1) that the correspondence is known, (2) that the initial message itself makes sense for the receiver. So the notion of Shannon's information applied to a perceptual system carries with it two elements of metaphysical knowledge: the relationship between sensor signals and externally defined properties of objects in the world, and the meaning of these objects.

In the robot example, the interaural delay between microphone signals is information about the source's angle in Shannon's sense. But to infer the source angle from that delay requires knowing the relationship between the quantities, which is metaphysical knowledge (it is not included in the sensor signals). Then knowing what the angle means requires some metaphysical knowledge of Euclidean geometry.

Thus what I have called the subjective structure of the world is not at all information in Shannon's sense. It is closer to the notion of scientific knowledge: universal statements about the world, as the laws of physics, rather than code words that stand for things in the world. The distinction has two important implications. One is that knowledge about the world is highly structured, contrary to Shannon's information, because it is in the form of laws. I will discuss this point in more detail in Section 4. The second implication is that such knowledge necessarily implies the notion of a sensory flow. Indeed, a law cannot exist in a single observation, it can only be seen through a flow of observations. This stands in contrast with a standard statistical learning framework in which an element of information is an image (seen as a vector of pixels). Since even the most elementary law is a relationship, it cannot involve a single image.

It is important to be fully aware of this distinction with Shannon's information, because the confusion seems to subtend a number of misunderstandings, in particular about Gibson's notion of information. Gibson considered that there is intrinsic information about the world in the "invariant structure" present in the sensory flow, as explained by this quote:

A great many properties of the array are lawfully or regularly variant with change of observation point, and this means that in each case a property defined by the law is invariant.

Gibson, 1972

This quote makes it clear that Gibson did not mean information in Shannon's sense, but truly in the sense of a law, that is, a universal statement about the sensory flow. The terms "invariant structure" stand for nothing else than a sensory law, as is made explicit in the quote above. It does not mean for example that some sensory signal is constant.

Subjective physics describe information about the world in the sense of structured knowledge in the form of laws, not in the sense of Shannon.

3.4 Noise, Finiteness, and Inference

It is assumed that the perceptual system can discover the subjective structure of the world, that is, using Gibson's terminology, it can "pick-up" the sensory and sensorimotor relationships. For example, it can observe that when a source produces sound, the two sound waves at the ears are delayed copies of each other. However, one may object that inferring a law from observations, which are necessarily finite and possibly noisy, is somewhat arbitrary: there are always an infinite number of laws that are consistent with a finite number of observations.

This is a relevant objection, but it is not an objection to the existence of the law: the argument emphasizes the difficulty for the perceptual system to infer the law. In fact, this and related questions are a major theme of the philosophy of knowledge (epistemology): how can knowledge be acquired? Subjective physics does not aim at answering this question, but it is assumed that knowledge can indeed be acquired by the perceptual system. Therefore I will summarize a few relevant arguments from philosophy of knowledge.

The argument that it is not possible to infer a universal law from a finite number of observations is the skeptic criticism of inductivism. It can be addressed by different means. One is "Occam's razor": the idea that among competing hypotheses, the most parsimonious one should be preferred (a principle for which statistical learning theory gives some justification). Another way of addressing the problem was provided by Karl Popper: first one makes a hypothesis, in the form of a universal law, and then the hypothesis is tested by future observations. Note that this view, falsificationism, does not actually suggest any particular method to produce the hypothesis in the first place, but for the problem at hand, it emphasizes that (1) knowledge is for predicting the future rather than for accounting for the past, (2) knowledge is useful if it has predictive power, even if it might be amended in the future. This is just a glimpse of a few relevant concepts from philosophy of knowledge. In fact, the relationship between perception and philosophy of knowledge is deep. For example, following the falsificationist account of knowledge, action could be seen as an experiment, chosen to test the current hypothesis about the subjective structure of the world [this is related to the idea of active learning (Cohn et al., 1996)].

To summarize this point: it may indeed be a difficult problem for a perceptual system to infer laws present in the sensorimotor signals, but these laws exist independently of the cognitive abilities of the perceptual system. The primary object of subjective physics is to describe these laws, without any particular claim about how the perceptual system could notice them. If it is assumed that a scientist can discover the laws of physics, then a perceptual system can grasp the laws of subjective physics.

4 SUBJECTIVE STRUCTURE

In this section, I will describe a few general concepts of subjective physics, taking again the example of the robot.

4.1 The Role of Action in Subjective Structure

As I previously argued, relationships can only be seen through a sensory flow, as opposed to a static pattern of inputs (an image). This means that something must change: either the organism moves (actions) or something in the world changes by itself. The latter can be a movement of an object or of an animal, or changes in acoustical pressure (a sound) or in illumination. An intermediate possibility is that the organism is made to move, for example, if she is sitting in the passenger's seat of a car.

The sensorimotor account of perception (O'Regan and Noë, 2001) emphasizes the relationship between action and the resulting sensory signals (or changes in sensory signals), and indeed I have observed that action must be involved to define any kind of spatial structure. In this section, I will examine the role of action in subjective structure. At first sight, such a sensorimotor relationship may be thought of as a mapping between actions and sensory signals. However, the robot example shows that this is a little too restrictive.

Action can be involved in the structure in five different ways.

1. First of all, it can be not involved at all. For example, if a sustained note of a musical instrument is played, a periodic sound is captured at each microphone. This periodicity structure exists independently of any movement, and it is unaffected by action.

2. It can also be involved negatively, that is, structure conditioned to the absence of action. For example, when a sound is produced by a source, two acoustical waves are captured at the two microphones that are delayed copies of each other: this structure exists as long as the robot does not turn its head. If the robot moves, the structure is changed. This is what gives its spatial character to this structure, which is otherwise sensory rather than motor.

3. Action can be involved as the cause of the sensory flow. For example, when a bird flies, there is a structure in the optic flow, which is related to the direction toward which the bird is flying. But this structure is purely sensory: action is not part of the structure but rather causing the structure (without movement, no optic flow). This is most closely related to Gibson's notion of "invariant structure": some sensory structure that is invariant with respect to some action (Gibson, 1986).

4. Action can be involved through proprioception. For example, the position of the robot's head is related to the interaural delay of the sound source, when the head moves. The proprioceptive signal is directly linked to action, rather than to the world. The notion of "sensorimotor contingency" is closest to this case.

5. Action can be seen as an event causing a change in structure or in a sensory signal. For example, a rotation of the head by a given angle changes the sensory structure from one interaural delay to another one. This is an event in the sense that it is transient (an action is not a flow), while the sensory structure is defined on a temporal flow. This can be seen as a mapping from (action, structure) to structure. Note how this is different from the previous case: action cannot be mapped to sensory structure but rather to a change in structure. This is a very important notion that is related to the mathematical notion of *group action*. It corresponds to what Henri Poincaré meant when claiming that the Euclidean structure of space derives from the relationship between movements and sensory inputs (Poincaré, 1968). In neuroscience, action defined in this sense corresponds to the notion of "efferent copy," the copy of motor commands that is available for the perceptual system.

4.2 Conditional and Unconditional Structure

There are two very different types of structure. One is the structure that is normally present in the world, which I will call *unconditional structure*. This is typically the relationship between actions and proprioceptive signals, for example signals that indicate the state of muscles or the relative position of different limbs. The same action normally results in the same change in proprioceptive signals, and the proprioceptive signals do not change unless an action is performed. On the other hand, when a source makes sound, it produces a sensorimotor structure that is *conditional* to the presence and direction of that source: the structure is indeed informative about the source precisely because it is conditional to it and to its properties. It can be said that this is what distinguishes the body from the external world.

I will give two anecdotic examples to help clarify this point. One usually feels her teeth as being part of the body—and in fact, in general, not feeling them at all. But when a tooth is removed and replaced by a dental implant of the same size, it initially feels very present and uncomfortable, as if there were an external body in the mouth. And indeed it is true that there is an external body in the mouth. Yet it occupies the same space as before and it is attached to the mouth as before, it simply has a slightly different shape that can be picked up by the tongue. The feeling is the same when a dental implant is renewed, so it does not have to do with the artificial nature of the tooth. And then after a few days, the implant feels like a normal tooth and it becomes difficult to distinguish the artificial tooth from the natural teeth. From the point of view of sensorimotor structure: before and after the tooth is changed, the structure is unconditional, but there is a transient change in structure at the time when the implant is inserted, which normally means that there is an external body in the mouth. The interesting point here is that this structure is carried by sensors, tactile receptors on the tongue that are not specifically proprioceptive. They are involved, for example, in the perception of taste. But depending on whether the structure is conditional or unconditional, it signals elements of the body or of the external world.

Another example is the vestibular system, which is involved in the perception of body balance. In the cochlea, there are receptors that sense head acceleration. The vestibular system also integrates multimodal information coming from motor and visual systems. Acceleration sensors qualify as proprioceptive, in the sense that they are normally only affected by self-generated movements: there is an unconditional sensorimotor structure. Interestingly, when there is a dysfunction of the vestibular system (due to a disease for example), so that this sensorimotor structure is disrupted, one feels nauseated. The standard interpretation of this fact is that nausea is a reflex response of the organism to a dysfunction of the vestibular system that is normally due to the ingestion of toxins: the organism then vomits to get rid of the toxins (Treisman, 1977). In other words: if the unconditional sensorimotor structure is disrupted then something must be wrong with the body.

Unconditional structure may also be defined in a statistical sense. For example, two nearby photoreceptors on the retina normally receive a similar amount of light. This simple observation provides topological relationships between receptors. In the same way, two neighboring inner hair cells in the cochlea transduce a similar displacement of the basilar membrane. More generally, a topology on sensors may be defined from the statistical structure of sensor data (specifically, from their correlations).

It is particularly important to characterize the unconditional structure, because conditional structure is defined with respect to it. That is, conditional structure is structure that is normally not observed: this is what makes it informative.

4.3 The Syntax of Subjective Structure

I have noted that structure must be understood in a much broader sense than just a mapping between actions and sensory inputs. This was clearly acknowledged by Gibson, who described the subjective structure of light ("ecological optics") in great detail. I will try to outline the main types and properties of subjective structure.

In the example of the robot, I noted that a sound produced by a source induces a particular structure, defined directly on the sensory inputs: an identity between the signal at one ear and the delayed signal at the other ear. This is an invariant structure in the sensory flow, where change is induced by the mechanical vibration of the sound source. As it is defined directly on the sensory inputs, one may call this structure "first-order." But the direction of the sound source is defined by the relationship between head position and interaural delay, that is, between head position and the first-order sensory structure. In this sense, it is second-order structure. This observation implies that subjective structure has two features: compositionality and hierarchy. Compositionality means that a relationship is defined between two or more constituents, and hierarchy means that the resulting relationship can be a constituent of another relationship (higher-order structure).

Another important feature is that subjective structure is contextual (or at least it can be). For example, consider a horizontal field of grass and a visual system looking at it. There are two textures, each of which qualifies as invariant visual structure: the sky, which is (more or less) uniform, and the grass, which is a statistically uniform visual texture on a perspective structure. Each of these structures is seen only in part of the visual field, for specific gaze angles. Thus there are two distinct structures, conditioned to some action

(eye movement) or some sensory input (proprioceptive information about eye position).

Thus, subjective structure is, in fact, very rich, much richer than a mapping between actions and sensory signals. It is compositional, hierarchical, and contextual: one may speak of the "syntax" of subjective structure.

4.4 Algebraic Properties of Structure

Subjective structure is, in fact, even richer: it can have an algebraic structure. The notion of "algebra" means that we are considering operations. This is precisely what an action is. Here we consider the notion that action is something that changes the sensory structure (case #5 in Section 4.1). Let us consider again the robot example. When a source makes a sound, a sensory structure with a particular interaural delay is observed. This structure is changed by a rotation of the head. This can be seen as a mapping from (rotation, delay) to delay, that is, from (action, structure) to structure. In other words, the set of actions acts (in the mathematical sense) on the set of structures. It can be seen that this action has the algebraic structure of a group action. Indeed, for every rotation there is an inverse rotation that brings the structure back to its original state; the combination of two rotations is another rotation. The group is, in fact, commutative. Note that this group structure exists even if rotations are not specified as angles or even as quantities linearly related to angles.

The existence of this algebraic structure in the subjective structure of the world was indeed noticed by Henri Poincaré, who remarked that movements and their relationships to sensory inputs provide us with the geometrical structure of the world.

Thus, there is a very rich subjective structure that exists independently of prior knowledge on behalf of the perceptual system. In principle, this structure can be captured by a perceptual system (provided appropriate cognitive abilities).

4.5 Analogy Versus Similarity

In this section, I want to touch on the notion of similarity. A standard notion of similarity is that defined mathematically in a metric space. For example, two sensory signals are considered similar if their difference is small. A notion of distance is defined on the space of signals, which provides a measure of similarity of any two signals. But there is another notion of similarity that is used in common language, for example in the statement: the heart is like a pump. Here the similarity is not implied in any metric sense. Rather, it is meant that the heart acts on the blood, a body fluid, in the same way as a pump acts on a fluid. The similarity is not about how the objects (heart and pump) look like in some representational space, but about the way they interact with other things. This is what we call an *analogy*. In his infamous criticism of artificial intelligence, Hubert Dreyfus noted that a great aspect of human intelligence is the faculty of analogy (Dreyfus, 1992). This notion is also related to Gibson's notion of affordances (Gibson, 1986), which are the ways with which we can interact with things in the world (eg, the ground affords standing, whereas a large volume of water affords swimming).

A famous question in philosophy of perception was formulated by William Molyneux in the 17th century. Imagine a man who was born blind, and who had learned to distinguish by touch between objects such as a sphere and a cube. If one day the man were given sight, would he be able to distinguish and name these two objects by sight alone? This question has generated considerable philosophical literature, and it is not my aim to answer it here (see Held et al., 2011 for an empirical study). I simply wish to point out that the question makes sense in terms of subjective physics. In vision, light rays reflected by an object are captured by photoreceptors. By moving the eye, the same photoreceptors capture light rays reflected by different parts of the object. In touch, tactile receptors capture mechanical signals at the interface between fingers and the object. A different part of the object is sampled when the fingers are moved across the object. There is some similarity in the subjective structure of the tactile and visual world: for example, for each movement of the eye, there is an opposite movement that makes sensory signals the same as they were prior to the first movement; the same is true for finger movements. This similarity is not metric: it is rather a set of properties that both subjective structures have in common. I propose to call this type of similarity between subjective structures "analogy."

The same physical object may produce analogous subjective structures in two different modalities. Indeed both touch and vision are about spatial configurations of surfaces. For example, a cube has sharp edges. This means that when the eye is moved across a face of the cube, the visual texture is uniform and then there is a discontinuity, which signals the edge. The same discontinuity occurs when a finger is swept across the face of the cube. In contrast, when the finger moves around a sphere, no such discontinuity occurs, as long as the finger is kept in contact with the surface. The situation is slightly different with vision, since there is also a visual boundary for a sphere, but it is a different type of boundary: when the eye is moved across the surface of a sphere, the visual texture progressively becomes denser as one gets closer to the boundary. Thus, there is a partial analogy between the two subjective structures induced by the presence of the same physical object.

Back to our robot, we may imagine a similar question in which the hearing robot is given visual sensors. Is there anything analogous between the sense of direction obtained through vision and through hearing? The sensors are very

different, and there is no such thing as an ITD in vision. However, one thing is identical: the rotation commands also act as a group action on the set of visual properties. Analogies may be useful for the perceptual system, as it may allow the system to learn new sensorimotor contingencies faster.

5 ADVANCED SUBJECTIVE PHYSICS OF A HEARING ROBOT

I will now describe the subjective physics of the hearing robot in a less idealized setting. It will illustrate the concepts I have previously presented in practical cases.

5.1 Two Ears

I consider again a robot with two ears, which are mounted on a head. Previously, I described a simplified physical situation in which the signals at the two ears are delayed copies of each other. This is not valid if there is an object between the two ears. The presence of the head produces intensity differences between the ears (it casts an "acoustic shadow"), but it also makes timing differences depend on frequency (Kuhn, 1977). The two signals are no longer delayed copies of each other. The physics of the situation is best described as follows: the source signal $S(t)$ is linearly filtered by two acoustical filters, which depend on sound direction (Fig. 7A). These filters are called head-related transfer functions (HRTFs) in the frequency domain, or head-related impulse responses (HRIRs) in the temporal domain. The monaural signals $S_L(t)$ and $S_R(t)$ are then described as the convolution of $S(t)$ with the two HRIRs:

$$S_L = \text{HRIR}_L(\theta) * S$$

$$S_R = \text{HRIR}_R(\theta) * S$$

where θ is the sound direction. This is the physical description of the situation. However the subjective physics is quite different, since only the monaural signals $S_L(t)$ and $S_R(t)$ are captured: neither the source signal $S(t)$ nor the set of HRIRs is directly captured. Both pieces of information constitute metaphysical knowledge for the perceptual system. When there was no head between the ears, one could describe one signal as a delayed copy of the other, but this is not possible anymore.

However, there is still some invariant structure in these two signals. In particular, there exist two filters F_L and F_R such that $F_L * S_L = F_R * S_R$. Indeed, this is true with $F_L = \text{HRIR}_R(\theta)$ and $F_R = \text{HRIR}_L(\theta)$ (Fig. 7B). It is also true with the pair of filters $F_L = U * \text{HRIR}_R(\theta)$ and $F_R = U * \text{HRIR}_L(\theta)$, for any linear filter U. Therefore, there exists a non-unique pair of filters such that $F_L * S_L = F_R * S_R$. This is a fact of subjective physics that does not include metaphysical knowledge. The perceptual

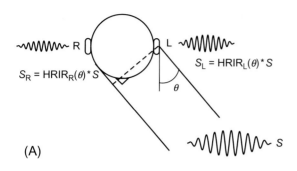

FIG. 7 Subjective structure of a binaural robot with sound diffraction. (A) The signal S_R at the right ear is the result of filtering the source signal S with a directional filter (HRIR for head-related impulse response). (B) As in the simplified case without diffraction, the two signals follow a law, which is falsified when a movement is produced. Here F_L and F_R are two filters, for example, $F_L = HRIR_R(\theta)$ and $F_R = HRIR_L(\theta)$ (or any filtered version of this filter pair). (C) Sound source location is specified by the relationship between a proprioceptive signal p and the sensory law followed by the auditory signals [characterized by a filter pair $F_L(p)$, $F_R(p)$].

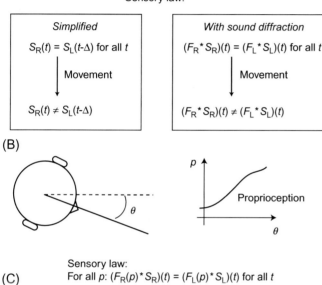

system cannot identify the HRIRs or source signal $S(t)$, but it can identify a particular pair of filters that satisfies the above-mentioned identity. This identity is an invariant, it holds as long as the sound has energy and it does not depend on the source $S(t)$ (at least for broadband sounds). Indeed it is related to spatial location through the identity $F_L * HRIR_L(\theta) = F_R * HRIR_R(\theta)$. This is an example of "invariant structure" in Gibson's terminology.

The identity is invalidated when the robot turns its head, which makes it a spatial invariant. Without a head, there is a relationship between proprioception (related to the head's angle) and interaural delay, which can be described by a real-valued function. The situation is slightly more complex here: there is a relationship between proprioception and sensory statements of the kind $F_L * S_L = F_R * S_R$. This can be summarized by a mapping from proprioceptive signal p to filter pair (F_L, F_R) (Fig. 7C). This mapping constitutes the subjective location of the sound source. The effect of actions on the structure is then identical to the idealized case with delays (the subjective structures are analogous in the sense proposed in Section 4.5).

This principle has been used in two state-of-the-art sound localization algorithms. One algorithm consists in calculating the convolutions $HRIR_R(\theta) * S_L$ and $HRIR_L(\theta) * S_R$ for all measured pairs of HRIRs, and identifying the pair that maximizes the correlation between the convolved signals (MacDonald, 2008). Another algorithm uses a similar technique, but refined in frequency bands (Durkovic et al., 2011). Note, however, that the two techniques rely on prior knowledge of the HRIRs.

This viewpoint has also been developed in a spiking neural model of sound localization (Goodman and Brette, 2010b). There the filters (F_L, F_R) are assumed to represent the auditory receptive fields of monaural neurons on both sides. The acoustical invariant $F_L * S_L = F_R * S_R$ is then reflected by a synchrony invariant: neurons with filters F_L and F_R fire in synchrony when the sound is presented at a specific location (see Brette, 2012 for a more general framework). Thus a pattern of synchrony is a signature of a particular invariant, and it must then be associated with the corresponding proprioceptive signal. In principle, this model does not require prior knowledge of the HRIRs, provided there is sufficient diversity in the neural filters (see Goodman and Brette, 2010a for a simplified learning procedure).

5.2 One Ear

I now describe the subjective physics of sound localization with a single ear. The ear is mounted on a head that acts as an acoustic shadow for the ear (Fig. 8A). Consequently, sound intensity at the ear depends on the position of the source relative to the ear. Physically, the relationship between the source signal $S(t)$ and the signal at the ear $X(t)$ is linear: $X(t) = a(\theta) \cdot S(t)$, where $a(\theta)$ is the attenuation for sound direction θ. In fact, more rigorously, the attenuation should be described as a filter: $X = a(\theta) * S$. However, I will just consider the simplified setting.

I will address two questions of subjective physics: (1) can there be a non-metaphysical notion of space, given that $S(t)$ is not observed and possibly non-stationary?, (2) what does "in front" mean?

If the source is on the left, then turning the robot's head to the left would make sound level increase (Fig. 8B). The

opposite observation can be made if the source is on the right. If the source is right in front, or in the back, then any movement makes sound level decrease. This seems to define three spatial concepts from a subjective viewpoint. The difficulty, of course, is that the source signal $S(t)$ is different before and after the movement. The level at the ear may change because of the movement or because the level of the source has changed (Fig. 8C).

This is a case where the concept of voluntary action is important. In Section 3.2, I explained that action is considered voluntary in the sense that it is not caused by sensory signals. On the contrary, action causes changes in sensory signals. The exact opposite would be a reflex (a motor command caused by sensory signals). Suppose the robot moves its head randomly between two positions $\theta - d\theta$ and $\theta + d\theta$ while the sound plays, and the squared signal (for example) is continuously measured: $X^2(t) = a(\theta(t))^2 \cdot S(t)^2$. Then, the expectation of X^2 given that the robot's head is at $\theta - d\theta$ is $a(\theta - d\theta)^2 \cdot E[S(t)^2]$ and similarly for the other position. It follows that the relative values of $a(\theta - d\theta)^2$ and $a(\theta + d\theta)^2$ can be observed (Fig. 8D). The same reasoning holds for the entire function $a(\theta)^2$, which can be observed up to a scaling factor. This function defines the subjective location of the sound.

This example was just meant to demonstrate that, because action is voluntary, it is possible for a perceptual system to extract spatial information in a situation that seems intrinsically ambiguous. Note that in contrast with the binaural case, inference is not possible without movement—unless the set of sounds is statistically constrained and statistical inference can be used.

What does "in front" mean? The source is "in front" when the level of the sound is maximal. The source is "in the back" when the level of the sound is minimal. This is a fine definition, assuming the notion of level is known. Specifically, the potentially problematic notion here is "maximal level": this implicitly requires that the value provided by the level sensor positively correlates with sound level. Without this implicit knowledge, an alternative definition can be proposed, by thinking of the moveable ear as an information-seeking device. Consider some omnidirectional background noise in addition to the source signal. The level of the source then acts on the signal-to-noise ratio, and therefore on the intelligibility of the source. Therefore, we may redefine "in front" as the spatial location of maximal intelligibility. Intelligibility requires us to also define the content of signals ("what"), not just their spatial location ("where"). For example, if source signals consist of pure tones of various frequencies and levels, intelligibility can be defined as the degree of predictability of the signals, which depends on the signal-to-noise ratio. This provides a definition of "in front" that is independent of the particular way level is transduced by sensors.

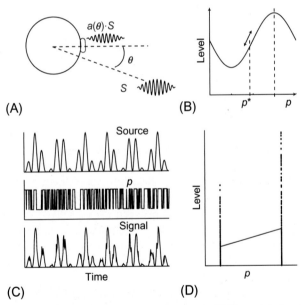

FIG. 8 Subjective physics of a robot with one ear. (A) The attenuation of the source signal at the ear is determined by the angle of the source relative to the ear. (B) The level at the ear varies with proprioceptive signal p, which is related to head angle in an unknown way. At position p^*, the source is said to be "on the right" of the robot because increasing p ("turning right") increases the level, for a constant source level. (C) If the source level is varying (top) while the head is moved (middle), then changes in level at the ear (bottom) cannot be directly attributed to source direction. Here the head was moved alternatively and at random between two positions around p^*. (D) If movements are independent from the source, as in (C) (middle), then on average the relative level at the two positions reflects the direction-dependent attenuation shown in (B) (*solid line* connects the two averages).

Other concepts can be defined, in relationship with the effect of infinitesimal movements:

- "in the axis of the ear" is when small movements have minimal impact on the signal (corresponding to front or back). This definition may be independent of intelligibility.
- "on the left" (resp. "on the right") is when a local movement in the clockwise (resp. anti-clockwise) direction decreases level or intelligibility.

6 SUBJECTIVE PHYSICS OF LIGHT, SOUND, AND TOUCH

6.1 Hearing Versus Seeing

Previously, I have chosen examples taken from auditory perception. Subjective physics applies to hearing, but also to vision and touch. In this section, I will try to describe analogies and differences between these sensory modalities. Physically, light is mediated by electromagnetic waves just as sound is mediated by acoustical waves. One can speak of the spectrum of light or of a sound, of light and sound diffraction, etc. Thus from the point of view of physics, there are many similarities. From the point of view of subjective physics, there are many differences, which I will try to outline.

The subjective physics of vision was described in great detail by James Gibson under the name "ecological optics" (Gibson, 1986). I will quickly summarize his view here. Illumination sources (the sun) produce light rays that are reflected by objects. More precisely, light is reflected by the surface of objects with the medium (air, or possibly water). What is available for visual perception are surfaces and their properties (color, texture, shape, etc.). Both the illumination sources and the surfaces in the environment are generally persistent. The observer can move, which changes the light rays received by the retina. These changes are highly structured because the surfaces persist, and this structure is informative of the surfaces in the environment. Thus visual subjective structure corresponds to the arrangement and properties of persistent surfaces. Persistence is crucial here, because it allows the observer to use its own movements to learn about the world.

On the other hand, sounds are produced by the mechanical vibration of objects. This means that sounds primarily convey information about volumes rather than surfaces. They depend on the shape, but also on the material and internal structure of objects (for example whether the object is full or empty). It also means that the information in sounds is about the source of the waves rather than their interaction with the environment, in contrast with vision. Crucially, contrary to vision, the observer cannot directly interact with sound waves, because a sound happens, it is not persistent. An observer can produce a sound wave, for example by hitting an object, but once the sound is produced there is no possible further interaction with it. The observer cannot move to analyze the structure of acoustic signals. The only available information is in the sound signal itself. In this sense, sounds are events (Casati and Dokic, 1998; O'Callaghan, 2010). There are of course some properties of sounds that are persistent: precisely the spatial properties (location of the sound source), as we have seen before. But the shape of an object is not specified by sound in the way it is specified by vision. In vision, the way the visual signals change when one moves around an object specifies the shape of the object. When the observer moves around the source of a sound, even a stationary one, the acoustical waves do not change is such a lawful way. This is not to say that the sound contains no information about shape. Indeed the structure of the acoustical signal is related to properties of the sounding object, in particular material and shape (Gaver, 1993). For example, the resonant modes are informative of the shape. However, the relationship between this structure and the three-dimensional shape of the object is metaphysical knowledge (if only auditory signals are available).

These observations highlight major differences between vision and hearing from the viewpoint of subjective physics, which go beyond the physical basis of these two senses (light waves and acoustic waves). Vision is the perception of persistent surfaces. Hearing is essentially the perception of mechanical events on volumes.

6.2 Touch

How about touch? As I previously mentioned when discussing Molyneux's problem, tactile perception is sometimes likened to vision in philosophy, by identifying the light ray impinging on a photoreceptor with a finger. This analogy also shows that the subjective structure of touch includes information about the spatial arrangement of surfaces, as vision, and the objects of touch are persistent (even though contact may not be). There are important differences with vision. There is no illumination source. Touch is a proximal sense that requires contact, whereas vision and hearing are distal senses. This means in particular that distance manifests itself in a different way in the subjective structure of touch: distance is defined not by the effect of movements on some continuous property of the sensory signals (level, visual solid angle), but by the movement necessary to make contact with the object. Touch is also about volume, or more precisely, about weight, and also about the type of material (soft/hard) in the relationship between hand movements and mechanical signals. Other aspects of active touch have been described by Gibson (1962).

There is also an analogy between sound and touch. I previously mentioned that sounds are about mechanical events on volumes, but this is incomplete. A mechanical event implies an interaction localized at the surface of the resonating object. This can be an impact, or more continuous interactions such as scratching or rolling. Information about the surface is then present in the temporal structure of the sound (Gaver, 1993). This is not about the spatial arrangement of the surface, but rather its texture. Such interactions result in mechanical waves that are tactile and auditory, possibly also visual. Therefore in the sensorimotor relationship between finger movement (or eye movement) and mechanical waves (or visual signals), textures produce an analogous subjective structure for hearing, touch, and vision.

7 THE MIND-WORLD BOUNDARY

The subjective structure of the world is determined by the interface between the perceptual system and the world, that is, by the set of sensors and possible actions on the world. This interface is traditionally located at the physical interface between the body and the external world. However, the exercise of subjective physics could as well be applied to less obvious interfaces, of which I will now give a few examples.

7.1 Subsumption Architectures

A subsumption architecture is a way to decompose complex behaviors into a set of layers, with each layer taking control

("subsuming") over the underlying layers. It was introduced in the field of behavior-based robotics by Rodney Brooks and colleagues (Brooks, 1986), as a strategy to develop autonomous robots of increasing complexity, but the concept may also apply to biology. For example, there is a reflex that makes the eye follow a slowly moving object appearing in the fovea. As far as subjective physics is concerned, this reflex may be considered as part of the perceptual system (one of the possible actions) or as part of the world, that is, as something that happens independently of the perceptual system. The subjective structure is different in both cases. If it is considered as being in the world, then motion of an object corresponds not to a displacement of the visual field, but to a change in proprioceptive signals (motion of the eye), together with a change in the background.

In the same way, actions may be considered as direct controls of the physical actuators, or as commands on structures that may issue complex motor sequences, possibly involving control mechanisms, as in servomotors, for example. A biological example is found in the locomotion of some birds like pigeons (Fig. 9A). When these animals walk, their head displays a characteristic forward and backward movement called "head-bobbing" (Necker, 2007). Each cycle consists of a rapid forward movement of the head (thrust phase) followed by a phase where the head is stable relative to the ground, that is, moving backward relative to the body so as to compensate for locomotion. It follows that the visual field changes

FIG. 9 The mind-world boundary. (A) Some birds move their head as they walk, in a characteristic fashion called "head-bobbing," such that the head is stationary relative to the ground except at discrete moments. It changes the effect of locomotion on the visual field. (B), The cochlea can be considered as a sensorimotor system, in which sensors are the afferent fibers capturing the vibration of the basilar membrane via the inner hair cells, and actuators are the efferent fibers influencing that vibration through the outer hair cells. (Adapted from Guinan, J.J., 2006. Olivocochlear efferents: anatomy, physiology, function, and the measurement of efferent effects in humans. Ear Hear. 27 (6) (December), 589–607. doi:10.1097/01.aud.0000240507.83072.e7.). (C) A neuron can be seen as a perceptual system, where the synapses are its sensors, its axon is its actuator, and all the other neurons together with the outside environment are "the world." (D) The same neuron can be seen as a perceptual system of a different kind, by redefining the sensors as molecular signal detectors (eg, calcium) and actions as expression or regulation of ionic channels and other elements of structure.

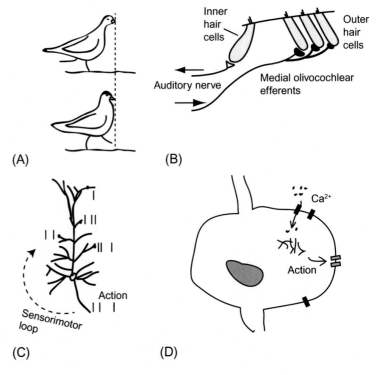

in almost discrete steps (during the thrust phase) when the body moves continuously.

7.2 The Cochlea as a Sensorimotor System

The cochlea is the organ of hearing. In the cochlea is the basilar membrane, which the acoustic wave received at the ear puts in motion. The stiffness varies from base to apex, and different places along that membrane are maximally sensitive to different frequencies. Thus the acoustic wave elicits a spatiotemporal pattern of vibration. Inner hair cells sit on the basilar membrane and transduce mechanical vibrations into electrical signals, which are in turn converted to spikes in the auditory nerve (Fig. 9B). The nerve then projects to various nuclei in the auditory brainstem. This process is generally depicted as a frequency analysis, as if a bank of bandpass filters were applied to the acoustic wave.

There are also other cells sitting on the basilar membrane, the outer hair cells. They are more numerous and they are thought to actively adjust the "gain" of the cochlea, that is, to be responsible for the dynamic compression that allows the 100 dB dynamic range of human hearing, and maybe to protect the cochlea against loud sounds. They receive inputs from olivocochlear efferents, with bodies in the superior olivary complex (also in the auditory brainstem), mostly on the contralateral side (Guinan, 2006). These are contacted by neurons in the cochlear nucleus, which themselves receive input from inner hair cells (through type I auditory nerve fibers). There are also (ipsilateral) olivocochlear efferent inputs onto inner hair cells. These are less numerous and seem to target the initiation site of type I auditory nerve fibers. I will focus on the most abundant ones, which target the outer hair cells.

Olivocochlear efferents can change the amplitude, but also the phase of the basilar membrane displacement in response to a sound (Cooper and Guinan, 2003). It would then be conceivable to examine the subjective physics of the cochlea seen as a sensorimotor system, where sensors are the afferent fibers (capturing the vibration of the basilar membrane), and actuators are the efferent fibers (influencing that vibration through the outer hair cells). There is then information about a sound in the relationship that it induces between the actions on outer hair cells and the inputs from the inner hair cells.

7.3 The Neuron as a Perceptual System

Electrical communication between neurons is directional: action potentials propagate along the axon of a neuron, from the initial segment near the cell body to the synaptic terminals, where they produce electrical changes in other neurons. Therefore one could consider the synapses of a neuron as its sensors, its axon as its actuator, and all the other neurons together with the outside environment as

"the world" (Fig. 9C). For the neuron seen as a perceptual system, spikes are actions that it produces on the world. The subjective physics of this system describes the relationship between spikes and their effects on the neuron's inputs. Somehow, this is what the neuron can "know" about the outside world.

The neuron could also be seen as a perceptual system of a different kind, by redefining the set of sensors and actions (Fig. 9D). Calcium is a universal signal in cells in general, and in neurons in particular (Berridge et al., 2000). It triggers a large number of processes such as the expression of ionic channels or morphological changes. The cell could then be seen as a perceptual system with these processes as the actions it can take, and calcium sensors as its sensors; sensors could also be voltage-activated calcium channels, for example. The subjective physics of this system describes the relationship between plastic changes (synaptic changes, ionic channel regulation) and their effect on calcium signals and other signals captured by the cell.

8 DISCUSSION

I have proposed to define subjective physics as the field of study that analyzes the structure of the sensory and sensorimotor relationships (laws) that are available to an organism embedded in an environment, given a set of sensors and actuators, without metaphysical knowledge about the world. I will first summarize the main concepts in this text, and then discuss the relevance of subjective physics to various fields.

8.1 Summary

Subjective physics describes a set of laws, in the same sense as physics describes a set of laws. The difference is that these laws are seen from the perspective of the perceptual system, the analog of the scientist. The perceptual system derives its knowledge of the world from its sensors and the actions it can produce—the "interface" with the world, in the same way as a scientist has access to measurements and can make experiments. This knowledge is a set of laws, which can take two forms: *weak knowledge*, the kind that has a predictive value about future sensory inputs; *strong knowledge*, the kind that applies to the effect of actions on sensory inputs. Any other form of knowledge is called *metaphysical knowledge*, in analog with Popper's demarcation criterion in philosophy of science. It encompasses all knowledge that an external observer might have, but that does not derive from the relationships between actions and sensors.

This set of laws is called the *subjective structure of the world*. It can be progressively analyzed at different depths of *ecological reduction*, which I defined as the suspension of prior knowledge about the world on behalf of the perceptual system: at the first level, we may consider that sensor data is specified in external terms (acoustical pressure); at the second level, we may consider that sensor data is specified in internal terms (transduced quantity); at the third level, we may consider that sensors are inhomogeneous.

A sensor is something that is modified by the world, and whose modification can be picked up by the perceptual system. An action is something that the perceptual system can do which produces modifications in the world and can ultimately affect sensors. In subjective physics, actions are considered *voluntary*, in the sense that any action can potentially be taken. A consequence is the existence of causality in the sensorimotor structure: causality can be distinguished from correlation because action can be taken or not taken, independently of sensory inputs. Action can be involved in the subjective structure in five different ways: not at all (sensory structure); structure conditioned to the absence of action; action as a cause of the sensory flow (without which there is no structure); through proprioception (sensorimotor contingencies); as an event causing a change in structure or in a sensory signal.

Subjective structure can be unconditional: a set of laws that are always satisfied. Or it can be conditional, that is, not normally observed (eg, induced by the presence of an object). Conditional structure is therefore defined with respect to unconditional structure (the reference). In contrast with classical notions of information (Shannon), which are unstructured, subjective structure is compositional, hierarchical, and contextual: one may speak of the "syntax" of subjective structure. Actions, seen as operations on the structure, define algebraic properties (eg, group action). This rich notion of structure provides a notion of analogy that is different from similarity between metric spaces: subjective structures are said to be analogous when they have a shared set of properties.

I have shown how to apply these concepts to two fictional examples involving a hearing robot. Applied to humans, subjective physics of light, sound and touch have analogies and differences, which go beyond their physical substrate (different types of waves). For example, vision is the perception of persistent surfaces, while hearing is essentially the perception of mechanical events on volumes.

The subjective structure of the world is determined by the interface between the perceptual system and the world, that is, by the set of sensors and possible actions on the world. As far as the exercise of subjective physics is concerned, any arbitrary interface can be defined. For example, the cochlea or a neuron can be seen as perceptual systems. In the first case, the "world" is the basilar membrane; in the second case, it is the rest of the brain and the external environment.

I will now discuss the relevance of subjective physics for various fields.

8.2 Subjective Physics and Computational Neuroscience

I started this essay with a quote from David Marr, a major figure in computational neuroscience: "If we are capable of knowing what is where in the world, our brains must somehow be capable of representing this information" (Marr, 1982). For a perceptual system, "knowing what is where in the world" corresponds to what Marr called the "computational level," what the system is supposed to do.

This text is largely an attempt to redefine the computational level of perceptual systems in a way that does not rely on objective descriptions made by an external observer. Instead, the computational level is described from the perspective of a perceiver embedded in its environment. This viewpoint departs from traditional approaches in computational neuroscience of perception, which follow the information-processing paradigm: the system is given some sensory inputs, for example a pair of acoustic signals, and it processes them into an output, which could be the estimated angle of the sound source. The output is given a meaning by the external observer, but for the perceptual system itself, it is nothing else than a real number. In this paradigm, the task of the system is to achieve a particular transformation between inputs and outputs.

This traditional approach implicitly assumes that the system has metaphysical knowledge about the world, in the sense that I proposed. If we do not want to make this assumption, then the traditional approach is not satisfying as a way to understand perceptual systems. In this case, the task of the perceptual system is radically different: it is not to achieve a particular transformation anymore, but to grasp the laws of subjective physics. The relevant analogy is not that of a machine applying a sequence of operations to inputs so as to produce an output, but rather that of a scientist who can make measurements and experiments, and who draws conclusions in the form of laws.

How should neural models of perceptual systems look, in the framework of subjective physics? In subjective physics, the basic element of perception is a law that sensory signals follow (in a broad sense, including proprioceptive signals), or "invariant structure" in the terms of Gibson. Therefore, one basic challenge for neural models of perceptual systems is to identify and respond to laws that unfold in time. There is some work in the field that is relevant to this theme. One relevant line of work is a set of learning algorithms based on the "slowness principle" (Földiák, 1991; Mitchison, 1991; Becker and Plumbley, 1996; Stone, 1996; Wiskott and Sejnowski, 2002). The idea is that characteristics of the world, for example the location of a sound source, vary more slowly than sensory signals that are caused by them (eg, acoustical signals), another way to express the idea that laws are invariant while the constituents of the laws are variable. The learning algorithms then consist in projecting the signal space into another space where projected signals vary as slowly as possible. These projected signals are then expected to be characteristics of laws.

Another relevant line of work is the idea that synchrony is a temporal invariant, and if spike trains are caused by sensory signals, then a particular pattern of synchrony in neural population reflects the occurrence of a particular sensory law (Brette, 2012; Goodman and Brette, 2010b). I defined the "synchrony receptive field" as the set of stimuli that elicit synchronous firing in a given group of neurons: it corresponds to a temporal invariant or law. A neuron that detects coincidences between these neurons then spikes when the stimulus follows that law.

These approaches are promising, but they only address a small subset of the concepts developed here. For example, computationally speaking, detecting that sensory signals follow a particular law is not the same as predicting the future of these signals, or predicting the effect of an action on future sensory signals. Perhaps more importantly, I have shown that the subjective structure of the world has syntax: it is compositional, hierarchical, and contextual; actions add an algebraic structure. These properties are ignored in standard representational theories in neuroscience because neural representations are intrinsically unstructured. In the concept of "neural assemblies," which is the mainstream assumption about how things we perceive are represented in the brain, any given object is represented by the firing of a given assembly of neurons. Therefore, the structure of such neural representations is the structure of subsets of a fixed set of elements: a neural assembly is a "bag of neurons," in the same way that search engines analyze the content of a web page as a "bag of words" with no relationships between the words. This weakness has, in fact, been observed many times in the past in the context of the "binding problem" (von der Malsburg, 1999): when two objects are present, they are represented by a merged assembly in which the identity of the two sub-assemblies is lost (the so-called superposition catastrophe). The identification of this weakness led to alternative propositions, such as using time as a signature of represented objects ["binding by synchrony" (Singer, 1999)]. Such representations are richer (in particular compositional and hierarchical), but still not as rich as subjective structure: sets of features can be represented, but not relations between features, apart from belonging to the same set. For example, the statement "action A changes sensory structure B to C" describes a relationship between B and C, labeled by A. This does not fit the neural assembly framework, even augmented with binding with synchrony. Other authors have proposed that taking into account the order of activation of neurons provides a syntax to neural firing, which may be a way to address this problem (Buzsáki, 2010).

Finally, even the most sophisticated representational theories still pose a problem to the viewpoint of subjective physics, because they leave the interpretation of the neural representations to some unspecified external observer. A simple way to solve this problem is to not have an interpretation stage. That is, instead of considering neural models that "represent" the world, one can consider neural models that produce actions in the world; in other words, models that are *autonomous*. This theme has been developed in behavioral robotics (Brooks, 1991) and in enactive philosophical theories of the mind (Thompson and Stapleton, 2009), but not very much in computational neuroscience. Yet it seems almost unavoidable that neural models of perception developed in the framework of subjective physics must be autonomous: it is a natural consequence of the attempt of removing metaphysical knowledge on behalf of the perceptual system.

8.3 Subjective Physics and Philosophy of Science

In this text, I have made a number of analogies with philosophy of science. This was not accidental. Starting from the observation that what a scientist can know about the world derives from her senses and the actions she can take in the world (experiments, possibly involving measurement devices), philosophy of science asks questions such as: What is knowledge? How can it be acquired? How can we distinguish between contradicting theories?

So there is, in fact, a formal analogy between these questions and corresponding questions in subjective physics, where sensors are the observation devices and actions are the kind of experiments the perceptual system can make. This is an interesting analogy, because concepts developed in philosophy of science are directly relevant to subjective physics. I will discuss a few of them.

The first remark is that science, like subjective physics, takes the form of universal statements or laws. A law is more than a collection of observations: it says something about observations that have not been made yet—this is what makes science useful. But how are laws formed? The naive view, classical inductivism, consists in collecting a large number of observations and generalizing from them. For example, one notes that all men she has seen so far have two legs, and concludes that all men have two legs. Unfortunately, inductivism cannot produce universal laws with certainty. It is well possible that one day you might see a man with only one leg. The problem is that there are always an infinite number of universal statements that are consistent with any finite set of observations.

Therefore, inductivism cannot guide the development of knowledge. Karl Popper, possibly the most influential philosopher of science of the 20th century, proposed to address this problem with the notion of *falsifiability* (Popper, 1959). What distinguishes a scientific statement from a metaphysical statement is that it can be *disproved* by an experiment. For example, "all men have two legs" is a scientific statement, because the theory could be disproved by observing a man with one leg. But "there is a God" is not a scientific statement. For any practical purpose, only scientific statements are useful, since a metaphysical statement can have no predictable impact on any of our experience, otherwise this would produce a test of that statement.

These concepts can be directly applied to subjective physics: knowledge about the world takes the form of laws; only those laws that could potentially be falsified in the future are useful from the perspective of the perceptual system; other kinds of laws can be considered "metaphysical."

I will not do an exhaustive review of philosophy of science, but I wish to point out that there are many other concepts that are directly relevant to subjective physics. I will mention a few of them. While Popper explains what a scientific statement is and how it can be tested, he does not explain the difficult part, which is how a scientific statement is made in the first place. From a logical point of view, there are an infinite number of possibilities given a finite set of observations. Which one should be chosen? A popular heuristic is "Occam's razor," that is, the idea that among competing hypotheses, the most parsimonious one should be preferred. This is a well-known concept in statistical learning theory, related to the problem of "overfitting": a simple law is more likely to generalize well than a complex one. But choosing the simple theory also means choosing a theory that is *not* consistent with observations. And indeed Post-Popperian philosophers and historians of science have argued that a scientific theory is not only a theory that *can* be falsified, it is a theory that *is* actually falsified (Lakatos et al., 1978; Kuhn, 1962; Feyerabend, 2010). This is made possible by treating falsifications of theories as anomalies, which can be explained by auxiliary hypotheses. For subjective physics, these concepts mean that consistency with observations is not the only criterion that a perceptual system should use to make laws—simplicity could be an additional one, or analogy with other laws (in the sense that I previously defined).

The most radical critics of Popper have made a remark that is highly relevant to subjective physics (Feyerabend, 2010; Kuhn, 1962). There is no such thing as an objective observation independent of any scientific theory. Observations are produced by scientists themselves, in the context of the theory they currently favor. Theories are not derived from observations, but rather there is a circular relationship between them. It is, in fact, implicit in Popper's exposition of falsifiability: a scientific theory should suggest a critical experiment that may falsify it; therefore, it drives future observations. In the same way, observations made by a perceptual system are not independent of that perceptual system. They depend on the actions taken by the system. This suggests that the formation of laws should not be thought of as a process of fitting a curve to a given set of data points, but rather as an active process in which observations are made so as to help the formation of laws. This relates to the concept of active learning (Cohn et al., 1996). Note that the fact that the relationship between observations and theories is circular does not imply that theories are arbitrary, since observations still depend on the world. But it does imply that the formation of theories (laws) depends on the history of the process.

To end this section, I may venture to propose that conversely, subjective physics may perhaps provide a simple conceptual framework in which to develop concepts of philosophy of science, instead of the more traditional framework of history of science.

8.4 Subjective Physics and Psychological Theories of Perception

There is a strong relationship between subjective physics and psychological theories of perception, mostly two of them: Gibson's ecological approach to perception (Gibson, 1986) and O'Regan's sensorimotor theory of perception (O'Regan, 2011). According to Gibson, the "information to be perceived" is the invariant structure in the sensory or sensorimotor flow, that is, the subjective structure. Gibson proposed informal descriptions of that structure mainly in vision, with the concept of the "optical array" (Gibson, 1972). He also wrote a less detailed account about active touch (Gibson, 1962), and his approach was applied to sounds by Gaver (1993).

A major Gibsonian theme is the notion of "affordances." Gibson considered that what we perceive in the world is affordances, a term he coined to designate the possibilities of action that an object in the world allows us to do. For example, a car is something that we can drive, and an opening in a cave is something we can go through. According to Gibson, the world is perceived through the actions we can do in it. It is these affordances that produce meaning for a particular organism, and the same object in the world can mean completely different things for organisms that act differently. Subjective physics may provide a framework to study affordances.

According to the sensorimotor theory of perception (O'Regan, 2011), what we perceive is the expected effect of our own actions on sensory signals. Subjective physics describes this expected effect. There are a number of differences with Gibson. One is that it grants an important role for inference in perception, which Gibson largely downplayed (he considered on the contrary that perception is direct). But what is inferred is a law (or sensorimotor contingency), not an objective "thing" in the world. A second important difference is that it refines the notion of invariant structure by acknowledging that the signals that the brain receives are only indirectly related to physical signals (light). Finally, in terms of subjective physics, the theory also proposes that the phenomenological structure of conscious perception reflects the subjective structure of the world.

An application of the sensorimotor theory of perception is sensory substitution, for example presenting the image of a camera or a sound through a tactile device (Kaczmarek et al., 1991). The theory predicts that sounds or images can be perceived through sensors that are different from those normally associated with the corresponding perceptual modality, provided that sensorimotor contingencies are preserved. What this text suggests is that sensory signals should be presented to the sensors in such a way that the substituted subjective structure is analogous to the original subjective structure, in the sense that I proposed in Section 4.5 (that the two structures have properties in common, eg, algebraic properties).

8.5 Subjective Physics, Robotics, and Neuromorphic Engineering

I will end this text on a short discussion of the relevance of subjective physics for robotics and neuromorphic engineering.

In robotics, the theme of subjective physics is connected to theories of embodiment, the idea that the body and its interaction with the environment are parts of the cognitive system. The traditional approach in robotics (and more generally in artificial intelligence) is to consider perception as a separate module whose function is to produce an objective representation of the world based on sensory inputs; another module takes decision and actions on the basis of that representation. There are alternative ideas in robotics, in which the robot learns to use its body and sensors, considered as an unknown "envelope," with a general-purpose algorithm, the "kernel" (Kaplan and Oudeyer, 2011). In terms of subjective physics, the envelope is the interface with the world and the kernel is the perceptual system. Thus subjective physics describes the structure of the world experienced by the robot through its envelope, which is to be discovered by the robot's kernel.

Some approaches in neuromorphic engineering use low-power analog circuits to model neurons (Indiveri et al., 2011). There are two motivations: to reproduce the computational abilities of biological neural networks, and to develop electronic devices that consume little power. Using low-power components comes at price: there is some tolerance in the properties of the electronic components, which make them partly unknown to the system. This issue corresponds to the notion that the relationship between physical stimuli and the activation of sensors is metaphysical knowledge for the system. Thus neuromorphic hardware must be designed to work in the absence of metaphysical knowledge of component properties (to some extent). Subjective physics describes what can still be known by the system under these constraints, and may suggest ways to deal with this issue, such as statistical homogenization, or including the unknown (but fixed) properties in the subjective structure.

This discussion has sketched a number of perspectives for the development of subjective physics along two main lines. One is to develop the description of subjective physics for human, animal, and artificial sensory systems. Another one is to develop the modeling of perceptual systems in the framework of subjective physics, an exciting challenge for computational neuroscience.

GLOSSARY

Action something that the perceptual system can do which produces modifications in the world. In turn, these modifications can affect sensors.

Ecological reduction the suspension of any prior knowledge about the world on behalf of the perceptual system, which precedes the analysis of the subjective structure of the world.

Eidetic variation varying the world, the interface, and the constraints on the perceptual system, so as to reveal their relationships with the subjective structure of the world.

Interface a set of sensors and possible actions on the world.

Metaphysical knowledge statements about the world that cannot be falsified with the available set of sensors and actuators.

Perceptual system the system that captures sensor data and performs actions on the world.

Sensor something that is modified by the world, and whose modification can be picked up by the perceptual system.

Sensor homogeneity the assumption that sensors have the same properties, for example, that different microphones provide values that have the same relationship to acoustical pressure.

Statistical homogenization a procedure by which sensor properties are made homogeneous, by tuning sensors so that their signals have identical statistics.

Subjective physics the field of study that analyzes the structure of the sensory and sensorimotor relationships that are available to an organism embedded in an environment, given a set of sensors and possible actions.

Subjective structure of the world the structure of the world, as it can be discovered by the perceptual system without a priori knowledge. It is made of the laws followed by sensor data and the relationships between actions and sensor data.

World the environment in which the perceptual system operates.

REFERENCES

Becker, S., Plumbley, M., 1996. Unsupervised neural network learning procedures for feature extraction and classification. Appl. Intell. 6 (3), 185–203. http://dx.doi.org/10.1007/BF00126625 (July 1).

Berridge, M.J., Lipp, P., Bootman, M.D., 2000. The versatility and universality of calcium signalling. Nat. Rev. Mol. Cell Biol. 1 (1), 11–21. http://dx.doi.org/10.1038/35036035. October.

Brette, R., 2012. Computing with neural synchrony. PLoS Comput. Biol. 8 (6).

Brooks, R.A., 1986. A robust layered control system for a mobile robot. IEEE J. Robot. Autom. 2 (1), 14–23. http://dx.doi.org/10.1109/JRA.1986.1087032.

Brooks, R.A., 1991. Intelligence without representation. Artif. Intell. 47 (1–3), 139–159. http://dx.doi.org/10.1016/0004-3702(91)90053-M. January.

Buzsáki, G., 2010. Neural syntax: cell assemblies, synapsembles, and readers. Neuron 68 (3), 362–385. http://dx.doi.org/10.1016/j.neuron.2010.09.023 (November 4).

Casati, Roberto, Dokic, Jérôme, 1998. La Philosophie Du Son. Jacqueline Chambon, Nîmes.

Cohn, D.A., Ghahramani, Z., Jordan, M.I., 1996. Active learning with statistical models. J. Artif. Intell. Res. 4, 129–145 arXiv:cs/9603104 (February 29). http://arxiv.org/abs/cs/9603104.

Cooper, N.P., Guinan, J.J., 2003. Separate mechanical processes underlie fast and slow effects of medial olivocochlear efferent activity. J. Physiol. 548 (1), 307–312. http://dx.doi.org/10.1113/jphysiol.2003.039081 (April 1).

Deneve, S., 2008. Bayesian spiking neurons I: inference. Neural Comput. 20 (1), 91–117. http://dx.doi.org/10.1162/neco.2008.20.1.91 (January).

Dreyfus, H.L., 1992. What Computers Still Can't Do: A Critique of Artificial Reason. MIT Press, New York.

Durkovic, M., Habigt, T., Rothbucher, M., Diepold, K., 2011. Low latency localization of multiple sound sources in reverberant environments. J. Acoust. Soc. Am. 130 (6), EL392–EL398. http://dx.doi.org/10.1121/1.3659146.

Feyerabend, P., 2010. Against Method. Verso, London.

Földiák, P., 1991. Learning invariance from transformation sequences. Neural Comput. 3 (2), 194–200. http://dx.doi.org/10.1162/neco.1991.3.2.194.

Gaver, W.W., 1993. What in the world do we hear?: an ecological approach to auditory event perception. Ecol. Psychol. 5 (1), 1–29. http://dx.doi.org/10.1207/s15326969eco0501_1.

Gibson, J.J., 1962. Observations on active touch. Psychol. Rev. 69, 477–491 (November).

Gibson, J.J., 1972. A theory of direct visual perception. In: The Psychology of Knowing. Gordon & Breach, New York.

Gibson, J.J., 1986. The Ecological Approach to Visual Perception. Routledge, Abingdon.

Goodman, D.F.M., Brette, R., 2010a. Learning to Localise Sounds with Spiking Neural Networks. Adv. Neural Inf. Process. Syst. 23, 784–792.

Goodman, D.F.M., Brette, R., 2010b. Spike-timing-based computation in sound localization. PLoS Comput. Biol. 6 (11) http://dx.doi.org/10.1371/journal.pcbi.1000993 (November 11).

Guinan, J.J., 2006. Olivocochlear efferents: anatomy, physiology, function, and the measurement of efferent effects in humans. Ear Hear. 27 (6), 589–607. http://dx.doi.org/10.1097/01.aud.0000240507.83072.e7 (December).

Held, R., Ostrovsky, Y., de Gelder, B., Gandhi, T., Ganesh, S., Mathur, U., Sinha, P., 2011. The newly sighted fail to match seen with felt. Nat. Neurosci. 14 (5), 551–553. http://dx.doi.org/10.1038/nn.2795.

Indiveri, G., Linares-Barranco, B., Hamilton, T.J., Etienne-Cummings, R., Delbruck, T., Liu, S.-C., Häfliger, P., et al., 2011. Neuromorphic silicon neuron circuits. Front. Neuromorphic Eng. 5, 73. http://dx.doi.org/10.3389/fnins.2011.00073.

Jeffress, L.A., 1948. A place theory of sound localisation. J. Comp. Physiol. Psychol. 41, 35.

Joris, P.X., Smith, P.H., Yin, T.C., 1998. Coincidence detection in the auditory system: 50 years after jeffress. Neuron 21 (6), 1235.

Kaczmarek, K.A., Webster, J.G., Bach-y-Rita, P., Tompkins, W.J., 1991. Electrotactile and vibrotactile displays for sensory substitution systems. IEEE Trans. Biomed. Eng. 38 (1), 1–16. http://dx.doi.org/10.1109/10.68204 (January).

Kaplan, F., Oudeyer, P.-Y., 2011. From hardware and software to kernels and envelopes: a concept shift for robotics, developmental psychology, and brain sciences. In: Neuromorphic and Brain-Based Robots. Cambridge University Press, New York. http://dx.doi.org/10.1017/CBO9780511994838.011.

Kuhn, T.S., 1962. The Structure of Scientific Revoutions. The University of Chicago Press, Chicago, IL.

Kuhn, G.F., 1977. Model for the interaural time differences in the azimuthal plane. J. Acoust. Soc. Am. 62 (1), 157–167. http://dx.doi.org/10.1121/1.381498.

Lakatos, I., Currie, G., Worrall, J., Lakatos, I., 1978. Philosophical Papers (of) Imre Lakatos. Cambridge University Press, Cambridge.

MacDonald, J.A., 2008. A localization algorithm based on head-related transfer functions. J. Acoust. Soc. Am. 123 (6), 4290–4296. http://dx.doi.org/10.1121/1.2909566.

Marr, D., 1982. Vision, first ed. W.H.Freeman & Co Ltd., New York.

Merleau-Ponty, M., 2002. Phenomenology of Perception, second ed. Routledge, London.

Mitchison, G., 1991. Removing time variation with the anti-hebbian differential synapse. Neural Comput. 3 (3), 312–320. http://dx.doi.org/10.1162/neco.1991.3.3.312 (September 1).

Necker, R., 2007. Head-bobbing of Walking Birds. J. Comp. Physiol. A. 193 (12), 1177–1183. http://dx.doi.org/10.1007/s00359-007-0281-3 (December 1).

O'Callaghan, C., 2010. Sounds: A Philosophical Theory. Oxford University Press, New York.

O'Regan, J.K., 2011. Why Red Doesn't Sound Like a Bell: Understanding the Feel of Consciousness: Understanding the Feel of Consciousness. Oxford University Press, New York.

O'Regan, J.K., Noë, A., 2001. A sensorimotor account of vision and visual consciousness. Behav. Brain Sci. 24 (05), 939–973. http://dx.doi.org/10.1017/S0140525X01000115.

Poincaré, H., 1968. La Science et l'Hypothèse. Flammarion, Paris.

Popper, K.R., 1959. The Logic of Scientific Discovery. Basic Books, Oxford.

Singer, W., 1999. Neuronal synchrony: a versatile code for the definition of relations? Neuron 24 (1), 49.

Stone, J.V., 1996. Learning perceptually salient visual parameters using spatiotemporal smoothness constraints. Neural Comput. 8 (7), 1463–1492 (October 1).

Thompson, E., Stapleton, M., 2009. Making sense of sense-making: reflections on enactive and extended mind theories. Topoi 28 (1), 23–30. http://dx.doi.org/10.1007/s11245-008-9043-2 (March 1).

Thompson, E., Palacios, A., Varela, F.J., 1992. Ways of coloring: comparative color vision as a case study for cognitive science. Behav. Brain Sci. 15 (01), 1–26. http://dx.doi.org/10.1017/S0140525X00067248.

Treisman, M., 1977. Motion sickness: an evolutionary hypothesis. Science 197 (4302), 493–495. http://dx.doi.org/10.1126/science.301659.

Von der Malsburg, C., 1999. The what and why of binding: the Modeler's perspective. Neuron 24 (1), 95.

Wiskott, L., Sejnowski, T.J., 2002. Slow feature analysis: unsupervised learning of invariances. Neural Comput. 14 (4), 715–770. http://dx.doi.org/10.1162/089976602317318938 (April).

Chapter 12

Contextual Emergence in Neuroscience

P. beim Graben*

Humboldt-University of Berlin, Berlin, Germany

1 INTRODUCTION

Science is regulated by methodological rules and idealizations (Kant, 1956, 1997; Primas, 1990). Perhaps the most important and uncontroversial of those assumptions is the existence of an *objective reality* that can be investigated by scientific technology in an *open loop* approach, where the system under study reveals its intrinsic properties through measurement results to a detached observer (Primas, 1977, 1990). This is basically the underlying prerequisite of current *big data* approaches, predominant, eg, in high energy physics, meteorology, corpus linguistics, and computational neuroscience. Using advanced data assimilation and analysis methods, scientists hope to get a compressed high-level description from a huge amount of detailed low-level data in such a way that the higher level description can be (at least in principle) completely reduced to the lower level description. This scenario of conventional open loop methodology is sketched in Fig. 1. Fig. 1A depicts the relationship between two levels of scientific description: a lower level L and a higher level H that can be strongly reduced to the former. Here, *strong reduction* is logically defined in that L bears necessary and sufficient conditions for H (Bishop and Atmanspacher, 2006), or, in other words, that the L-description implies the H-description (sufficiency) and vice versa (necessity), as indicated by the arrows. On the other hand, Fig. 1B displays the relationship between a system under study and its observer. The system feeds the detached observer with (big) measurement data.

The picture illustrated in Fig. 1 is the methodological ideal of classical physics, where the lowest level comprises the first principles of Newtonian mechanics and Maxwellian electrodynamics and all higher level accounts, such as thermodynamics or geometrical optics can be strongly reduced to those descriptions that do not refer to any observer. Yet this scientific ideal became questioned in the early 20th century through the rise of relativity, quantum mechanics, and cybernetics (beim Graben, 2011), where observation conditions (eg, relative motion, complementary measurements, or feedback control) explicitly had to be taken into account by general system theory. Instead of the open loop, *closed-loop*

methodologies were demanded (Potter et al., 2014). In particular, the concept of strong reduction appeared inappropriate for clarification of interlevel dependencies.

In order to elucidate the intricate situation of modern epistemology, we consider more closely the paradigmatic example of the relationship between Newtonian point mechanics (L) and thermodynamics (H). A gas in a container at thermal equilibrium is regarded as a huge collection of microscopic particles that freely move around, collide with each other, and bounce the container's boundaries in a deeply elastic manner. The so-called *microstate* of the gas at some time is an instantaneous listing of all positions and velocities of every particle, or likewise a point (ie, a *pure* state) in a high-dimensional vector space, called the *phase space*. The coordinates of this individual point are then given by the entries of the given list. This phase space representation is thus an individual description for the lower level, henceforth L_i, as every particle can be identified either from the list or from a projection of the phase space vector. Physical quantities, such as a particle's momentum or its energy, that are (at least principally) measurable are called *observables*. These are mappings from a microstate to the real numbers of possible measurement results. Because the outcome of such a measurement is always the same number for a given microstate, there is no observable variability at the individual description, a situation characterized through the fact that microstates are *dispersion-free* (in every observable) (beim Graben and Atmanspacher, 2006, 2009).

However, individual descriptions are epistemically unfeasible for the large number of particles involved. Therefore, physicists employ statistical descriptions L_s in statistical mechanics where a statistical state is a probability distribution over phase space such that the value of the distribution density in a certain phase space region gives the likelihood to find the system in that area. These distribution functions are called *statistical states* and are required for the calculation of macroscopically measurable quantities such as pressure, volume, or temperature through expectation values. Yet, computing such an expectation value for one of the lower level observables yields different results for different realizations of one given statistical state. Thus, arbitrary statistical states are

Closed Loop Neuroscience. http://dx.doi.org/10.1016/B978-0-12-802452-2.00012-3

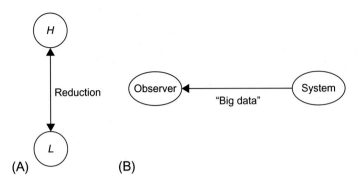

FIG. 1 Conventional *open loop* scientific methodology. (A) In the case of *strong reduction*, the lower level description *L* is necessary and sufficient for a higher level description *H* of a system under study. (B) In the case of *big data* approaches, the system under study delivers large amounts of measurement results to the observer.

generally not dispersion-free with respect to the system's observables. Moreover, the space of all possible statistical states is a mathematical "monster, which contains myriads of unphysical states" (Primas, 1990) that are rarely realized in a particular physical setup. For these reasons, lower level statistical descriptions L_s are generally unsuitable for higher level descriptions *H*. One needs therefore some criterion to select statistical states that are appropriate for a particular higher level description such as thermodynamics.

For the latter case, such selection criterion is given by the phenomenal concept of *thermal equilibrium*, which is operationalized by the so-called *Zeroth Law* of thermodynamics. The Zeroth Law establishes an equivalence relation over thermal systems, or likewise, over statistical states. If two systems are in thermal equilibrium with another third system, they are in thermal equilibrium with each other as well. Interestingly, thermal equilibrium states are characterized by one particular macroscopic observable, namely *temperature*, which does not belong to the conceptual domain of lower level statistical mechanics (Takesaki, 1970; Primas, 1998; Bishop and Atmanspacher, 2006). This observable and the corresponding thermal equilibrium states are emergent at the higher level description from the underlying lower level description. Moreover, different systems in thermal equilibrium are not distinguishable by measuring their temperature, they are epistemically equivalent with each other (beim Graben and Atmanspacher, 2006, 2009).

Finally, one tries to identify thermal equilibrium states with their statistical counterparts from the lower level of description. It turns out that the huge space of largely unphysical statistical states has to be considerably restricted to the so-called Kubo-Martin-Schwinger (KMS) states (Haag et al., 1967, 1974; Primas, 1998; Bishop and Atmanspacher, 2006), which implement three important stability conditions at the lower level: (1) KMS states are stationary, ie, expectation values and higher statistical moments of observables do not change in the course of time; (2) they are ergodic, ie, KMS states are structurally stable under small parametric perturbations; (3) they are mixing, ie, KMS states are asymptotically uncorrelated. The latter property additionally implies that KMS states are *relatively pure*, meaning that they are approximately dispersion-free in the higher level observables (Haag et al., 1974; Atmanspacher and beim Graben, 2007). Thus, thermal equilibrium states behave at the higher level description approximately as microstates do at the lower level description. For that reason they are called *macrostates*, thereby establishing an individual description for higher level properties H_i.

In contrast to several claims in philosophy (eg, Rorty, 1965; Levine, 1983), thermodynamics is not strongly reducible to point mechanics via statistical mechanics in an open loop fashion. It rather emerges from point mechanics and statistical mechanics by the selection of particularly suitable macrostates through an observer in a *closed-loop methodology* as shown in Fig. 2. These macrostates become operationally accessible through the phenomenal *context* of thermal equilibrium and implement sufficient conditions for the higher level through *stability*

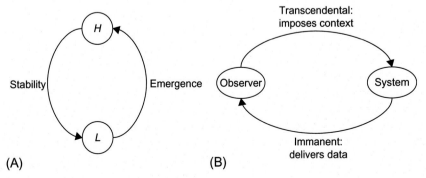

FIG. 2 Alternative *closed-loop* scientific methodology for the case of *contextual emergence*. (A) The lower level description *L* is necessary, but not sufficient for a higher level description *H* of a system under study; sufficient conditions are only defined on the higher level description and implemented as stability criteria at the lower level. (B) The *immanent arc* where the system under study delivers measurement results to the observer becomes augmented by the *transcendental arc*, where the observer imposes a *contingent context*, defining the *relevant properties*.

criteria at the lower level. The general situation where the lower level only provides necessary but not sufficient conditions for the higher level, while these are introduced by a contingent context has been named *contextual emergence* by Bishop and Atmanspacher (2006) and Atmanspacher and Bishop (2007) (see also Atmanspacher and beim Graben, 2009 for a review).

Fig. 2 illustrates the closed-loop methodology of contextual emergence where the open arcs from Fig. 1 get closed by their respective counterparts. In Fig. 2A the higher level description *H* emerges from the lower level description *L*, which provides only necessary conditions. On the other hand, the sufficient conditions that are introduced from a contextually given higher level description implement stability conditions similar to the KMS conditions for the contextual emergence of thermal equilibrium states. The loop between levels *L* and *H* is closed through this "downward confinement" of stability (Atmanspacher, 2015). Similarly, Fig. 2B shows that the system under study still delivers measurement results to the observer that we may call the *immanent arc* here. However, the observer is no longer detached from the observed system as in the classical open loop ideal, but linked back to the system by what we could call the *transcendental arc*, alluding to Kant's conception of *transcendental phenomenology* as the study of the preconditions of the possibility of experience as residing in the epistemic constitution of the experiencer (Kant, 1956, 1997) (cf. Meixner, 2003; Lyre, 2006; beim Graben, 2011 for recent interpretations).[1] Applied to the framework of contextual emergence, the transcendental arc is realized by a contingent context imposed by an observer. By implementing stability conditions, the context defines the relevant properties that can actually be observed and allows the interpretation of the data collected along the immanent arc (Primas, 1998; Bishop and Atmanspacher, 2006). In this sense contextual emergence can be regarded as a closed-loop methodological "response clamp" (Wallach, 2013).

In the remainder of this contribution, I elucidate the significance of contextual emergence for contemporary closed-loop neuroscience (Atmanspacher and Rotter, 2008; Potter et al., 2014) by means of four selected examples. I start with some tutorial on neural observables and microstates with particular emphasis on the role of contextual topologies. My second example is devoted to the contextual emergence of structure and function of ion channels. And finally, I discuss two examples for the emergence of neural macrostates. First in large-scale neural networks, second in neurophysiological time series.

2 CONTEXTUAL EMERGENCE IN NEUROSCIENCE

What are lower and higher level descriptions of neural systems? A single neuron regarded as a point model could constitute a lower level for a large-scale neural network comprising the higher level. However, regarded as a spatially extended cell with a multitude of membrane compartments, imbedded ion channels, receptors, and other proteins, it constitutes a higher level based on the lower level of molecular biology. Obviously, already answering the question above requires the selection of a context that defines the relevant properties for a neuroscientific study and neglects its irrelevant details. In this section I discuss three examples of differently selected contexts and how relevant properties are contextually emergent upon such a choice.

2.1 Neural Observables and Microstates

Already a single neuron is a biological system of perplexing complexity. Its state and behavior can be described by a multitude of variables and parameters, eg, the number of receptors imbedded in a certain patch of its cell membrane, the electric potential of this patch, or the different ionic currents flowing through this patch. The activity of the neuron itself could be characterized by the number of spikes emitted in some time interval, or by the average spike rate, or by the spatial distribution of the electric potential across the entire cell membrane. Other important variables could be the concentration of some sort of ions or of a certain type of a neuromodulator molecule in the extracellular space. As far as these variables can be quantitatively measured, we call them *neural observables*. There are several functional dependencies between different neural observables, eg, the spike rate depends on the membrane potential at the axon hillock, which in turn depends on the distribution of membrane potential along the dendritic tree. The latter again depends on the concentration of neurotransmitters in the synaptic clefts, and so on. Thus, there is a minimal number *n* of independent neural observables that are called *state variables*.

Again, state variables can be selected according to contextual constraints: In a large-scale neural network simulation the representation of a single neuron should be as parsimonious as possible and therefore, one would probably use only one state variable, describing a neuronal point model (see Section 2.4). Suitable choices could be the membrane potential at the axon hillock, *v*, or the spike rate *r*. In a medium-scale neural network simulation, however, one might be interested in the biologically more realistic Hodgkin-Huxley firing dynamics (Hodgkin and Huxley, 1952), which require four state variables for a single neuron, namely the membrane potential *v* and the gating variables *h,n, m* of voltage-gated sodium and potassium

---

1. The epistemological term "transcendental" must not be confused with "transcendent," which refers to the "things-in-itself" in Kant's terminology. The ontic transcendent is beyond any scientific exploration.

channels (see Section 2.2). Also in the resulting model, a single neuron is regarded as a point without spatial extension. Spatial properties become relevant as soon as one is interested in the propagation of electric fields in extracellular space, which can either be incorporated through multicompartmental neuron models (Einevoll et al., 2013) with about 100,000 state variables, namely the membrane potential of each compartment, or likewise, through point models augmented with extracellular observation equations (beim Graben and Rodrigues, 2013). Thus, the selected n independent state variables of a single neuron model comprise a point in an n-dimensional state space, to which we refer to as the *neural phase space* $X \subset \mathbb{R}^n$ in analogy to Newtonian point mechanics. A point $x \in X$ is equivalently referred to as a *neural microstate*.

Let us restrict for the remainder of this section to a point model with one-dimensional (compact) phase space $X \subset \mathbb{R}$, assuming that the relevant state variable is scaled to some real interval, eg, $X = [-\pi, \pi]$ (which could be a suitably gauged membrane potential v). Then, all dependent observables become real-valued functions $f : X \to \mathbb{R}$, such that the neuron's firing probability, eg, is obtained as $p = h(v)$ with the continuous sigmoidal activation function

$$h(x) = \frac{1}{1 + e^{-\gamma(x-\theta)}}, \tag{1}$$

parameterized through gain $\gamma > 0$ and firing threshold $\theta > 0$. From the firing probability, we can determine the spike rate through multiplication with the maximal firing rate r_{max}, ie, $r = r_{max} p = r_{max} h(v)$. In this way, all interesting observables can be generated by either multiplying or adding different functions of the state variables. Therefore, the set of neural observables constitutes a (function) *algebra* that can be conveniently taken as the space of continuous functions over the neural phase space, $C(X)$. Let $f(x)$ and $g(x)$ be two continuous

functions over X, then functions $f(x) + g(x)$ and $f(x)g(x)$ belong to the algebra $C(X)$ as well (and as $g(x) = -i(x)$, or $g(x) = 1/i(x)$, for $i(x) \neq 0$ also subtraction and division are well defined).

Using function addition and multiplication, we can also compute polynomials in our observable algebra such as

$$f_m(x) = \sum_{k=0}^{m} a_k (x - x_0)^k \tag{2}$$

with coefficients $a_k \in \mathbb{R}$ and $x_0 \in X$. This construction entails an interesting question: Does the limit $f(x) = \lim_{m \to \infty} f_m(x)$ exist in the observable algebra $C(X)$? In order to tackle this question we have to introduce a topology over $C(X)$. This is usually achieved through the maximum norm

$$\|f\|_\infty = \max_{x \in X} (f(x)) \tag{3}$$

for $f \in C(X)$. This norm induces a particular notion of convergence, which is called *uniform convergence*. We say that a series of functions $f_m \in C(X)$ converges uniformly against a limit function $f \in C(X)$ if

$$\lim_{m \to \infty} \|f_m - f\|_\infty = 0, \tag{4}$$

that is, $\lim_{m \to \infty} f_m(x) = f(x)$ independently from $x \in X$, or likewise, uniformly in an ε-strip around f as displayed in Fig. 3A for the convergence of the particular series

$$f_m(x) = \sum_{k=0}^{m} (-1)^k \frac{x^{2k+1}}{(2k+1)!} \tag{5}$$

which is the Taylor series of the sine function $f(x) = \sin x$.

Fig. 3A illustrates the uniform convergence of the series (5) against the sine function $f(x) = \sin x$ in a strip with $\varepsilon = 0.1$. The first three partial sums of Eq. (5), shown in blue, green,

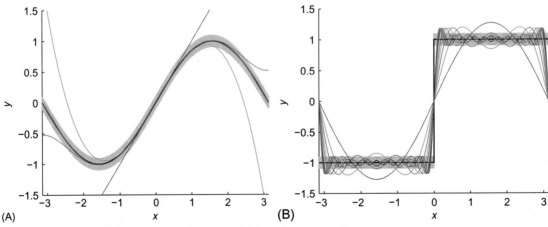

FIG. 3 Contextually relevant properties emerge from a change of topology. (A) Strong uniform convergence of power series against a sine function (Taylor expansion). (B) Weak quadratic mean convergence of sine functions against a periodic rectangular function (Fourier expansion). Both plots depict the first 10 iterates (1st blue, 2nd green, 3rd red, 4th turquoise, 5th magenta, 6th brown, 7th gray, 8th blue, 9th green, 10th red) converging against the limit function (bold black). Additionally, the strip of uniform convergence ($\varepsilon = 0.1$) for the Taylor series is shown in both plots for comparison.

and red, are not yet confined to the gray strip, but the remaining seven are so completely.

The maximum norm (3) equips the neural observable algebra $C(X)$ with a *strong norm topology*, which is closed with respect to this topology, hence it is a so-called Banach algebra of *intrinsic neural observables*, comprising our lower level description.

As far as we are only interested in the firing probability of our neural point model, we could apply the continuous sigmoidal activation function (1). But next we would like to know whether the neuron in state $x(t)$ at time t will actually emit a spike or not. To answer this question, we had to employ the noncontinuous Heaviside step function, or some derived observable. Let us use instead of the Heaviside function, the related sign function

$$\operatorname{sgn} x = \begin{cases} -1 & : x < 0 \\ 0 & : x = 0 \\ +1 & : x > 0 \end{cases} \quad (6)$$

such that $\operatorname{sgn}(x - \theta) = 1$ indicates a spike, when the membrane potential is larger than firing threshold θ, and $\operatorname{sgn}(x - \theta) = -1$ indicates no spike. This new spiking observable $f(x) = \operatorname{sgn} x$ does not belong to our algebra of intrinsic neural observables as it is not continuous. Therefore, $f(x)$ is neither infinitely often differentiable and does not even possess a Taylor series of converging polynomials. How can we obtain this crucial observable?

The answer is through contextual emergence. Higher level observables, to which we refer as to *contextual neural observables* henceforth, are introduced through a context by weakening the topology of the lower level algebra of intrinsic neural observables. Instead of uniform convergence in the strong intrinsic norm topology, we consider convergence in quadratic mean in the following. To this aim, we first have to extend the domain of our observables, ie, the phase space X, to the space of real numbers $\mathbb{R}$ through periodic continuation. Then we can define the Fourier series

$$f_m(x) = \frac{4}{\pi} \sum_{k=0}^{m} \frac{\sin(2k+1)x}{2k+1} \quad (7)$$

which converge to the periodic continuation of the sign function over $X = [-\pi, \pi]$ in quadratic mean

$$\lim_{m \to \infty} \|f_m - f\|_2 = 0 \quad (8)$$

with the *weak norm topology*

$$\|f\|_2 = \sqrt{\frac{1}{2\pi} \int_{-\pi}^{\pi} |f(x)|^2 \, dx}. \quad (9)$$

The convergence of the Fourier series in quadratic mean is illustrated in Fig. 3B. As for the example of uniform convergence in Fig. 3A the first ten partial sums of the Fourier series (7) are plotted in different colors. The convergence is

obviously not uniform, as none higher iterate becomes confined to the gray ε-strip. Moreover, we recognize that the higher iterates become even more oscillatory around the discontinuities of the sign function. These must be damped out by the averaging in the quadratic mean norm (9).

Using the weaker topology, new contextual observables emerge at a higher level of description from the lower level of intrinsic observables. But what can be identified with the implemented stability conditions? In order to clarify this point we slightly rephrase definition (9)

$$\|f\|_2 = \sqrt{\int_{-\pi}^{\pi} |f(x)|^2 \frac{dx}{2\pi}}. \quad (10)$$

Written in this way, we can generalize the uniform Lebesgue measure $dx/2\pi$ by introducing a weight function $\rho(x)$ through

$$\|f\|_\rho = \sqrt{\int_{-\pi}^{\pi} |f(x)|^2 \rho(x) dx}. \quad (11)$$

This resembles an expectation value functional

$$\langle f \rangle_\rho = \int_X f(x) \rho(x) \, dx \quad (12)$$

in statistical mechanics, where the weight functions ρ define statistical states. By selecting one particular state ρ from the huge set of all possible states, which is the function space of essentially bounded and normalized functions over X, we choose a suitable contextual topology. In our example above, the quadratic mean topology is selected through the uniform probability distribution function over $X = [-\pi, \pi]$

$$\rho(x) = \frac{1}{2\pi}. \quad (13)$$

Only this particular choice defines the Fourier series and thereby stable convergence of continuous functions against step functions in quadratic mean. We may interpret this as the implementation of a stability criterion. Most weight functions ρ that could be selected for a contextual topology would be useless for establishing stable notions of convergence.

Interestingly, the uniform probability distribution functions such as Eq. (13) also play an important role in statistical mechanics. They realize the ideal of unbiased estimation, maximizing entropy under the constraint of clamping the value of some contextual observables (Wallach, 2013). Therefore, uniform distributions correspond to microcanonical ensembles, an important precursor for the derivation of canonical ensembles as stable thermal equilibrium states. In this sense, the selection of the quadratic mean topology through a stable statistical reference state closes the loop between the lower level of intrinsic and the higher level of contextual neural observables.

Through contextual emergence, we obtain a larger algebra of contextual neural observables from the underlying algebra of intrinsic neural observables. The emergent algebra contains discontinuous functions such as the Heaviside and sign step functions. These step functions exhibit another important property that can be exploited for the contextual emergence of macrostates from microstates below. Looking at two different microstates $x, y \in X$ and computing their values under a step function f, yields, as f is not *injective*, $f(x) = f(y)$ (beim Graben and Atmanspacher, 2009). Therefore, the microstates are indistinguishable by measuring f, or, in other words, they are epistemically equivalent with respect to f (beim Graben and Atmanspacher, 2006, 2009). Collecting all microstates together that are epistemically equivalent to each other, yields an equivalence class in phase space and the family of all those equivalence classes partitions the whole phase space X. Moreover, also topologies can be introduced as families of (either open or closed) subsets of a phase space X. Therefore, a partition always induces a coarse-grained topology as well.

Finally, contextual emergence is also related to the possibility of separate time scales of a dynamical system. Using a fine-grained topology corresponding to some partition, the system spends little time within each cell and proceeds fast to the next transition. By contrast, a coarse-grained partition could be constructed in such a way that the system dwells in one cell for larger portions of time. There are several segmentation methods (Froyland, 2005; Gaveau and Schulman, 2006; beim Graben and Hutt, 2013) for analyzing time series, eg, from neurophysiological experiments through partitioned phase spaces (Lehmann et al., 1987; Wackermann et al., 1993; Hutt and Riedel, 2003; Allefeld et al., 2009; beim Graben and Hutt, 2015). We see in the next sections how partitioned phase spaces are used to exploit contextual emergence in neuroscience.

2.2 Ion Channels

In the famous Hodgkin-Huxley model (Hodgkin and Huxley, 1952), a single spiking neuron is described by four state variables v, h, m, n with v as membrane potential of the axon, h, m as kinetic variables of voltage-gated sodium ion channels and n the kinetic variable of voltage-gated potassium ion channels (Hille, 2001). The Hodgkin-Huxley equations are given as

$$I = C\frac{dv}{dt} + n^4 \bar{g}_{K^+}(v - E_{K^+}) + m^3 h \bar{g}_{Na^+}(v - E_{Na^+})$$
$$+ g_l(v - E_l) \tag{14}$$

$$\frac{dn}{dt} = \alpha_n(1-n) - \beta_n n \tag{15}$$

$$\frac{dm}{dt} = \alpha_m(1-m) - \beta_m m \tag{16}$$

$$\frac{dh}{dt} = \alpha_h(1-h) - \beta_h h, \tag{17}$$

where Eq. (14) is a conductivity equation for v with membrane capacitance C, maximal conductivities $\bar{g}$ and reversal potentials E of potassium (K^+) and sodium (Na^+) ions, as well as leakage conductivity g_l, and membrane resting potential E_l. The current I is injected into the cell, eg, through synaptic receptors. Eqs. (15)–(17) are master equations for the stochastic channel kinetics with opening rates α and closing rates β for variables n, m, and h, respectively.

The structure and behavior of ion channels has been intensively studied using closed-loop patch clamp electrophysiology, X-ray crystallography and molecular genetics (Hille, 2001). The crucial class of potassium channels, eg, is comprised of tetrameric proteins that are imbedded into the cell membrane like a cylindrical tube with a pore connecting intracellular and extracellular space. This pore may be opened or closed depending on the actual value of the membrane potential v.

2.2.1 Molecular Structure

What are the lower level and higher levels of description for an ion channel? The higher level is certainly comprised by the functional role of a channel as a voltage-gated pore through the neuron's cell membrane. This function is determined by the molecular structure given, eg, as four KCNA1 subunits, each one consisting of 495 amino acids (Hille, 2001). At the lower level, the channel is accordingly described as a quantum mechanical many particle system. Thus, structure and dynamics of an ion channel are described with a Schrödinger equation

$$i\frac{\partial}{\partial t}\Psi(\xi, x, t) = H(\xi, x)\Psi(\xi, x, t) \tag{18}$$

with $\Psi(\xi, x, t)$ as the molecular wave function, depending on the positions of all electrons ξ, of all atomic nuclei x, and time t and $H(\xi, x)$ as the corresponding Hamiltonian operator, ie,

$$H(\xi, x) = T_e + T_n + V_{ee}(\xi) + V_{nn}(x) + V_{en}(\xi, x) \tag{19}$$

with total electronic kinetic energy T_e, total nuclear kinetic energy T_n, electronic interaction potential V_{ee}, nuclear interaction potential V_{nn}, and the interaction potential of electrons and nuclei V_{en} (Born and Oppenheimer, 1927; Hagedorn, 1980; Primas, 1981). Interestingly, the macromolecular Schrödinger equation (18) is only necessary, but not sufficient for explaining molecular structure and function because the wave function $\Psi(\xi, x, t)$ represents a highly entangled state between electrons and nuclei. The molecule is therefore a "whole" not comprising any "parts" such as subunits, membrane domains or even a pore (Primas, 1998). Moreover, all chemical isomers of a molecule are described by the same Schrödinger equation, which therefore depends only on the chemical sum formula.

For the tetrameric potassium channel, this formula is approximately[2] $C_{6000}H_{14000}N_{2000}O_{4000}$ such that the lower level quantum mechanical description is not sufficient for representing the crucial structure and function of ion channels.

A first step in identifying the sufficient conditions for the emergence of molecular structure was achieved as early as in 1927 by Born and Oppenheimer (1927). In their famous "approximation" they make essentially an asymptotic expansion of the molecule's Hamiltonian (19) in the singular perturbation parameter $\epsilon = (\mu/m)^{1/4}$ with electron mass μ and (reduced) nuclear mass m (Primas, 1981, 1998; Bishop and Atmanspacher, 2006). The singular limit $\epsilon \to 0$ corresponds to infinite nuclear mass $m \to \infty$ when the movement of nucleons can be completely neglected in comparison with electronic dynamics. Then, the stationary wave function factorizes

$$\Psi(\xi, x) = \varphi(\xi|x)\psi(x), \qquad (20)$$

where $\varphi(\xi|x)$ becomes the electronic wave function for a given rigid nuclear frame and $\psi(x)$ the wave function of the nuclei. The latter obeys a reduced Schrödinger equation

$$(T_n + e(x))\psi(x) = E_n\psi(x) \qquad (21)$$

that is solved through a variational problem for $e(x) = V_{nn} + E_e$ where E_e are the energy eigenvalues of the electrons and E_n of the nuclei. Hence, the nuclei assume a rigid nuclear frame that is determined by the minimum of the electronic energies E_e. On the other hand, the positions of the nuclei x become represented as real parameters instead of quantum mechanical operators, thus as *classical variables* that emerge from a purely quantum mechanical lower level description at the higher level.

A careful conceptual analysis of the so-called Born-Oppenheimer approximation reveals that it is essentially another instance of contextual emergence (Primas, 1981, 1998; Bishop and Atmanspacher, 2006). The lower level quantum mechanical description of a molecule bears only necessary conditions for the higher level description of its structure and function, eg, as an ion channel in neurodynamics. The sufficient conditions are introduced by a contingent context of *molecular equilibrium*, being implemented as a stability criterion at the lower level through the decorrelation condition (20), which eventually leads to a breakdown of holistic quantum entanglement and the contextual emergence of a molecule as a classical object with definite structure and function.

2.3 Channel Kinetics

The contextual emergence of molecular structure is the first step for the description of the voltage-gated potassium ion

channel in the Hodgkin-Huxely model of a spiking neuron. However, the crucial characterization through a Markov process with master equation (15) is not achieved yet. This requires another step of contextual emergence.

The Born-Oppenheimer approximation results in first order in a Schrödinger equation for the electron hull in the potential of a rigid nuclear frame. In second and fourth order it also describes oscillations of the nuclei through their equilibrium positions and rotations of the nuclear frame. Taking all orders of the singular perturbation together, the movement of the nuclei is described by a stochastic process (Primas, 1981; Hagedorn, 1986). Thus, the coordinates of the nuclei become time-dependent random variables $\Xi(t) \in X$ in nuclear configuration space X, which are governed by a hierarchy of probability density functions $\rho_n(x_1, t_1; x_2, t_2; \ldots; x_n, t_n)$, $n \in \mathbb{N}$ such that $\rho_1(x, t)dx$, eg, is the probability to find a configuration around $x \in X$ at time t, and $\rho_2(x_1, t_1; x_2, t_2)d^2x$ as the probability to find a configuration around x_1 at time t_1 and later around x_2 at time $t_2 > t_1$ (van Kampen, 1992). The densities are positive and normalized

$$\int_X \rho_n(x_1, t_1; x_2, t_2; \ldots; x_n, t_n)\, d^n x = 1 \qquad (22)$$

for all consecutive times $t_1, t_2, \ldots, t_n$.

This stochastic description comprises the necessary lower level condition for the desired higher level description. In order to obtain the sufficient conditions for contextual emergence, we compare the given description with the paradigmatic example of thermal equilibrium in Section 1. Thus, a realization x of some molecular configuration comprises an individual microstate, while the multitude of probability densities ρ_n can be identified with a statistical state at the lower level of description. Emergent individual macrostates would be those statistical states that correspond in some way to thermal equilibrium KMS states. This correspondence can be explicated in analogy to Section 1 as follows: meaningful molecular macrostates would be stationary, ergodic, and mixing statistical states. In the framework of stochastic dynamics, these conditions are implemented through stationary Markov processes where densities ρ_n factorize into products of ρ_1 and ρ_2 only. This important *Markov property* supplies the sufficient decorrelation condition for the contextual emergence of stable macrostates in analogy to Boltzmann's "Stoßzahlansatz," implementing the crucial *molecular chaos assumption* for the derivation of his transport equation by the factorization of a two-point correlation function into the product of two one-point correlation functions (Boltzmann, 1872). This contextual idealization is the essential reason for the emergence of nonlinear dynamics in physical kinetics (Bishop, 2008; Primas, 1990).

For a Markov process, the probability to find the system in configuration x at time t after preparation in initial

---

2. For this approximation I assume that the KCNA1 subunit consists of 495 alanine molecules only.

condition $p(x, t_0) = \delta(x - x_0)$ is given by a continuous master equation (van Kampen, 1992)

$$\frac{\partial p(x,t)}{\partial t} = \int_X w(x, y)p(y, t) - w(y, x)p(x, t)\, dy, \quad (23)$$

where $w(x,y)$ is the transition rate from state y into state x. This restriction to stationary Markov processes is not yet sufficient for the contextual emergence of molecular function such as ion channel kinetics. Before we are able to implement the remaining stability conditions, we introduce the crucial coarse-graining through a contextual observable as in Section 2.1.

The macroscopically relevant property of an ion channel is the configuration of its pore, being either open or closed. This can be formalized through a contextual observable $f : X \rightarrow \{0,1\}$ such that $f(x) = 0$ if the channel's configuration x determines the closed pore and $f(x) = 1$ the open pore. Thus, the space of configurations is partitioned into two disjoint subsets $C, O \subset X$ such that $C \cup O = X$ and $C \cap O = \varnothing$ and two configurations $x, y \in X$ are equivalent with respect to f, ie, $x \sim_f y$ if either $f(x) = f(y) = 0$, when $x, y \in C$ or $f(x) = f(y) = 1$, when $x, y \in O$ (beim Graben and Atmanspacher, 2006; Atmanspacher and beim Graben, 2009).

Integrating (23) over the set O yields the idealized open probability of the ion channel through the discretized master equation

$$\begin{aligned}\frac{dp_O(t)}{dt} &= \int_O dx \int_X dy\, w(x,y)p(y,t) - w(y,x)p(x,t)\\ \frac{dp_O(t)}{dt} &= \alpha(1 - p_O(t)) - \beta p_O(t),\end{aligned} \quad (24)$$

ie, basically Eq. (15), with $\alpha = \int_O dx \int_C dy\, w(x,y)$ as the transition rate from the closed into the open state and β vice versa. However, this theoretical finding is at variance with experimental evidence about potassium channels as expressed by the proper result $p_O = n^4$ from the Hodgkin-Huxley model. Thus, our contextual coarse-graining was actually too coarse. By contrast, empirical data suggest a five state Markov process comprising four closed pore configurations $C_1 - C_4$ and an open one O as depicted in Fig. 4 (Hille, 2001).

The transition graph of the Hodgkin-Huxley potassium channel in Fig. 4 illustrates two important conditions for the contextual emergence of higher level equilibrium states. First the transition graph is connected, ie, every node can be reached from every other node through a path of sufficient length. This corresponds to ergodicity of the

corresponding Markov process. Second, besides states C_1 and O every node is source of two arrows, rendering the Markov process approximately mixing. These properties guarantee the existence of a unique stationary probability distribution that is asymptotically approached from every initial condition, in close analogy to thermal equilibrium states (Haken, 1983).

The coarse-grained configurations $C_1 - C_4$, O correspond to disjoint regions in configuration space X again, yielding a discrete master equation

$$\frac{dp(t)}{dt} = W \cdot p(t) \quad (25)$$

with transition rate matrix

$$W = \begin{pmatrix} -4\alpha & \beta & 0 & 0 & 0 \\ 4\alpha & -3\alpha - \beta & 2\beta & 0 & 0 \\ 0 & 3\alpha & -2(\alpha + \beta) & 3\beta & 0 \\ 0 & 0 & 2\alpha & -\alpha - 3\beta & 4\beta \\ 0 & 0 & 0 & \alpha & -4\beta \end{pmatrix} \quad (26)$$

and probabilities $p = (p_1, p_2, p_3, p_4, p_5)^T$ such that $p_i = \Pr(C_i)$ for $1 \le i \le 4$ and $p_5 = \Pr(O)$. In order to make the connection to the Hodgkin-Huxley model, we have to interpret $p_5 = n^4$. The closed state probabilities are then determined through the normalization condition

$$1 = p_1 + p_2 + p_3 + p_4 + p_5 = (n + (1 - n))^4. \quad (27)$$

Evaluating the binomial yields $p_1 = (1-n)^4$, $p_2 = 4n(1-n)^3$, $p_3 = 6n^2(1-n)^2$, and $p_4 = 4n^3(1 - n)$. Finally we combine Eqs. (25) and (26) with those expressions and obtain the desired master equation (15) of the Hodgkin-Huxley model.

Interestingly, the last assumption $p_5 = n^4$ has a straightforward interpretation as another sufficient stability condition. It requires the decorrelation of four independently moving gating charges. Thus we have shown for the second time that the breakdown of correlations plays an important role for the contextual emergence of molecular structure and function in neuroscience.

2.4 Neural Macrostates

Large-scale neural networks such as an entire brain consist of billions of neurons. Their description requires statistical methods similar to those developed in statistical physics where macroscopic properties are contextually emergent from stable microscopic descriptions. In this section I discuss this kind of closed-loop scenario by means of two illustrative examples: Amari's approach to *statistical neurodynamics* (Amari, 1974; Amari et al., 1977) and the segmentation of electroencephalographic time series in cognitive neuroscience (beim Graben and Hutt, 2015).

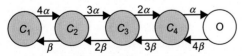

FIG. 4 Transition graph of the Hodgkin-Huxley potassium channel, exhibiting four closed states $C_1 - C_4$ (indicated in gray) and one open state O.

2.4.1 Large-Scale Neural Networks

The phase space of a single Hodgkin-Huxley neuron is spanned by four real variables. Correspondingly, the phase space of a large-scale neural network of N Hodgkin-Huxley neurons has dimension $d = 4N$. Contrastingly, a single leaky-integrator neuron is only described by its membrane potential v, so that N of those units span an N-dimensional phase space. In this section we assume for the sake of simplicity that a single neuron can be described by only one relevant variable, and consider large-scale neural networks of such units with phase space $X \subset \mathbb{R}^N$. The dynamics of the network is generally described by a differential equation

$$\frac{\mathrm{d}x(t)}{\mathrm{d}t} = g(x(t)) \tag{28}$$

for $x(t) \in X$ as the activation vector of the entire network. In accordance with the paradigmatic example of contextual emergence, we call the vector $\boldsymbol{x} \in X$ the *microstate* of the system comprising its lower level description.

The equation of movement (28) allows for different forms, eg, the leaky-integrator model

$$g(x(t)) = -x + W \cdot h(x) \tag{29}$$

with synaptic weight matrix W and a vector-valued activation function h defined component-wise as in Eq. (1) (beim Graben and Kurths, 2008). Eq. (28) is formally solved through the system's phase space flow Φ^t, parameterized by the elapsed time t,

$$x(t) = \Phi^{t-t_0}(x(t_0)) \tag{30}$$

when the dynamics starts with initial condition $x(t_0)$. This microstate dynamics characterizes the necessary conditions for contextual emergence of neural macrostates.

In order to introduce the higher level of description, we consider a family of contextual observables $f_i : X \to \mathbb{R}$ defining the relevant macroscopic properties, eg, average membrane potential, local field potential or cerebral blood flow, which are provided through suitable observation models (beim Graben and Rodrigues, 2013; Hämäläinen et al., 1993; Stephan et al., 2004). Gathering the images of a microstate x under all observables $F(x) = (f_i(x))$ we obtain a vector $y = F(x)$ in another space, called the observation space $Y = F(X)$ (Birkhoff and von Neumann, 1936), constituting the higher level description. Sometimes these observables can be regarded as mean fields

$$F(x) = \sum_i f_i(x), \tag{31}$$

where the sum extends over a population of N neurons and f_i denotes a projector of the neural phase space X onto its ith coordinate axis measuring the microscopic activation of the ith neuron. Clearly, F is not injective as the terms in the sum in Eq. (31) can be arranged in arbitrary order. Therefore, two neural activation vectors $x, y \in X$ can lead to the same value $F(x) = F(y)$ (eg, when $f_i(x) - \epsilon = f_j(x) + \epsilon$, $i \neq j$), so that they are indistinguishable by means of F and, therefore, epistemically equivalent (Atmanspacher and beim Graben, 2007). Thus, $y = F(x_1) = F(x_2)$ for different points $x_1, x_2 \in X$. The preimage $R = F^{-1}(y) = \{x \in X | F(x) = y\}$ is then a subset of X. Then we may construct a probability density function $\rho(x)$ in phase space that has R as essential support and we can therefore identify regions R with statistical states ρ. This delivers the statistical description at the lower level.

Next we study the impact of the dynamics upon statistical states. To this aim, we have to apply the flow Φ^t to each individual point in R and obtain $\Phi^t(R)$, another region in phase space X. Observing these sets by means of the observable F yields eventually a set $F(\Phi^t(R)) \in Y$ as shown in Fig. 5.

Fig. 5 also demonstrates that we can construct a mapping

$$\widetilde{\varphi}^t(y) = (F \circ \Phi^t \circ F^{-1})(y) \tag{32}$$

that brings an individual point $y \in Y$ to a set $\widetilde{\varphi}^t(y) \subset Y$, thereby constituting a nondeterministic dynamics in observation space (beim Graben et al., 2009). The crucial question, under which prerequisites this observation space dynamics becomes deterministic has been addressed some 40 years ago by Amari and coworkers in their framework of statistical neurodynamics (Amari, 1974; Amari et al., 1977) and recently resumed by beim Graben et al. (2009) to prove the equivalence of their approach with contextual emergence.

Instead of repeating our main results from beim Graben et al. (2009), I present a slightly modified argumentation in the following. Looking again at Fig. 5 we recognize that a sufficient condition for the emergence of a deterministic macrostate dynamics in observation space

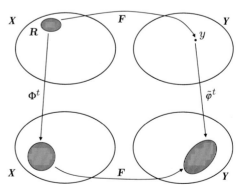

FIG. 5 Construction of a nondeterministic dynamics $\widetilde{\varphi}^t$ in the space of macroscopic observables. Upper row: the preimage of a point y under the observable F is an equivalence class of microstates, $R = F^{-1}(y) \subset X$. Bottom row: The temporal iterates of epistemically equivalent microstates, $\Phi^t(R)$, belong to a set $\widetilde{\varphi}^t(y) = F(\Phi^t(R)) \subset Y$ in observation space.

$$y(t) = \varphi^{t-t_0}(y(t_0)) \tag{33}$$

with a one-to-one mapping φ^t is that the flow must be contracting such that $\Phi^t(F^{-1}(y)) \subset F^{-1}(\varphi^t(y))$. This reflects essentially Amari's macrostate criterion for random neural networks (Amari, 1974; Amari et al., 1977; beim Graben et al., 2009). In this case, the set $\Phi^t(R)$ would be completely contained in another region of epistemically equivalent microstates that are mapped onto their dynamic image $\varphi^t(y)$. Contracting flows are well-known for the class of attractor neural networks (Amit, 1989; Hopfield, 1982) that possess asymptotically stable fixed point or limit torus attractors. However, a much more general class are flows that are either expanding or hyperbolic. For this case, we assume for simplicity, that, after fixing t, the regions of epistemically equivalent microstates provide a finite partition of the phase space X into disjoint cells A_i. We may further assume that this partition is a so-called Markov partition (Sinai, 1968; Ruelle, 1989). For expanding maps a Markov partition is defined through the decorrelation condition

$$A_i \cap \Phi^t(A_j) \neq \varnothing \Longrightarrow A_i \subset \Phi^t(A_j) \tag{34}$$

ie, images of partition cells do not partially overlap with other cells, or, in other words, the flow preserves cell boundaries (for hyperbolic maps things are much more involved). In any case, a Markov partition entails that the dynamics of statistical states, ie, regions in X, becomes a Markov process and in the thermodynamic limit $N \to \infty$ we may assume that this dynamics is described by a continuous state master equation (23) in X (cf. with neural *population density models* (Omurtag et al., 2000; de Kamps, 2003; Chizhov et al., 2007), with mean-field models (Faugeras et al., 2008), and neural field theory (Coombes et al., 2014)).

Under these assumptions, we compute the expectation of y under the statistical states $p(x, t)$ as in Eq. (12), insert into the master equation as in van Kampen (1992) and obtain

$$\frac{dy(t)}{dt} = \frac{d}{dt}\langle F(x) \rangle_p = \int_X F(x)\frac{dp(x,t)}{dt}\,dx = \langle a_1(F(x)) \rangle_p \tag{35}$$

with the jump moments

$$a_m(x) = \int_X (y-x)^m w(y,x)\,dy. \tag{36}$$

From Eq. (35) we obtain an emergent macrostate dynamics

$$\frac{dy(t)}{dt} = a_1(y(t)) \tag{37}$$

in two distinguished cases: firstly, when the first jump moment a_1 is a linear function, but secondly and more importantly, when the variances and all higher statistical moments of y are almost vanishing for the solution states

of the master equation. In this case, a_1 can be expanded into a power series with only the linear term contributing to Eq. (37). Remarkably, this case finds its analogue in our paradigmatic example of contextual emergence of thermal equilibrium states where thermal macrostates are relatively pure, and hence almost dispersion-free. As this results from the mixing property of Markov processes, we find a sufficient decorrelation condition also for the contextual emergence of neural macrostates.

2.4.2 Segmentation of Neurophysiological Time Series

Macroscopic observables for brain activity are, eg, electrophysiological mass potentials such as local field potentials (LFP), electrocorticograms (ECoG) and the electroencephalogram (EEG). In cognitive neuroscience, EEG measurements are often conducted according to the event-related potential (ERP) paradigm. In analyzing time series of neurophysiological observables, one is often interested to separate fast time scales that are considered to be irrelevant from slow time scales where essential transitions take place. To that aim several segmentation techniques have been proposed, based, eg, on cluster analysis (Lehmann et al., 1987; Wackermann et al., 1993; Hutt and Riedel, 2003), on spectral clustering (Froyland, 2005; Gaveau and Schulman, 2006; Allefeld et al., 2009), or on recurrence analysis (beim Graben and Hutt, 2013, 2015). Using spectral clustering, Allefeld et al. (2009) have argued in favor of contextual emergence of EEG-segments by identifying gaps in the eigenvalue spectrum of estimated Markov transition matrices. The resulting segments, reflecting metastable EEG topographies, could be called "brain microstates" after a suggestion of Lehmann et al. (1987) (cf. also Wackermann et al., 1993), referring to the even higher level of description used in psychology and cognitive neuroscience. However, in our present approach these microstates are macrostates emerging from the lower level description of suitably selected observation spaces. This case that emergent properties at a higher level description, such as neural macrostates, serve as microstates of another lower level description in a contextual hierarchy is crucial for the notion of *relative onticity* (Atmanspacher and Kronz, 1999), (cf. beim Graben, 2014 for another example).

In the following, I present another reanalysis of an ERP experiment on the processing of ungrammaticalities in German (Frisch et al., 2004) who examined processing differences for different violations of lexical and grammatical rules. Here we focus only on the contrast between a so-called *phrase structure violation* (38b), indicated by the asterisk, in comparison to grammatical control sentences (38a)

Im Garten wurde oft **gearbeitet** und . . .

In the garden was often **worked** and . . .   (38a)

"Work was often going on in the garden . . ."

* Im Garten wurde am **gearbeitet** und . . .

In the garden was on − the **worked** and . . .   (38b)

"Work was on − the going on in the garden . . ."

Sentences as (38b) are ungrammatical in German because the preposition *am* is followed by a past participle instead of a noun. A correct continuation could be, eg, *im Garten wurde am **Zaun** gearbeitet* ("work was going on in the garden at the fence") with *am Zaun* ("at the fence") as an admissible constituent.

The ERP study was conducted with 17 subjects in a visual word-by-word presentation paradigm. Subjects were presented with 40 structurally identical examples per condition (ie, 40 trials). The critical word in all conditions was the past participle printed in bold font above. EEG and additional electrooculogram (EOG) for controlling eye-movement were recorded with a 64 electrode montage from which 59 channels were measuring EEG proper.

The lower level description for the contextual emergence of ERP segments is the observation space spanned by those 59 recording electrodes such that the multivariate time series for the experimental conditions (38a) and (38b) are represented as trajectories $y(t)$ exploring this 59-dimensional space Y in the course of time t. These data are subjected to the recurrence grammar analysis of beim Graben and Hutt (2013, 2015), where segments are identified as *recurrence domains* partitioning the observation space Y. Then we can introduce a contextual observable $S : Y \to \{0, 1, 2, ..., K\}$ such that

$$S(y(t)) = k \text{ if } y(t) \in \text{recurrence domain } k \qquad (39)$$

(where $k = 0$ indicates transient regimes).

In contrast to previous approaches (beim Graben and Hutt, 2013, 2015), I optimize the partitioning in such a way that $S(y(t))$ should be a series of 0's followed by a series of 1's when the system dwells in the first recurrence domain, followed by a series of 0's, again for a transient period, then followed by a series of 2's when the system enters the second recurrence domain, thereafter followed

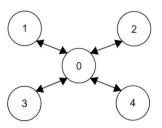

FIG. 6   Sketch of the transition graph of a Markov chain for optimal recurrence grammar.

by a series of 0's, indicating another transient, and so on. For this optimal encoding, the coarse-grained behavior would be described by the transition graph of a Markov chain shown in Fig. 6 (beim Graben et al., 2016).

Fig. 6 reveals that for an optimal segmentation of a time series into, say, four recurrence domains, the transient state 0 acts as a hub that must be passed through for every transition from one of the recurrence domains 1–4 into another one. The Markov chain characterized by the graph in Fig. 6 is obviously ergodic and mixing, thus satisfying the stability conditions for the contextual emergence of neural macrostates. The new algorithm developed by beim Graben et al. (2016) measures the distance between a given segmentation and the optimal one described by the Markov chain Fig. 6 being minimized.

The results of this optimization procedure are plotted in Fig. 7.

The upper panels of Fig. 7 show the multivariate ERP time series for the correct condition (38a) (left) and for the phrase structure violation condition (38b) (right) as colored traces, one for each EEG electrode. Both plots start 200 ms before stimulus presentation at $t = 0$. In both conditions similar N100/P200 ERP complexes are evoked. Furthermore, condition (38b) exhibits a large positive P600 ERP at central and posterior electrodes around 600 ms after stimulus presentation.

The emergent segmentation of the ERP time series into optimized recurrence domains is depicted in lower panels of Fig. 7, such that identical segments (Lehmann's "brain microstates") are plotted in the same color (the transient regime 0 in dark blue). During the prestimulus interval

FIG. 7   Grand average event-related brain potentials and optimal segmentation into recurrence domains of a language-processing experiment in German. (Left panels) Correct condition (38a). (Right panels) Phrase structure violation condition (38b). (Upper panels) grand average ERPs over 17 subjects for all EEG channels, eg, C5 blue, T7 green, FC5 red. Each trace showing one recording channel. (Bottom panels) Emergent segmentation into optimized recurrence domains.

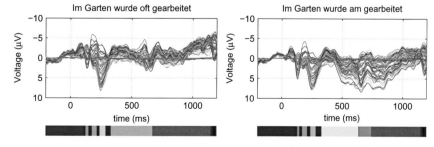

and until the first 100 ms both conditions exhibit the same resting state behavior, followed by a sequence of phasic N100/P200 complexes. Most importantly, the correct condition (38a) (left) shows only two segments between 300 and 1000 ms, whereas the ungrammatical condition (38b) (right) displays three recurrence domains, indicating the presence of the P600 component.

3 DISCUSSION

Contextual emergence is an important methodological closed-loop approach in the sciences. Its paradigmatic examples from theoretical physics are the emergence of thermal equilibrium states and of molecular structure through the introduction of weak contextual topologies and their implementation as stability conditions. Guided by these principles, I discussed four applications of contextual emergence in the neurosciences. First, I showed that relevant neural observables may belong to different classes of function algebras and that the transition from continuous functions to physiologically important step functions involves a transition from a strong norm topology to a weak contextual topology. Second, starting from the paradigmatic case of contextual emergence of molecular structure according to the Born-Oppenheimer approximation, I argued that the description of neurophysiological functioning of ion channels requires contextual emergence over two distinct steps from a quantum mechanical treatment of a molecule as a many particle system over a stochastic treatment of molecular configurations to a coarse-grained description of functional states. In the last two examples, I discussed contextual emergence of macroscopic descriptions in large-scale neural networks and in experimental data in close analogy to the treatment of thermodynamics.

In all examples, higher level properties appeared to be contextually emergent from lower level descriptions by the introduction of relevance criteria through contextual observables and the implementation of stability conditions that essentially involve statistical decorrelation principles. The first example, the emergence of firing step functions from continuous intrinsic neural observables, required the relaxation of their strong norm topology into weak quadratic mean topology through the introduction of a uniformly distributed reference state, in analogy to the microcanonical ensemble in statistical physics. The second example, the emergence of structure and function of ion channels, was established over a hierarchy of two levels. In a first step, molecular structure emerged via the Born-Oppenheimer singular perturbation method from many particle quantum mechanics through the breakdown of entanglement correlations. In a second step, the functionally relevant properties of open and close configurations emerged from a coarse-graining of a general stochastic process toward a stationary, ergodic, and mixing Markov process. And the last two examples, deterministic macroscopic dynamics and its relevant slow time scales, emerged from contextual observables for large-scale neural networks through the decorrelation in phase space and in time.

Contextual emergence also found other applications in cognitive neuroscience and in the philosophy of mind by further closing the loop between the observer and the system being observed in Fig. 2B. Based on ideas about *phenomenal families* that partition a putative mental phase space (Chalmers, 2000), Atmanspacher and beim Graben (2007) showed that appropriately defined contextual observables induce a partitioning of the underlying neural phase space discussed in Section 2.4. Applying basically the same decorrelation techniques as above, one could speak about the contextual emergence of mental states from neurodynamics (Atmanspacher and beim Graben, 2007; Atmanspacher, 2015). In another recent study, beim Graben (2014), I argued that Dennett's operationalistic treatment of the *intentional stance* (Dennett, 1978, 1989) can be regarded as a route of contextual emergence through a many-level hierarchy in analogy to that of fluid dynamics (Bishop, 2008, 2012). And finally, beim Graben and Potthast (2012) discuss the contextual emergence of symbolic computation in neural field automata through the restriction of the space of statistical states to that of uniform distributions that are compatible with a partitioned space of microstates.

In sum, contextual emergence is a promising closed-loop methodology that may find many other applications in neurodynamics in the near future.

ACKNOWLEDGMENTS

I am gratefully indebted to Anja Hahne, Angela Friederici, and Stefan Frisch for their courtesy to present another reanalysis of their ERP experiment and to Axel Hutt and Robert Bishop for helpful comments and suggestions.

REFERENCES

Allefeld, C., Atmanspacher, H., Wackermann, J., 2009. Mental states as macrostates emerging from EEG dynamics. Chaos 19, 015102.

Amari, S.I., 1974. A method of statistical neurodynamics. Kybernetik 14, 201–215.

Amari, S.I., Yoshida, K., Kanatani, K.I., 1977. A mathematical foundation for statistical neurodynamics. SIAM J. Appl. Math. 33 (1), 95–126.

Amit, D.J., 1989. Modeling Brain Function. The World of Attractor Neural Networks. Cambridge University Press, Cambridge (MA).

Atmanspacher, H., 2015. Contextual emergence of mental states. Cogn. Process. 16 (4), 359–364.

Atmanspacher, H., beim Graben, P., 2007. Contextual emergence of mental states from neurodynamics. Chaos Complex. Lett. 2 (2/3), 151–168.

Atmanspacher, H., beim Graben, P., 2009. Contextual emergence. Scholarpedia 4 (3), 7997.

Atmanspacher, H., Bishop, R.C., 2007. Stability conditions in contextual emergence. Chaos Complex. Lett. 2 (2/3), 139–150.

Atmanspacher, H., Kronz, F., 1999. Relative onticity. In: Atmanspacher, H., Amann, A., Müller-Herold, U. (Eds.), On Quanta, Mind and Matter. Hans Primas in Context, Fundamental Theories of Physics. Kluwer, Dordrecht, pp. 273–294.

Atmanspacher, H., Rotter, S., 2008. Interpreting neurodynamics: concepts and facts. Cogn. Neurodyn. 2 (4), 297–318.

beim Graben, P., 2011. Naphtas Visionen. Perspektivität in der Naturwissenschaft. In: Knaup, M., Müller, T., Spät, P. (Eds.), Post-Physikalismus. Alber, Freiburg/Br, pp. 122–141.

beim Graben, P., 2014. Contextual emergence of intentionality. J. Conscious. Stud. 21 (5–6), 75–96.

beim Graben, P., Atmanspacher, H., 2006. Complementarity in classical dynamical systems. Found. Phys. 36 (2), 291–306.

beim Graben, P., Atmanspacher, H., 2009. Extending the philosophical significance of the idea of complementarity. In: Atmanspacher, H., Primas, H. (Eds.), Recasting Reality. Wolfgang Pauli's Philosophical Ideas and Contemporary Science. Springer, Berlin, pp. 99–113.

beim Graben, P., Hutt, A., 2013. Detecting recurrence domains of dynamical systems by symbolic dynamics. Phys. Rev. Lett. 110 (15), 154101.

beim Graben, P., Hutt, A., 2015. Detecting event-related recurrences by symbolic analysis: applications to human language processing. Proc. R. Soc. Lond. A373, 20140089.

beim Graben, P., Kurths, J., 2008. Simulating global properties of electroencephalograms with minimal random neural networks. Neurocomputing 71 (4), 999–1007.

beim Graben, P., Potthast, R., 2012. Implementing Turing machines in dynamic field architectures. In: Bishop, M., Erden, Y.J. (Eds.), Proceedings of AISB12 World Congress 2012—Alan Turing 2012, Volume Fifth AISB Symposium on Computing and Philosophy: Computing, Philosophy and the Question of Bio-Machine Hybrids. pp. 36–40.

beim Graben, P., Rodrigues, S., 2013. A biophysical observation model for field potentials of networks of leaky integrate-and-fire neurons. Front. Comput. Neurosci. 6(100).

beim Graben, P., Barrett, A., Atmanspacher, H., 2009. Stability criteria for the contextual emergence of macrostates in neural networks. Netw. Comput. Neural Syst. 20 (3), 178–196.

beim Graben, P., Sellers, K.K., Fröhlich, F., Hutt, A., 2016. Optimal estimation of recurrence structures from time series. EPL 114, 38003.

Birkhoff, G., von Neumann, J., 1936. The logic of quantum mechanics. Ann. Math. 37 (4), 823–843.

Bishop, R.C., 2008. Downward causation in fluid convection. Synthese 160 (2), 229–248.

Bishop, R.C., 2012. Fluid convection, constraint and causation. Interf. Focus 2 (1), 4–12.

Bishop, R.C., Atmanspacher, H., 2006. Contextual emergence in the description of properties. Found. Phys. 36 (12), 1753–1777.

Boltzmann, L., 1872. Weitere Studien über das Wärmegleichgewicht unter Gasmolekülen. Sitzungsberichte der kaiserlichen Akademie der Wissenschaften, mathematisch-naturwissenschaftliche Klasse, 66:275–370. 1 (2003), 262–349. English translation: "Further Studies on the Thermal Equilibrium of Gas Molecules", History of Modern Physical Sciences.

Born, M., Oppenheimer, R., 1927. Zur Quantentheorie der Molekeln. Ann. Phys. 389 (20), 457–484.

Chalmers, D.J., 2000. What is a neural correlate of consciousness? In: Metzinger (2000), Chap. 2, pp. 17–39.

Chizhov, A.V., Rodrigues, S., Terry, J.R., 2007. A comparative analysis of a firing-rate model and a conductance-based neural population model. Phys. Lett. A 369 (1–2), 31–36.

Coombes, S., beim Graben, P., Potthast, R., Wright, J., 2014. Neural Fields. In: Theory and Applications. Springer, Berlin.

de Kamps, M., 2003. A simple and stable numerical solution for the population density equation. Neural Comput. 15 (9), 2129–2146.

Dennett, D.C., 1978. Intentional Systems, Chap. 1, pages 3–22. MIT Press, Cambridge, MA, pp. 162–183 Reprinted from J. Philos. 68 (4), 87–106, 1971.

Dennett, D.C., 1989. True Beliefers: The Intentional Strategy and Why it Works. MIT Press, Cambridge, MA, pp. 13–35. Chap. 2.

Einevoll, G.T., Kayser, C., Logothetis, N.K., Panzeri, S., 2013. Modelling and analysis of local field potentials for studying the function of cortical circuits. Nat. Rev. Neurosci. 14 (11), 770–785.

Faugeras, O.D., Touboul, J.D., Cessac, B., 2008. A constructive mean-field analysis of multi population neural networks with random synaptic weights and stochastic inputs. Front. Comput. Neurosci. 3, 1.

Frisch, S., Hahne, A., Friederici, A.D., 2004. Word category and verb-argument structure information in the dynamics of parsing. Cognition 91 (3), 191–219.

Froyland, G., 2005. Statistically optimal almost-invariant sets. Phys. D 200, 205–219.

Gaveau, B., Schulman, L.S., 2006. Multiple phases in stochastic dynamics: geometry and probabilities. Phys. Rev. E 73, 036124.

Haag, R., Hugenholtz, N.M., Winnink, M., 1967. On the equilibrium states in quantum statistical mechanics. Commun. Math. Phys. 5, 215–236.

Haag, R., Kastler, D., Trych-Pohlmeyer, E.B., 1974. Stability and equilibrium states. Commun. Math. Phys. 38, 173–193.

Hagedorn, G.A., 1980. A time dependent Born-Oppenheimer approximation. Commun. Math. Phys. 77 (1), 1–19.

Hagedorn, G.A., 1986. High order corrections to the time-dependent Born-Oppenheimer approximation I: smooth potentials. Ann. Math. 124 (3), 571–590.

Haken, H., 1983. Synergetics. An Introduction, Volume 1 of Springer Series in Synergetics, Springer, Berlin.

Hämäläinen, M., Hari, R., Ilmoniemi, R.J., Knuutila, K., Lounasmaa, O.V., 1993. Magnetoencephalography—theory, instrumentation, and applications to noninvasive studies of the working human brain. Rev. Mod. Phys. 65 (2), 413–497.

Hille, B., 2001. Ion Channels of Excitable Membranes. Sinauer, Sunderland.

Hodgkin, A.L., Huxley, A.F., 1952. A quantitative description of membrane current and its application to conduction and excitation in nerve. J. Physiol. 117, 500–544.

Hopfield, J.J., 1982. Neural networks and physical systems with emergent collective computational abilities. Proc. Natl. Acad. Sci. USA 79 (8), 2554–2558.

Hutt, A., Riedel, H., 2003. Analysis and modeling of quasi-stationary multivariate time series and their application to middle latency auditory evoked potentials. Phys. D 177 (1–4), 203–232.

Kant, I., 1956. Kritik der reinen Vernunft. Felix Meiner, Hamburg. English translation: Critique of Pure Reason, by P. Guyer and A. Wood. Cambridge University Press, Cambridge, MA (1999).

Kant, I., 1997. Metaphysische Anfangsgründe der Naturwissenschaft, Philosophische Bibliothek. Felix Meiner Verlag, Hamburg. English translation: Metaphysical Foundations of Natural Science by M. Friedman. Cambridge University Press, Cambridge, MA (2004).

Lehmann, D., Ozaki, H., Pal, I., 1987. EEG alpha map series: brain microstates by space-oriented adaptive segmentation. Electroencephalogr. Clin. Neurophysiol. 67, 271–288.

Levine, J., 1983. Materialism and qualia: the explanatory gap. Pac. Philos. Q. 64 (4), 354–361.

Lyre, H., 2006. Kants metaphysische Anfangsgründe der Naturwissenschaft: gestern und heute. Deut. Z. Philos. 54, 1–16.

Meixner, U., 2003. Die Aktualität Husserls für die moderne Philosophie des Geistes. In: Meixner, U., Newen, A. (Eds.), Seele, Denken, Bewusstsein: Zur Geschichte der Philosophie des Geistes. Walter de Gruyter, Berlin.

Omurtag, A., Knight, B., Sirovich, L., 2000. On the simulation of large populations of neurons. J. Comput. Neurosci. 8 (1), 51–63.

Potter, S.M., Hady, A.E., Fetz, E.E., 2014. Closed-loop neuroscience and neuroengineering. Front. Neural Circuits 8, 115.

Primas, H., 1977. Theory reduction and non-Boolean theories. J. Math. Biol. 4, 281–301.

Primas, H., 1981. Chemistry, Quantum Mechanics and Reductionism. Number 24 in Lecture Notes in Chemistry. Springer, Berlin.

Primas, H., 1990. Mathematical and philosophical questions in the theory of open and macroscopic quantum systems. In: Miller, A.I. (Ed.), Sixty-Two Years of Uncertainty: Historical, Philosophical and Physics Inquiries into the Foundation of Quantum Mechanics. Plenum Press, New York, pp. 233–257.

Primas, H., 1998. Emergence in exact natural sciences. Acta Polytech. Scand. 91, 83–98.

Rorty, R., 1965. Mind-body identity, privacy, and categories. Rev. Metaphys. 19 (1), 24.

Ruelle, D., 1989. The thermodynamic formalism for expanding maps. Commun. Math. Phys. 125, 239–262.

Sinai, Y.G., 1968. Construction of Markov partitions. Funct. Anal. Appl. 2 (3), 245–253.

Stephan, K.E., Harrison, L.M., Penny, W.D., Friston, K.J., 2004. Biophysical models of fMRI responses. Curr. Opin. Neurobiol. 14, 629–635.

Takesaki, M., 1970. Disjointness of the KMS-states of different temperatures. Commun. Math. Phys. 17, 33–41.

van Kampen, N.G., 1992. Stochastic Processes in Physics and Chemistry. Elsevier, Amsterdam.

Wackermann, J., Lehmann, D., Michel, C.M., Strik, W.K., 1993. Adaptive segmentation of spontaneous EEG map series into spatially defined microstates. Int. J. Psychophysiol. 14 (3), 269–283.

Wallach, A., 2013. The response clamp: functional characterization of neural systems using closed-loop control. Front. Neural Circuits 7, 5.

Part II

Experimental Axis

Chapter 13

Closed-Loop Methodologies for Cellular Electrophysiology

J. Couto*,‡, **D. Linaro***,‡, **R. Pulizzi*** and **M. Giugliano***,†,‡,§

**University of Antwerp, Antwerpen, Belgium,* †*Neuro-Electronics Research Flanders, Leuven, Belgium,* ‡*University of Sheffield, Sheffield, United Kingdom,* §*Brain Mind Institute, EPFL, Lausanne, Switzerland*

1 INTRODUCTION

Many components of the nervous system interact and operate in closed-loop. For instance, the opening and closing of a voltage-gated ion channel embedded in a biological lipid bilayer are influenced by the local electrical potential established across the bilayer. For excitable cells, such as neurons, the opening of ion channels affects the total flow of ions permeating the cell membrane and the intracellular charge distribution, which in turn determines the electrical depolarization or hyperpolarization of the membrane, thus closing the loop (Hille and others, 2001).

This example illustrates a general mechanism by which the state of (or the input to) a biophysical system often depends on the state itself (or its output), widely employed in nervous systems, across different spatial and temporal scales, from molecules to interconnected brain areas (Andersen et al., 2006; Sherman and Guillery, 2013).

Explicitly addressing the features of the biological systems under investigation, cellular electrophysiology adopted closed-loop protocols and technological approaches since its early days. For instance, the invention in the 1950s of electronic voltage-clamp feedback amplifiers was instrumental for dissecting the ionic bases of the action potential (Hodgkin and Huxley, 1952). Despite its simplicity, such an invention constitutes the most successful closed-loop technique, and is still routinely used worldwide for the study of intrinsic and synaptic ionic currents.

The design of similar methods has been inspired by automatic regulation techniques, first developed in engineering, and continued during the last decades for the study of biological systems. Today, we distinguish between methods aiming at (i) "breaking" the existing closed-loop operation of a biological system, and those aiming at (ii) artificially "replacing" or "mimicking" it. Voltage-clamp belongs to the first category, as it decouples temporal interdependencies between biophysical variables (ie, membrane potential and membrane currents) allowing the experimenter to study the dynamics of voltage-dependent ion channels in open loop. The general idea of the methods belonging to the former category can be summarized as a strategy to maintain or constrain one observable to a certain state over time (ie, as by a "clamp" in a mechanical system), while reading out the dynamical properties of the other parts of the system.

The second category is represented by methods interfacing a real biological system to a physical analog of a (artificial) biophysical component, or to a virtual portion of the environment. The most prominent example is the dynamic-clamp (Robinson and Kawai, 1993; Sharp et al., 1993), by which the biological effects of synthetic ion conductances can be studied in a living cell, upon automated regulation of an ionic current injected externally.

What these techniques have in common is the general principle of operation: one or more biophysical quantities are recorded over time and employed to continuously regulate an externally applied stimulus. Earlier implementations of closed-loop protocols relied on custom-made analog electronic circuits (eg, as a four-quadrant analog multiplier) and connected to a conventional electrophysiological amplifier (Robinson and Kawai, 1993; Yarom, 1991). Unfortunately, this approach lacks flexibility and makes experiments hard to reproduce across laboratories without considerable technical expertise. In recent years however, the increasing computational power of inexpensive personal computers (PCs), equipped with data acquisition boards or dedicated electronic digital signal processors, has led to a surge in software-based implementations of closed-loop protocols. These novel solutions have overcome most of the limitations of dedicated analog hardware, being easily reconfigurable upon software programming. In general, closed-loop experiments using regular PCs require the computer to run software that

Closed Loop Neuroscience. http://dx.doi.org/10.1016/B978-0-12-802452-2.00013-5

FIG. 1 Software designs for closed-loop experiments using a personal computer. (A) Graphical representation of a cycle in closed-loop protocols: at each iteration, the PC monitors one or more experimental quantities, employed to update the command to be applied to the biological system. (B) Schematic description of *hard* and *soft* real-time design approaches.

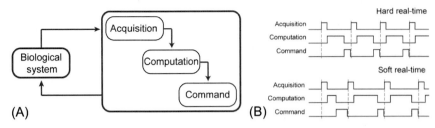

(a) iteratively monitors the state of the biological system at precisely specified time intervals and (b) reacts accordingly. It should be thus clear that all closed-loop techniques involve some form of real-time operation, within a repetitive cycle (a)-(b)-(a)-(b)-(a)-..., respecting precise temporal constraints. In more details, this cycle of operations must comprise three major steps (Pinto et al., 2001), as summarized in Fig. 1A:

1. the acquisition of the values of a set of experimental variables;
2. the computation of quantities required to accordingly adapt the stimulus;
3. the action on the biological system upon synthesis and injection of the stimulus.

This cycle has to run continuously and for the whole duration of the experiment. Two main approaches have been developed and described in the literature, in order to meet real-time constraints required by the experimental design: (i) a *hard* approach that employs strictly constrained time intervals, and (ii) a *soft* approach that instead employs variable time intervals. As its name suggests, the former requires that at each iteration of the cycle all the computations necessary for step (2) be completed by the PC before the next iteration starts (Fig. 1B, *top*). The latter approach instead compensates for unpredictable time intervals by adapting accordingly the step (2) of each cycle (Fig. 1B, *bottom*).

Therefore, the *hard* approach necessarily requires the computer to run a special operating system, able to perform real-time operations—such as real-time Linux, QNX Neutrino, or FreeRTOS—and ensuring that the closed-loop cycle is completed within each iteration. On the contrary, the *soft* approach may use conventional operating systems but offers no guarantee on timing. It requires the implementation (often via software) of additional logic to compensate for the actual elapsed iteration time and to properly implement the closed-loop paradigm.

The required real-time implementation is ultimately determined by the experiment: protocols requiring sub-millisecond precision usually require *hard* real-time, whereas protocols reacting to slowly changing variables can be equally implemented by means of *soft* real-time systems.

In the following, we focus on recent developments of closed-loop paradigms for cellular electrophysiology. We first describe how these can be used to synthetically recreate neurobiological components, spanning a diversity of spatial scales ranging from ion channels to background synaptic inputs, induced by large neuronal networks.

We then illustrate how closed-loop techniques are used at the single-cell level to constrain (ie, clamp) higher-order features of neuronal activity such as the firing rate, for the sake of indirectly measuring otherwise inaccessible quantities underlying cellular excitability. Finally, we show how such a concept can be extended to large populations of neurons, investigating the mechanisms underlying the emergence of network excitability.

2 SYNTHETIC RECREATION OF REALISTIC BIOLOGICAL MECHANISMS AND SYNAPTIC INPUTS

Conventional single-cell intracellular electrophysiological experiments often consist in probing the response of a neuron to an external stimulus, delivered, for example, as a current waveform $I(t)$ injected by the experimenter. This approach has been instrumental in elucidating countless neuronal features over the last decades, such as active and passive membrane intrinsic properties across cell types and brain areas, and more recently synaptic transmission and its short- and long-term plasticity, thought to underlie learning and memory.

However, the pioneering work by Hodgkin and colleagues (Hodgkin et al., 1952; Hodgkin and Huxley, 1952) demonstrated that changes in the membrane ionic permeability arise from voltage-dependent gating of selective ionic conductances, which ultimately underlies the initiation of an action potential, as explained by a set of elegant experiments and by a mathematical model. Subsequently, the introduction of the patch-clamp experimental technique (Sakmann and Neher, 1984) allowed accurately measuring the ionic currents associated with conductance changes in very small patches of cell membranes, fully disclosing the existence of discrete and microscopic ion channels with stochastic behavior (Neher and Sakmann, 1976).

Ultimately, this body of analytic work made it clear that the intrinsic and synaptic electrical potentials emerge from changes in membrane permeability and not from ionic currents per se. This sparked huge interest for the quantitative investigation of excitability and neuronal computation by means of mathematical models and computer simulations, but later it also prompted the introduction of an experimental synthetic approach. Known as dynamic-clamp, in its simple form this approach allows the study of the consequences of artificial changes in membrane permeability beyond computer simulations, while employing living cells (Robinson and Kawai, 1993; Sharp et al., 1993). This is accomplished by measuring the membrane potential $V(t)$ of a neuron and computing the current $I(t)$ to be injected, using the Ohmic relation

$$I(t) = g(t)(E_{rev} - V(t)), \qquad (1)$$

where E_{rev} is the Nernst potential of the ionic species the synthetic conductance is permeable to and $g(t)$ represents a time-varying conductance to be inserted (artificially) into the cell membrane. The main departure from traditional current-clamp stimulation lies therefore in the more faithful recreation by dynamic-clamp of the biophysical mechanisms underlying the flow of ions (and therefore of currents) through the membrane. For instance, when recreating virtual synaptic inputs to a neuron, dynamic-clamp allows reproducing the shunting effect on the cell membrane expected during the activation of real inhibitory synapses. The same effect simply cannot be achieved under conventional current-clamp stimulation, where only a transient hyperpolarization of the membrane can be induced upon emulating the activation of an inhibitory synapse. Since its introduction more than 20 years ago, the dynamic-clamp technique and its extensions have been employed in several studies.

Dynamic-clamp has been instrumental in identifying the contributions of voltage-gated ion channels to specific features of both the sub- and supra-threshold membrane potential dynamics. While pharmacological manipulation allows us to selectively abolish the net effects of a certain ionic conductance, thanks to the dynamic-clamp one can later synthetically reintroduce an artificial conductance, with known voltage-dependency, and compare the emerging features to control conditions. In fact, as the kinetics and the dependency on the voltage of the artificial conductance are entirely known being specified by the experimenter, dynamic-clamp offers a tool for causative investigation with high flexibility and richness. A particularly interesting case study is the work by Dorval and White (2005): using dynamic-clamp, they provided the first direct experimental evidence of the functional importance of random fluctuations associated with membrane permeability and arising from the stochastic gating of ion channels (White et al., 2000). Investigating the integrative properties of entorhinal stellate neurons,

the authors tested the effect of a selective pharmacological antagonist on the perithreshold spontaneous oscillations of the membrane potential of stellate cells (Fig. 2A). Since the oscillations were greatly reduced in the presence of the blocker, this experiment demonstrates that the activation of a specific ion channel (ie, voltage-gated persistent sodium channels) is necessary for the emergence of the oscillations. Nonetheless, the question on which specific mechanisms participate to the generation of the oscillations remains unanswered. However, Dorval and White artificially inserted in neurons virtual persistent sodium currents, under the presence of the pharmacological blocker. The effects of reintroducing synthetic channels with subtle different features could be now explicitly examined, addressing in particular whether or not the spontaneous current fluctuations due to the stochastic nature of membrane permeability had any detectable effects on the emerging oscillations. Interestingly, only the presence of synthetic ion channels with current fluctuations restored the oscillations to a degree comparable to the control conditions (Fig. 2B and C). Deterministic, noise-free recreation of persistent sodium channels could not therefore explain the emergence of the oscillations, suggesting that a rather limited number of channels were necessary and sufficient to ensure robust oscillations. Additionally, the recreated random fluctuations in membrane permeability could be causally linked to the time locking of individual action potentials fired in response to weak sinusoidal current stimuli, which emulated the ongoing inputs received by the same class of cells during theta activity in vivo.

With regards to the injection of artificial synaptic conductances in living neurons by dynamic-clamp, the recreation of in vivo-like synaptic bombardment gathered great interest in the last decade. A similar regime of activity could then be recreated realistically even in an in vitro slice preparation, which notoriously lacks any spontaneous synaptic activity (Destexhe et al., 2001). Complementing numerical studies that used exclusively mathematical models, the experimental application of dynamic-clamp provided in this context a first quantitative understanding of how a large number of synaptic inputs modify the intrinsic properties of cortical cells. For instance, Destexhe and collaborators (Destexhe et al., 2001; Fellous et al., 2003) showed that somatic injection of a synthetic conductance is sufficient to reproduce effects observed in vivo, such as a depolarized membrane potential and a three to fivefold reduction in input resistance. In addition, dynamic-clamp further paved the way for a series of studies examining in vivo-like background conditions in a highly controlled slice preparation, tackling research questions that would have otherwise been intractable, given the greater complexity and shorter duration of single-cell recordings in the intact brain.

FIG. 2 Applications of dynamic-clamp in in vitro cellular electrophysiology. In stellate cells of the enthorinal cortex, pharmacological blockade (by Riluzole) of persistent sodium currents reduces theta oscillations in the membrane potential (A). Oscillations can be restored by reintroducing, by dynamic-clamp, synthetic sodium currents but only when their kinetics is stochastic (*SDE*): a deterministic model (*ODE*) fails to restore oscillations (B and C), proving a functional role for ion channel random flickering. The activation rate of background synthetic synaptic inputs, recreated by dynamic-clamp, modulates the slope (ie, the gain) of the static frequency-current relationship in cortical pyramidal neurons (D–F). As recreated background synaptic inputs are combined with reactive-clamp synthetic input reverberation, transiently sustained cortical firing activity (G) can be induced and studied in vitro in single neurons. The duration of the sustained firing epochs depends on the number of spikes fired during the cue phase (H). *(Panels A–C are reproduced, with permission, from Dorval, A.D., White, J.A., 2005. Channel noise is essential for perithreshold oscillations in entorhinal stellate neurons. J. Neurosci. 25(43), 10025–10028. doi:10.1523/JNEUROSCI.3557-05.2005. © (2005) Society for Neuroscience. Panels D–F are reproduced, with permission, from Chance, F.S., Abbott, L.F., Reyes, A.D., 2002. Gain modulation from background synaptic input. Neuron 35(4), 773–782. http://www.ncbi.nlm.nih.gov/pubmed/12194875. © (2002) Cell Press. Parts G–H are reproduced, with permission, from Fellous, J.-M., Sejnowski, T.J., 2003. Regulation of persistent activity by background inhibition in an in vitro model of a cortical microcircuit. Cereb. Cortex 13(11), 1232–1241. doi:10.1093/cercor/bhg098. © (2003) Oxford University Press.)*

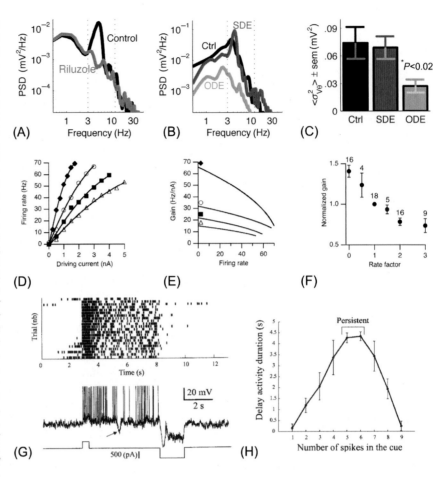

In this category, two studies in particular are worth mentioning. The first one (Chance et al., 2002) investigated experimentally how the degree of background synaptic input modulates the gain of the stationary input-output response curve of cortical neurons. Crucially, the virtual synaptic background activity recreated by Reyes and collaborators (Chance et al., 2002) allowed them to study in the details the so-called *balanced state* (Shadlen and Newsome, 1994), hypothesized for the in vivo operation of cortical neurons and defined when the average excitatory and inhibitory components of the total incoming synaptic inputs cancel each other out. While maintaining such a balance, dynamic-clamp allowed Reyes and collaborators to alter proportionally the activation rate of presynaptic inputs, varying only the total conductance load of the cell under investigation. Under these conditions, the frequency-current relationship of the cell, quantified upon an additional DC current injection, could be shown to depend on the overall background activity, acting by divisive and not additive effects (Fig. 2D–F). This experimental result provided for the first time a possible cellular correlate for the modulation of the gain of single-neuron responses, among the key computational principles observed throughout the central nervous system (see Salinas and Thier, 2000 for a review).

Another work that leveraged on the in vitro recreation of synthetic background synaptic inputs is by Fellous and Sejnowski (2003). While the recreation of realistic synaptic input conductance constitutes in itself a form of closed-loop stimulation, the authors introduced an additional type of feedback protocol, named reactive-clamp, mimicking the effect of a local (virtual) cortical microcircuit triggered by the activity of the neuron under examination. In fact, the "reaction" consisted in the injection of an additional train of a fixed number of excitatory and inhibitory synaptic inputs, in response to the online detection of an action potential fired by the neuron being stimulated. Combining dynamic- and reactive-clamp, the authors were able to show how reverberatory incoming synaptic activity, caused by the local cortical microcircuitry, is capable of producing sustained firing that outlasts the stimulation (Fig. 2G). In response to a brief transient depolarization of the neuron under investigation, for example, imitating an excitatory sensory cue in a delay match-to-sample task, the reactive-clamp induced a realistic self-sustained positive feedback reaction, whose features were similar to the firing activity observed in prefrontal cortex during working memory tasks.

The advantage of having both background synaptic activity and local microcircuit reverberations under direct (synthetic) control allowed dissecting the mechanisms underlying the self-sustained activity. Interestingly, the authors discovered that the duration of the sustained firing epoch was dependent on the number of action potentials emitted in the cue part, as well as on the interplay of excitatory and inhibitory feedback conductances (Fig. 2H). This suggests that the balance between fast feedback inhibition and slow feedback excitation, largely mediated by glutamatergic NMDA receptors, might be directly responsible for sustained activity in prefrontal cortex during working memory tasks.

As we have seen, using an in vitro neurobiological preparation, in which artificial components are reintroduced to mimic the authentic behavior of the in vivo, intact system, offers unparalleled advantages of a greater control on both artificial and real parts of the whole system. This, however, comes with a tradeoff: often, the recreated activity matches only certain aspects of the real one. In addition, sensory inputs and motor behaviors cannot be emulated in completely faithful ways. Notwithstanding these limitations, often this type of approach, known as bio-artificial hybrid, provides insight into the functioning of parts of the nervous system, which can be tested later, if necessary, in a more complete biological preparation.

The concept of hybrid networks dates back more than two decades, when Yarom showed how a network composed of a neuron from the inferior olivary nucleus coupled to four damped electronic oscillators generates sustained oscillations in response to a triggering event (Yarom, 1991). This is the result of the interplay between the intrinsic properties of each unit (ie, unstable oscillatory behavior) and network properties, in the form of electrical coupling between the units.

Another example of the advantages of an increased control over several parts of the same experimental preparation is represented by the studies on synchronization of neuronal networks. Indeed, the artificial components of a hybrid network (neurons and/or synapses) can be modified to allow testing of different hypotheses on the functional role of specific phenomena. Rabinovich and collaborators (Nowotny et al., 2003) used a hybrid network composed of a pacemaker model neuron coupled to a biological

neuron of the *Aplysia* abdominal ganglion. Plastic or static synthetic synapses were used alternatively to connect together the real and artificial neurons, clarifying the role played by synaptic plasticity in the entrainment of action potentials fired in the hybrid network. Along similar lines, Netoff and collaborators (Netoff et al., 2005) showed how single-cell properties in the rat hippocampus determine synchronized oscillations in networks composed of real neurons coupled together by artificial synapses.

Additionally, hybrid neural networks have been used to investigate how feedback inhibition regulates the gating of information through the thalamus (Le Masson et al., 2002). This work was notable in that it combined simulated sensory input and simulated feedback inhibition onto a real thalamic relay cell. The authors demonstrated how modulation of the gain of the thalamic inhibitory feedback loop can gate the flow of information through the thalamus, thus effectively "disconnecting" the cerebral cortex from the incoming sensory inputs. As illustrated in Fig. 3A, the authors showed that when the inhibitory feedback was strong enough, the temporal correlation between sensory inputs and thalamic outputs decreased substantially compared to the lack of feedback inhibition. Additionally, a modulation of single-cell excitability by local application of noradrenaline was observed to be sufficient to increase the correlation between sensory inputs and principal cells' outputs, thus offering a microcircuit explanation of how single-cell and synaptic properties interact to drive the network across distinct activity states: from sleep to arousal.

As a last example of hybrid neural networks, Calabrese and collaborators (Sorensen et al., 2004) examined the role of the hyperpolarization-activated inward current (I_h) in the rhythmic bursting dynamics observed in the interneurons of two ganglia of the leech heart. The authors first showed that mutual inhibitory coupling of a regular firing heart interneuron and a regular firing artificial model neuron causes both cells to change their firing pattern into a stable regular bursting, for strong enough inhibitory coupling. This antiphasic bursting mode is characteristic of the leech heart interneuron pacemaker network: the activity of one neuron inhibits that of its partner, which in turn inhibits the former, once intrinsic currents overcome the inhibitory drive it receives and restore the firing behavior, thus restarting the cycle. In the case of this two-cell network, the I_h intrinsic current provides a mechanism for the termination of the silent phase in each neuron. As shown in Fig. 3B, the hyperpolarization of the membrane potential caused by the incoming inhibitory drive produces a gradual increase of the hyperpolarization-activated membrane conductance: once the associated I_h reaches a sufficient level of intensity, it drives spiking in the neuron that was silent, causing it to inhibit the other cell, which will thus enter the silent phase. An additional demonstration of this mechanism was

FIG. 3 Sample implementations of hybrid networks. The efficacy of feedback inhibition regulates the correlation between sensory inputs and output firing of thalamic relay cells (A), in a hybrid network. Hyperpolarization-activated ion currents are responsible for the intrinsic bursting behavior of a hybrid half-center oscillator (B), with its oscillation duty-cycle modulated by the maximal ionic conductance of such currents (C). *(Part A reproduced, with permission, from Le Masson, G., Renaud-Le Masson, S., Debay, D., Bal, T., 2002. Feedback inhibition controls spike transfer in hybrid thalamic circuits. Nature, 417(6891), 854–858. doi:10.1038/nature00825. © (2002) Nature Publishing Group. Parts B and C reproduced, with permission, from Sorensen, M., DeWeerth, S., Cymbalyuk, G., Calabrese, R.L., 2004. Using a hybrid neural system to reveal regulation of neuronal network activity by an intrinsic current. J. Neurosci. 24(23), 5427–5438. doi:10.1523/JNEUROSCI.4449-03.2004. © (2004) Society for Neuroscience.)*

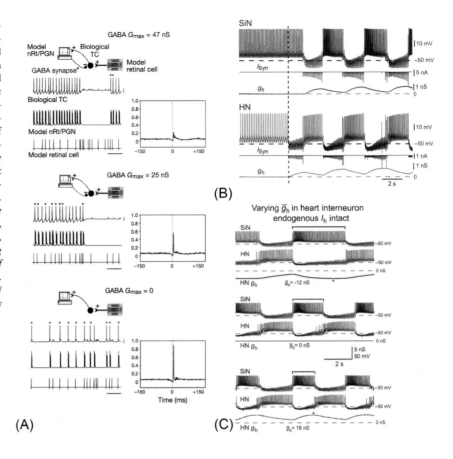

provided by altering the dynamics of I_h in the artificial model neuron and by supplementing the naturally present I_h in the real cell with additional synthetic components, whose dynamics could be modulated during the experiment. These manipulations showed (Fig. 3C) that increasing the maximal hyperpolarization-activated conductance in one neuron causes a decrease in the duty-cycle of the other cell, dissecting complex network interactions in closed-loop.

Overall, the contributions to the literature described so far highlight the breadth of possible uses of closed-loop techniques in cellular electrophysiology, to synthetically introduce biophysically realistic inputs in reduced preparations as well as intrinsic components, with the ultimate goal of exploring specific aspects of neuronal and synaptic function in reduced preparations.

3 FEEDBACK REGULATION OF THE FIRING RATE IN SINGLE NEURONS

A common feature shared by the methodologies reviewed so far is the requirement of reacting to changes (ie, in the membrane potential) on the millisecond timescale. However, the dynamical properties of neuronal responses vary on several other time scales, prompting the development of closed-loop techniques particularly suited for longer time scales. An area of research that has received

significant attention concerns the study, at level of single neurons, of the interplay between ongoing activity and spiking dynamics. In fact, regardless of the computation performed, all neurons integrate inputs from different sources, a process that is inevitably affected by their ongoing electrical activity. While this may be the result of synaptic inputs, it is sometimes endogenously generated (ie, spontaneous firing). Furthermore, how synaptic inputs are integrated depends on the history of emission of action potentials of the neuron. For instance, if an action potential has been generated very recently, the neuron will be almost insensitive to further inputs. On the contrary, a cell whose membrane potential is close to its spiking threshold is likely to be highly sensitive even to the weakest synaptic potential.

It is therefore essential to quantify how spiking history affects the integration of synaptic inputs. This can be done by delivering perturbations to the neuron while carefully monitoring its spiking history. However, the fact that neuronal excitability and spiking history are strongly coupled, so that the (lack of) emission of an action potential at a certain moment in time will affect the cell's excitability and vice versa (Gal et al., 2010), poses experimental challenges to neurophysiologists.

A conventional procedure for coping with this interdependency is known as *trial selection*: after performing a very large number of repeated trials, the experimenter selects the subset of trials that meet a certain constraint depending on the spiking history, while discarding the remainder of the repetitions. In this context, closed-loop control techniques have been employed to dramatically reduce the number of necessary repetitions, by clamping the history of spiking activity to a desired level before delivering perturbations (ie, test stimulus). A successful example of such method is the feedback regulation of neuronal firing rate to clamp the baseline activity of a cell (Couto et al., 2015; Linaro et al., 2014; Miranda-Domínguez et al., 2010; Miranda-Dominguez and Netoff, 2013; Netoff et al., 2005a,b; Sorensen et al., 2004). This is largely based on a continuous estimate of the error between actual firing rate and its desired target level. In the case of an intracellular protocol, described here, the error is then used to proportionally adapt the amplitude of the injected current.

The method is highlighted in Fig. 4A, where intracellular access allows simultaneously recording the membrane potential and injecting a current. A simple real-time operating algorithm is in charge of estimating the neuronal firing rate (eg, by monitoring the membrane potential and threshold crossing), as well as of comparing the aforementioned to a target rate, thus estimating the instantaneous error signal. This is finally employed to update the amplitude of the current injected to the cell. Because an accurate measurement of the firing rate requires a long observation window, an online estimate can be used in alternative by means of a sliding averaging window. In detail, this can be implemented in real-time as an iterative formula that is updated after the detection of each action potential (Linaro et al., 2014; Wallach, 2013):

$$\widetilde{F}_k = \mathrm{ISI}_k{}^{-1} \cdot \left(1 - e^{\frac{-\mathrm{ISI}_k}{\tau}}\right) + \widetilde{F}_{k-1} \cdot e^{\frac{-\mathrm{ISI}_k}{\tau}}, \qquad (2)$$

where $\widetilde{F}_k$ is the estimate of the firing frequency after the kth spike, τ is a time constant representing how much the previous history will influence the current running estimate, and ISI_k is the kth interspike interval, that is, the time difference between the kth and the $(k-1)$th action potential.

Such an estimate is then compared to the target value and the resulting error, $\overline{\overline{e}}_k = \widetilde{F}_k - F_{\mathrm{target}}$, is used in a feedback controller. To this end, any feedback controller can be employed: when using a Proportional-Integral controller (PI), the injected current at the kth spike takes the form:

$$I_k = g_{\mathrm{p}} \cdot \overline{\overline{e}}_k + g_i \cdot \sum_{i=0}^{k} \overline{e}_i, \qquad (3)$$

where g_{p} and g_i are known in the automatic control literature as proportional and integral gains, respectively. Netoff and collaborators (Miranda-Domínguez et al., 2010) describe a procedure to obtain the optimal values for those coefficients.

This feedback regulation technique has proven very valuable to measure neuronal dynamical response

properties that require stationary firing rates, such as the phase response curve (PRC) (Netoff et al., 2005b; Schultheiss et al., 2011). Considering the neuron as an ideal oscillator, the PRC quantifies the effect of an external perturbation on the cycle of an oscillator and readily allows quantitative predictions on network phenomena. Netoff and collaborators (Netoff et al., 2005b), for example, employed a firing rate feedback controller to maintain a stationary firing rate, while using dynamic-clamp to mimic the activation of a time-varying membrane conductance at precise times, with respect to the cells' interspike interval.

An important reason to adopt a similar feedback regulation of the ongoing neuronal activity is the severe errors in the PRC estimate caused by nonstationaries in the firing rate. Our laboratory employed a similar method to explore the dependency of the PRC on different firing rates (Couto et al., 2015), across a wide range (20–150 spikes/s), and allowing the computation of up to 12 PRCs for the same cell. Indeed, the use of open-loop methods under the same conditions would be prohibitive, due to the temporal constraints for stable experimental recording conditions. Additionally, the same method was employed with success to perform optimal experimental sampling of response phases using pseudorandom intervals (Couto et al., 2015; Netoff et al., 2005a).

Firing rate feedback regulation proves to be very useful when delivery of an external stimulus must occur at a particular phase of the firing cycle. To exemplify the high level of precision achievable with this method, Fig. 4B shows an overlay of 250 repeated trials in which a cerebellar Purkinje neuron was injected with a recreated synaptic conductance to elicit complex spike firing as in an earlier experiment by Häusser and coworkers (Davie et al., 2008). The feedback controller was used to maintain the cell's firing rate around 40 spikes/s, while allowing a conductance synaptic input to be delivered (every six interspike intervals) using dynamic clamp. As the firing rate was clamped to a stationary value by the controller, it was possible to deliver the synaptic input precisely in the middle of the interspike interval.

The feedback regulation of the firing rate does not only refer to stationary target rates. Fig. 4C shows an example of a target firing rate increasing linearly through time (Linaro et al., 2014). In this case, slowly increasing the target rate can be used to rapidly estimating the frequency-current relationship of the cell. As the firing rate is estimated in real time (*blue trace*, Fig. 4C) and the amplitude of the injected current is updated after each action potential (*red trace*, Fig. 4C), almost a continuum of firing rates versus current amplitudes (*gray circles*, Fig. 4C) is generated, accurately deriving the frequency-current curve. Such an estimate is equivalent to what is obtained in open loop, stimulating the cell by discrete current steps of increasing amplitudes (Fig. 4C, *red points on the right plot*) but is obtained in a fraction of the time.

FIG. 4 Feedback regulation of neuronal firing rate. Sketch of an experimental setup (A) for closed-loop intracellular control of the firing rate of neuron, including an online spike detector and a running estimator of the spike rate. The error between the actual and the target rates is employed by the controller to adapt the amplitude of injected current. This setup can be employed to clamp the firing rate of a cerebellar Purkinje neuron (B), during external stimulus delivery at precise times within the firing cycle. Firing rate control (*blue*) is disabled during the interval of stimulus delivery. The stimulus mimics a synaptic input eliciting a complex spike (Davie et al., 2008). The feedback regulation of the firing rate can also be employed as a fast method to estimate the cell's frequency-current (*F-I*) curve (C), when the target rate is linearly increased in time (*green*). The black trace is the membrane voltage recorded from a cortical layer 5 pyramidal neuron during firing rate regulation, while red and blue traces represent the injected current and the estimated firing rate, respectively. The right panel compares the conventional iterative measures of the *F-I* curve (*red circles*) and those obtained in closed-loop (*gray circles*). (Panel A is adapted, with permission, from Couto, J., Linaro, D., De Schutter, E., Giugliano, M., 2015. On the firing rate dependency of the phase response curve of rat purkinje neurons in vitro. PLoS Comput. Biol. 11(3), e1004112. doi:10.1371/journal. pcbi.1004112. © (2015) Public Library of Science. Panels B and C are adapted, with permission, from Linaro et al. © (2014) Elsevier.)

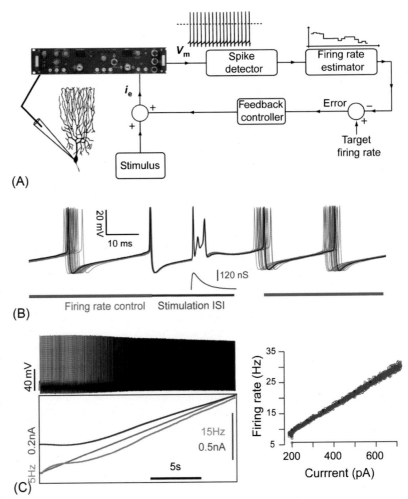

4 FEEDBACK REGULATION OF NEURONAL RESPONSE VARIABILITY

So far, we focussed on closed-loop methods applied to individual, isolated neurons, describing first dynamic-clamp and its variations, and how it could be used to recreate synthetic intrinsic and synaptic conductances into living cells or simple hybrid microcircuits. We next discussed how to regulate higher-order features of neuronal excitability, such as regular firing, characterized by longer time scales than ion channels. In this section, we continue a progression with respect to time scales and features of neuronal excitability, considering the closed-loop control of neuronal response variability. The degree by which a neuron is excitable, and thus it can emit an action potential in response to a stimulus, depends on the previous firing history and on the intensity of the stimulus. Therefore the variability in responding or not to a repeated stimulation depends on the neuronal excitability, which in turn is affected by whether or not an action potential has been generated shortly in response to the stimulus. Such interplay between

variability and excitability makes it desirable to design a closed-loop protocol able to decouple those two components, revealing underlying mechanisms characterized by complex spatial and temporal dynamics.

Excitability depends, in fact, not only on intrinsic neuronal features, such as distribution of ion channels (Kole and Stuart, 2012), but also on the distribution and strength of synaptic contacts across its cellular morphology and on the activation of specific presynaptic populations (Turrigiano, 1999). On the other hand, history dependency has been shown to be particularly rich, with a relative refractoriness over a variety of time scales, ranging from milliseconds to several hours (Gal and Marom, 2013; Gal et al., 2010).

As in voltage-clamp, the control signal required to constrain the neuronal membrane potential contains rich indirect information on otherwise inaccessible quantities, and being able to regulate the likelihood of a response to a stimulus (for instance, in terms of percentage of success out of the total number of repetitions) may disentangle cause and effect and provide access to hidden biophysical quantities.

Marom and collaborators (Wallach et al., 2011) were the first to devise a closed-loop strategy that enabled direct observation of changes in neuronal excitability over multiple time scales and named *neuronal response-clamp*. The broad aim of this technique was to gain direct insight into internal variables, upon explicit regulation of a given observable, such as the probability of spike emission in response to a repeated electrical stimulation. A simple proportional-integral-derivative controller could be employed to adapt continuously the amplitude of the extracellular stimuli. Fig. 5A and B illustrates how a desired level of response variability could be held constant or modulated sinusoidally in time over more than 10 min of continuous stimulation. The control signal disclosed temporal correlations over a variety of time scales and unexpected fluctuations, offering a rather direct readout of previously neglected features of neuronal excitability (Fig. 5D).

It was the first time that an unexpectedly rich heterogeneity of time scales and stereotypical neuronal response properties (Fig. 5C) could be studied experimentally.

While Marom and collaborators first considered the excitability properties, emerging in a pharmacologically isolated neuron, the significance of their discovery was much deeper. When experimentally dealing with an in vitro recurrent network of neurons, and observing the population mean firing rates over time instead of the membrane potential of individual cells, one can also define a measure of response variability and excitability at the network level. In fact, the transient emergence of

synchronized spiking activity throughout the network can also be seen as an all-or-none phenomenon, such as in the case of action potentials for a single neuron (Wallach and Marom, 2012). While the control signals obtained in closed-loop response-clamp experiments with isolated neurons provides access to intracellular effective properties, the control signal obtained while constraining multiple neuron participating in network activity offers unparalleled access on how incoming synaptic activity modulates the excitability of individual cells, enabling the study of extrinsically generated complex dynamics. As shown in Fig. 6, the authors studied how the instantaneous temporal evolution of an effective excitability threshold of each clamped neuron was modulated during synchronous network events. Employing an array of 60 substrate-integrated microelectrodes, the spontaneous spiking activity of all unclamped electrodes (yellow dots) and the firing of the two clamped neurons (blue and purple dots) can be represented simultaneously (Fig. 6A).

As the control signals related to multiple cells were averaged through time and referred to the time of occurrence of a spontaneous network synchronization event (Fig. 6B), the effective evolution of neuronal excitability could be probed directly, highlighting unexpected complex excitability dynamics sustaining the emergence of network synchronization, as well as heterogeneity across neurons.

In summary, the neuronal response-clamp is a valuable tool for exploring hidden neuronal dynamics, both intrinsic

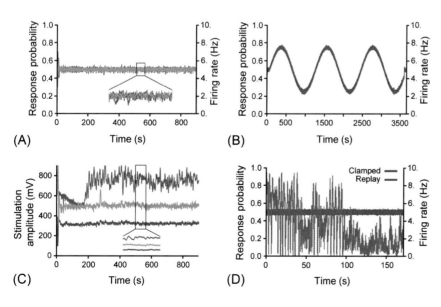

(A)   (B)   (C)   (D)

FIG. 5 The *response-clamp* technique in isolated neurons. Examples from three cultured neurons dissociated from the rat neocortex and pharmacologically isolated (A; *blue, red, yellow traces*). The variability of their firing responses was directly tested upon continuous extracellular electrical pulsed stimulation, repeated every 0.1 s for 15 min. By a closed-loop controller, appropriately adapting the stimulation intensity, the variability of the neuronal responses could be regulated by the experimenter and fixed to 50% (A), or slowly modulated in time (B). The sequence of stimulus intensities, required for clamping the response variability in (A), discloses an extremely rich repertoire of fluctuations (C), unexpectedly characterized by long-lasting temporal correlations. The open-loop replay of the control signal in the same neurons (D) leads to large fluctuations in the response variability, which is no longer *clamped*. (*Figure adapted, with permission, from Wallach, A., Eytan, D., Gal, A., Zrenner, C., Marom, S., 2011. Neuronal response clamp. Front. Neuroeng. 4, 3. doi:10.3389/fneng.2011.00003.* © (2011) Frontiers Media S.A.)

and extrinsic, illustrating how closed-loop control of neuronal system might be the sole means to decouple complex biophysical interdependencies.

Besides extracellular electrical stimulation, optogenetics (Boyden et al., 2005) has also been applied as a

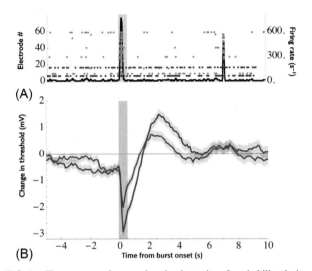

FIG. 6 The *response-clamp* probes the dynamics of excitability during spontaneous network synchronization. A large in vitro network of cortical neurons is coupled to an array of 60 substrate-integrated microelectrodes, each detecting spontaneous firing and episodic network-wide synchronization (A), denoted by the shaded box. Only by means of closed-loop response-clamp is it possible to probe the effective excitability threshold, averaged across repeated instances of spontaneous network synchronization events and aligned to their peak (B). *(Figure adapted, with permission, from Wallach, A., Marom, S., 2012. Interactions between network synchrony and the dynamics of neuronal threshold. J. Neurophysiol. 107(11), 2926–2936. doi:10.1152/jn.00876.2011. © (2012) American Physiological Society.)*

control signal to modulate neural activity in closed loop (Grosenick et al., 2015). Recently, Potter and collaborators (Newman et al., 2015) introduced the first optogenetic feedback control system, named *optoclamp*, allowing noninvasive firing rate control, both in vivo and in vitro. By means of a simple *on-off* controller, *optoclamp* was shown to constrain across a very long time interval the collective firing rate of a large neuronal network in vitro (Fong et al., 2015). The noninvasive character of the control stimuli and the very long duration of the experiments allowed the design of an experimental setting to discriminate between glutamatergic synaptic currents, mediated by AMPA or by NMDA receptors, in the context of homeostatic plasticity regulation. Similar to the *response-clamp*, the *optoclamp* employs a PI controller to activate multiple wide-field light sources simultaneously and thus exploit the working of distinct opsins (Fig. 7A). For instance, when the firing rate of the entire neuronal network was constrained to a range of desired fixed values (Fig. 7B), the controller was effective and accurate up to rates of 10 spikes/s.

The same concept could be applied in vivo (Fig. 7C), as the authors demonstrated controlling the firing rate of thalamocortical neurons in 75% of the cases. In particular, while tracking error was higher than in vitro due to the lower number of units recorded simultaneously by extracellular electrical recordings, also in vivo the error grew for increasing values of the target firing rate.

It is interesting to note that *optoclamp* builds on recently developed software tools, proposed for closed-loop experiments, developed since 2012 by Potter and collaborators (Newman et al., 2012). As for the real-time approach of

FIG. 7 Regulating network activity by optical stimulation. Overall sketch of the closed-loop system employed in (Newman et al., 2015) (A). Raw voltage traces recorded at each of 60 substrate-integrated microelectrodes are processed to adapt the control signal to the LED light source, every 10 ms. Sample experiment where the network firing rate is dynamically regulated over a period of 60 s in the range 0–10 Hz in 1 Hz steps (B). When a similar experimental design was translated to in vivo in ventral posteromedial thalamic nucleus (C), detection of extracellular spikes (mean and SD of a waveform, vertical bar 100 μV, horizontal bar 1 ms) was processed to adapt the photo stimulation delivered by an implanted fiber optic, and similarly succeeded constraining the firing rate (*black traces*) to four different targets levels (*red traces*). *(Figure adapted with permissions from Newman, J.P., Fong, M., Millard, D.C., Whitmire, C.J., Stanley, G.B., Potter, S.M., 2015. Optogenetic feedback control of neural activity. eLife 4, 1–24. doi:10.7554/eLife.07192. © (2015) eLife Sciences.)*

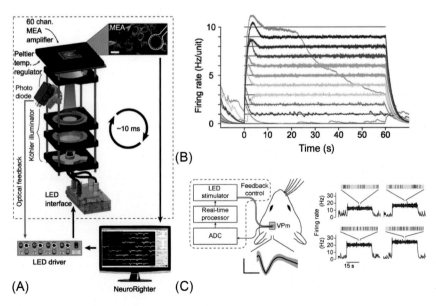

response-clamp, open-source software[1] was publicly disseminated enabling real-time and bidirectional electrophysiological interactions with large neuronal networks, relying on conventional acquisition hardware boards. The software implements the feedback control presented earlier by Potter and collaborators (Wagenaar et al., 2005): the on-off control law is given by

$$V_k(t + \triangle T) = V_k(t) - \alpha V_k \left(\frac{\langle f(t) \rangle}{f^*} - 1 \right), \qquad (4)$$

where V_k is the intensity of extracellular voltage-controlled electrical stimulation delivered at microelectrode k, $\langle f(t) \rangle$ is the average network firing rate over a 2 s running window, f^* is the target firing rate, $\triangle T$ is the update period of the feedback loop, and α defines the time constant of the feedback controller.

Compared to previous closed-loop approaches (Wagenaar et al., 2005; Wallach et al., 2011), this software platform represents a step forward, given the use of a graphical user interface and of an application programming interface, simplifying the development of custom real-time experimental protocols.

In this section we reviewed recent developments in the study of neuronal excitability, focussing on closed-loop techniques involving extracellular electrical stimulation or photoactivation as command signals. The future will likely see the improvement of the existing software platforms, and the implementation of more advanced real-time spike sorting algorithms to be used in real-time experimental design. More complex nonlinear control algorithms will also be adopted, for refining the feedback control as already explored in other fields (eg, Grosenick et al., 2015).

5 CONCLUSIONS

We reviewed and discussed a selection of recent closed-loop techniques proposed in the field of cellular electrophysiology. We followed a bottom-up approach, starting from single-cell methods aiming at recreating synthetic biophysical components, acting on rather short time scales (eg, the opening and closing of individual ion channels) and then moved forward to discuss control techniques relevant for higher-order neuronal features, such as the emission of action potentials and its response variability at the level of single neurons and of networks.

By these techniques, it has been possible to gain significant insights into the working of neurobiological systems that would have been impossible to gain with conventional approaches. For instance, the discovery of the role of stochastic ion channel in the generation of membrane potential oscillations (Dorval and White, 2005) linked two

description levels of neuronal complexity together: the microscopic stochastic fluctuations of individual channels (Colquhoun, 1995) and the mesoscopic emergence of network oscillations (Buzsaki, 2006).

Analogously, the impact of balanced background synaptic inputs on the response gain of cortical neurons (Chance et al., 2002) offered a possible cellular mechanism for a computational primitive widespread across levels of organization in the nervous system (Salinas and Thier, 2000).

In addition, disentangling the distinct role of glutamatergic synaptic currents mediated by AMPA and NMDA synaptic transmission in homeostatic plasticity, among the most studied mechanisms of plasticity and stability maintenance in neuronal circuits (Fong et al., 2015), had long remained inaccessible to conventional techniques.

Closed-loop methodologies ultimately brought efficiency and simplicity to the design of electrophysiological experiments. For instance, the use of firing rate feedback regulation boosted the accuracy and greatly reduced the time required to estimate PRCs and their dependency on the firing rate (Couto et al., 2015).

In conclusion, the last few decades witnessed the development of novel software tools for closed-loop experiments. Interestingly, the scientific community produced mostly open-source options, focussing on flexibility and ease of use. Yet, conceptual challenges remain, but future clever experimental designs will likely bring within reach previously impossible experiments, with simplicity and elegance.

REFERENCES

Andersen, P., Morris, R., Amaral, D., Bliss, T., O'Keefe, J., 2006. The Hippocampus Book. Oxford University Press, USA.

Boyden, E.S., Zhang, F., Bamberg, E., Nagel, G., Deisseroth, K., 2005. Millisecond-timescale, genetically targeted optical control of neural activity. Nat. Neurosci. 8, 1263–1268. http://dx.doi.org/10.1038/nn1525.

Buzsaki, G., 2006. Rhythms of the Brain. Oxford University Press, Inc., New York. http://www.caam.rice.edu/~yad1/miscellaneous/References/Neuroscience/Papers/Hippocampus/RythmsOfTheBrainBuszaki.pdf

Chance, F.S., Abbott, L.F., Reyes, A.D., 2002. Gain modulation from background synaptic input. Neuron 35 (4), 773–782. http://www.ncbi.nlm.nih.gov/pubmed/12194875.

Colquhoun, D., Hawkes, A.G., 1995. The principles of the stochastic interpretation of ion-channel mechanism. In: Sakmann, B., Neher, E. (Eds.), Single-Channel Recordings. second ed. Plenum Press, New York, pp. 397–482. http://dx.doi.org/10.1007/978-1-4419-1229-9 (Chapter 18).

Couto, J., Linaro, D., De Schutter, E., Giugliano, M., 2015. On the firing rate dependency of the phase response curve of rat purkinje neurons in vitro. PLoS Comput. Biol.. 11(3), e1004112. http://dx.doi.org/10.1371/journal.pcbi.1004112.

---

1. https://sites.google.com/site/neurorighter/.

Davie, J.T., Clark, B.A., Häusser, M., 2008. The origin of the complex spike in cerebellar purkinje cells. J. Neurosci. 28 (30), 7599–7609. http://dx.doi.org/10.1523/JNEUROSCI.0559-08.2008.

Destexhe, A., Rudolph, M., Fellous, J.M., Sejnowski, T.J., 2001. Fluctuating synaptic conductances recreate in vivo-like activity in neocortical neurons. Neuroscience 107 (1), 13–24. http://www.pubmedcentral.nih.gov/articlerender.fcgi?artid=3320220&tool=pmcentrez&rendertype=abstract.

Dorval, A.D., White, J.a., 2005. Channel noise is essential for perithreshold oscillations in entorhinal stellate neurons. J. Neurosci. 25 (43), 10025–10028. http://dx.doi.org/10.1523/JNEUROSCI.3557-05.2005.

Fellous, J.-M., Sejnowski, T.J., 2003. Regulation of persistent activity by background inhibition in an in vitro model of a cortical microcircuit. Cereb. Cortex 13 (11), 1232–1241. http://dx.doi.org/10.1093/cercor/bhg098.

Fellous, J.M., Rudolph, M., Destexhe, A., Sejnowski, T.J., 2003. Synaptic background noise controls the input/output characteristics of single cells in an in vitro model of in vivo activity. Neuroscience 122 (3), 811–829. http://dx.doi.org/10.1016/j.neuroscience.2003.08.027.

Fong, M., Newman, J.P., Potter, S.M., Wenner, P., 2015. Upward synaptic scaling is dependent on neurotransmission rather than spiking. Nat. Commun. 6, 1–11. http://dx.doi.org/10.1038/ncomms7339.

Gal, A., Marom, S., 2013. Entrainment of the intrinsic dynamics of single isolated neurons by natural-like input. J. Neurosci. 33 (18), 7912–7918. http://dx.doi.org/10.1523/JNEUROSCI.3763-12.2013.

Gal, A., Eytan, D., Wallach, A., Sandler, M., Schiller, J., Marom, S., 2010. Dynamics of excitability over extended timescales in cultured cortical neurons. J. Neurosci. 30 (48), 16332–16342. http://dx.doi.org/10.1523/JNEUROSCI.4859-10.2010.

Grosenick, L., Marshel, J.H., Deisseroth, K., 2015. Closed-loop and activity-guided optogenetic control. Neuron 86 (1), 106–139. http://dx.doi.org/10.1016/j.neuron.2015.03.034.

Hille, B., et al., 2001. Ion Channels of Excitable Membranes, vol. 507. Sinauer Sunderland, Massachusetts.

Hodgkin, A.L., Huxley, A.F., 1952. A quantitative description of membrane current and its application to conduction and excitation in nerve. J. Physiol. 117, 500–544. http://dx.doi.org/10.1016/S0092-8240(05)80004-7.

Hodgkin, A.L., Huxley, A.F., Katz, B., 1952. Measurement of current-voltage relations in the membrane of the giant axon of Loligo. J. Physiol. 116, 424–448.

Kole, M.H.P., Stuart, G.J., 2012. Signal processing in the axon initial segment. Neuron 73 (2), 235–247. http://dx.doi.org/10.1016/j.neuron.2012.01.007.

Le Masson, G., Renaud-Le Masson, S., Debay, D., Bal, T., 2002. Feedback inhibition controls spike transfer in hybrid thalamic circuits. Nature 417 (6891), 854–858. http://dx.doi.org/10.1038/nature00825.

Linaro, D., Couto, J., Giugliano, M., 2014. Command-line cellular electrophysiology for conventional and real-time closed-loop experiments. J. Neurosci. Methods 230C, 5–19. http://dx.doi.org/10.1016/j.jneumeth.2014.04.003.

Miranda-Dominguez, O., Netoff, T.I., 2013. Parameterized phase response curves for characterizing neuronal behaviors under transient conditions. J. Neurophysiol. 109 (9), 2306–2316. http://dx.doi.org/10.1152/jn.00942.2012.

Miranda-Domínguez, O., Gonia, J., Netoff, T.I., 2010. Firing rate control of a neuron using a linear proportional-integral controller. J. Neural Eng. 7 (6), 66004. http://stacks.iop.org/1741-2552/7/i=6/a=066004?key=crossref.d44e95b7aaa072b7b3740ffd9fe21e66.

Neher, E., Sakmann, B., 1976. Single-channel currents recorded from membrane of denervated frog muscle fibres. Nature 260 (5554), 799–802. http://dx.doi.org/10.1038/260799a0.

Netoff, T.I., Acker, C.D., Bettencourt, J.C., White, J.A., 2005a. Beyond two-cell networks: experimental measurement of neuronal responses to multiple synaptic inputs. J. Comput. Neurosci. 18 (3), 287–295. http://eutils.ncbi.nlm.nih.gov/entrez/eutils/elink.fcgi?dbfrom=pubmed&id=15830165&retmode=ref&cmd=prlinks.

Netoff, T.I., Banks, M.I., Dorval, A.D., Acker, C.D., Haas, J.S., Kopell, N., White, J.a., 2005b. Synchronization in hybrid neuronal networks of the hippocampal formation. J. Neurophysiol. 93 (3), 1197–1208. http://dx.doi.org/10.1152/jn.00982.2004.

Newman, J.P., Zeller-Townson, R., Fong, M.-F., Arcot Desai, S., Gross, R.E., Potter, S.M., 2012. Closed-loop, multichannel experimentation using the open-source neurorighter electrophysiology platform. Front. Neural Circuits 6, 98. http://dx.doi.org/10.3389/fncir.2012.00098.

Newman, J.P., Fong, M., Millard, D.C., Whitmire, C.J., Stanley, G.B., Potter, S.M., 2015. Optogenetic feedback control of neural activity. eLife 4, 1–24. http://dx.doi.org/10.7554/eLife.07192.

Nowotny, T., Zhigulin, V.P., Selverston, A.I., Abarbanel, H.D.I., Rabinovich, M.I., 2003. Enhancement of synchronization in a hybrid neural circuit by spike timing dependent plasticity. J. Neurosci. 23 (30), 9776–9785. http://www.jneurosci.org/cgi/content/full/23/30/9776.

Pinto, R.D., Elson, R.C., Szücs, A., Rabinovich, M.I., Selverston, A.I., Abarbanel, H.D.I., 2001. Extended dynamic-clamp: controlling up to four neurons using a single desktop computer and interface. J. Neurosci. Methods 108 (1), 39–48. http://dx.doi.org/10.1016/S0165-0270(01)00368-5.

Robinson, H.P., Kawai, N., 1993. Injection of digitally synthesized synaptic conductance transients to measure the integrative properties of neurons. J. Neurosci. Methods 49 (3), 157–165.

Sakmann, B., Neher, E., 1984. Patch clamp techniques for studying ionic channels in excitable membranes. Annu. Rev. Physiol. 46, 455–472. http://dx.doi.org/10.1146/annurev.physiol.46.1.455.

Salinas, E., Thier, P., 2000. Gain modulation: a major computational principle of the central nervous system. Neuron 27 (1), 15–21. http://scholar.google.com/scholar?hl=en&btnG=Search&q=intitle:Gain+Modulation+:+A+Major+Computational+Principle+of+the+Central+Nervous+System#0.

Schultheiss, N.W., Prinz, A.A., Butera, R.J., 2011. Phase Response Curves in Neuroscience: Theory, Experiment, and Analysis. Springer Science & Business Media, New York.

Shadlen, M.N., Newsome, W.T., 1994. Noise, neural codes and cortical organization. Curr. Opin. Neurobiol. 4 (4), 569–579. http://dx.doi.org/10.1016/0959-4388(94)90059-0.

Sharp, A.A., O'Neil, M.B., Abbott, L.F., Marder, E., 1993. Dynamic-clamp: computer-generated conductances in real neurons. J. Neurophysiol. 69 (3), 992–995. http://dx.doi.org/10.1016/0166-2236(93)90004-6.

Sherman, S.M., Guillery, R.W., 2013. Functional Connections of Cortical Areas: A New View from the Thalamus. MIT Press, Cambridge, MA.

Sorensen, M., DeWeerth, S., Cymbalyuk, G., Calabrese, R.L., 2004. Using a hybrid neural system to reveal regulation of neuronal network activity by an intrinsic current. J. Neurosci. 24 (23), 5427–5438. http://dx.doi.org/10.1523/JNEUROSCI.4449-03.2004.

Turrigiano, G.G., 1999. Homeostatic plasticity in neuronal networks: the more things change, the more they stay the same. Trends Neurosci. 22 (5), 221–227. http://dx.doi.org/10.1016/S0166-2236(98)01341-1.

Wagenaar, D.A., Madhavan, R., Pine, J., Potter, S.M., 2005. Controlling bursting in cortical cultures with closed-loop multi-electrode stimulation. J. Neurosci. 25 (3), 680–688. http://dx.doi.org/10.1523/JNEUROSCI.4209-04.2005.

Wallach, A., 2013. The response clamp: functional characterization of neural systems using closed-loop control. Front. Neural Circuits 7, 5. http://dx.doi.org/10.3389/fncir.2013.00005.

Wallach, A., Marom, S., 2012. Interactions between network synchrony and the dynamics of neuronal threshold. J. Neurophysiol. 107 (11), 2926–2936. http://dx.doi.org/10.1152/jn.00876.2011.

Wallach, A., Eytan, D., Gal, A., Zrenner, C., Marom, S., 2011. Neuronal response clamp. Front. Neuroeng. 4, 3. http://dx.doi.org/10.3389/fneng.2011.00003.

White, J.A., Rubinstein, J.T., Kay, A.R., 2000. Channel noise in neurons. Trends Neurosci. 23 (3), 131–137. http://www.pubmedcentral.nih.gov/articlerender.fcgi?artid=3279159&tool=pmcentrez&rendertype=abstract.

Yarom, Y., 1991. Rhythmogenesis in a hybrid system—interconnecting an olivary neuron to an analog network of coupled oscillators. Neuroscience 44 (2), 263–275. http://dx.doi.org/10.1016/0306-4522(91)90053-Q.

Chapter 14

Bidirectional Brain–Machine Interfaces

M. Semprini, F. Boi and A. Vato

Italian Institute of Technology, Rovereto, Italy

1 INTRODUCTION

Brain–machine interfaces (BMIs) mediate the interaction between the brain and the external world by establishing an artificial channel of communication. Typical examples include systems aiming to restore or enhance missing or defective motor functions (Mussa-Ivaldi and Miller, 2003; Fagg et al., 2007; Daly and Wolpaw, 2008; Hatsopoulos and Donoghue, 2009). In such scenarios, the brain is connected through the interface to an artificial end effector, such as a robotic arm, a cursor on a screen, or a mobile wheelchair. Within the interface, information can travel in two directions: from the brain to an end effector (to convey information about the intended movement—*efferent BMI*) or from an end effector to the brain (to convey information about current state of the end effector or about the environment—*afferent BMI*). A bidirectional BMI can perform these two operations at the same time establishing a two-way communication channel between brain and devices, which is crucial for restoring motor functions after paralysis.

Examples are given by recent demonstrations where decoded neural activities from cortical motor areas are used to control the movement of artificial effectors, such as robotic arms (Chapin et al., 1999; Hochberg et al., 2006; Ojakangas et al., 2006; Velliste et al., 2008). In these paradigms, subjects, like rats, monkeys, or humans, learn to activate the recorded neurons and to guide external devices.

They represent undoubtedly one of the greatest achievements of neuroscience and neuroengineering, but also draw attention to two major limits in the current state of BMIs.

One limit arises from the constraints imposed by the visual system, which has processing delays ranging between 100 and 200 ms (Thorpe et al., 1996). These systems are based on visual control and the sensory information about the controlled device is purely visual. The delay imposed by the processing of such sensory information is in sharp contrast with the physiological control of the limbs, as the brain can guide the arms and the legs without looking at them, based mainly on proprioceptive signals originating from the joints, tendons, and muscles.

A second limit concerns the need for BMI users to dedicate continuous attention while performing the task. This is, again, unlike movement physiology: when opening a door it is not necessary to focus the attention on the motions of the fingers wrapping around the handle.

For these reasons, we proposed a novel experimental paradigm by creating a bidirectional brain–machine interface by setting up a decoding and an encoding interface, which generate a dynamic control policy in the form of a force field. We aimed at emulating the operation of the spinal cord, as the prime biological interface between the brain and the musculoskeletal apparatus that in vertebrates combines brain instructions and sensory information, and organizes patterns of muscle forces (Shadmehr et al., 1993; Mussa-Ivaldi et al., 1994).

In this chapter we describe the components of a typical bidirectional BMI, explaining the technological limits and bottlenecks and we illustrate different decoding algorithms to interpret the neural signals and techniques to encode sensory information. We also describe state of the art BMIs, and we finally illustrate the bidirectional BMI system developed in our lab that aims to overcome the already introduced limits of current systems.

From now on we refer to bidirectional closed-loop BMI simply as *BMI*, therefore monodirectional or open loop BMI systems can be considered as "special cases" of the described BMI.

It must also be noted that we are here referring to invasive BMI that makes use of neural signals recorded via intracortical or intracranial electrodes. This implies that subjects need to undergo brain surgery. We therefore exclude from our dissertation the so-called brain–computer interface (BCI) systems that are mainly focussed on noninvasive signal extraction techniques, such as electroencephalography, and are traditionally aimed at providing subjects with the ability to verbally communicate through a dedicated speller device (Millán et al., 2010; Reza et al., 2012). Recently, other noninvasive or less invasive signal extraction techniques have also been adopted in the BCI field and used for different applications: not only as word processors, but also to brain control

Closed Loop Neuroscience. http://dx.doi.org/10.1016/B978-0-12-802452-2.00014-7

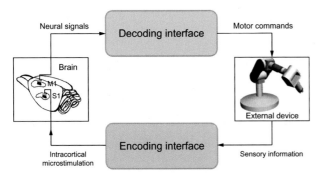

FIG. 1 Overview of a classical bidirectional brain–machine interface. The connection between the brain and an external device is established by a decoding interface that transforms the neural signals into motor commands, and an encoding interface that transduces the information collected by the device interacting with the environment into suitable signals to be delivered directly into the brain.

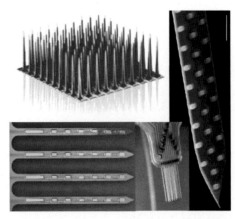

FIG. 2 Different examples of intracortical electrodes used in several BMI systems to record the neural activity or to deliver patterns of electrical stimuli.

a steering wheelchair, or interact with web browsers (Nicolas-Alonso and Gomez-Gil, 2012).

2 THE ESSENTIAL ELEMENTS OF A BMI

Two main blocks compose a BMI: a decoding and an encoding interface that permit the communication between the brain and an external device (Fig. 1).

2.1 Decoding Interface

The *decoding interface* transforms brain signals into commands that can be interpreted by the artificial device. It is composed by a *recording unit*, a *processing unit*, and a *decoder*.

2.1.1 Recording Unit

The type of recording electrode may vary depending on the neural signals that need to be extracted. Invasive BMIs record single-unit (SU), multi-unit (MU), and local field potential (LFP) neural signals by means of intracortical electrodes or multielectrode arrays (Fig. 2).

The *headstage* is a small amplifier directly connected to the electrodes, which represents the first stage of amplification of the collected signals. After this first stage, the signals travel either wired or wireless reaching a *preamplifier* that amplifies and filters the different components of the neural signals. Finally they reach the core unit of the *acquisition system* to be digitalized and, of course, depending on the acquisition system, differences may occur, but the output result is that neural data are digitalized and filtered into the frequency bands of interest.

2.1.2 Processing Unit

LFPs are captured by filtering the raw signals typically between 1 and 250 Hz, while spikes, representing SU and

MU activities, are obtained by setting the filter parameters between 300 Hz and 5 kHz. Spike signals are then generally processed by an algorithm that first performs spike detection and then clusters spikes recorded from the same electrode into groups with similar features (Quiroga et al., 2004). Different techniques and methods are available and several commercial neural recording systems are equipped with such spike sorting algorithms embedded.

2.1.3 Decoder

Once the neural signals have been conveniently filtered, it is necessary to extract the features to control the external device to perform the desired task. For example, if the goal is to move a robotic arm, the decoder needs to estimate movement directions and translate them into joint positions that the robotic arm needs to assume. Clearly, depending on the type of device or end effector to be controlled, and also on the brain area from which signals are recorded, different techniques will be used (Fig. 3).

As an example, we can mention the decoder developed by Schwartz and colleagues (Velliste et al., 2008) and used by a monkey to control a mechanized arm replica in a self-feeding task. In this experiment, the monkey was implanted with intracortical microelectrodes arrays in the primary motor cortex and the device to control was a robotic arm with 5 degrees of freedom performing 3D movements. The decoder used a simple algorithm based on the population vector algorithm (PVA) already used in cursor-control experiments. This algorithm is based on the observations of Georgopoulos and his colleagues (1982) that showed that the neural activity recorded from the motor cortex was tuned by the movement directions of the monkey's hand while performing a reaching task. Several units showed a single preferred direction: when the monkey was performing the movements in that particular direction, the unit was firing

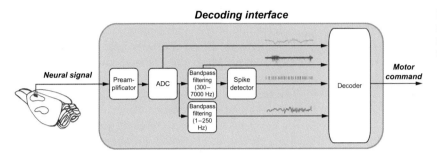

FIG. 3 Detailed representation of a BMI decoding interface able to process the recorded neural signals and to extract the features needed to control an external device.

at the maximum level. The researchers were able to predict the direction of the movement by adding together the weighted neural activity of several units by obtaining a vector (ie, *population* vector) whose direction and magnitude were instantaneously highly correlated with the velocity of the movement. This population vector was used by Schwartz and colleagues to predict the trajectories of the monkey's arm in a reaching task and in different drawing tasks (Schwartz, 1992). The decoder used to control the robotic arm reported in Velliste et al. (2008) was based on an updated version of the PVA and was used to extract the three velocity dimensions of the endpoint of a robotic arm and to control the aperture of the gripper fingers. With this simple linear decoding algorithm, the monkey was able to continuously control the movement of a prosthetic hand by modulating the velocity and the direction of the robotic endpoint in a self-feeding task. This example confirms how crucial the design of a decoder is in order to develop a robust and performing BMI, and this issue has been addressed by many research groups in the last 15 years (Wessberg et al., 2000; Serruya et al., 2002; Taylor et al., 2002; Carmena et al., 2003; Musallam et al., 2004; Orsborn et al., 2014).

Recently, researchers have been exploring the possibility of increasing the performances and robustness of these decoders by also engaging the concept of neural adaptation, indicating with this expression, the changes that occur in the neural activity due to the learning of the controlled prosthetic device (Carmena, 2013; Orsborn et al., 2014; Shenoy and Carmena, 2014).

2.2 Encoding Interface

The *encoding interface* allows us to communicate the sensory information captured by the device interacting with the environment to the brain. In invasive BMIs feedback information is communicated to the brain through *intracortical microstimulation* (ICMS), trains of electrical pulses that are delivered directly to specific brain regions. The encoding interface is composed by a *transducer unit,* an *encoder*, and a *feedback generator unit* (Fig. 4).

2.2.1 Transducer Unit

The transducer unit is composed by a set of hardware and software that translates the analog sensor outputs mounted on the external device into digital values. It is not a simple analog-to-digital converter (ADC), because, for example, the transducer coupled with a motor sensor that takes as input the current/voltage coming from the sensor, is capable of providing more complex information such as motor position, velocity, acceleration, or torque.

2.2.2 Encoder

The goal of the encoder is to translate the information coming from the transducer into a representative and comprehensible feedback stimulus to provide to the subject connected to the BMI.

The encoding interface feedback stimuli can be subdivided into *natural* and *artificial* stimuli. Natural stimuli are composed by signals that the subject can easily detect and recognize by using his/her natural sensory system

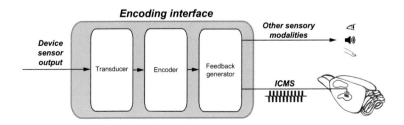

FIG. 4 Detailed representation of a BMI encoding interface designed to transform the information collected by the controlled device into feedback signals suitable for the brain.

(ie, visual, auditory, or kinesthetic stimuli). *Artificial stimuli* are generated to bypass the natural sensory apparatus, ie, by delivering a series of electrical pulses directly to the sensory cortex eliciting different artificial sensations.

2.2.3 Feedback Generator Unit

The feedback generator unit triggers the onset of the stimuli whose parameters such as amplitude, duration, and frequency have been previously set by the encoder.

The feedback generator physically converts the digital information passed by the encoder into analog signals that can be both delivered directly to the brain (*ICMS*) or felt naturally by the subject (sounds, light, vibration).

3 BMI SYSTEMS: STATE OF THE ART

In the last 15 years, several reviews have been published describing the advancement of BMI systems and the potential of these devices to help millions of people suffering from neurological injuries and diseases (Mussa-Ivaldi and Miller, 2003; Musallam et al., 2004; Lebedev and Nicolelis, 2006; Mussa-Ivaldi et al., 2010; Jackson and Zimmermann, 2012; Moxon and Foffani, 2015).

The majority of these systems use natural visual system as sensory feedback to close the loop in real time. In this section, we are interested in describing some examples of closed-loop BMIs in which the feedback sensory channel is artificial and represented by patterns of electrical stimulation.

One of the first examples is represented by the experiment of Mussa Ivaldi and colleagues (Reger et al., 2000) who connected the brain of a lamprey to a mobile robot. They established a two-way communication channel by placing two recording and two stimulating electrodes in the extracted brain of a sea lamprey placed and maintained at a constant temperature in a recording chamber. The two stimulating electrodes were placed on left and right side of midline, along the axons of vestibular pathways. The two recording electrodes were placed on left and right side of the brainstem's midline. The recorded spiking activity from the two electrodes was transformed by a simple decoder into driving signals to control the speed of the two wheels of a small mobile robot moving in an arena. On the other side, the robot was equipped with a set of optical sensors, and the light-intensity signals detected by these sensors were converted by an encoder in patterns of electrical stimulation with different frequency delivered to the brain through the two stimulation electrodes. The result was that the robot exhibited different behaviors such as following the movements of a shining light (ie, positive phototaxis)

or moving away from the light source (ie, negative phototaxis) according to the configuration of the motor and the sensory interfaces. This is considered one of the first implementations of a BMI in closed-loop between a brain and a robotic device.

This technique of using patterns of ICMS delivered into the brain as an artificial sensory channel has been explored in BMI systems by several research groups in rats and monkeys (Romo et al., 1998; Fitzsimmons et al., 2007; Fridman et al., 2010; Semprini et al., 2012; Zaaimi et al., 2013).

The first realization of the integration of this artificial sensory channel in real time into a BMI is considered the experiment carried out by (O'Doherty and colleagues 2011) with behaving monkeys. They reported the operation of a BMI called a brain–machine–brain interface that controls the movement of an artificial actuator and, at the same time, uses ICMS to deliver artificial tactile feedback into the primary somatosensory cortex of the subjects. The monkeys implanted with multielectrode arrays in the primary motor cortex (M1) and in the primary somatosensory cortex (S1) were trained to perform an exploration task of virtual objects by using a computer cursor, or a virtual image of their arm (ie, an avatar). They performed this task by controlling a joystick with the left hand (ie, *hand control* modality) to move the actuator over the target object to obtain the reward, or by controlling the movement of the actuator by modulating the activity of right-hemisphere M1 neurons recorded by the implanted arrays (ie, *brain control* modality). The researchers were able to use ICMS of the hand representation in the somatosensory cortex to communicate directly to the brain the different texture of the virtual object that the monkeys were "touching" with their virtual arm. The textures for different objects were represented with different frequency of the stimulation patterns delivered through the electrodes in S1. In this experiment, the main challenge for the researchers was to integrate in real time the ICMS feedback signal into the system without interrupting the functioning of the BMI decoder; indeed ICMS artifact masks the recorded neural activity for 5–10 ms after each stimulation pulse. To overcome this problem, they multiplexed the recorded activity with the stimulation signal with a 20 Hz clock signal, obtaining a continuous online control of the virtual arm without interferences.

These examples show the importance of integrating an artificial sensory channel in a closed-loop system and represent the proof of concept toward the development of motor prosthetic devices equipped with sensors that are able to collect information about the environment to be delivered directly to the brain by using different techniques such as ICMS.

4 DYNAMIC BMI: A BIDIRECTIONAL BMI THAT EMULATES THE SPINAL CORD

Motor actions have the apparently paradoxical property of being at once automatic and consciously accessible to voluntary control. However, such behavior is not implemented in traditional BMIs: indeed, in order to control the external device, BMI users need to constantly and consciously regulate their behavior. Our group recently proposed a novel BMI system, called *dynamic BMI* (dBMI) (Vato et al., 2012), which aims to overcome this limit. The key idea of this study is to program a bidirectional BMI for generating control policies in the form of force fields acting on the controlled external device, using rats as subjects. This study was conceived by Ferdinando Mussa-Ivaldi and stems from his seminal work on spinal cord and movement representation (Bizzi et al., 1991).

A bidirectional interface can, in principle, be programmed to implement a pattern of neural stimuli and responses capable of approximating a desired behavior of the controlled system. Mathematically, this process corresponds to translating the behavior of the neural system into a control policy that maps the current observed state of the controlled system into a corresponding action. This concept is closely related to earlier evidence that spinal interneurons organize muscles into synergy groups whose mechanical outputs are force fields acting upon the limbs. In fact, a simple mechanism of vector summation is capable of generating a repertoire of control policies out of a small set of nonlinear force fields (Mussa-Ivaldi et al., 2010).

Earlier studies in frogs (Mussa-Ivaldi et al., 1994), rats (Tresch and Bizzi, 1999), and cats (Lemay and Grill, 2004) have revealed the presence of circuits within the spinal cord that transform brain commands into the activation of several limb muscles. The force generated by a muscle on a limb varies depending on muscle length, on state of motion, and on the position of the limb (Kandel et al., 2000). When the spinal cord activates an ensemble of muscles in response to a cortical command, the net mechanical outcome is a spatial pattern of forces—a *force field*—that sets the limb in motion. This simplified description is consistent with the general view (Schöner and Kelso, 1988; Hatsopoulos, 1996) that higher brain centers are not concerned with specifying at each instant the position of the limb. Instead, they instruct the spinal cord to achieve movement goals, such as reaching for a point in space (Pesaran et al., 2008).

Force fields afford a simple mathematical foundation for this view of motor control: by imposing a force field, the control system specifies a goal and regulates the interactions between the limb and the environment. For example, when the brain issues the command to reach for an object with the hand, the spinal cord activates the arm muscles to produce a field of forces converging upon the object. Then, the hand will move toward the object along different trajectories depending on its starting location. The spatial structure of a force field describes a control policy, by specifying the forces to be generated throughout the reachable space in response to unexpected perturbations.

This perspective was adopted for developing our dBMI. This dBMI borrows a local portion of cortical tissue to emulate the generation of force fields by the spinal cord.

The BMI algorithm implements a neuromechanical translation by acting on two levels:

1. *The motor interface*, a decoder that maps recorded output activity into a force vector;
2. *The sensory interface*, an encoder that maps the state of the device into a pattern of stimuli.

Recording and stimulation take place on two different, but connected areas to investigate whether brain plasticity can produce an adaptation to the interface as the subject gradually learns how to use the connections between the two areas to control the BMI. In fact, this marriage of the nervous system with artificial devices offers an unparalleled opportunity to acquire knowledge about neural computation and plasticity (Reger et al., 2000) while opening a path for restoring functions lost to accident or disease.

In a dynamically shaped interface, the external neural input sets an initial condition and the dynamic field, absent other influences, determines the ensuing trajectory. This approach frees the user from the need to guide the connected device instant by instant. At the same time, however, the subject is able to perform a continuous control, thus guiding the device through arbitrary paths.

Experiments were first run on anesthetized animals implanted with two arrays of electrodes in the whisker representation areas of primary motor (M1) and primary sensory (S1) cortices. In this implementation, the physical external device is replaced with a virtual mass moving in a viscous medium in a 2D space (Vato et al., 2012).

Fig. 5 shows the outline of the dBMI experiment.

Two microwire arrays are placed into the motor (M1) and sensory cortical regions (S1) of the vibrissal system of a rat (Fig. 5A). Activity recorded from M1 (Fig. 5B) is converted by the motor interface (Fig. 5C) into a force. The force moves the point mass in the virtual space according to the law of the dynamical system (Fig. 5D). The position of the point mass is encoded by the sensory interface (Fig. 5E) into a pattern of electrical stimuli (Fig. 5F) delivered to S1.

FIG. 5 Outline of the dynamic BMI. We placed two 16-channel microwire arrays in the vibrissa motor (M1) and sensory areas (S1) of a rat brain cortex (A). The recorded evoked activity (B) is used by the motor interface (C) to generate a force vector from the first two principal components. The obtained force vector is applied to a simulated dynamical system (D). The sensory interface maps each point in the field into the corresponding stimulation pattern (E). The four electrical stimulation patterns are set by specifying a pair of electrodes in stimulating array (F), adapted from Vato et al. (2012) with permission from the authors.

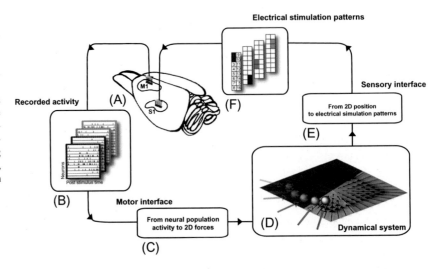

5 dBMI ALGORITHM

Activity recorded from M1 is converted by the dBMI algorithm into a force vector in a bidimensional space, requiring a dimensionality reduction from N to 2, N being the number of recorded units. The virtual mass then moves according to this force. Every point of the bidimensional space is associated with a particular stimulus, so the new position of the virtual mass determines which stimulus has to be delivered on S1.

In order to run the dBMI, a calibration phase is needed: activity in M1 is recorded while different stimulation patterns occur one at the time; this procedure is repeated R times, R being the number of calibration cycles.

Data is then organized in a matrix in which each row contains the sequence of binned neural response from different units. Principal component analysis (PCA) is performed and scores of the first two PC are scaled to fit the space of the virtual mass and used as reference system for it.

Mean responses to every stimulus pattern are calculated and plotted as force vectors in the new bidimensional space so that every point in the space is associated with the stimulus corresponding to the closer calibration vector.

Following is a detailed description of how the algorithm operates.

5.1 Calibration of the Interface

Before running the algorithm, motor and sensory interfaces need a calibration procedure that establishes the form of the force field operating on the controlled dynamical system.

First, a set of key parameters needs to be defined:

- N: number of recorded neurons as established by spike sorting;
- T: number of time-intervals (bins) recorded per each neuron;

- S: number of stimulation patterns (the stimulus vocabulary); and
- R: number of repetitions of each stimulus pattern during the calibration procedure.

During the calibration, each stimulation pattern is repeated R times and, accordingly, $R \times N$ neural responses are recorded. Each response is an array of T values: the number of spikes in each bin. The calibration responses are then represented as $R \times N$ N-dimensional vector functions:

$$\psi_s^r(t) = \begin{bmatrix} \psi_{s,1}^r(t) \\ \psi_{s,2}^r(t) \\ \dots \\ \psi_{s,N}^r(t) \end{bmatrix} \quad (s = 1, \dots, S;\ r = 1, \dots, R;\ t = 1, \dots, T)$$

(1)

From these calibration responses, the responses obtained from the repetitions of each stimulus are averaged, to extract S mean responses

$$\varphi_s(t) = \frac{1}{R} \sum_{r=1}^{R} \psi_s^r(t)$$

(2)

Following the same notation, a neural response vector is a N-dimensional vector function

$$v(t) = \begin{bmatrix} v_1(t) \\ v_2(t) \\ \dots \\ v_N(t) \end{bmatrix} \quad (t = 1, \dots, T)$$

(3)

The inner product of two neural responses is defined by extension over time bins and units of the Euclidean inner product:

$$\langle v | g \rangle = \sum_{n=1}^{N} \sum_{t=1}^{T} v_n(t) \cdot g_n(t)$$

(4)

The S mean calibration responses form a set of basis fields—an extension of the concept of basis vectors (Mussa-Ivaldi, 1992) and are used to approximate all recorded neural responses. In particular, each calibration response is approximated as a sum of mean responses:

$$\psi_s^r(t) \approx \sum_{i=1}^{S} d_{s,i}^r \cdot \varphi_i(t) \tag{5}$$

To derive the combination coefficients $d_{s,i}^r$ the inner product of each side of Eq. (5) with each basis function must be taken. This leads to S equations

$$\Psi_s^r = \Phi \cdot d_s^r \tag{6}$$

where

$$\begin{aligned}[\Psi_s^r]_i &= \langle \varphi_i | \psi_s^r \rangle \\ [\Phi]_{i,j} &= \langle \varphi_i | \varphi_j \rangle \\ (i,j &= 1, \ldots, S)\end{aligned} \tag{7,8}$$

Eq. (6) can be solved for d_s^r provided that $\det(\Phi) \neq 0$ (if the projection matrix is singular, one can use a pseudoinverse. But this does not seem to be a likely situation and was not encountered with any datasets). With this, each calibration response is mapped, respectively, into an S-dimensional vector

$$d_s^r = \begin{bmatrix} d_{s,1}^r \\ d_{s,2}^r \\ \ldots \\ d_{s,S}^r \end{bmatrix} \tag{9}$$

Each response corresponds to a d-vector and vice versa, each d-vector corresponds to a unique approximation of the response (the likelihood that two distinct signals map onto the same d-vector is vanishingly small). Therefore, the S-dimensional d_s^r vectors are taken as representations of the individual neural responses obtained after applying each stimulus.

5.2 Motor Interface

To calibrate the motor interface, we used the first two principal components of PCA: these capture the greatest amount of variance in the set of the $S \times R$ calibration vectors, d_s^r. These two components are two S-dimensional arrays that form the rows of the $2 \times S$ projection matrix

$$W = \begin{bmatrix} w_{1,1} & w_{1,2} & \cdots & w_{1,S} \\ w_{2,1} & w_{2,2} & \cdots & w_{2,S} \end{bmatrix} \tag{10}$$

This operator defines the bidimensional plane with maximum variance over the set of S stimuli. The next step of the calibration procedure involves stretching the matrix as to match the range of variation of the x and y components of the force vectors over the desired force field domain:

$$\overline{W} = \sigma \cdot W \tag{11}$$

The gain σ is a 2×2 diagonal matrix that scales the bidimensional projections of the calibration recordings to cover the range of the desired force field, $F = K(\rho)$. The field establishes a correspondence between the position, r, of the controlled object—in this first implementation a point mass—and a resulting force $F = [F_x, F_y]^T$. Here, the additional hypothesis that this field is invertible is made, which means that there is a function $x = K^{-1}(F)$ mapping force vectors to corresponding positions. This is obviously the case if the field is linear, as in $F = K \cdot (r - r_0)$ and the stiffness matrix is nonsingular. However, the requirement of invertibility can be relaxed to a local and continuous form. The two projection matrices, Φ and $\overline{W}$, and the mean calibration responses, $\phi_i(t)$, to all the stimuli generate a map from the data collected during the experiment to a corresponding bidimensional force vector

$$F = W \cdot \Phi^{-1} \begin{bmatrix} \langle \varphi_1 | \nu \rangle \\ \langle \varphi_2 | \nu \rangle \\ \ldots \\ \langle \varphi_S | \nu \rangle \end{bmatrix} \tag{12}$$

This is a linear filter that operates in real time.

5.3 Sensory Interface

The sensory interface maps the instantaneous position of the controlled object onto one of the stimulation patterns in the calibration vocabulary.

This sensory interface performs a look-up operation:

$$\hat{s} = \underset{i \in \{1, \ldots, S\}}{\arg \min} (\|\rho - \xi_i\|) \tag{13}$$

that picks up the stimulus, $\hat{s}$, corresponding to the "calibration site" $\xi_{\hat{s}}$ that is closest to the current position x of the controlled object. The calibration sites $\{\xi_1, \ldots \xi_S\}$ are the S locations:

$$\xi_i = K^{-1}(F(\varphi_i)) \tag{14}$$

where $F(\varphi_i)$ is the force derived by Eq. (12) from the average response, ϕ_i, to the ith stimulus in the vocabulary.

As an example, suppose that during the calibration of the motor interface the stimulus s_A is set to produce a force vector F_A. In the desired force field, F_A is associated to a location x_A. Then, the sensory interface sets the correspondence from all the points in the neighborhood of x_A to s_A. This is effectively a look-up table, based on a nearest-neighbor interpolation.

5.4 dBMI Procedure

The working procedure of the dBMI can be summarized as follows:

- The force field establishes a mechanical correspondence between positions and forces;
- The motor interface establishes a correspondence between recorded activities and forces;
- The sensory interface looks for the stimulus that generates a force attached to a position that is nearest—among all stimuli—to the current position of the controlled object.

It is important to keep in mind that each time a stimulus is repeated, the neural response produces a different force. This difference is due to background activities that interact with the activities induced by the stimulus.

In the alert animal, these activities may also contain a voluntary component. Therefore, while the field produced by a noiseless preparation is a piecewise linear approximation of the desired field, the actual field is an additive superposition of this approximation—established by the interface—and of a variable component induced by background noise and voluntary activity.

5.5 Simulation of the Dynamical System Interacting With Neural Activity

In our first dBMI implementation, the device interacting bidirectionally with neural activity was a simulated point mass in a viscous medium. The activities recorded from the neural tissue and decoded by the motor interface were supplied as input to a dynamic simulation, and the position of the simulated point mass was fed to the sensory interface to determine the stimulus delivered to the neural preparation. This section provides details about the simulation procedures.

Typical values for the mass and viscosity were $M = 10$ kg and $B = 15$ Ns/m. A linear force field $F = K \cdot r$, results in the linear differential equation

$$M \cdot \ddot{x} + B \cdot \dot{x} + K \cdot x = 0 \qquad (15)$$

so, with an isotropic stiffness of $K = 4$ N/m, the ideal system driven by the noiseless linear field is slightly over-damped

$$\left(\zeta = 1.19, \text{ with } \zeta = \frac{B}{2 \cdot M \cdot \omega_0} \text{ and } \omega_0 = \sqrt{\frac{K}{M}} \right).$$

As the interface implements a piecewise constant approximation of the linear field, $F = \bar{K}(x)$, corrupted by random background activity, the stability properties afforded by the desired continuous field can only be considered as an optimal limit.

This first realization of the interface had some notable limitations. One is that the control law generates an output force in response to a position input. In a more complete system, the input should convey not only position, but state information, that is position and velocity. Here, the derivative component of the controller is a fixed property, expressed by the term $B \cdot \dot{x}$ in the dynamics equation.

Another obvious simplification is in the choice of a point mass ($M \cdot \ddot{x}$) for controlled object. A mechanical arm is generally characterized by a nonlinear differential equation. However, the second-order linear ordinary differential equation (15) is used in robotics to represent the error dynamics of nonlinear systems controlled by proportional-derivative (PD) methods (Spong et al., 2006):

$$M \cdot \ddot{e} + B \cdot \dot{e} + K \cdot e = 0 \qquad (16)$$

with $e(t) \triangleq x - x_D(t)$, ($x_D(t)$ is a desired trajectory). In this framework, the PD control law can be reformulated as

$$M \cdot \ddot{x} + B \cdot \dot{x} + K \cdot (x - \rho(t)) = 0 \qquad (17)$$

where $\rho(t) = [M \cdot \ddot{x}_D + B \cdot \dot{x}_D + K \cdot x_D] \cdot K^{-1}$ is a time-varying function to be supplied by the voluntary input to the interface.

In this case, the dBMI would provide stability to a desired movement in a way analogous to the combined influence on limb movements of muscle mechanics and feedback mechanisms of the spinal cord. Therefore, while the form of Eq. (15) is quite simple, it also expresses a fundamental mathematical representation for control.

5.6 Online Operation of the dBMI

The online operation of the dBMI is summarized by the following algorithm:

1. The experimenter places the point mass at a starting position inside the workspace.
2. The sensory interface determines the stimulus to be delivered at that position, based on the nearest calibration site, according to Eq. (13), the stimulus is delivered.
3. The motor interface decodes the ensuing neural activity and derives the force vector, F, to be applied to the point mass, according to Eq. (12).
4. The next position is computed by integrating the equation $M \cdot \ddot{x} + B \cdot \dot{x} + F = 0$ for a time interval t (typically 1 s); note that this integration time is time in the formal physical units used to simulate the dynamical system, and it does not have to be interpreted as a real-time value for the operation of the dBMI; in fact, the simulation of the dynamical system took only a fraction of millisecond in real time and did not limit in any way the intervals between electrical stimulation, which are the real "clock" for the operation of the online dBMI.
5. The process is repeated from step 2 until the point mass reaches the zero-force zone surrounding the equilibrium point; this condition may not be encountered if the field is unstable or if it has a circulating pattern; in those

cases, the algorithm is set to stop after a preset number of repetitions (typically 30).

6 dBMI WITH ANESTHETIZED ANIMALS

The process presented in the previous paragraph has been tested on anesthetized rats. During these experiments, the subject was implanted with a couple of microwire arrays: one placed in the whisker motor cortex (M1, recording site) and the other in the whisker somatosensory cortex (S1, stimulating site).

By performing stimulation along four different electrodes of the microwire array placed in S1, four different repeatable evoked neural activities have been recorded from the array sited in M1.

Raster plots (Fig. 6A) and poststimulus time histogram (PSTH) of the evoked activity (Fig. 6B) show the difference in terms of firing rate and activation timings of the recorded neural population in response to different stimulation patterns.

After calibration of the dBMI, the motor interface was able to translate each evoked activity into a bidimensional force. We grouped the decoded forces belonging to the same stimulation and displayed them as direction arrows (Fig. 7). Arrows belonging the same group are very close to each other (this represents the fact that the response evoked by a stimulus is highly repeatable) and

simultaneously forces belonging to different groups point in almost orthogonal directions (this means that different stimulations evoked different recognizable spiking activity).

Once the dBMI was calibrated, we also performed a series of runs in which the virtual point mass was randomly initialized each time in a different starting position with the goal to reach the equilibrium region. The point mass was piloted just by the decoded activity evoked by the proper stimulations depending on the current position of the point mass, as established by the sensory interface.

Looking at Fig. 8 it is possible to appreciate the fact that starting from whichever initial position, the dBMI was able to reach the target region. Subject implied less time, thus performing straighter trajectories, when the point mass belongs a sensory region in which forces (Fig. 7) are closer to each other (eg, stimulation pattern 1, yellow region) while in other regions (eg, stimulation pattern 4, blue region) the resulting path is more uncertain because of the bigger spread of force directions. By performing this operation many times it was demonstrated that the BMI has a convergence rate closer to 70%. To further prove the efficacy of this new BMI concept, we performed a test in which the forces are decoded starting from neural activity evoked by random stimulation nonrelated to the point mass state. In this case the BMI reached a convergence rate lower than 10%.

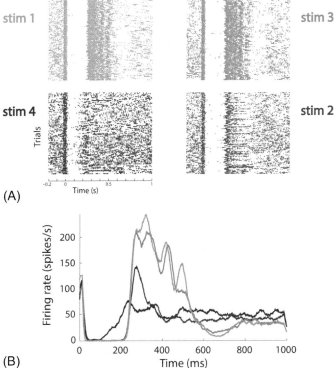

(A)

(B)

FIG. 6 (A) Raster plots for each stimulation pattern of the spiking stimulus-evoked activity of a single unit; (B) poststimulus time histograms (PSTH) of neural evoked responses of selected neuron. The color code represents four different stimulation patterns, adapted from Vato et al. (2012) with permission from the authors.

FIG. 7 The output of the motor interface collected during the calibration procedure is represented as force vectors (*black arrows*) while the calibration forces are represented by *red arrows*, adapted from Vato et al. (2012) with permission from the authors.

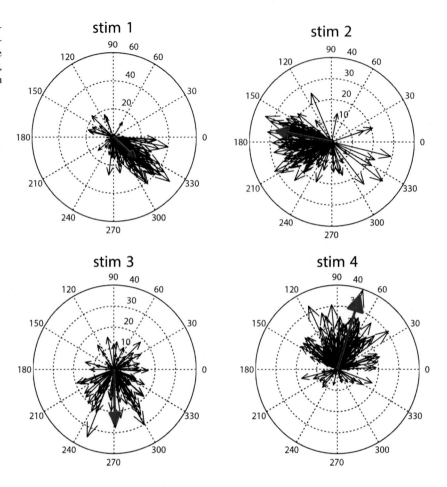

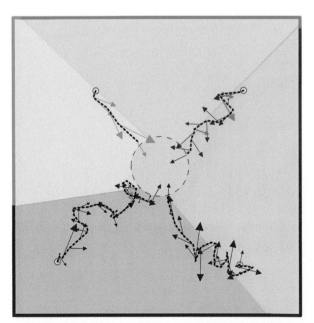

FIG. 8 Example of a family of trajectories generated by running the BMI from different random initial positions of the bidimensional workspace, adapted from Vato et al. (2012) with permission from the authors.

7 dBMI WITH ALERT ANIMALS

In previous section we generated a family of trajectories designed by the controlled simulated object starting from different positions and converging upon a selected equilibrium point. Our group also explored the possibility of programming the neural interface by using nonlinear force fields to overcome the limitation of the preliminary algorithm that could only approximate invertible-force fields (Szymanski et al., 2011).

We now wish to demonstrate that during the performing movement, a volitional command issued by the user can modify the force field produced by the interface. To verify this, we need to run experiments with alert animals. We therefore developed a novel experimental set-up which will allow us to understand how the brain of alert rats interacts with a real dynamical system and to explore which field parameters are accessible and modifiable by volitional commands (9).

In this experiment involving alert animals, the neural interface is designed to control the movements of a pellet dispenser with the goal of placing the small cart in

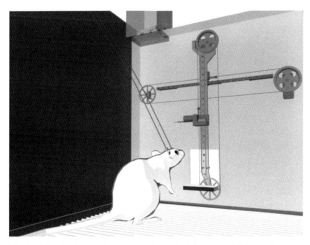

FIG. 9 The freely moving rat, implanted with two multielectrode arrays, is placed in a box with a wall equipped with a small cart connected to the pellet dispenser and controlled by two servomotors. To get a reward, the rat needs to position the cart close to the window on the see-trough Plexiglas sheet by modulating its neural activity. A mechanism formed by two perpendicular rails allows the cart to span a squared area of 38×38 cm, adapted from Boi et al. (2015) with permission from the authors.

correspondence of the slot through which the rats get the reward. This task is less abstract than controlling a computer cursor on a screen (Nekovarova and Bures, 2006) because the controlled object coincides with the reward dispenser and the rats are encouraged to get the reward in the shortest time. In this particular implementation, the subjects need to bring the cart close to the slot by modulating the neural activity of the motor cortex in response to the electrical microstimulation of the sensory cortex (Fig. 9).

We ran pilot tests of this apparatus with awake animals and showed that freely moving rats can modulate their brain activity to control the position on a vertical plane of a small cart connected to the pellet dispenser (Boi et al., 2015). We therefore believe that this experimental framework will allow us to explore the existence and the role of additive volitional neural commands in modifying the automatic responses already obtained in the anesthetized experiments. We expect to be able to quantify the effects of this additional input on the behavior of the controlled object, by registering a significant modification of the trajectories generated by this novel dBMI.

This research will offer an interesting opportunity to develop a new family of BMIs to provide patients with severe motor and sensory disabilities with systems capable of performing actions without the need of a constant online visual supervision.

ACKNOWLEDGMENTS

We are grateful to F.A. Mussa-Ivaldi, S. Panzeri, F.D. Szymanski, E. Maggiolini, and L. Fadiga for their precious collaboration on earlier work relevant to this chapter, to L. Taverna for graphical work, and to D. Torrazza for the design and development of the rat box. This work was supported by the *SI-CODE* project of the Future and Emerging Technologies (FET) program within the Seventh Framework Programme for Research of the European Commission, under FET-Open grant number: FP7-284553.

REFERENCES

Bizzi, E., Mussa-Ivaldi, F.A., Giszter, S., 1991. Computations underlying the execution of movement: a biological perspective. Science (New York, N.Y.) 253 (5017), 287–291.

Boi, F., Semprini, M., Mussa-Ivaldi, F.A., Panzeri, S., Vato, A., 2015. A bidirectional brain-machine interface connecting alert rodents to a dynamical system. In: Engineering in Medicine and Biology Society (EMBC), 2015 Annual International Conference of the IEEE, Milan.

Carmena, J.M., 2013. Advances in neuroprosthetic learning and control. PLoS Biol. 11 (5).

Carmena, J.M., Lebedev, M.A., Crist, R.E., O'Doherty, J.E., Santucci, D.M., Dimitrov, D.F., Patil, P.G., Henriquez, C.S., Nicolelis, M.A., 2003. Learning to control a brain-machine interface for reaching and grasping by primates. PLoS Biol. 1 (2), E42.

Chapin, J.K., Moxon, K.A., Markowitz, R.S., Nicolelis, M.A., 1999. Real-time control of a robot arm using simultaneously recorded neurons in the motor cortex. Nat. Neurosci. 2 (7), 664–670.

Daly, J.J., Wolpaw, J.R., 2008. Brain-computer interfaces in neurological rehabilitation. Lancet Neurol. 7 (11), 1032–1043.

Fagg, A.H., Hatsopoulos, N.G., de Lafuente, V., Moxon, K.A., Nemati, S., Rebesco, J.M., Romo, R., Solla, S.A., Reimer, J., Tkach, D., Pohlmeyer, E.A., Miller, L.E., 2007. Biomimetic brain machine interfaces for the control of movement. J. Neurosci. 27 (44), 11842–11846.

Fitzsimmons, N.A., Drake, W., Hanson, T.L., Lebedev, M.A., Nicolelis, M.A., 2007. Primate reaching cued by multichannel spatiotemporal cortical microstimulation. J. Neurosci. 27 (21), 5593–5602.

Fridman, G.Y., Blair, H.T., Blaisdell, A.P., Judy, J.W., 2010. Perceived intensity of somatosensory cortical electrical stimulation. Exp. Brain Res. 203 (3), 499–515.

Georgopoulos, A.P., Kalaska, J.F., Caminiti, R., Massey, J.T., 1982. On the relations between the direction of two-dimensional arm movements and cell discharge in primate motor cortex. J. Neurosci. 2 (11), 1527–1537.

Hatsopoulos, N.G., 1996. Coupling the neural and physical dynamics in rhythmic movements. Neural Comput. 8 (3), 567–581.

Hatsopoulos, N.G., Donoghue, J.P., 2009. The science of neural interface systems. Neuroscience 32 (1), 249–266.

Hochberg, L.R., Serruya, M.D., Friehs, G.M., Mukand, J.A., Saleh, M., Caplan, A.H., Branner, A., Chen, D., Penn, R.D., Donoghue, J.P., 2006. Neuronal ensemble control of prosthetic devices by a human with tetraplegia. Nature 442 (7099), 164–171.

Jackson, A., Zimmermann, J.B., 2012. Neural interfaces for the brain and spinal cord—restoring motor function. Nat. Rev. Neurol. 8 (12), 690–699.

Kandel, E.R., Schwartz, J.H., Jessell, T.M., 2000. Principles of Neural Science. McGraw-Hill, New York.

Lebedev, M.A., Nicolelis, M.A., 2006. Brain-machine interfaces: past, present and future. Trends Neurosci. 29 (9), 536–546.

Lemay, M.A., Grill, W.M., 2004. Modularity of motor output evoked by intraspinal microstimulation in cats. J. Neurophysiol. 91 (1), 502–514.

Millán, J.D., Rupp, R., Müller-Putz, G.R., Murray-Smith, R., Giugliemma, C., Tangermann, M., Vidaurre, C., Cincotti, F., Kübler, A., Leeb, R., Neuper, C., Müller, K.R., Mattia, D., 2010. Combining brain-computer interfaces and assistive technologies: state-of-the-art and challenges. Front. Neurosci. 4.

Moxon, K.A., Foffani, G., 2015. Brain-machine interfaces beyond neuro-prosthetics. Neuron 86 (1), 55–67.

Musallam, S., Corneil, B.D., Greger, B., Scherberger, H., Andersen, R.A., 2004. Cognitive control signals for neural prosthetics. Science 305 (5681), 258–262.

Mussa-Ivaldi, F.A., 1992. From basis functions to basis fields: vector field approximation from sparse data. Biol. Cybern. 67 (6), 479–489.

Mussa-Ivaldi, F.A., Miller, L.E., 2003. Brain-machine interfaces: computational demands and clinical needs meet basic neuroscience. Trends Neurosci. 26 (6), 329–334.

Mussa-Ivaldi, F.A., Giszter, S.F., Bizzi, E., 1994. Linear combinations of primitives in vertebrate motor control. Proc. Natl. Acad. Sci. U. S. A. 91 (16), 7534–7538.

Mussa-Ivaldi, F.A., Alford, S.T., Chiappalone, M., Fadiga, L., Karniel, A., Kositsky, M., Maggiolini, E., Panzeri, S., Sanguineti, V., Semprini, M., Vato, A., 2010. New perspectives on the dialogue between brains and machines. Front. Neurosci. 4, 44.

Nekovarova, T., Bures, J., 2006. Spatial decisions in rats based on the geometry of computer-generated patterns. Neurosci. Lett. 394 (3), 211–215.

Nicolas-Alonso, L.F., Gomez-Gil, J., 2012. Brain computer interfaces, a review. Sensors (Basel, Switzerland) 12 (2), 1211–1279.

O'Doherty, J.E., Lebedev, M.A., Ifft, P.J., Zhuang, K.Z., Shokur, S., Bleuler, H., Nicolelis, M.A., 2011. Active tactile exploration using a brain-machine-brain interface. Nature 479 (7372), 228–231.

Ojakangas, C.L., Shaikhouni, A., Friehs, G.M., Caplan, A.H., Serruya, M.D., Saleh, M., Morris, D.S., Donoghue, J.P., 2006. Decoding movement intent from human premotor cortex neurons for neural prosthetic applications. J. Clin. Neurophysiol. 23 (6), 577–584.

Orsborn, A.L., Moorman, H.G., Overduin, S.A., Shanechi, M.M., Dimitrov, D.F., Carmena, J.M., 2014. Closed-loop decoder adaptation shapes neural plasticity for skillful neuroprosthetic control. Neuron 82 (6), 1380–1393.

Pesaran, B., Nelson, M.J., Andersen, R.A., 2008. Free choice activates a decision circuit between frontal and parietal cortex. Nature 453 (7193), 406–409.

Quiroga, R.Q., Nadasdy, Z., Ben-Shaul, Y., 2004. Unsupervised spike detection and sorting with wavelets and superparamagnetic clustering. Neural Comput. 16 (8), 1661–1687.

Reger, B.D., Fleming, K.M., Sanguineti, V., Alford, S., Mussa-Ivaldi, F.A., 2000. Connecting brains to robots: an artificial body for studying the computational properties of neural tissues. Artif. Life 6 (4), 307–324.

Reza, F.-R., Brendan, Z.A., Christoph, G., Eric, W.S., Sonja, C.K., Andrea, K., 2012. P300 brain computer interface: current challenges and emerging trends. Front. Neuroeng. 5, 14.

Romo, R., Hernandez, A., Zainos, A., Salinas, E., 1998. Somatosensory discrimination based on cortical microstimulation. Nature 392 (6674), 387–390.

Schöner, G., Kelso, J.A., 1988. A synergetic theory of environmentally-specified and learned patterns of movement coordination. I. Relative phase dynamics. Biol. Cybern. 58 (2), 71–80.

Schwartz, A.B., 1992. Motor cortical activity during drawing movements: single-unit activity during sinusoid tracing. J. Neurophysiol. 68 (2), 528–541.

Semprini, M., Bennicelli, L., Vato, A., 2012. A parametric study of intracortical microstimulation in behaving rats for the development of artificial sensory channels. In: Conference proceedings: … Annual International Conference of the IEEE Engineering in Medicine and Biology Society. IEEE Engineering in Medicine and Biology Society. Annual Conference 2012, pp. 799–802.

Serruya, M.D., Hatsopoulos, N.G., Paninski, L., Fellows, M.R., Donoghue, J.P., 2002. Instant neural control of a movement signal. Nature 416 (6877), 141–142.

Shadmehr, R., Mussa-Ivaldi, F.A., Bizzi, E., 1993. Postural force fields of the human arm and their role in generating multijoint movements. J. Neurosci. 13 (1), 45–62.

Shenoy, K.V., Carmena, J.M., 2014. Combining decoder design and neural adaptation in brain-machine interfaces. Neuron 84 (4), 665–680.

Spong, M.W., Hutchinson, S., Vidyasagar, M., 2006. Robot Modeling and Control. John Wiley and Sons, Hoboken, NJ. ISBN: 978-0-471-64990-8.

Szymanski, F.D., Semprini, M., Mussa-Ivaldi, F.A., Fadiga, L., Panzeri, S., Vato, A., 2011. Dynamic brain-machine interface: a novel paradigm for bidirectional interaction between brains and dynamical systems. Conf. Proc. IEEE Eng. Med. Biol. Soc. 2011, 4592–4595.

Taylor, D.M., Tillery, S.I., Schwartz, A.B., 2002. Direct cortical control of 3D neuroprosthetic devices. Science 296 (5574), 1829–1832.

Thorpe, S., Fize, D., Marlot, C., 1996. Speed of processing in the human visual system. Nature 381 (6582), 520–522.

Tresch, M.C., Bizzi, E., 1999. Responses to spinal microstimulation in the chronically spinalized rat and their relationship to spinal systems activated by low threshold cutaneous stimulation. Exp. Brain Res. 129 (3), 401–416.

Vato, A., Semprini, M., Maggiolini, E., Szymanski, F.D., Fadiga, L., Panzeri, S., Mussa-Ivaldi, F.A., 2012. Shaping the dynamics of a bidirectional neural interface. PLoS Comput. Biol. 8(7).

Velliste, M., Perel, S., Spalding, M.C., Whitford, A.S., Schwartz, A.B., 2008. Cortical control of a prosthetic arm for self-feeding. Nature 453 (7198), 1098–1101.

Wessberg, J., Stambaugh, C.R., Kralik, J.D., Beck, P.D., Laubach, M., Chapin, J.K., Kim, J., Biggs, S.J., Srinivasan, M.A., Nicolelis, M.A., 2000. Real-time prediction of hand trajectory by ensembles of cortical neurons in primates. Nature 408 (6810), 361–365.

Zaaimi, B., Ruiz-Torres, R., Solla, S.A., Miller, L.E., 2013. Multi-electrode stimulation in somatosensory cortex increases probability of detection. J. Neural Eng. 10 (5), 056013.

Chapter 15

Adaptive Brain Stimulation for Parkinson's Disease

M. Beudel

University of Groningen, Groningen, The Netherlands

1 INTRODUCTION

Parkinson's Disease (PD) is the second most common neurodegenerative disorder in which patients gradually develop motor and nonmotor symptoms (van Rooden et al., 2011). It affects roughly 1% of the 65 yo population and 4–5% of the 85 yo population (Van Den Eeden et al., 2003) and its cardinal motor symptoms are bradykinesia, tremor, rigidity, and postural imbalance. Examples of nonmotor symptoms are depression, hallucinations, dementia, and drooling. In its first years, PD can, most often, be managed in satisfying ways with dopaminergic medication. However, after approximately 5 years of disease progression, medication often no longer works sufficiently and often induces side effects (Hauser et al., 2006). One of the most debilitating and frequently occurring (paradoxical) side effects are Levodopa Induced Dyskinesia's (LIDs), in which patients develop debilitating involuntary movements in addition to their usual paucity of movements.

In these more progressed disease stages, deep brain stimulation (DBS) is one of the additional therapeutical options in PD. With DBS, electrodes that provide small electrical pulses are inserted into nuclei deep in the brain and connected to a battery that is implanted subcutaneously, most often under the clavicle. In PD, these electrodes are typically inserted in the subthalamic nucleus (STN), or the Internal Segment of the globus pallidus (GPi). However, in case of a tremor-dominant PD, the Ventral Intermediate nucleus of the thalamus (Vim) is used as a target as well. Next to these targets, other targets like the Pedunculopontine Nucleus (PPN) (Fasano et al., 2015) and basal nucleus of Meynert (NBM) (Kuhn et al., 2015) are currently explored as a potential target for, respectively, axial signs and cognition, but their clinical indication and effectiveness still need to be established.

DBS can substantially improve motor symptoms, reduce dopaminergic medication, and improve quality of life in PD (Deuschl et al., 2006; Williams et al., 2010), which makes it the most effective treatment for advanced PD to date. However, although DBS has been applied for more than 25 years, there are still limitations in terms of efficacy, side effects, and efficiency. At present, the effect of chronically applied DBS is, on average, only an improvement of around 45% (eg, Odekerken et al., 2013) on the motor items of the Unified Parkinson's Disease Rating Scale (UPDRS III) OFF dopaminergic medication. Furthermore, there is even evidence that DBS can, paradoxically, worsen motor functioning by not only influencing pathological, but also physiological, neural activity (Brittain et al., 2014; Chen et al., 2006). Next to this, the potential of DBS is often limited due to stimulation-induced side effects like Stimulation-Induced Dysarthria (SID) and neuropsychiatric symptoms. Finally, the capacity of the currently implanted batteries is still limited. For this reason, battery replacement surgery needs to take place every few years.

At present, DBS is applied in such a way that the small electrical pulses, with an average current of approximately 2.5 V, are applied in an uninterrupted way with a frequency of approximately 130 Hz and a characteristic pulse width of 60 µs. Although there are many stimulation parameters that can be adjusted, like stimulation voltage and frequency, pulse duration, contact, and polarity and that there are relatively new stimulation techniques like interleaved stimulation, the common denominator of the current stimulation algorithms is that they stimulate continuously and do not respond to external stimuli: they are *nonadaptive*. In theory, DBS could work more effectively with fewer side effects and be more efficient were it only to stimulate when necessary. This type of stimulation is called adaptive DBS (aDBS) or closed-loop stimulation.

The most elementary form of adaptive stimulation is an adjustment of stimulation parameters by a clinician based on a limited clinical effect or the occurrence of side effects. In theory, a frequent clinical evaluation and a manual adjustment of stimulation parameters are already a form of adaptive stimulation. In fact, more frequent parameters

Closed Loop Neuroscience. http://dx.doi.org/10.1016/B978-0-12-802452-2.00015-9

213

adjustments improve DBS efficacy (Moro et al., 2006). However, the temporal interval over which parameters could be modified is in the magnitude of days to years. On a shorter temporal interval there are currently two strategies to *adapt* DBS treatment on a sub-circadian interval. In the first one, patients can choose between two different stimulation settings or can switch off their stimulator at night. The advantage of the first option is that patients can adjust the stimulator settings themselves with an external programmer based on their current activities and preferences. A well-known example of this first option is that patients switch between a stimulator setting with more efficacy *and* more side effects like SID when they are planning to walk, but not talk, and vice versa. A well-known example of the second option is that patients can switch off the DBS system at night in order to save battery life.

These very crude adaptive strategies either require the expertise of a professional or an intervention of the patient him or herself, and have a temporal resolution that is lower than the typical fluctuation of PD symptoms. To apply adaptive stimulation with a higher temporal resolution, measurements and adjustments need to be automatized. In PD, these measurements could consist of electrical signals like local field potentials (LFPs) that could be derived from the nonstimulating DBS electrodes, but also from other sites in the nervous system and also from non-neural signals, like accelerometer signals.

In this chapter the neurophysiological perturbations of the cardinal PD symptoms at different levels of the nervous system will be reviewed. After this, the modulation of these signals by (deep) brain stimulation will be reviewed and an assessment of whether these signals could be developed as biomarkers for adaptive (deep) brain stimulation will be performed. Finally, a review of the first data on adaptive brain stimulation will be given and a road-map toward clinical applicability will be sketched. This chapter will primarily focus on adaptive *DBS*, since DBS is the only established neuromodulation treatment in PD as yet. Although, other experimental neuromodulation techniques will be reviewed as well, this will be primarily from a scientific perspective.

2 BIOMARKERS IN PARKINSON'S DISEASE SUITABLE FOR APPLYING ADAPTIVE BRAIN STIMULATION

In the last decades, our knowledge on the pathophysiological basis of PD has increased dramatically. The main landmarks of this journey have been the discovery of the loss of dopaminergic neurons in substantia nigra, the alterations of dopaminergic projections to the striatum (Parent and Parent, 2010), and the expression of Levy bodies throughout the central nervous system (Braak et al., 2003).

Despite this extensive pathophysiological knowledge, the relation between the pathology at a cellular level and clinical symptomatology is far less established. In between these descriptive layers, namely the molecular and clinical layers, the neurophysiological alterations occur and lead to clinical symptoms.

With the development of stereotactic neurosurgery, not only a new opportunity to lesion or stimulate deep brain nuclei became available, but also a new opportunity to do human recordings of the deep brain nuclei was born. The first intra-cranial recordings in humans were described by Albe Fessard et al. and Guiot and Brion in the 1950s and 1960s (Albe Fessard et al., 1963). With this recording method, it became possible to look at spontaneous neural activity in different deep nuclei and look at their modulations during movement and the application of medication or stimulation. Further sophistication of these recordings in the late 1960s, using electrodes with smaller tip sizes (around 2–3 μm), made it possible to actually record single-units (ie, single neurons). When these single-units synchronize their oscillatory activity, a rhythmic population activity, the LFP is generated. These LFPs can be measured with the electrode tip size of actual DBS electrodes (around 2 mm).

In PD, the previously mentioned alterations of dopaminergic projections toward the striatum result in changes in neural firing patterns in the Cortio Basal-Ganglia Thalamo Cortical (CBGTC) loop (Alexander et al., 1986). This CBGTC loop model is a leading anatomical model that explains the subsequent processing of cortical information toward the basal ganglia and further downstream the thalamus, and finally back to the cortex. Although the anatomical accuracy of the model is validated by many subsequent anatomical tracer studies (eg, Haber et al., 2000), its assumption that PD symptoms are caused by a change in neural discharge rate have not been experimentally confirmed (Hammond et al., 2007, Rivlin-Etzion et al., 2006). In fact, it seems most probable that PD symptoms are not caused exclusively by changes in excitatory or inhibitory firing rate, but by a change of patterning of (group) neural burst activity (Chan et al., 2011; Kühn et al., 2005). In PD, the most distinct synchronized (LFP) neural activity are oscillations, ie, "bursts," in the beta (13–30 Hz) band that are present at various sites in the CBGTC loop (Kühn et al., 2006).

2.1 Biomarkers for Bradykinesia and Rigidity

At present there is substantial experimental evidence showing that the power of synchronized LFP beta power is anti-kinetic in nature. Beta power decreases prior and during movements in PD patients (Cassidy et al., 2002) but this decrement is smaller when PD patients are withdrawn from their dopaminergic medication and are in a bradykinetic state (Doyle et al., 2005). Furthermore, when DBS is applied to the STN, beta power decreases in the GPi combined with a clinical improvement in bradykinesia (Brown et al., 2004). After this, when DBS is provided at 20 Hz (the frequency roughly in the middle of the beta range), bradykinesia deteriorates (Chen et al., 2007). At another level of the CBGTC loop, the cortex, beta synchronization over central motor areas correlates with motor impairment (Silberstein et al., 2005b). This cortical beta synchronization decreases when either DBS is switched ON or dopaminergic medication is provided (Silberstein et al., 2005b). From the opposite side, an excess of beta power might be related to dyskinesias in PD (Silberstein et al., 2005a, see the following section).

Although the experimental evidence mentioned before support a correlation between beta power and an antikinetic state in PD, there are several other, and often independent (Stochl et al., 2008), motor phenomena in PD like tremor and axial symptoms. Beta power is not correlated with PD tremor, which might be more related to (low) gamma power (Beudel et al., 2015) and does not correlate with the total UPDRS III score OFF medication (eg, Weinberger et al., 2006). This might implicate that beta power only correlates with that portion of the bradykinetic-rigid state that can be reversed with dopaminergic medication or DBS (Eusebio and Brown, 2009). This, however, does not mean that beta power cannot be used as a biomarker for aDBS but the importance of gaining more insight in more subtle aspects of beta oscillations, instead of the power in the whole beta range, and developing other biomarkers for controlling aDBS.

2.2 Spectrally Distinct Beta Oscillations

In the first human aDBS study of Little et al. (see next section, Little et al., 2013a) the peak power of beta oscillations in a range between 13 and 35 Hz was used as a biomarker. Beta oscillations can be divided into low (13–20 Hz) and high (21–35 Hz) beta. Although magneto encephalography and LFP studies show especially cortical-subcortical coherence in the high beta range (Litvak et al., 2011) and LFP recordings show especially modulation of low beta power after the application of dopaminergic medication (Priori et al., 2004) or a correlation with severity with Parkinsonism in the untreated state (Lopez-Azcarate et al., 2010), it has not yet been elucidated whether there is a difference between low or high beta power as optimal biomarker for aDBS. In the study of Little et al. (2013a) six of the eight subjects had a target frequency in the high beta range and the same held for low beta. The average target frequency was 22 Hz. In future experiments, this putative difference can be elucidated by comparing the efficiency of aDBS based on the power of low with that of high beta power.

2.3 Power Change and Phase of Beta Oscillations

Besides high and low beta, the volatility of beta power might potentially serve as a biomarker for aDBS as well. The reason for this is that coefficient of variation (CV) of high beta power in the OFF dopaminergic medication is significantly, inversely, correlated with UPDRS III scores, independent of spectral amplitude and with the change in UPDRS III scores after the application of dopaminergic medication (Little et al., 2012). One can think of applying aDBS in such a way that aDBS is applied at those moments that a pathological synchronous state (de Hemptinne et al., 2015) is entered for too long.

Apart from the amplitude or the change in its amplitude, beta oscillations can also be characterized by their phase, which also might be a potential biomarker for aDBS. In the Parkinsonian STN, the severity of bradykinesia and rigidity correlates with the spatial extend of beta synchrony (Pogosyan et al., 2010). This was discovered by successively assessing the phase coherence of two different locations within and surrounding the STN. In theory, one could think of disturbing the contact to contact phase coherence by selectively applying a pulse to one contact when two, or more, contacts are pathologically synchronized. Such stimulation algorithm was recently proposed by Azodi-Avval and Gharabaghi (2015) and supported by another recent finding that showed phase locking of beta oscillations in the STN and GPi (Cagnan et al., 2015). In this last study, certain phases of phase locking supported further coupling resulting in increased amplitude but also other phases resulted in the amplitude being suppressed and coupling terminated. By steering aDBS to those phases, a very selective form of stimulation could be applied (Cagnan et al., 2015).

2.4 Broad Gamma and High-Frequency Oscillations

Next to beta oscillations, recent findings indicate that information processing in the basal ganglia not only involves activities within a certain frequency bands but also interactions between frequency bands. One important example of this cross-frequency coupling, in PD, is Phase Amplitude Coupling (Canolty and Knight, 2010). In the last years a vast amount of experimental evidence has become available that showed that the phase of beta oscillations is correlated with the amplitude of broad gamma oscillations ($\pm$50–200 Hz). As with the power of beta LFPs, PAC also decreased in relation to movements (Yanagisawa et al., 2012).

In PD, an increased PAC between motor cortex beta phase and broad gamma amplitude was detected that reversed after the application of STN DBS (de Hemptinne et al., 2013). Since nonParkinsonian patients have a lower and more volatile beta to broad gamma PAC, this excessive coupling state in which the primary motor cortex is restricted to a monotonous pattern of coupling, might be used as another biomarker for aDBS. The main advantages of this method would be that the recordings are subdural and don't penetrate the cortex and are performed at a different location in the nervous system than where the stimulation is actually given. Further support for this idea came from a very recent paper by the same group, showing that the DBS-induced reduction in beta to broad gamma PAC was correlated with motor improvement (de Hemptinne et al., 2015). Ongoing research by the same group currently involves the chronic implantation of subdural grids over the primary cortex and are connected to a new device that is able to simultaneously stimulate and record LFP's (see Section 5).

Next to the broad gamma (50–200) Hz oscillations in PD, yet another distinct spectral peak is present in the range between 200 and 400 Hz (Yang et al., 2014). These oscillations are coined High-Frequency Oscillations (HFOs) and show, contrarily to beta oscillations, a movement-related increase that is also more prominent after the application of dopaminergic medication (Litvak et al., 2012). Furthermore, the peak HFO frequency changes from around 250 to 340 Hz after the application of dopaminergic medication (Lopez-Azcarate et al., 2010) and the strength of HFO PAC was correlated with the UPDRS III score OFF medication. However, it is not yet known which UPDRS III items are especially well correlated and to what extent DBS influences HFO PAC, although electrodes that show greater HFO PAC turned out to be more likely to be the contacts that were clinically effective (Yang et al., 2014). At present, more extensive knowledge about the effects of DBS on HFO PAC and its volatility needs to be established before this can be translated toward a potential biomarker for adaptive stimulation.

2.5 Adaptive Stimulation Characteristics

Finally, an even more adaptive way of stimulation might be possible in which the stimulation algorithm is not predetermined but determined on the effect of different stimulation algorithms on a biomarker (Heldman et al., 2016; Su et al., 2015). By doing this, the optimal stimulation strategy can be obtained by only setting the boundaries in which DBS should take place and determining the biomarker on which it should. In practice this would mean that the stimulation device is actively searching for the best stimulation parameters, eg, stimulation fraction, to suppress, for example, LFP beta power or clinical parameters like tremor and bradykinesia.

2.6 Other Symptom-Based Biomarkers

Most of the potential biomarkers discussed in the previous section are all related to bradykinesia and rigidity. At present all the human and nonhuman experimental evidence in favor of the applicability in PD has focussed on either biomarkers or outcome measurements that were related to bradykinesia and rigidity. However, in PD, several other debilitating motor phenomena are present. The three most important are tremor, dyskinesias, and freezing (of gait) and other axial signs. Although there is no experimental evidence for the application of aDBS to selectively control these symptoms, hypothetical stimulation algorithms will be discussed. Furthermore, the indirect effect of aDBS on nonmotor symptoms will be briefly discussed.

2.7 Tremor

As described in the first sub-section, tremor and bradykinesia-rigidity are two independently occurring motor phenomena in PD (Stochl et al., 2008) and don't share the same oscillatory alterations. Contrary to bradykinesia-rigidity, tremor induces a suppression of beta oscillatory activity. Next to this decrease of beta activity, also increased neural activity at the tremor frequency has been observed in the CBGTC loop in PD tremor (Hirschmann et al., 2013). In the last paper, an adaptive stimulation algorithm with the power of the dominant tremor frequency and its first harmonic, plus a decrease in beta activity was proposed as a potential biomarker for aDBS. However, given the conflicting existence of beta activity related to bradykinesia-rigidity and tremor is an argument against using beta spectral LFP power in patients with both prominent tremor and bradykinesia-rigidity. In addition to this, recent findings indicate that when DBS decreases tremor amplitude, a decrease in low gamma (31–45 Hz) activity occurs (Beudel et al., 2015), which further illustrates the richness of the STN oscillatory activity.

Contrary to the proposed spectral density-based aDBS, PD tremor might also be treated with aDBS based on phase-interference stimulation techniques. Recent evidence from such an approach came from a study in which a noninvasive technique, Transcranial Direct Current Stimulation (TDCS), was able to reduce PD tremor by delivering pulses at "phase cancellation phases" based on peripheral tremor phase (Brittain et al., 2013). However, this phenomenon was not seen in another study in which DBS was provided at the tremor frequency (Cagnan et al., 2014). A crucial remark on this last study is that DBS was provided at the tremor frequency, but not consistently at a certain tremor phase (Cagnan et al., 2014). Future tremor tracking studies should assess whether this is the case.

2.8   Dyskinesias

Although DBS has a dramatic effect on the treatment of dyskinesias in PD, in 2–4% of the patients experience (STN and not GPi) DBS induced dyskinesias with a temporal relation with their remaining dopaminergic medication (Sriram et al., 2014). These dyskinesias were recently coined "brittle dyskinesias" (Sriram et al., 2014). Since PD patients are usually OFF their dopaminergic medication during DBS surgery, not so much is known about the neurophysiological hallmark of dyskinesias. In one study, this subject was addressed (Alonso-Frech et al., 2006). The main finding was that in the transition from OFF to ON dopaminergic state in PD patients experiencing dyskinesias, beta activity was reduced and 4–10 Hz activity increased. This finding was further strengthened by the fact that it was limb-specific. Interestingly, such a low-frequency spectral peak is also seen in the LFP power spectrum of dystonia patients undergoing DBS surgery of the GPi (Barow et al., 2014), and stimulation at 5 Hz induced involuntary choreiform movements in PD patients undergoing DBS surgery (Liu et al., 2002). For these reasons, STN and GPi low-frequency oscillations might, contrary to beta oscillations, be a pro-kinetic oscillation. Whether these low-frequency oscillations could serve as a biomarker is not established yet. In theory, they could serve as a secondary biomarker for the hyperkinetic state. However, the main expectation of aDBS would be that it is able to prevent a hyperkinetic state by stimulating less after dopaminergic medication has reduced beta activity. This hypothesis was supported by the recently published case report of Rosa et al. (2015) (see next section) in which less stimulation was provided after dopaminergic medication was provided and less dyskinesias were observed compared to continuous DBS (cDBS).

2.9   Freezing and Other Axial Features

After 10–15 years, it is not limb-bradykinesia-rigidity, but axial motor features that dominate the motor phenotype of PD (Hely et al., 2005). Examples of these features are gait disturbances, postural imbalance, and freezing of gait (FOG). Contrary to "appendicular" motor sings, axial symptoms respond less well to DBS (Fasano et al., 2015). Recently, a new target that might be more effective in treating axial symptoms, the PPN, has been investigated (Fasano et al., 2015). Interestingly, recent findings (Fraix et al., 2013) showed decreased 5–12 Hz activation in the PPN when patients were unable to step because of severe FOG. When patients were able to walk, ON dopaminergic

medication, this 5–12 Hz activity increased and beta activity decreased (Androulidakis et al., 2008).

Such an imbalance between higher and lower frequencies shows a strong resemblance with the findings in Parkinson's dyskinesia. Interestingly, 10 Hz stimulation of the PPN turned out to have a better effect on gait than conventional high-frequency stimulation (Mazzone et al., 2005). Unfortunately, no data is present yet on the modulation of oscillatory activity by DBS yet. Furthermore, it would be interesting to see how gait parameters, eg, phase and amplitude as in tremor, relate to oscillatory activity in the PPN. Given this limited data, the only feasible biomarkers for PPN aDBS up to know would be the 5–12 Hz power or the ratio between higher and lower frequencies.

2.10   Cognitive and Limbic Symptoms

At present, one of the main limitations of DBS in PD is that although motor symptoms improve and dopaminergic medication can be lowered, certain nonmotor symptoms persist and tend to worsen. One plausible mechanism for this is that DBS especially stimulates the motor loop, and to a lesser extent, the cognitive and limbic CBGTC loop (Alexander et al., 1986). One of the most striking examples of this is apathy that can neutralize benefits in quality of life after STN DBS (Martinez-Fernandez et al., 2015). In theory, aDBS might lead to an improvement of this apathy since dopaminergic drugs can be decreased to a lesser extent since the stimulation will switch off when a hyper dopaminergic, ie, dyskinetic, state is reached. This interesting hypothesis is difficult to test in the acute postoperative setting but with the dawn of new implantable devices (see Section 5) these nonmotor symptoms can be assessed over longer periods.

3   CURRENT NONHUMAN AND HUMAN EXPERIMENTAL EVIDENCE FOR aDBS IN PD

The first experimental evidence of the successful application of aDBS in PD comes from nonhuman primates (Rosin et al., 2011). In this study, two African green monkeys were rendered Parkinsonian by the application of the neurotoxin MPTP (1-methyl-4-fenyl-1,2,3,6-tetrahydropyridine). The monkeys were implanted with two electrodes in the GPi and four in the ipsilateral primary motor cortex. Although aDBS was provided in different settings, the most successful was the version in which spike activity was detected in M1 and 7 pulses were provided subsequently in the GPi with a delay of 80 ms. The idea behind this 80 ms delay was that this coincides with the cycle length of the frequency in which coherence between the cortex and basal ganglia is observed, given the dominant frequency of 12.5 Hz. The most important finding of this aDBS study was that motor velocity increased significantly compared to 130 Hz cDBS, despite less overall stimulation. Next to this, spike and oscillatory activity between 4 and 7 Hz and 9 and 15 Hz within the GPi all decreased significantly compared to 130 Hz cDBS.

Based on this proof-of-principle study, human aDBS studies were set up. These studies are, for obvious reasons, all performed during patient care and have several limitations compared to the nonhuman primate study of Rosin et al. These limitations are both in the spatial and in the temporal domain. In DBS surgery, only depth electrodes, and no cortical grids are implanted. For this reason, no cortical feedback signal could be used. Next to this, the implanted DBS electrodes have a tip size of approximately 2 mm and are not able to record single action potentials as biomarker. For this reason, the current experiments are limited to depth electrodes and LFPs.

At present, two studies and one case report in which PD patients have been treated with aDBS have been published. Their first study was published in 2013 by Little et al. (2013a) and described eight PD patients in their immediate postoperative phase after they had undergone the electrode placement in the STN. In this study, different forms of stimulation, including aDBS, were applied unilaterally for a short period up to 5 min, aDBS was applied based on the amount of beta (13–35 Hz) peak power between two DBS electrodes (either 0–2 or 1–3) that was continuously analyzed using a moving average filter. The stimulation trigger threshold was determined in such a way that the device was stimulating for approximately 50% of the time while remaining clinically effective. The stimulation itself consisted of 130 Hz monopolar stimulation with a pulse duration of 100 μs with an average voltage of 2.1 V.

The most important finding of this study was that there was a clinical improvement that was quantified by blinded and unblinded UPDRS III score items of 50% and 66%, respectively. This was significantly better than the effect of cDBS. Furthermore, there was decrease of 56% of stimulation time compared to aDBS. aDBS was also significantly more effective than random intermittent stimulation which was ineffective in the blinded ratings. Finally, there were no side effects other than those that were also encountered in cDBS like transient paresthesias.

This proof of principle study showed that aDBS is technically feasible, that it can be applied in a safe way and that it turned out to be highly effective. Although this landmark paper was of crucial importance for the development of aDBS, there were still some major limitations for translating these findings toward clinical applicability. One of the limitations of this study was that aDBS was only applied unilaterally while DBS is most often applied bilaterally since bilateral symptoms don't respond equally effective to unilateral stimulation (Kim et al., 2009). Although there is an interaction between the oscillations in both STNs, this

interaction turned out to be weak (Little et al., 2013b). Furthermore, patients were only tested on a very small subset (three) of the (31/34) UPDRS III items (bradykinesia, rigidity, and tremor) for a very short period. This is a crucial point since axial signs and speech disturbances can actually deteriorate with the application of DBS. Next to this, the stimulation parameters that were applied for all types of stimulation in this study (aDBS, cDBS, and random stimulation) were different in terms of pulse configuration, active contact and pulse duration compared to those that are applied in clinical practice, which might have underestimated the effect of cDBS. Another point, inherent to the study, was that the stimulation fraction was set to 50% without a true rationale. Finally, no side effects were formally assessed and the study was conducted in the immediate postoperative phase which would result in the so-called "stun" or micro-lesion effect in which patients temporarily clinically improve due to the lesions effect on the deep brain nuclei of the DBS procedure itself (Rosa et al., 2010; Tykocki et al., 2013).

In a follow-up study (Little et al., 2015), aDBS was applied in a bilateral way with independent LFP recording and stimulation. Next to this, the full UPDRS III score, including axial sings and walking was assessed and the interaction of aDBS with dopaminergic medication was assessed as well (Little et al., 2015). In line with the unilateral study, aDBS was titrated on the amount of beta (13–35 Hz) peak power between two DBS contact points (either 0–2 or 1–3). The two independent stimulation trigger thresholds were determined in such a way that the device was stimulating for approximately 50% of the time while remaining clinically effective. The stimulation itself consisted of 130 Hz monopolar stimulation with a pulse duration of 100 μs with an average voltage of 3.0 V. The main results of this study were that aDBS significantly reduced the UPDRS III score compared to no stimulation (average 43% reduction) with less than half of its energy consumption, and that the application of dopaminergic medication further reduced the energy consumption whilst staying effective (Little et al., 2015).

Although no other study on aDBS is published up to now, a case report on aDBS was published in 2015 in which the patient was able to freely move during the application of aDBS, so more UPDRS III, including axial, items could be assessed (Rosa et al., 2015). In this report, a single PD patient is described after STN implantation in the immediate postoperative period ON and OFF stimulation. During two 120 min intervals, the patient was clinically assessed every 20 min in a cDBS and an aDBS condition using the UPDRS III and a dyskinesia rating scale. aDBS was applied based on the amount of power in the lowest beta range (13–17 Hz) and not provided in an ON-OFF fashion but in a linear way so that more 13–17 Hz activity resulted in a higher stimulation voltage and vice versa. The main results of this case study were that both aDBS and cDBS improved the axial symptoms, but aDBS improved bradykinesia better than cDBS. Furthermore, aDBS improved dyskinesia's more than cDBS did in this patient in the ON medication state. Given the $N = 1$ character of this report, its most important contribution is that aDBS can now be tested in a more ecological setting, which is another step toward a further, more externally valid, studies.

4 STIMULATION ALGORITHMS

As in cDBS, many stimulation algorithms can be applied in aDBS. It is good to realize that many of the stimulation parameters in cDBS have their origin based on empirical findings and have not been systematically investigated. In the aDBS studies that have been published up to now (Little et al., 2013a, 2015; Rosa et al., 2015), high-frequency stimulation, with regular pulses with a fixed inter-pulse interval, were given. In theory, aDBS could also be given with single pulses based on certain biomarker characteristics (eg, Rosin et al., 2011). In future head to head studies based on high-frequency stimulation, it will be, in the first instance, most straightforward to use as much similar parameters for aDBS and cDBS as possible. In the next section, the most important stimulation characteristics will be reviewed.

In the first aDBS study, a stimulation frequency of 130 Hz was used. This is in line with the most frequently applied stimulation frequency in chronically implanted PD patients. Although, there is a question to be answered whether low-frequency stimulation (60 Hz) might work better in treating axial symptoms in PD (Moreau et al., 2008), there is little rationale to apply a different stimulation frequency yet. The most commonly applied pulse duration of cDBS is that of 60 µs, whereas the duration that was used in the first study was 100 µs. In the case report on aDBS, a pulse duration of 60 µs was used, which was tolerated and shown to be effective in the single patient. To harmonize the amount of applied current in future head to head aDBS-cDBS studies, it is most intuitive to apply a 60 µs pulse duration in the initial future studies. In line with this, it is essential to apply the same pulse configuration in aDBS and cDBS. In the first aDBS study, a charge balanced symmetrical pulse waveform was applied, whereas the pulse waveform that is most often clinically applied is charge balanced, but asymmetrical. Also for this parameter it is important to harmonize between aDBS and cDBS.

Next to these temporal aspects of stimulation, there is currently an issue with aDBS that is more difficult to deal with and that is the location and polarity of stimulation. With cDBS it is possible to stimulate with every DBS contact in various configurations, eg, monopolar, double monopolar, bipolar, or double bipolar. With the present

DBS leads, that typically have no more than four contacts, the options for applying aDBS are more restricted and limited to the two electrodes that are not involved in the LFP recordings. Conducting a balanced experiment with this limited amount of stimulation contact options makes it difficult to apply cDBS in the optimal way and challenges head to head comparisons. However, these issues could be dealt with, with new DBS leads that have an increased number of contacts, eg, the new Sapiens electrode with 32 contacts (Bour et al., 2015).

Finally, the amount of stimulation that is provided with either cDBS or aDBS can be changed according to the different algorithms for determining the amount of stimulation. In general, there are two different approaches to changing the amount of stimulation in an adaptive way: a binary approach and a scalar approach. Both have their advantages and disadvantages. In the first, aDBS study, a binary approach was used in which stimulation was applied during approximately 50% of the time. One disadvantage of the on-off approach is that it requires a ramping period of approximately 250 ms to overcome paresthesias that can occur after quickly switching the stimulation ON and OFF. One advantage of a binary approach is that it can deal with a nonlinear character of beta power in which pathological beta (synchronization) bursts can be suppressed, whereas the physiological beta activity, with a lower power, are not suppressed. The advantage of using a scalar approach is that no ramping period is required and changes in stimulation amplitude occur more gradually. The disadvantage of the (linear) scalar approach, as applied in the case report of Rosa et al., is that the aDBS is virtually never switched off, which might also result in suppression of physiological beta activity. One way to overcome this would be to apply a nonlinear approach, which could, in practice, be quite similar to the binary approach. Irrespective of the binary or scalar approach, the optimal amplitude of stimulation voltage is not established yet for aDBS. In the study on aDBS an average voltage of 2.1 V was applied and in the case report on aDBS a stimulation voltage between 0 and 2 V was applied. This is in the lower range of voltages that are applied early in the course of cDBS stimulation. This seems to be a rational voltage to start with, but it could be that even higher voltages are tolerated when not continuously applying stimulation or lower voltages are equally effective.

To summarize, even at the present level of DBS sophistication many stimulation parameters could be adjusted in many different ways. To move forward in translating aDBS toward clinical applicability, it is of crucial importance to carefully consider each deviation from current stimulation techniques given the additional difficulties in comparing aDBS with cDBS.

To a similarly extend these issues also holds for the other aDBS approach in which single pulses, instead of continuous pulses, are applied. However, given the smaller amount of pulses delivered, the amount of current per pulse can possibly be higher.

5 HARDWARE DEVELOPMENTS

At present, aDBS has only been applied in externalized patients in their immediate postoperative phase. Although studies in a later phase, eg, during battery replacement surgery, can take place and circumvent the well-known problems related to the immediate postoperative phase, eg, the stun-effect, the ultimate goal is to provide aDBS in an implanted device. At present, many companies are developing DBS systems with adaptive functionality. One example toward an implanted system is the Activa PC+S implanted neurostimulator from Medtronic (Ryapolova-Webb et al., 2014) that can establish a connection with an external computer via an investigational research tool, the Nexus-D system to apply adaptive stimulation.

The main challenges from an engineering point of view are that the implantable device needs to be small, reliable, and energy efficient. With a very limited availably of clinical data on the application of aDBS, it is difficult to determine the necessary capacities of an implantable DBS device with adaptive capacity as yet. Since the current devices are already capable of providing all different kinds of stimulation, the main engineering challenges concern the recording and processing of biomarkers, either from the brain, the muscular system, or from movement sensors. Based on the findings of ongoing studies on aDBS, the optimal (minimal) sampling frequency and resolution for establishing adaptive stimulation will be established which will set the boundaries of the engineering challenge. Given the relatively slow transition from OFF to ON state, one can imagine that it might even be possible to only determine a biomarker every few minutes. However, experimental data is needed to back-up these ideas.

6 TRANSLATIONAL ASPECTS

After the initial studies of the application of aDBS, a plethora of new studies should be considered to translate aDBS toward clinical applicability. Apart from the character of the intervention or endpoint(s), the crucial importance is that aDBS is studied in a well-controlled way. In this section, a road-map of development for the current experimental aDBS based on the amplitude of LFP beta peak-power toward clinical applicability will be sketched.

To translate the neurophysiological alterations described in the previous section toward clinical applicability, certain criteria should be met. The ideal biomarker would be both sensitive and specific for disease states. Furthermore, its fluctuation should have a fixed temporal relation with the disease state, and it should be detected with a minimum of energy. It is a matter of debate whether the biomarker should have a causal relation with the disease state or can even be an epiphenomenon (Little and Brown, 2012). This is not yet relevant for the current stimulation paradigms but might be for future, phase-based, adaptive stimulation paradigms (Brittain et al., 2014).

Given the vast amount of fundamental work especially, and the very recent first clinical experiments in line with the even more recent hardware developments, there seems to be a promising future for aDBS in PD. Based on the current pace of developments, it might be possible that aDBS can be clinically applied within the next 5 years.

REFERENCES

Albe Fessard, D., Arfel, G., Guiot, G., Derome, P., Herran dela, Korn, H., et al., 1963. Characteristic electric activities of some cerebral structures in man. Ann. Chir. 17, 1185–1214.

Alexander, G.E., DeLong, M.R., Strick, P.L., 1986. Parallel organization of functionally segregated circuits linking basal ganglia and cortex. Annu. Rev. Neurosci. 9, 357–381.

Alonso-Frech, F., Zamarbide, I., Alegre, M., Rodriguez-Oroz, M.C., Guridi, J., Manrique, M., et al., 2006. Slow oscillatory activity and levodopa-induced dyskinesias in Parkinson's disease. Brain 129, 1748–1757.

Androulidakis, A.G., Mazzone, P., Litvak, V., Penny, W., Dileone, M., Gaynor, L.M.F.D., et al., 2008. Oscillatory activity in the pedunculopontine area of patients with Parkinson's disease. Exp. Neurol. 211, 59–66.

Azodi-Avval, R., Gharabaghi, A., 2015. Phase-dependent modulation as a novel approach for therapeutic brain stimulation. Front. Comput. Neurosci. 9, 26.

Barow, E., Neumann, W.-J., Brücke, C., Huebl, J., Horn, A., Brown, P., et al., 2014. Deep brain stimulation suppresses pallidal low frequency activity in patients with phasic dystonic movements. Brain. awu258.

Beudel, M., Little, S., Pogosyan, A., Ashkan, K., Foltynie, T., Limousin, P., et al., 2015. Tremor reduction by deep brain stimulation is associated with gamma power suppression in Parkinson's disease. Neuromodulation 18, 349–354.

Bour, L.J., Lourens, M.A.J., Verhagen, R., de Bie, R.M.A., van den Munckhof, P., Schuurman, P.R., et al., 2015. Directional recording of subthalamic spectral power densities in Parkinson's disease and the effect of steering deep brain stimulation. Brain Simul. 8 (4), 730–741.

Braak, H., del Tredici, K., Rüb, U., de Vos, R.A.I., Jansen Steur, E.N.H., Braak, E., 2003. Staging of brain pathology related to sporadic Parkinson's disease. Neurobiol. Aging 24, 197–211.

Brittain, J.-S., Probert-Smith, P., Aziz, T.Z., Brown, P., 2013. Tremor suppression by rhythmic transcranial current stimulation. Curr. Biol. 23, 436–440.

Brittain, J.-S., Sharott, A., Brown, P., 2014. The highs and lows of beta activity in cortico-basal ganglia loops. Eur. J. Neurosci. 39, 1951–1959.

Brown, P., Mazzone, P., Oliviero, A., Altibrandi, M.G., Pilato, F., Tonali, P.A., et al., 2004. Effects of stimulation of the subthalamic area on oscillatory pallidal activity in Parkinson's disease. Exp. Neurol. 188, 480–490.

Cagnan, H., Little, S., Foltynie, T., Limousin, P., Zrinzo, L., Hariz, M., et al., 2014. The nature of tremor circuits in parkinsonian and essential tremor. Brain. awu250.

Cagnan, H., Duff, E.P., Brown, P., 2015. The relative phases of basal ganglia activities dynamically shape effective connectivity in Parkinson's disease. Brain 138, 1667–1678.

Canolty, R.T., Knight, R.T., 2010. The functional role of cross-frequency coupling. Trends Cogn. Sci. 14, 506–515.

Cassidy, M., Mazzone, P., Oliviero, A., Insola, A., Tonali, P., Di Lazzaro, V., et al., 2002. Movement-related changes in synchronization in the human basal ganglia. Brain 125, 1235–1246.

Chan, V., Starr, P.A., Turner, R.S., 2011. Bursts and oscillations as independent properties of neural activity in the parkinsonian globus pallidus internus. Neurobiol. Dis. 41, 2–10.

Chen, C.C., Brücke, C., Kempf, F., Kupsch, A., Lu, C.S., Lee, S.T., et al., 2006. Deep brain stimulation of the subthalamic nucleus: a two-edged sword. Curr. Biol. 16, R952–R953.

Chen, C.C., Litvak, V., Gilbertson, T., Kühn, A., Lu, C.S., Lee, S.T., et al., 2007. Excessive synchronization of basal ganglia neurons at 20 Hz slows movement in Parkinson's disease. Exp. Neurol. 205, 214–221.

de Hemptinne, C., Ryapolova-Webb, E.S., Air, E.L., Garcia, P.A., Miller, K.J., Ojemann, J.G., et al., 2013. Exaggerated phase-amplitude coupling in the primary motor cortex in Parkinson disease. Proc. Natl. Acad. Sci. 110, 4780–4785.

de Hemptinne, C., Swann, N.C., Ostrem, J.L., Ryapolova-Webb, E.S., San Luciano, M., Galifianakis, N.B., et al., 2015. Therapeutic deep brain stimulation reduces cortical phase-amplitude coupling in Parkinson's disease. Nat. Neurosci.

Deuschl, G., Schade-Brittinger, C., Krack, P., Volkmann, J., Schäfer, H., Bötzel, K., et al., 2006. A randomized trial of deep-brain stimulation for Parkinson's disease. N. Engl. J. Med. 355, 896–908.

Doyle, L.M.F., Kühn, A.A., Hariz, M., Kupsch, A., Schneider, G.-H., Brown, P., 2005. Levodopa-induced modulation of subthalamic beta oscillations during self-paced movements in patients with Parkinson's disease. Eur. J. Neurosci. 21, 1403–1412.

Eusebio, A., Brown, P., 2009. Synchronisation in the beta frequency-band—the bad boy of Parkinsonism or an innocent bystander? Exp. Neurol. 217, 1–3.

Fasano, A., Aquino, C.C., Krauss, J.K., Honey, C.R., Bloem, B.R., 2015. Axial disability and deep brain stimulation in patients with Parkinson disease. Nat. Publ. Group 11, 98–110.

Fraix, V., Bastin, J., David, O., Goetz, L., Ferraye, M., Benabid, A.-L., et al., 2013. Pedunculopontine nucleus area oscillations during stance, stepping and freezing in Parkinson's disease. PLoS One. 8.

Haber, S.N., Fudge, J.L., McFarland, N.R., 2000. Striatonigrostriatal pathways in primates form an ascending spiral from the shell to the dorsolateral striatum. J. Neurosci. 20, 2369–2382.

Hammond, C., Bergman, H., Brown, P., 2007. Pathological synchronization in Parkinson's disease: networks, models and treatments. Trends Neurosci. 30, 357–364.

Hauser, R.A., McDermott, M.P., Messing, S., 2006. Factors associated with the development of motor fluctuations and dyskinesias in Parkinson disease. Arch. Neurol. Psychiatry 63, 1756–1760.

Heldman, D.A., Pulliam, C.L., Urrea Mendoza, E., Gartner, M., Giuffrida, J.P., Montgomery, E.B., et al., 2016. Computer-guided deep brain stimulation programming for Parkinson's disease. Neuromodulation 19 (2), 127–132.

Hely, M.A., Morris, J.G.L., Reid, W.G.J., Trafficante, R., 2005. Sydney Multicenter Study of Parkinson's disease: non-L-dopa-responsive problems dominate at 15 years. Mov. Disord. 20, 190–199.

Hirschmann, J., Hartmann, C.J., Butz, M., Hoogenboom, N., Ozkurt, T.E., Elben, S., et al., 2013. A direct relationship between oscillatory subthalamic nucleus-cortex coupling and rest tremor in Parkinson's disease. Brain 136, 3659–3670.

Kim, H.-J., Paek, S.H., Kim, J.-Y., Lee, J.-Y., Lim, Y.H., Kim, D.G., et al., 2009. Two-year follow-up on the effect of unilateral subthalamic deep brain stimulation in highly asymmetric Parkinson's disease. Mov. Disord. 24, 329–335.

Kühn, A.A., Trottenberg, T., Kivi, A., Kupsch, A., Schneider, G.-H., Brown, P., 2005. The relationship between local field potential and neuronal discharge in the subthalamic nucleus of patients with Parkinson's disease. Exp. Neurol. 194, 212–220.

Kühn, A.A., Kupsch, A., Schneider, G.-H., Brown, P., 2006. Reduction in subthalamic 8–35 Hz oscillatory activity correlates with clinical improvement in Parkinson's disease. Eur. J. Neurosci. 23, 1956–1960.

Kuhn, J., Hardenacke, K., Lenartz, D., Gruendler, T., Ullsperger, M., Bartsch, C., et al., 2015. Deep brain stimulation of the nucleus basalis of Meynert in Alzheimer's dementia. Mol. Psychiatry 20 (3), 353–360.

Little, S., Brown, P., 2012. What brain signals are suitable for feedback control of deep brain stimulation in Parkinson's disease? Ann. N. Y. Acad. Sci. 1265, 9–24.

Little, S., Pogosyan, A., Kühn, A.A., Brown, P., 2012. β band stability over time correlates with Parkinsonian rigidity and bradykinesia. Exp. Neurol. 236, 383–388.

Little, S., Pogosyan, A., Neal, S., Zavala, B., Zrinzo, L., Hariz, M., et al., 2013a. Adaptive deep brain stimulation in advanced Parkinson disease. Ann. Neurol. 74, 449–457.

Little, S., Tan, H., Anzak, A., Pogosyan, A., Kühn, A., Brown, P., 2013b. Bilateral functional connectivity of the Basal Ganglia in patients with Parkinson's disease and its modulation by dopaminergic treatment. PLoS One 8.

Little, S., Beudel, M., Zrinzo, L., Foltynie, T., Limousin, P., Hariz, M., et al., 2015. Bilateral adaptive deep brain stimulation is effective in Parkinson's disease. J. Neurol. Neurosurg. Psychiatr. http://dx.doi.org/10.1136/jnnp-2015-310972.

Litvak, V., Jha, A., Eusebio, A., Oostenveld, R., Foltynie, T., Limousin, P., et al., 2011. Resting oscillatory cortico-subthalamic connectivity in patients with Parkinson's disease. Brain 134, 359–374.

Litvak, V., Eusebio, A., Jha, A., Oostenveld, R., Barnes, G., Foltynie, T., et al., 2012. Movement-related changes in local and long-range synchronization in Parkinson's disease revealed by simultaneous magnetoencephalography and intracranial recordings. J. Neurosci. 32, 10541–10553.

Liu, X., Ford-Dunn, H.L., Hayward, G.N., Nandi, D., Miall, R.C., Aziz, T.Z., et al., 2002. The oscillatory activity in the Parkinsonian subthalamic nucleus investigated using the macro-electrodes for deep brain stimulation. Clin. Neurophysiol. 113, 1667–1672.

Lopez-Azcarate, J., Tainta, M., Rodriguez-Oroz, M.C., Valencia, M., Gonzalez, R., Guridi, J., et al., 2010. Coupling between beta and high-frequency activity in the human subthalamic nucleus may be a pathophysiological mechanism in Parkinson's disease. J. Neurosci. 30, 6667–6677.

Martinez-Fernandez, R., Pelissier, P., Quesada, J.-L., Klinger, H., Lhommée, E., Schmitt, E., et al., 2015. Postoperative apathy can neutralise benefits in quality of life after subthalamic stimulation for

Parkinson's disease. J. Neurol. Neurosurg. Psychiatr. http://dx.doi.org/10.1136/jnnp-2014-310189.

Mazzone, P., Lozano, A., Stanzione, P., Galati, S., Scarnati, E., Peppe, A., et al., 2005. Implantation of human pedunculopontine nucleus: a safe and clinically relevant target in Parkinson's disease. Neuroreport 16, 1877–1881.

Moreau, C., Defebvre, L., Destée, A., Bleuse, S., Clement, F., Blatt, J.L., et al., 2008. STN-DBS frequency effects on freezing of gait in advanced Parkinson disease. Neurology 71, 80–84.

Moro, E., Poon, Y.-Y.W., Lozano, A.M., Saint-Cyr, J.A., Lang, A.E., 2006. Subthalamic nucleus stimulation: improvements in outcome with reprogramming. Arch. Neurol. Psychiatry 63, 1266–1272.

Odekerken, V.J.J., van Laar, T., Staal, M.J., Mosch, A., Hoffmann, C.F.E., Nijssen, P.C.G., et al., 2013. Subthalamic nucleus versus globus pallidus bilateral deep brain stimulation for advanced Parkinson's disease (NSTAPS study): a randomised controlled trial. Lancet Neurol. 12, 37–44.

Parent, M., Parent, A., 2010. Substantia nigra and Parkinson's disease: a brief history of their long and intimate relationship. Can. J. Neurol. Sci. 37, 313–319.

Pogosyan, A., Yoshida, F., Chen, C.C., Martinez-Torres, I., Foltynie, T., Limousin, P., et al., 2010. Parkinsonian impairment correlates with spatially extensive subthalamic oscillatory synchronization. Neuroscience 171, 245–257.

Priori, A., Foffani, G., Pesenti, A., Tamma, F., Bianchi, A.M., Pellegrini, M., et al., 2004. Rhythm-specific pharmacological modulation of subthalamic activity in Parkinson's disease. Exp. Neurol. 189, 369–379.

Rivlin-Etzion, M., Marmor, O., Heimer, G., Raz, A., Nini, A., Bergman, H., 2006. Basal ganglia oscillations and pathophysiology of movement disorders. Curr. Opin. Neurobiol. 16, 629–637.

Rosa, M., Marceglia, S., Servello, D., Foffani, G., Rossi, L., Sassi, M., et al., 2010. Time dependent subthalamic local field potential changes after DBS surgery in Parkinson's disease. Exp. Neurol. 222, 184–190.

Rosa, M., Arlotti, M., Ardolino, G., Cogiamanian, F., Marceglia, S., Di Fonzo, A., et al., 2015. Adaptive deep brain stimulation in a freely moving parkinsonian patient. Mov. Disord. 30, 1003–1005.

Rosin, B., Slovik, M., Mitelman, R., Rivlin-Etzion, M., Haber, S.N., Israel, Z., et al., 2011. Closed-loop deep brain stimulation is superior in ameliorating parkinsonism. Neuron 72, 370–384.

Ryapolova-Webb, E., Afshar, P., Stanslaski, S., Denison, T., de Hemptinne, C., Bankiewicz, K., et al., 2014. Chronic cortical and electromyographic recordings from a fully implantable device: preclinical experience in a nonhuman primate. J. Neural Eng. 11, 016009.

Silberstein, P., Oliviero, A., Di Lazzaro, V., Insola, A., Mazzone, P., Brown, P., 2005a. Oscillatory pallidal local field potential activity inversely correlates with limb dyskinesias in Parkinson's disease. Exp. Neurol. 194, 523–529.

Silberstein, P., Pogosyan, A., Kühn, A.A., Hotton, G., Tisch, S., Kupsch, A., et al., 2005b. Cortico-cortical coupling in Parkinson's disease and its modulation by therapy. Brain 128, 1277–1291.

Sriram, A., Foote, K.D., Oyama, G., Kwak, J., Zeilman, P.R., Okun, M.S., 2014. Brittle dyskinesia following STN but not GPi deep brain stimulation. Tremor Other Hyperkinet. Mov. (N.Y.) 4, 242.

Stochl, J., Boomsma, A., Růžička, E., Brozova, H., Blahus, P., 2008. On the structure of motor symptoms of Parkinson's disease. Mov. Disord. 23, 1307–1312.

Su, F., Wang, J., Deng, B., Wei, X.-L., Chen, Y.-Y., Liu, C., et al., 2015. Adaptive control of Parkinson's state based on a nonlinear computational model with unknown parameters. Int. J. Neural Syst. 25, 1450030.

Tykocki, T., Nauman, P., Koziara, H., Mandat, T., 2013. Microlesion effect as a predictor of the effectiveness of subthalamic deep brain stimulation for Parkinson's disease. Stereotact. Funct. Neurosurg. 91, 12–17.

Van Den Eeden, S.K., Tanner, C.M., Bernstein, A.L., Fross, R.D., Leimpeter, A., Bloch, D.A., et al., 2003. Incidence of Parkinson's disease: variation by age, gender, and race/ethnicity. Am. J. Epidemiol. 157, 1015–1022.

van Rooden, S.M., Colas, F., Martinez-Martin, P., Visser, M., Verbaan, D., Marinus, J., et al., 2011. Clinical subtypes of Parkinson's disease. Mov. Disord. 26, 51–58.

Weinberger, M., Mahant, N., Hutchison, W.D., Lozano, A.M., Moro, E., Hodaie, M., et al., 2006. Beta oscillatory activity in the subthalamic nucleus and its relation to dopaminergic response in Parkinson's disease. J. Neurophysiol. 96, 3248–3256.

Williams, A., Gill, S., Varma, T., Jenkinson, C., Quinn, N., Mitchell, R., et al., 2010. Deep brain stimulation plus best medical therapy versus best medical therapy alone for advanced Parkinson's disease (PD SURG trial): a randomised, open-label trial. Lancet Neurol. 9, 581–591.

Yanagisawa, T., Yamashita, O., Hirata, M., Kishima, H., Saitoh, Y., Goto, T., et al., 2012. Regulation of motor representation by phase-amplitude coupling in the sensorimotor cortex. J. Neurosci. 32, 15467–15475.

Yang, A.I., Vanegas, N., Lungu, C., Zaghloul, K.A., 2014. Beta-coupled high-frequency activity and beta-locked neuronal spiking in the subthalamic nucleus of Parkinson's disease. J. Neurosci. 34, 12816–12827.

Chapter 16

Closed-Loop Neuroprosthetics ☆

A. Gharabaghi

Eberhard Karls University, Tuebingen, Germany

1 INTRODUCTION

Modulating the neural activity of the brain in a state-dependent manner for therapeutic purposes has received increasing attention. These closed-loop paradigms adapt the timing of stimulation on the basis of online recorded physiological markers and may thereby improve both the efficiency and therapeutic efficacy of standard neuromodulatory interventions, such as deep brain stimulation in Parkinson's disease (Little et al., 2013; Rosin et al., 2011).

By introducing this concept into the field of neurorehabilitation, even functional restoration of persistent deficits after stroke, which represents an unmet goal in many patients despite intensive rehabilitation programs, may come within reach. Previous experimental findings in animal models suggest that activity-dependent stimulation of the nervous system may strengthen neuronal connectivity by inducing Hebbian-like plasticity (Jackson et al., 2006; Rebesco et al., 2010; Guggenmos et al., 2013).

Thus, closed-loop neuroprosthetics represents an emerging field at the interface between neuroscience, neuroengineering, neurotechnology, neurocomputing, neurology, and neurosurgery. Accordingly, the terminology in this field has been quite diverse up till now.

2 TERMINOLOGY

2.1 Closing the Loop

When the term *closed-loop* is used in the context of stimulation, it is usually contrasted with *open-loop* interventions, which apply input to the nervous system in a predefined, continuous way independent of the brain's current state. As soon as neural stimulation is linked to output signals, spanning from single neuron or neural population activity to movement and behavior, a broad spectrum of terms is used in parallel to *closed-loop* to characterize this approach: *recurrent, triggered, controlled, responsive, bidirectional, state-dependent, activity-dependent, adaptive,*

feedback, autonomous, or *intelligent.* Currently, there is no systematic differentiation between these terms, which are often used on the basis of personal preferences.

Recording and stimulation sites of closed-loop systems might be either in the immediate vicinity less than millimeters apart, or quite distant from each other by bridging several neurofunctional systems. While a temporal contingency between input and output is required, lasting usually from milliseconds to seconds, the parameters of the stimulation itself are often predefined and not determined by the recorded response, unlike classical concepts of control engineering.

2.2 Neuroprosthetics

While different terms are used to describe the concept of closed-loop interventions, the opposite phenomenon holds true for the term *neuroprosthetics.* The very same term might be used to define quite diverse devices and tools such as (i) a classical prosthesis, ie, the replacement of a missing limb, which is controlled by neural signals, or to describe (ii) systems that are assisting impaired, but still existing, parts of the body similar to a robotic orthosis or exoskeleton, or even for (iii) devices that are connected to neural structures only and not to the periphery for pure recording and/or stimulation purposes.

Neuroprosthetics has predominantly been associated with neural implants, but indirect connections to neural structures, eg, via electroencephalography (EEG), have recently been termed this way as well. Therefore, many different devices meant to support, replace, or augment biological functionality could be summarized under this umbrella term; this may also include other, mutually overlapping terms such as brain–computer interfaces, brain–machine interfaces, or brain–robot interfaces. These devices may be differentiated on the basis of their invasiveness or by the specific characteristics of their respective peripheral actuators, eg, reaching from visual feedback on a computer screen to multidimensional control of a robotic reaching movement. Thereby, these neural interfaces, which have primarily been implemented for recording

Closed Loop Neuroscience. http://dx.doi.org/10.1016/B978-0-12-802452-2.00016-0

purposes, eg, to control external assistive tools, may turn into stimulation devices as well, by indirectly modulating the brain physiology via the sensory input and feedback signals generated by the controlled actuators. Thus, neuroprosthetic devices may close the loop on different functional system levels.

2.3 Assistance Versus Restoration

A better understanding of the interaction between brain and technology within recurrent feedback loops has led to a paradigm shift in the field of closed-loop neuroprosthetics. Past research was dominated by the concept of *assistive* technology, which intends to compensate for lost function, eg, by controlling external devices such as upper extremity prostheses or wheelchairs. More recent approaches focus on *rehabilitative* technology that aims at restoring function by providing tools for strengthening disturbed connections or facilitating motor learning within training environments, eg, with rehabilitation robots.

This change in goals necessitates conceptual modifications as well, even when applying the same brain-interface-based technology. Research in the field of *assistive* neuroprosthetics has focussed on maximizing the speed and classification accuracy of this technology in differentiating brain states for high-dimensional control of external actuators, ie, on influencing the *outside world*. In contrast, *restorative* neuroprosthetics aims at achieving neuroplastic changes either by (i) directly strengthening neural connections according to Hebbian mechanisms, when the brain is particularly responsive to external input, or by (ii) operant conditioning of brain states, which are regarded as relevant for neurorestoration, via neurofeedback and reinforcement learning. Thus, these approaches intend to modulate the *inside world* of the brain. Such an ambitious objective, however, requires a better understanding of the underlying physiology and interaction between the human brain and the applied technology.

3 BRAIN–TECHNOLOGY INTERACTION

Closing the sensorimotor loop and providing motor imagery-related proprioceptive/haptic feedback with a brain–robot interface facilitates—in both healthy subjects and stroke patients—the decoding of those brain states, which are controlling the neuroprosthetic device (Gomez-Rodriguez et al., 2011). Therefore, both healthy subjects and stroke survivors show similar patterns, but different aptitudes, of controlling three-dimensional robotic assistance for reaching movements with a multi-joint exoskeleton during motor imagery-related brain self-regulation (Brauchle et al., 2015). However, using these neuroprosthetic training devices based on brain self-regulation and neurofeedback is challenging, ie, characterized by a significant association with the experience of frustration, even for young healthy subjects, when they apply hand robots with straight forward opening/closing functions only (Fels et al., 2015). In order to optimize motor learning with this closed-loop technology, knowledge about the participant's individual cognitive load is essential to adapt the difficulty of the intervention to the respective mental resources (Bauer and Gharabaghi, 2015a). Adjustment of the training difficulty of this rehabilitation technology, eg, by adapting the classifier threshold of restorative brain-technology interfaces, has the potential to improve reinforcement learning, particularly in participants with limited capability for brain self-regulation, eg, following stroke (Bauer and Gharabaghi, 2015b). More specifically, adaptive threshold-setting based on the participant's perceived mental effort has been shown to facilitate reinforcement learning with restorative neuroprosthetics for brain state-dependent neurofeedback (Bauer et al., 2015a).

Complementary neuroprosthetic approaches apply gravity-compensating multi-joint exoskeletons together with virtual reality feedback (Grimm et al., 2016) and adaptive closed-loop neuromuscular stimulation (Grimm and Gharabaghi, 2016) to enhance the upper extremity movement range of severely impaired stroke patients; this strategy provides assistance as needed while preserving the engagement of the participant.

4 CORTICAL NETWORK PHYSIOLOGY

Closing the loop with a brain–robot neuroprosthesis bridges the gap between the cortical networks and abilities of motor imagery and motor execution. They may be predicted, moreover, with high specificity from resting state cortical connectivity patterns *before* the intervention (Bauer et al., 2015b). The brain self-regulation *during* the closed-loop intervention is regulated by the coupling of distant cortical areas. This physiological pattern allows differentiating between good and poor performance of volitional brain modulation when using these neuroprostheses (Vukelić et al., 2014). The comparison between different closed-loop approaches revealed that proprioceptive feedback with a brain–robot interface was more suitable than visual feedback with a brain–computer interface to entrain the motor learning network during neuroprosthetic training, thereby resulting in better volitional control of regional brain activity (Vukelić and Gharabaghi, 2015a). Furthermore, this intervention has modulated the cortical connectivity of *subsequent* resting state networks in a spatially selective and frequency-specific way, thus conforming to a Hebbian-like sharpening concept (Vukelić and Gharabaghi, 2015b).

5 CORTICO-MUSCULAR CONNECTIVITY

Projecting navigated transcranial magnetic stimulation sites on the gyral anatomy allows for precise group analysis, by decreasing inter-subject variability of cortical motor maps, and provides thereby, a valuable tool for monitoring the functional topography of cortico-muscular connectivity in humans (Kraus and Gharabaghi, 2015). Applying this technique, a complex pattern of modulated corticospinal excitability following the application of a brain–robot neuroprosthesis could be detected (Kraus et al., 2015). More specifically, this intervention induces a re-distribution of corticospinal excitability with a decrease of excitability in the hand area of the primary motor cortex, which controlled the neuroprosthesis, but an increase of excitability in the surrounding somatosensory and premotor cortex via synchronization of neuronal firing. The increased excitability in the somatosensory area correlated, moreover, with the amount of brain-self regulation, ie, beta-band desynchronization, achieved by the participants (Kraus et al., 2015). Pilot data suggested that the very same closed-loop intervention, ie, neuroprosthetic brain–robot feedback of sensorimotor beta-band desynchronization, facilitated the reinforcement learning of frequency-specific brain self-regulation—when performed repetitively for several weeks, while continuously adapting the training difficulty—and led to muscle-specific motor improvement in chronic stroke (Naros and Gharabaghi, 2015).

6 BRAIN STATE-DEPENDENT STIMULATION

Bi-hemispheric transcranial electric stimulation prior to training with assistive rehabilitation technology may improve both motor learning and consolidation of exoskeleton-based arm movements (Naros et al., 2016a). Such a general priming of subsequent rehabilitation exercises may, however, not be sufficient to restore motor function in severely affected patients with persistent deficits. Such conditions might necessitate more targeted—in time and space—strengthening of weakened corticospinal connectivity. Applying single pulse stimulation during rehabilitation training may be one suitable way to achieve this goal (Edwardson et al., 2013). Determining the optimal time point for neuroplastic modulation during an exercise could, however, be highly dependent on the executed task. This suggests that the corresponding brain state may represent the actual physiological measure to be considered for the timing of activity-dependent cortical stimulation. For this purpose, neuroprosthetic brain-interfaces seem to offer suitable tools for novel activity-dependent stimulation paradigms based on intrinsic brain activity, ie, providing brain state-dependent stimulation or even pairing cortical with afferent input by robotic devices (Gharabaghi, 2015) or neuromuscular electrical stimulation (Royter and Gharabaghi, 2016). Along these lines, it has been demonstrated that the increased

modulation range in the beta-band correlates in a frequency-specific way with motor gains following neuroprosthetic learning and may, thus, serve as the biomarker for state-dependent interventions (Naros et al., 2016b). Accordingly, closed-loop transcranial magnetic stimulation, when applied during beta-band desynchronization induced a robust increase of corticomuscular excitability of the somatosensory *and* primary motor cortex, an effect that was not observed when the same number and pattern of stimuli were applied independent of the brain state (Kraus et al., 2016). Closing the loop between intrinsic brain state, cortical stimulation, and proprioceptive input in a neuroprosthetic brain–robot environment may thereby provide a novel neurorehabilitation strategy for stroke patients lacking residual hand function. Pilot data in the persistent motor deficit condition in chronic stroke suggests that this closed-loop neuroprosthetic approach may induce corticospinal plasticity, ie, turning the absence of motor evoked responses (MEP < 50 μV) at baseline into the presence of MEP in the ipsilesional primary motor cortex hand representation (Gharabaghi et al., 2014a).

7 BI-DIRECTIONAL NEURAL INTERFACE

Neuroprosthetic approaches based on neural implants may—thanks to their proximity to the neural signal source—be able to surmount difficulties of non-invasive techniques, such as EEG, related to a characteristically low spatial resolution, low signal-to-noise ratio because of signal attenuation caused by the skull, possible contamination by muscle artifacts or external electrical activity (Gharabaghi et al., 2014b). We have recently proposed a new neuroprosthetic technique for human application, which is less invasive than the classical implanted approaches with subdural grids (Yanagisawa et al., 2011, 2012; Wang et al., 2013) or even brain penetrating electrodes (Hochberg et al., 2012; Collinger et al., 2013). This novel approach entailed the application of *epidural* electrocorticography (eECoG) to decode volitional brain activity in patients with locked-in syndrome suffering from amyotrophic lateral sclerosis (Bensch et al., 2014), with chronic pain as a result of upper limb amputation (Gharabaghi et al., 2014c), and with hemiparesis following subcortical (Gharabaghi et al., 2014d) and cortical stroke (Gharabaghi et al., 2014b). The eECoG technique may record proprioceptive input to the brain even in severe neurodegenerative conditions (Murguialday et al., 2011) and detect cognitive processing in the absence of voluntary muscle control (Bensch et al., 2014). This approach is also suitable for long-term monitoring of state-dependent dominant frequencies and capturing physiological biomarkers of neuromodulatory interventions (Martens et al., 2014). It may, furthermore, serve as a bi-directional interface for both brain signal recording and electrical stimulation revealing mutually overlapping cortical representations, even in the absence of volitional hand movement (Gharabaghi et al., 2014c). Moreover, epicortical electrical stimulation allows disentangling with high-precision cortical areas, which are relevant for different behavior within the same functional domain (Becker et al., 2013). The simultaneous application of these features has become possible by using methods for spectral estimation in the presence of stimulation after-effects, thereby allowing combined recording and stimulation for brain state-dependent modulation (Walter et al., 2013). Such closed-loop neural implants may detect and train brain activity with epidural brain–robot neuro-prostheses even when the cortical physiology is distorted following severe brain injury, eg, extended stroke lesions, allowing for reinforcement learning of preserved neural networks (Gharabaghi et al., 2014b). Furthermore, this technique has demonstrated to decode in severely para-lyzed chronic stroke patients up to seven hand movement intentions in the absence of volitional hand control (Spüler et al., 2014). Finally, eECoG-based closed-loop neuroprosthetic training may facilitate reinforcement learning of specific brain-states during short training periods; pilot data suggests that this may result in limited functional improvement even in the severely affected chronic stroke condition (Gharabaghi et al., 2014d). However, none of the closed-loop techniques, which are currently available, has led to functionally relevant movement restoration in chronic stroke patients with per-sistent deficits.

The technological developments in the field of closed-loop neuroprosthetics provide us with powerful tools and options for neural modulation. Translating restorative con-cepts into effective clinical practice poses, however, an ongoing challenge. How and where should we stimulate? What are the appropriate physiological markers for guiding the stimulation? Should we apply direct and spatially spe-cific or rather indirect and more global modulation of the brain and corticospinal connections?

For addressing these questions, we need to explore the methodological opportunities in a meaningful way in order to achieve real benefits for our patients. This will neces-sitate a better neurophysiological understanding of the underlying mechanisms and a continuous exchange of the different disciplines contributing to this field.

ACKNOWLEDGMENTS

AG was supported by grants from the German Research Council [DFG GH 94/2-1, DFG EC 307], from the Federal Ministry of Education and Research [BFNT 01GQ0761, BMBF 16SV3783, BMBF 0316064B, BMBF 16SV5824], and from the European Union [ERC 227632].

REFERENCES

Bauer, R., Gharabaghi, A., 2015a. Estimating cognitive load during self-regulation of brain activity and neurofeedback with therapeutic brain-computer interfaces. Front. Behav. Neurosci. 9, 21.

Bauer, R., Gharabaghi, A., 2015b. Reinforcement learning for adaptive threshold control of restorative brain-computer interfaces: a Bayesian simulation. Front. Neurosci. 9, 36.

Bauer, R., Fels, M., Royter, V., Raco, V., Gharabaghi, A., 2015a. Adaptive threshold-setting based on mental effort facilitates reinforcement learning of restorative brain-robot interfaces for brain state-dependent neurofeedback (in revision).

Bauer, R., Fels, M., Vukelić, M., Ziemann, U., Gharabaghi, A., 2015b. Bridging the gap between motor imagery and motor execution with a brain–robot interface. Neuroimage 108, 319–327.

Becker, H.G., Haarmeier, T., Tatagiba, M., Gharabaghi, A., 2013. Elec-trical stimulation of the human homolog of the medial superior tem-poral area induces visual motion blindness. J. Neurosci. 33 (46), 18288–18297.

Bensch, M., Martens, S., Halder, S., Hill, J., Nijboer, F., Ramos, A., Birbaumer, N., Bogdan, M., Kotchoubey, B., Rosenstiel, W., Schölkopf, B., Gharabaghi, A., 2014. Assessing attention and cog-nitive function in completely locked-in state with event-related brain potentials and epidural electrocorticography. J. Neural Eng. 11(2).

Brauchle, D., Vukelić, M., Bauer, R., Gharabaghi, A., 2015. Brain state-dependent robotic reaching movement with a multi-joint arm exoskeleton: combining brain-machine interfacing and robotic reha-bilitation. Front. Hum. Neurosci. 9, 564.

Collinger, J.L., Wodlinger, B., Downey, J.E., Wang, W., Tyler-Kabara, E.C., Weber, D.J., McMorland, A.J.C., Velliste, M., Boninger, M.L., Schwartz, A.B., 2013. High-performance neuropros-thetic control by an individual with tetraplegia. Lancet 381 (9866), 557–564.

Edwardson, M.A., Lucas, T.H., Carey, J.R., Fetz, E.E., 2013. New modal-ities of brain stimulation for stroke rehabilitation. Exp. Brain Res. 224 (3), 335–358.

Fels, M., Bauer, R., Gharabaghi, A., 2015. Predicting workload profiles of brain–robot interface and electromyographic neurofeedback with cor-tical resting-state networks: personal trait or task-specific challenge? J. Neural Eng. 12 (4), 046029.

Gharabaghi, A., 2015. Activity-dependent brain stimulation and robot-assisted movements for use-dependent plasticity. Clin. Neurophysiol. 126 (5), 853.

Gharabaghi, A., Kraus, D., Leão, M.T., Spüler, M., Walter, A., Bogdan, M., Rosenstiel, W., Naros, G., Ziemann, U., 2014a. Coupling brain-machine interfaces with cortical stimulation for brain-state dependent stimulation: enhancing motor cortex excitability for neurorehabil-itation. Front. Hum. Neurosci. 8, 122.

Gharabaghi, A., Naros, G., Khademi, F., Jesser, J., Spüler, M., Walter, A., Bogdan, M., Rosenstiel, W., Birbaumer, N., 2014b. Learned self-regulation of the lesioned brain with epidural electrocorticography. Front. Behav. Neurosci. 8, 429.

Gharabaghi, A., Naros, G., Walter, A., Roth, A., Bogdan, M., Rosenstiel, W., Mehring, C., Birbaumer, N., 2014c. Epidural electro-corticography of phantom hand movement following long-term upper-limb amputation. Front. Hum. Neurosci. 8, 285.

Gharabaghi, A., Naros, G., Walter, A., Grimm, F., Schuermeyer, M., Roth, A., Bogdan, M., Rosenstiel, W., Birbaumer, N., 2014d. From assistance towards restoration with epidural brain-computer interfacing. Restor. Neurol. Neurosci. 32 (4), 517–525.

Grimm, F., Gharabaghi, A., 2016. Closed-loop neuroprosthesis for reach-to-grasp assistance: combining adaptive multi-channel neuromuscular stimulation with a multi-joint arm exoskeleton. Front. Neurosci. 10, 284. http://dx.doi.org/10.3389/fnins.2016.00284.

Grimm, F., Naros, G., Gharabaghi, A., 2016. Compensation or restoration: closed-loop feedback of movement quality for assisted reach-to-grasp exercises with a multi-joint arm exoskeleton. Front. Neurosci. 10, 280. http://dx.doi.org/10.3389/fnins.2016.00280.

Gomez-Rodriguez, M., Peters, J., Hill, J., Schölkopf, B., Gharabaghi, A., Grosse-Wentrup, M., 2011. Closing the sensorimotor loop: haptic feedback facilitates decoding of motor imagery. J. Neural Eng. 8 (3), 036005.

Guggenmos, D.J., Azin, M., Barbay, S., Mahnken, J.D., Dunham, C., Mohseni, P., Nudo, R.J., 2013. Restoration of function after brain damage using a neural prosthesis. Proc. Natl. Acad. Sci. 110 (52), 21177–21182.

Hochberg, L.R., Bacher, D., Jarosiewicz, B., Masse, N.Y., Simeral, J.D., Vogel, J., Haddadin, S., Liu, J., Cash, S.S., van der Smagt, P., Donoghue, J.P., 2012. Reach and grasp by people with tetraplegia using a neurally controlled robotic arm. Nature 485 (7398), 372–375.

Jackson, A., Mavoori, J., Fetz, E.E., 2006. Long-term motor cortex plasticity induced by an electronic neural implant. Nature 444 (7115), 56–60.

Kraus, D., Gharabaghi, A., 2015. Projecting navigated TMS sites on the Gyral anatomy decreases inter-subject variability of cortical motor maps. Brain Stimul. 8 (4), 831–883.

Kraus, D., Naros, G., Bauer, R., Leão, M.T., Ziemann, U., Gharabaghi, A., 2015. Brain–robot interface driven plasticity: distributed modulation of corticospinal excitability. NeuroImage 125, 522–532.

Kraus, D., Naros, G., Bauer, R., Leão, M.T., Ziemann, U., Gharabaghi, A., 2016. Brain state-dependent transcranial magnetic closed-loop stimulation controlled by sensorimotor desynchronization induces robust increase of corticospinal excitability. Brain Stimul. 9 (3), 415–424. http://dx.doi.org/10.1016/j.brs.2016.02.007.

Little, S., Pogosyan, A., Neal, S., Zavala, B., Zrinzo, L., Hariz, M., Foltynie, T., Limousin, P., Ashkan, K., FitzGerald, J., Green, A.L., Aziz, T.Z., Brown, P., 2013. Adaptive deep brain stimulation in advanced Parkinson disease. Ann. Neurol. 74 (3), 449–457.

Martens, S., Bensch, M., Halder, S., Hill, J., Nijboer, F., Ramos-Murguialday, A., Schoelkopf, B., Birbaumer, N., Gharabaghi, A., 2014. Epidural electrocorticography for monitoring of arousal in locked-in state. Front. Hum. Neurosci. 8, 861.

Murguialday, A.R., Hill, J., Bensch, M., Martens, S., Halder, S., Nijboer, F., Schoelkopf, B., Birbaumer, N., Gharabaghi, A., 2011. Transition from the locked in to the completely locked-in state: a physiological analysis. Clin. Neurophysiol. 122 (5), 925–933.

Naros, G., Gharabaghi, A., 2015. Reinforcement learning of self-regulated β-oscillations for motor restoration in chronic stroke. Front. Hum. Neurosci. 9, 391.

Naros, G., Geyer, M., Koch, S., Mayr, L., Ellinger, T., Grimm, F., Gharabaghi, A., 2016a. Enhanced motor learning with bilateral transcranial direct current stimulation: impact of polarity or current flow direction? Clin. Neurophysiol. 127 (4), 2119–2126. http://dx.doi.org/10.1016/j.clinph.2015.12.020.

Naros, G., Naros, I., Grimm, F., Ziemann, U., Gharabaghi, A., 2016b. Reinforcement learning of self-regulated sensorimotor β-oscillations improves motor performance. Neuroimage 2 (134), 142–152. http://dx.doi.org/10.1016/j.neuroimage.2016.03.016.

Rebesco, J.M., Stevenson, I.H., Körding, K.P., Solla, S.A., Miller, L.E., 2010. Rewiring neural interactions by micro-stimulation. Front. Syst. Neurosci. 4, 39.

Rosin, B., Slovik, M., Mitelman, R., Rivlin-Etzion, M., Haber, S.N., Israel, Z., Vaadia, E., Bergman, H., 2011. Closed-loop deep brain stimulation is superior in ameliorating parkinsonism. Neuron 72 (2), 370–384.

Royter, V., Gharabaghi, A., 2016. Brain state-dependent closed-loop modulation of paired associative stimulation controlled by sensorimotor desynchronization. Front Cell Neurosci. 10 (10), 115. http://dx.doi.org/10.3389/fncel.2016.00115.

Spüler, M., Walter, A., Ramos-Murguialday, A., Naros, G., Birbaumer, N., Gharabaghi, A., Rosenstiel, W., Bogdan, M., 2014. Decoding of motor intentions from epidural ECoG recordings in severely paralyzed chronic stroke patients. J. Neural Eng. 11 (6).

Vukelić, M., Bauer, R., Naros, G., Naros, I., Braun, C., Gharabaghi, A., 2014. Lateralized alpha-band cortical networks regulate volitional modulation of beta-band sensorimotor oscillations. Neuroimage 87, 147–153.

Vukelić, M., Gharabaghi, A., 2015a. Oscillatory entrainment of the motor cortical network during motor imagery is modulated by the feedback modality. NeuroImage 111, 1–11.

Vukelić, M., Gharabaghi, A., 2015b. Self-regulation of circumscribed brain activity modulates spatially selective and frequency specific connectivity of distributed resting state networks. Front. Behav. Neurosci. 9, 181.

Walter, A., Murguialday, A.R., Spüler, M., Naros, G., Leão, M.T., Gharabaghi, A., Rosenstiel, W., Birbaumer, N., Bogdan, M., 2013. Coupling BCI and cortical stimulation for brain-state-dependent stimulation: methods for spectral estimation in the presence of stimulation after-effects. Front. Neural Circuit. 7, 6.

Wang, W., Collinger, J.L., Degenhart, A.D., Tyler-Kabara, E.C., Schwartz, A.B., Moran, D.W., Wodlinger, B., Vinjamuri, R.K., Ashmore, R.C., Kelly, J.W., Boninger, M.L., 2013. An electrocorticographic brain interface in an individual with tetraplegia. PLoS One. 8(2).

Yanagisawa, T., Hirata, M., Saitoh, Y., Goto, T., Kishima, H., Fukuma, R., Yokoi, H., Kamitani, Y., Yoshimine, T., 2011. Real-time control of a prosthetic hand using human electrocorticography signals: technical note. J. Neurosurg. 114 (6), 1715–1722.

Yanagisawa, T., Hirata, M., Saitoh, Y., Kishima, H., Matsushita, K., Goto, T., Fukuma, R., Yokoi, H., Kamitani, Y., Yoshimine, T., 2012. Electrocorticographic control of a prosthetic arm in paralyzed patients. Ann. Neurol. 71 (3), 353–361.

Chapter 17

Closed-Loop Stimulation in Emotional Circuits for Neuro-Psychiatric Disorders

A.S. Widge*,† and C.T. Moritz‡

*Massachusetts General Hospital, Charlestown, MA, United States, †Picower Institute for Learning & Memory, Massachusetts Institute of Technology, Cambridge, MA, United States, ‡University of Washington, Seattle, WA, United States

1 INTRODUCTION

Electrical stimulation of the nervous system is a valuable clinical tool. Therapies such as vagus nerve stimulation (VNS) for epilepsy and deep brain stimulation (DBS) for Parkinson's disease are standards of care for neurologic disorders (Ben-Menachem, 2002; Bronstein and Tagliati, 2011). Electrical stimulation of the spinal cord, peripheral nerves, and muscles can improve movement or restore sensation after nerve or spinal cord injury or limb loss (Ho et al., 2014; Moritz et al., 2008; Tan et al., 2014). VNS, DBS, and related technologies have also been tested for psychiatric disorders, but the results have not been as promising. Despite early success in open-label human studies, these technologies have not shown consistent clinical benefit. A fundamental problem is that they treat mental disorders as static entities that can be managed with constant, open-loop therapy. In that context, these failures may not be surprising—it is more surprising that open-loop works so well for movement disorders. This apparent failure highlights the potential opportunity for a closed-loop approach for psychiatric disorders.

In this chapter, we review the state of the art and challenges of psychiatric neurostimulation. The major barrier to closed-loop psychiatric stimulation is a lack of reliable symptom biomarkers. We review work that suggests paths to biomarker discovery. Finally, we consider the hardware that will be needed to detect and respond to those signals. Existing platforms should be adequate for the first generation of closed-loop psychiatric devices, but new approaches may expand clinical prospects.

2 PSYCHIATRIC BRAIN STIMULATION— STATE OF THE ART AND CHALLENGES

Across diagnoses, medications have limited efficacy, generally 30–50% (Gaynes et al., 2009; Jonas et al., 2013; Pittenger and Bloch, 2014). Talk therapy can do better,

particularly in the anxiety disorders (Ellard et al., 2010; Jonas et al., 2013; Powers et al., 2010). Unfortunately, well-trained therapists are difficult for many patients to access. Even when standard treatments help, the overall morbidity and mortality of mental illness have not improved in decades (Cuthbert and Insel, 2013; Insel, 2009, 2010). More recently, the neuroscientific community's conception of mental illness has shifted. Thought leaders now argue that mental disorders represent disrupted information processing in distributed circuits (Cuthbert and Insel, 2013; Etkin, 2012a; Etkin and Wager, 2007; Insel, 2010, 2012). Most biologic therapies, particularly medication, do not act directly on these circuits. They globally alter brain function, perhaps incidentally achieving a desirable effect in pathologic networks. Electrical and magnetic brain stimulation, by contrast, can focally modulate many putative nodes in the networks of mental illness (Rosa and Lisanby, 2012). The most specific of these is arguably DBS, which can specifically access the deep nuclei believed to drive emotional experiences.

At the turn of the 21st century, multiple investigators tested DBS in the limbic circuit, either following classical lesion targets for obsessive-compulsive disorder (OCD) (Nuttin et al., 1999), or identifying new targets with functional imaging in major depressive disorder (MDD) (Mayberg et al., 2005). Early results were promising, with response rates up to 90% (Greenberg et al., 2010; Holtzheimer et al., 2012; Malone et al., 2009). Just a few years later, well-controlled clinical trials of DBS for MDD failed to meet clinical endpoints (Dougherty et al., 2015; Morishita et al., 2014). VNS for MDD did little better, failing to separate from sham stimulation in a key trial (Rush et al., 2005). Transcranial magnetic stimulation (TMS) for MDD separated from placebo, but was not as effective in treatment-resistant patients (George et al., 2010). As of this writing, the only invasive brain stimulation treatment available in the US market is open-loop DBS for OCD, which holds an FDA Humanitarian Device

Closed Loop Neuroscience. http://dx.doi.org/10.1016/B978-0-12-802452-2.00017-2

Exemption due to the very small number of patients who qualify (Garnaat et al., 2014).

Those difficulties have multiple roots. The anatomy is heterogenous; standard image-guided targeting may not reach the right structure in many patients (Riva-Posse et al., 2014). There are powerful placebo effects of high-technology/high-cost procedures that may obscure true clinical effects. The largest problem, however, is that none of the studies provably engaged any functional target. While electrical energy was clearly delivered into the limbic circuit, that delivery followed *ad hoc* clinical rules. There was neither measurement of the brain's response nor therapy titration to achieve such response (Widge et al., 2015b; Widge and Dougherty, 2015). Hence, the case for closed-loop brain stimulation, and especially DBS, in psychiatry. Until target engagement is clearly shown, trials may continue to fail because clinicians cannot be sure that they are accurately targeting the pathologic circuits.

3 CLOSED-LOOP SYSTEMS AS THE NEXT STEP FORWARD

We define a closed-loop stimulator as an implanted device that senses the brain's electrical activity through one or more electrodes (sensing) and delivers electrical stimulation through the same electrode or at a different site (actuation). In Fig. 1, we illustrate a version where the feedback loop is provided by a controller and algorithm that automatically adjust stimulation based on observed electrical signals. This "detect and treat" paradigm has succeeded in epilepsy (Morrell, 2011). A simpler version would be "human in the loop," where the sensors provide a clinician with feedback on the results of his/her manual stimulation

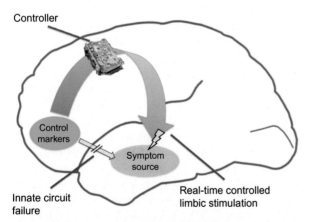

FIG. 1 Schematic of closed-loop implantable neurostimulation for psychiatric indications. The core deficit is a disconnection or dysfunction in an emotion regulatory circuit (*broken arrow*). The closed-loop controller senses an electrical biomarker corresponding to that deficit, possibly from cortex, then delivers neurostimulation to a deep structure based on a real-time algorithm.

adjustment. While this may not achieve fine-grained symptom control, it would still be a major advance (Widge et al., 2015a).

3.1 Clinical Efficacy for Time-Varying Symptoms

Closed-loop psychiatric DBS could have substantial advantages. First, it likely would produce better symptom control. In Parkinson's disease, therapy that self-titrated based on neural signals produced better motor outcomes in monkeys and humans (Little et al., 2013; Rosin et al., 2011). Closed-loop stimulation may also be critical for tracking the relatively rapid symptom changes of mental disorders. Panic, anxiety, obsessions, and mood all can change with little warning. In many disorders, patients experience brief, but intense symptom flares (American Psychiatric Association, 2013). It might be possible to deliver such a high "dose" of open-loop stimulation that patients experience very few acute flares, but this is not a true solution. It would be the electrical equivalent of keeping a patient on high doses of sedatives, and might cause the same unpleasant emotional numbing (Widge et al., 2014). A closed-loop system, on the other hand, should be able to track and tightly control symptoms without "overtreating."

3.2 Patient Tolerability

Closed-loop limbic stimulation may also reduce side effects. Beyond the possibility of emotional numbing, psychiatric DBS causes hypomania in roughly 50% of patients (Goodman et al., 2010; Greenberg et al., 2010; Haq et al., 2010; Widge et al., 2015c). That syndrome, characterized by impulsive actions, distractibility, and irritability can be highly detrimental to patients' functioning. If DBS systems could deliver less total energy via closed-loop operation, they could widen the "therapeutic window" and reduce the incidence of hypomania. Lowering the total energy also lengthens battery life. Even rechargeable DBS batteries require surgical replacements, and increasing the time between changes would improve patients' quality of life.

On the other hand, patients may have difficulties with closed-loop limbic DBS. If the system does suppress symptoms in real time, this implies that it can drive a patient's emotional state to an arbitrary setpoint. In other words, the patient would only be "allowed" to feel a prespecified range of emotions, and that specification would be under a clinician's control. This may be considered "mind control," since the clinician can effectively deny the patient his/her most positive or negative experiences. In some situations, that paternalism is ethical under the principle of nonmaleficence. For instance, patients who have experienced hypomania sometimes request DBS increases that would provoke that state, because they enjoy the mild euphoria (and do not recognize the profound negative consequences). This is similar to the impulse-control and dopamine-dysregulation problems encountered with DBS for Parkinson's (Samuel et al., 2015). In other situations, a well-meaning physician could inadvertently blunt a patient's emotional experience, cutting him/her off from the fullness of life. The boundary between these scenarios is not always clear-cut. Below, we discuss ways in which this may require us to rethink closed-loop strategies in psychiatry, either through giving control to the patient or through emphasizing DBS as an assistive technology to enhance self-regulation.

3.3 Long-Term Neuroplasticity to Repair Diseased Circuits

A closed-loop system links the activity of one brain area to another. If that linkage is correctly designed, it cannot only correct abnormal network fluctuations, but also "re-wire" circuits into their healthy state. The best-known examples are in motor control, using a system called the Neurochip (Fig. 2). This programmable electronic system detects events (a single neuron firing or a marked change in the local field potential (LFP)) and triggers brief stimulation (Mavoori et al., 2005; Zanos et al., 2011). When two neurons in macaque motor cortex were coupled together for 48 h through a Neurochip, the recorded (Nrec) cell became connected to the stimulated (Nstim) cell in such a way that driving Nrec could produce Nstim's motor output (Jackson et al., 2006). The plasticity lasted for many days after this relatively brief intervention. Subsequent studies extended this result to corticospinal connections, although they did not find the same long-duration plasticity (Nishimura et al., 2013). As we discuss further below, this plasticity approach may be directly applicable to the problem of closed-loop limbic stimulation, particularly for anxiety disorders.

4 PSYCHIATRIC BIOMARKER DEVELOPMENT IS A CRITICAL BARRIER

The largest barrier to closed-loop limbic stimulation is not hardware or control algorithms; it is sensing. Despite extensive investigations, there is no electrophysiologic signature that closely or reliably tracks the symptoms of any known mental disorder. Small studies have identified many candidates, but those studies often fail to replicate in independent cohorts (Whelan and Garavan, 2013; Widge et al., 2013). Single studies have identified possible response signatures for DBS for MDD at Brodmann area 25 (Broadway et al., 2012), and one for ventral capsule/ventral striatum (VC/VS) DBS for OCD (Figee et al., 2013). The former will be difficult to replicate, as its investigators have almost

exclusive access to patients with DBS at their target; the latter was a neuro-imaging study that is not yet correlated to electrophysiology. Thus, much work remains to identify markers that can be sensed and controlled. Two distinct strategies may break this barrier. Both arise from the same postulate: if there is no known marker of disorder-specific symptoms, then disorders and diagnoses may be the wrong framework for the search.

5 OVERCOMING BIOMARKERS—CROSS-DIAGNOSTIC NEUROSCIENCE APPROACHES

Psychiatric diagnoses are heterogeneous and often unreliable. MDD, one of the most common, had an interrater reliability (k) of 0.25 in field trials of the most recent diagnostic manual (Regier et al., 2013). Other diagnoses were more reliable, but many still showed $k < 0.4$. Atop this, each categorical diagnosis likely represents multiple neurobiologic entities (Cuthbert and Insel, 2013; Insel and Wang, 2010). If that is true, then our failure to find biomarkers for loop closure is no surprise. If we perform imaging or EEG studies of five different disorders and average the results together, the most likely outcome is zero. That is, unfortunately, what the psychiatric neuroscience community has done. In contrast, leaders at the National Institutes of Health have proposed a new framework: studying not disorders, but cross- or trans-diagnostic "Research Domain Criteria" (Cuthbert and Insel, 2013; Insel and Wang, 2010). These objectively measurable behaviors should link more tightly to human and animal neuroscience, and thus should be more amenable to computational modeling (Wang and Krystal, 2014; Widge et al., 2015b).

For psychiatric DBS, disorder-based studies might be used to seed that cross-diagnostic neuroscience. Most DBS targets have been studied for a single disorder (Mayberg et al., 2005; Schlaepfer et al., 2013). The notable exception is VC/VS, a site derived from the lesion target for capsulotomy (Greenberg et al., 2010; Nuttin et al., 1999). That single lesion treats both OCD and MDD (Dougherty et al., 2002; Yang et al., 2013). DBS at VC/VS also can improve both MDD and OCD, although the MDD randomized trial was statistically nonsignificant (Dougherty et al., 2015; Greenberg et al., 2010; Malone et al., 2009). If a single target is effective in both disorders, then there is likely a common neural substrate for at least some cases. By studying VC/VS DBS patients across diagnoses, we may identify useful biomarkers.

MDD and OCD are both characterized by "stuck" thinking (American Psychiatric Association, 2013). Top-down executive control is essential for "unsticking" oneself and flexibly changing behavior patterns, thus mental flexibility deficits may drive treatment response in MDD

(Etkin et al., 2015). A network of structures linked to executive function is anatomically different in multiple psychiatric diagnoses when compared to controls (Goodkind et al., 2015). Further, DBS at the VC/VS target is marked by a rapid "rebound" to a pathologic state when stimulation stops (Ewing and Grace, 2013; Ooms et al., 2014). Therefore, if we "stress" VC/VS DBS patients with an executive function task in the stimulator on/off state, we may identify a signature that defines the clinical response. Fig. 3 shows a preliminary example of such a marker. We tested a mixed group of MDD ($n=10$) and OCD ($n=2$) patients with DBS on and off while they performed the Multiple Source Interference Task (MSIT) under 60-channel EEG recording. This Stroop-like task requires frequent

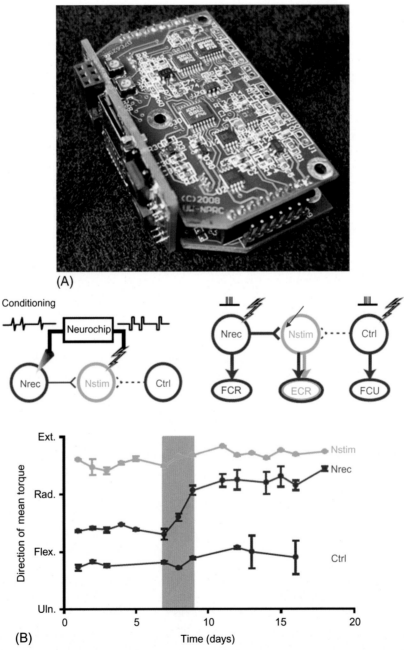

FIG. 2 (A) Photograph of the "Neurochip" closed-loop neurostimulation and plasticity induction platform (scale bar 1 cm). (B) Neurochip-induced plasticity. Driving stimulation at one cortical site (Nstim) based on spikes recorded at a nearby electrode (Nrec) for 48 h (*gray bar*) results in robust connectivity changes within the brain. These are likely driven by strengthened synaptic connections (*red arrow*). Ext, extension; Rad, radial; Fle, flexion; Uln, ulnar. (*From Jackson, A., Mavoori, J., Fetz, E.E., 2006. Long-term motor cortex plasticity induced by an electronic neural implant. Nature 44, 56–60. http://dx.doi.org/10.1038/nature05226, reprinted with permission from Nature Publishing Group.*)

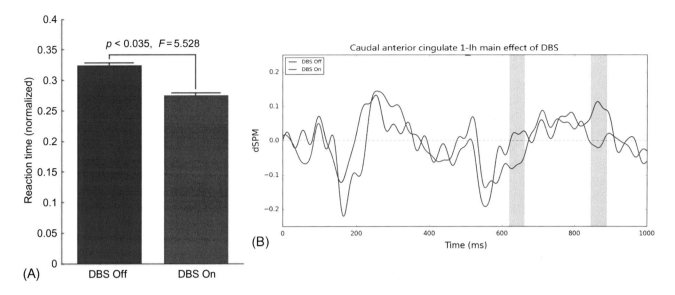

FIG. 3 Example of identifying neural and behavioral biomarkers of cross-diagnostic response to DBS. (A) Reaction time with DBS ON/OFF in the Multi-Source Interference Task (MSIT). Patients became significantly faster with neurostimulation at ventral capsule/ventral striatum (VC/VS), implying that DBS at this target alters cognitive processing. (B) Example event-related potentials (ERPs) from left dorsal anterior cingulate (dACC), a structure known to be engaged during MSIT performance. In a cluster-corrected permutation ANOVA, stimulus-locked ERPs showed significant DBS main effects at two time windows, 620–660 ms ($p < 0.039$) and 845–890 ms ($p < 0.007$). Bilateral dlPFC and right dACC showed similar effects at different time points.

response inhibition and rule switching (Bush and Shin, 2006). Patients showed a marked reaction time decrease (better performance) with DBS ON ($F = 5.53$, $df = 1$, $p < 0.035$ for main effect of DBS in multiway ANOVA). Further, a permutation-based ANOVA on the event-related potentials (ERPs) showed a significant main effect of DBS in several areas known to be linked to the MSIT, including the dorsolateral prefrontal cortex (PFC) and the dorsal anterior cingulate (dACC) bilaterally (Widge et al., 2015b,d). Fig. 3B illustrates one example, based on cluster mass statistics from a nonparametric F-test (Maris and Oostenveld, 2007). The left dACC ERP showed significant differences centered on 635 ms ($p < 0.039$) and 867 ms ($p < 0.007$). These are preliminary findings, not a conclusive biomarker, but they demonstrate that a trans-diagnostic approach is feasible.

6 OVERCOMING BIOMARKERS—INTENTION AND BRAIN-COMPUTER INTERFACES

Another common feature of many psychiatric syndromes is a desire to be free of symptoms. Disorders are often ego-dystonic: patients know they are acting irrationally, but feel unable to suppress or avoid responding to overwhelming emotions (American Psychiatric Association, 2013). They have difficulty with top-down emotion regulation, which is linked to descending connections from PFC to various nodes of the limbic circuit (Drevets, 1999; Etkin, 2012a,b;

Etkin and Wager, 2007; Milad and Rauch, 2012; Pitman et al., 2012; Price and Drevets, 2012). Deficits in emotion regulation can thus be seen as a problem of circuit connectivity between PFC and downstream structures. This is analogous to the problem of spinal cord injury, where top-down signals from primary motor cortex (M1) cannot reach the distal spinal cord or muscles. As described earlier, those deficits may be partly corrected by carefully timed recording and stimulation (Jackson et al., 2006; Mondello et al., 2014; Nishimura et al., 2013). A related line of work has shown that paralyzed patients or animals still have effective control signals in M1, and those signals can be used to control robotic actuators (Collinger et al., 2013; Hochberg et al., 2012; Ifft et al., 2013; Moritz et al., 2008). It may not be necessary to identify a specific biomarker for each psychiatric disorder. We could instead create a system that was under the patient's volitional control, using technologies almost identical to these motor brain-computer interfaces (BCIs). In that model, when the patient experienced undesired symptoms, he/she would alter the firing of PFC neurons. Since PFC is a key source of natural regulatory signals it is a logical starting point for an emotion-regulating BCI. That change in activity would be detected by the controller, which would increase or decrease neurostimulation according to the patient's command. Note that such a controller would not need to directly measure single-unit firing; the change in neural activity should also be reflected in LFP power, especially in higher frequency bands (Ojemann et al., 2013).

We recently demonstrated a proof-of-concept of this system, using a rodent model (Widge and Moritz, 2014). As an analog of a human patient using PFC to control an implanted limbic stimulator, four adult Long-Evans rats were implanted with 16-wire recording electrodes in PFC at the border of infralimbic and prelimbic cortex, and a bipolar stimulating electrode in medial forebrain bundle (MFB). While the latter is relevant as a human DBS target (Schlaepfer et al., 2013), we chose it because of its value in operant behavior paradigms. Brief MFB stimulation is reinforcing and can shape behavior (Carlezon and Chartoff, 2007; Olds, 1958). That reinforcement is believed to occur by creating a positive hedonic state, which may be similar to the effect of clinical psychiatric DBS.

We demonstrated that animals could learn the contingency between changes in their prefrontal activity and a desirable emotional experience, and then control that PFC activation to trigger stimulation. As an intermediate paradigm, we trained the rats to control their PFC firing to drive an auditory feedback signal (Gage et al., 2005; Koralek et al., 2012; Marzullo et al., 2006). As shown in Fig. 4A, activity of a single unit recorded from PFC was converted to a feedback tone, which was then played back to the rat. Neurons were not preselected for their relation to MFB activity, nor was the BCI in any way pretuned. Rather, at the start of each session, a unit was selected and mapped into the BCI with fixed parameters. By hearing the feedback cursor, feeling the positive effects of brain stimulation, and repeatedly experiencing the close temporal relation between the two, the animals were operantly shaped and motivated to take control of the stimulator (Fig. 4B).

All four rats successfully controlled the PFC BCI to trigger MFB stimulation. To demonstrate that stimulator activations were due to animals' intentional behavior (as opposed to stochastic neural fluctuations), we used real-time "catch" trials without audio feedback and offline "bootstrap" permutations (Fig. 5). Over all animals ($n = 4$) and experimental days ($n = 41$), rats were more successful at acquiring the target when they received audio feedback ($p < 0.006$, one-tailed paired-sample t-test; Fig. 5A). They were also faster at acquiring targets (Fig. 5B). The normalized time to a target acquisition was faster on actual cursor trials than in real-time catch or offline bootstrap trials ($p < 0.02$, two-tailed Kolmogorov-Smirnov test). Across animals, 80% of PFC recording sites supported control, suggesting that the technique can generalize broadly.

This approach may be clinically useful in anxiety and trauma-related disorders, where patients show difficulties with regulating their own fear and their responses to fear. Those deficits are found in generalized anxiety (Etkin and Wager, 2007), OCD (Dougherty et al., 2004; Milad and Rauch, 2012), and depression (Drevets, 1999; Greening et al., 2013). They have been linked to a failure of extinction learning—the ability to learn new memories indicating that a feared stimulus (eg, trauma-associated cues) no longer signals danger. Fear regulation depends in part on descending "pro-extinction" signals from ventromedial PFC to the amygdala (Etkin et al., 2011; Likhtik and Paz, 2015). Basolateral amygdala stimulation appears to reduce PTSD (post-traumatic stress disorder)-like behaviors in conditioned rats (Stidd et al., 2013). Thus, a clinical realization might allow patients to temporarily

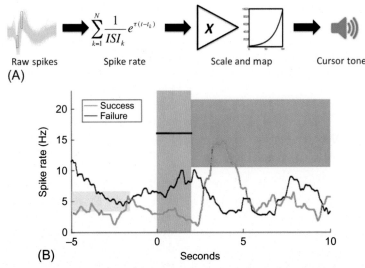

FIG. 4 Prefrontal operant brain-computer interface (BCI) for rodents. (A) BCI schematic. Spiking of a single PFC unit was integrated to an instantaneous rate estimate, and then mapped onto a varying-frequency auditory feedback tone. (B) Trial schematic. Rats maintained activity of the PFC unit at baseline (*yellow*), then heard a cue tone (*gray*). They had up to 10 s to increase the unit's activity (spike rate) and hold it within a target window (*red*). Successful unit control (*green curve*) earned reinforcing brain stimulation. Unsuccessful control (*black curve*) resulted in a time-out until the next trial opportunity.

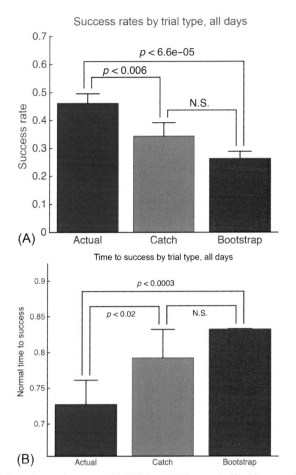

Success rates by trial type, all days

Time to success by trial type, all days

FIG. 5 Rat performance with PFC-based BCI control of limbic neurostimulation. Both graphs represent pooled data across all days when neurostimulator use exceeded chance; for details see Widge and Moritz (2014). "Actual" trials involved an auditory feedback signal, "Catch" provided neither feedback nor brain stimulation reward, and "Bootstrap" chance performance was computed offline based on temporal permutation of trials. (A) Number of trials successfully completed. (B) Time to success. When the BCI cursor was available, animals obtained more successes ($p < 0.006$, one-tailed t-test) and reached the target more quickly ($p < 0.02$, K-S test of distributions). This suggests that they were aware of the feedback and actively using it to reach an intentional goal of receiving brain stimulation. The two chance metrics did not differ, confirming their validity.

"turn down" amygdala activity by increasing vmPFC firing. This is what patients effectively learn to do during exposure psychotherapy, wherein they learn new safety memories that compete with their anxiety habits. Similar to the examples of motor system plasticity above, it could also be a powerful rehabilitative tool. When a patient wants quick anxiety relief under present treatment paradigms, his/her most likely approach is to take an anxiolytic medication, usually a benzodiazepine. This involves little use of PFC or emotion regulation skills, and may actually impair learning of those skills (Makkar et al., 2010). By contrast, our proposed "emotional prosthesis" approach provides anxiety relief only if the patient "exercises" PFC. If done

enough times, this should promote positive Hebbian plasticity in the fear regulation circuits, improving the patient's ability to self-regulate even without neurostimulation (Widge et al., 2014). Anecdotally, some DBS psychiatrists have opined that even open-loop DBS works through this mechanism, allowing the patients to benefit more strongly from psychotherapy and other rehabilitative interventions.

This intention-oriented approach also may solve major limitations of other closed-loop designs. First, it addresses the concerns of patient tolerability and "mind control." We propose a system in which neurostimulation is always under the patient's direct control. The clinician may set safety limits, but stimulation onset (and the resulting altered emotional experience) would always depend on the patient's volitional mental command. Second, some volitional component may be necessary in any closed-loop device for mood and anxiety regulation. A series of experiments with "locked in" human patients demonstrated "extinction of thought"— the notion that if a patient's volitional intent becomes disconnected from what actually happens to him/her, the capacity for self-regulation or intentional action can atrophy (Kübler and Birbaumer, 2008). In theory, if a psychiatric closed-loop DBS were not under a patient's direct control, it could disconnect a patient's emotional intentions (what he/she wants to feel at that moment) from the experienced emotional state. This could effectively atrophy his/her emotional regulation capacity, just as anxiolytic medications can. The marker-based approaches discussed may need to be implemented as a "hybrid BCI," with two or more algorithms sharing stimulator control (Widge et al., 2014). The cross-diagnostic biomarker algorithm would determine the type and amount of stimulation, while an intention-decoding algorithm would act as a "gate," enabling or disabling stimulation based on his/her present wishes. This shared control could also be a valuable fail-safe that could speed regulatory approval of closed-loop devices.

7 INVASIVE AND NONINVASIVE HARDWARE CONSIDERATIONS

The algorithms and approaches above will only be useful to patients when embodied in devices that they can actually receive. The only commercially available closed-loop brain stimulator depends on a specific cortical recording configuration that would not support the schemas discussed earlier (Morrell, 2011). A major DBS manufacturer has demonstrated a "sensing" DBS platform that can record from the human brain and stream those signals out to clinicians (Rouse et al., 2011). In animal models of epilepsy, that same device has been shown capable of running closed-loop algorithms and controlling LFP features (Afshar et al., 2013; Stypulkowski et al., 2014). The control algorithms we have proposed for psychiatric stimulation are similar and should be implementable on that platform. New clinical stimulators optimized for closed-loop approaches are under development via the Defense Advanced Research Projects Agency (DARPA; Miranda et al., 2015). Preliminary specifications for at least one of those systems are available (Bjune et al., 2015; Wheeler et al., 2015). By contrast, the algorithms for closed-loop psychiatric stimulation are not clinically mature. They require further animal testing, and several groups have demonstrated suitable open hardware platforms (Rolston et al., 2010; Siegle et al., 2015; Wu et al., 2015).

While we have mainly considered DBS and other invasive technologies, non- or minimally invasive approaches to closed-loop limbic control would be desirable. They could reduce the surgical risk and enable more widespread and rapid human testing of innovative therapies. The challenge is that almost all limbic structures lie in the deep brain. TMS can only activate the first 1 cm of cortex (Deng et al., 2013, 2014). Magnetic stimulation can specifically be targeted to deep tissues by coupling neurons to magnetic nanoparticles (Chen et al., 2015). This requires a small fluid injection and far fewer indwelling parts. Focussed ultrasound (FUS) is able to penetrate to the deep brain without affecting overlying tissues. Preliminary animal and human studies suggest that low-intensity FUS can increase or decrease neural activity, in focussed brain regions, without heating or other damage (Legon et al., 2014; Mehić et al., 2014; Tufail et al., 2011). If FUS transducers can be modified to reliably focus their energy in limbic regions, this could become a noninvasive variant of DBS.

8 CONCLUSIONS

Brain stimulation for emotional and mental disorders has recently faced challenges. A number of open-loop studies have found much smaller effects than expected, and some have failed outright. Those difficulties, however, make a clear case for attempting closed-loop stimulation to reconnect the prefrontal and limbic circuits that underpin mental disorders. Closing the feedback loop, particularly for DBS, should lead to stimulators that are more effective, more energy-efficient, and potentially applicable to a much larger range of mental illnesses. If done correctly, it could even strengthen the pathways whose dysregulation is believed to produce those illnesses. Closed-loop stimulation in movement circuits has been shown to induce helpful neuroplasticity, and the same should be possible in emotional circuits.

One key challenge for the field is defining the biomarkers—the neural signals that can be sensed and used to titrate stimulation. As we have shown, new research in this area is changing the way we search for those markers. By orienting neuroscience across diagnostic boundaries, and by leveraging technologies developed for closed-loop motor prosthetics, it may be possible to bring closed-loop psychiatric technologies closer to human use even while biomarkers are still being refined. We have presented some preliminary results showing that both directions are feasible, and outlined a possible agenda for future work. Further, we have linked those results to the capabilities of current and near-future neurostimulation hardware, showing that there is a viable path for clinical translation. The future for closed-loop techniques in psychiatric neuromodulation appears quite bright.

REFERENCES

Afshar, P., Khambhati, A., Carlson, D., Dani, S., Lazarewicz, M., Cong, P., Denison, T., 2013. A translational platform for prototyping closed-loop neuromodulation systems. Front. Neural Circuits 6, 117. http://dx.doi.org/10.3389/fncir.2012.00117.

American Psychiatric Association, 2013. Diagnostic and Statistical Manual of Mental Disorders (DSM), fifth ed. American Psychiatric Publishing, Arlington, VA.

Ben-Menachem, E., 2002. Vagus-nerve stimulation for the treatment of epilepsy. Lancet Neurol. 1 (8), 477–482. http://dx.doi.org/10.1016/S1474-4422(02)00220-X.

Bjune, C.K., Marinis, T.F., Brady, J.M., Moran, J., Wheeler, J., Sriram, T.S., Eskandar, E.N., 2015. Package architecture and component design for an implanted neural stimulator with closed loop control. In: Engineering in Medicine and Biology Society (EMBC). 2015 37th Annual International Conference of the IEEE. IEEE, pp. 7825–7830. Retrieved from, http://ieeexplore.ieee.org/xpls/abs_all.jsp?arnumber=7320206.

Broadway, J.M., Holtzheimer, P.E., Hilimire, M.R., Parks, N.A., DeVylder, J.E., Mayberg, H.S., Corballis, P.M., 2012. Frontal theta cordance predicts 6-month antidepressant response to subcallosal cingulate deep brain stimulation for treatment-resistant depression: a pilot study. Neuropsychopharmacology 37 (7), 1764–1772. http://dx.doi.org/10.1038/npp.2012.23.

Bronstein, J., Tagliati, M., 2011. Deep brain stimulation for Parkinson disease: an expert consensus and review of key issues. Arch. Neurol. 68 (2), 165. http://dx.doi.org/10.1001/archneurol.2010.260.

Bush, G., Shin, L.M., 2006. The multi-source interference task: an fMRI task that reliably activates the cingulo-frontal-parietal cognitive/attention network. Nat. Protoc. 1 (1), 308–313. http://dx.doi.org/10.1038/nprot.2006.48.

Carlezon, W.A., Chartoff, E.H., 2007. Intracranial self-stimulation (ICSS) in rodents to study the neurobiology of motivation. Nat. Protoc. 2 (11), 2987–2995. http://dx.doi.org/10.1038/nprot.2007.441.

Chen, R., Romero, G., Christiansen, M.G., Mohr, A., Anikeeva, P., 2015. Wireless magnetothermal deep brain stimulation. Science 347 (6229), 1477–1480. http://dx.doi.org/10.1126/science.1261821.

Collinger, J.L., Wodlinger, B., Downey, J.E., Wang, W., Tyler-Kabara, E.C., Weber, D.J., Schwartz, A.B., 2013. High-performance neuroprosthetic control by an individual with tetraplegia. Lancet 381 (9866), 557–564. http://dx.doi.org/10.1016/S0140-6736(12)61816-9.

Cuthbert, B.N., Insel, T.R., 2013. Toward the future of psychiatric diagnosis: the seven pillars of RDoC. BMC Med. 11 (1), 126. http://dx.doi.org/10.1186/1741-7015-11-126.

Deng, Z.-D., Lisanby, S.H., Peterchev, A.V., 2013. Electric field depth–focality tradeoff in transcranial magnetic stimulation: simulation comparison of 50 coil designs. Brain Stimul. 6 (1), 1–13. http://dx.doi.org/10.1016/j.brs.2012.02.005.

Deng, Z.-D., Lisanby, S.H., Peterchev, A.V., 2014. Coil design considerations for deep transcranial magnetic stimulation. Clin. Neurophysiol. 125 (6), 1202–1212. http://dx.doi.org/10.1016/j.clinph.2013.11.038.

Dougherty, D.D., Baer, L., Cosgrove, G.R., Cassem, E.H., Price, B.H., Nierenberg, A.A., Rauch, S.L., 2002. Prospective long-term follow-up of 44 patients who received cingulotomy for treatment-refractory obsessive-compulsive disorder. Am. J. Psychiatry 159 (2), 269–275. http://dx.doi.org/10.1176/appi.ajp.159.2.269.

Dougherty, D.D., Rauch, S.L., Deckersbach, T., Marci, C., Loh, R., Shin, L.M., Fava, M., 2004. Ventromedial prefrontal cortex and amygdala dysfunction during an anger induction positron emission tomography study in patients with major depressive disorder with anger attacks. Arch. Gen. Psychiatry 61 (8), 795–804. http://dx.doi.org/10.1001/archpsyc.61.8.795.

Dougherty, D.D., Rezai, A.R., Carpenter, L.L., Howland, R.H., Bhati, M.T., O'Reardon, J.P., Malone Jr., D.A., 2015. A randomized sham-controlled trial of deep brain stimulation of the ventral capsule/ventral striatum for chronic treatment-resistant depression. Biol. Psychiatry 78 (4), 240–248. http://dx.doi.org/10.1016/j.biopsych.2014.11.023.

Drevets, W.C., 1999. Prefrontal cortical-amygdalar metabolism in major depression. Ann. N. Y. Acad. Sci. 877 (1), 614–637. http://dx.doi.org/10.1111/j.1749-6632.1999.tb09292.x.

Ellard, K.K., Fairholme, C.P., Boisseau, C.L., Farchione, T.J., Barlow, D.H., 2010. Unified protocol for the transdiagnostic treatment of emotional disorders: protocol development and initial outcome data. Cogn. Behav. Pract. 17 (1), 88–101. http://dx.doi.org/10.1016/j.cbpra.2009.06.002.

Etkin, A., 2012a. Functional neuroimaging of major depressive disorder: a meta-analysis and new integration of baseline activation and neural response data. Am. J. Psychiatr. 169 (7), 693–703. http://dx.doi.org/10.1176/appi.ajp.2012.11071105.

Etkin, A., 2012b. Neurobiology of anxiety: from neural circuits to novel solutions? Depress. Anxiety 29 (5), 355–358. http://dx.doi.org/10.1002/da.21957.

Etkin, A., Wager, T.D., 2007. Functional neuroimaging of anxiety: a meta-analysis of emotional processing in PTSD, social anxiety disorder, and specific phobia. Am. J. Psychiatr. 164 (10), 1476–1488. http://dx.doi.org/10.1176/appi.ajp.2007.07030504.

Etkin, A., Egner, T., Kalisch, R., 2011. Emotional processing in anterior cingulate and medial prefrontal cortex. Trends Cogn. Sci. 15 (2), 85–93. http://dx.doi.org/10.1016/j.tics.2010.11.004.

Etkin, A., Patenaude, B., Song, Y.J.C., Usherwood, T., Rekshan, W., Schatzberg, A.F., Williams, L.M., 2015. A cognitive–emotional biomarker for predicting remission with antidepressant medications: a report from the iSPOT-D trial. Neuropsychopharmacology 40 (6), 1332–1342. http://dx.doi.org/10.1038/npp.2014.333.

Ewing, S.G., Grace, A.A., 2013. Long-term high frequency deep brain stimulation of the nucleus accumbens drives time-dependent changes in functional connectivity in the rodent limbic system. Brain Stimul. 6 (3), 274–285. http://dx.doi.org/10.1016/j.brs.2012.07.007.

Figee, M., Luigjes, J., Smolders, R., Valencia-Alfonso, C.-E., van Wingen, G., de Kwaasteniet, B., Denys, D., 2013. Deep brain stimulation restores frontostriatal network activity in obsessive-compulsive disorder. Nat. Neurosci. 16, 386–387. http://dx.doi.org/10.1038/nn.3344.

Gage, G.J., Ludwig, K.A., Otto, K.J., Ionides, E.L., Kipke, D.R., 2005. Naïve coadaptive cortical control. J. Neural Eng. 2 (2), 52.

Garnaat, S.L., Greenberg, B.D., Sibrava, N.J., Goodman, W.K., Mancebo, M.C., Eisen, J.L., Rasmussen, S.A., 2014. Who qualifies for deep brain stimulation for OCD? Data from a naturalistic clinical sample. J. Neuropsychiatry Clin. Neurosci. 26 (1), 81–86. http://dx.doi.org/10.1176/appi.neuropsych.12090226.

Gaynes, B., Warden, D., Trivedi, M.H., Wisniewski, S., Fava, M., Rush, A.J., 2009. What did STAR*D teach us? Results from a large-scale, practical, clinical trial for patients with depression. Psychiatr. Serv. 60 (11), 1439–1445. http://dx.doi.org/10.1176/appi.ps.60.11.1439.

George, M.S., Lisanby, S.H., Avery, D., McDonald, W.M., Durkalski, V., Pavlicova, M., Sackeim, H.A., 2010. Daily left prefrontal transcranial magnetic stimulation therapy for major depressive disorder: a sham-controlled randomized trial. Arch. Gen. Psychiatry 67 (5), 507–516. http://dx.doi.org/10.1001/archgenpsychiatry.2010.46.

Goodkind, M., Eickhoff, S.B., Oathes, D.J., Jiang, Y., Chang, A., Jones-Hagata, L.B., Etkin, A., 2015. Identification of a common neurobiological substrate for mental illness. JAMA Psychiatry 72 (4), 305. http://dx.doi.org/10.1001/jamapsychiatry.2014.2206.

Goodman, W.K., Foote, K.D., Greenberg, B.D., Ricciuti, N., Bauer, R., Ward, H., Okun, M.S., 2010. Deep brain stimulation for intractable obsessive compulsive disorder: pilot study using a blinded, staggered-onset design. Biol. Psychiatry 67 (6), 535–542.

Greenberg, B., Gabriels, L., Malone, D., Rezai, A., Friehs, G., Okun, M., Nuttin, B., 2010. Deep brain stimulation of the ventral internal capsule/ventral striatum for obsessive-compulsive disorder: worldwide experience. Mol. Psychiatry 15 (1), 64–79. http://dx.doi.org/10.1038/mp.2008.55.

Greening, S.G., Osuch, E.A., Williamson, P.C., Mitchell, D.G.V., 2013. The neural correlates of regulating positive and negative emotions in medication-free major depression. Social Cogn. Affect. Neurosci. 9 (5), 628–637. http://dx.doi.org/10.1093/scan/nst027.

Haq, I.U., Foote, K.D., Goodman, W.K., Ricciuti, N., Ward, H., Sudhyadhom, A., Okun, M.S., 2010. A case of mania following deep brain stimulation for obsessive compulsive disorder. Stereotact. Funct. Neurosurg. 88 (5), 322–328. http://dx.doi.org/10.1159/000319960.

Ho, C.H., Triolo, R.J., Elias, A.L., Kilgore, K.L., DiMarco, A.F., Bogie, K., Mushahwar, V.K., 2014. Functional electrical stimulation and spinal cord injury. Phys. Med. Rehabil. Clin. N. Am. 25 (3), 631–654. http://dx.doi.org/10.1016/j.pmr.2014.05.001.

Hochberg, L.R., Bacher, D., Jarosiewicz, B., Masse, N.Y., Simeral, J.D., Vogel, J., Donoghue, J.P., 2012. Reach and grasp by people with tetraplegia using a neurally controlled robotic arm. Nature 485 (7398), 372–375. http://dx.doi.org/10.1038/nature11076.

Holtzheimer, P.E., Kelley, M.E., Gross, R.E., Filkowski, M.M., Garlow, S.J., Barrocas, A., Mayberg, H.S., 2012. Subcallosal cingulate deep brain stimulation for treatment-resistant unipolar and bipolar depression. Arch. Gen. Psychiatry 69 (2), 150–158. http://dx.doi.org/10.1001/archgenpsychiatry.2011.1456.

Ifft, P.J., Shokur, S., Li, Z., Lebedev, M.A., Nicolelis, M.A.L., 2013. A brain-machine interface enables bimanual arm movements in monkeys. Sci. Transl. Med. 5 (210), 210ra154. http://dx.doi.org/10.1126/scitranslmed.3006159.

Insel, T.R., 2009. Disruptive insights in psychiatry: transforming a clinical discipline. J. Clin. Invest. 119 (4), 700–705. http://dx.doi.org/10.1172/JCI38832.

Insel, T.R., 2010. Faulty circuits. Sci. Am. 302 (4), 44–51. http://dx.doi.org/10.1038/scientificamerican0410-44.

Insel, T.R., 2012. Next-generation treatments for mental disorders. Sci. Transl. Med. 4 (155), 155ps19. http://dx.doi.org/10.1126/scitranslmed.3004873.

Insel, T.R., Wang, P.S., 2010. Rethinking mental illness. J. Am. Med. Assoc. 303 (19), 1970–1971. http://dx.doi.org/10.1001/jama.2010.555.

Jackson, A., Mavoori, J., Fetz, E.E., 2006. Long-term motor cortex plasticity induced by an electronic neural implant. Nature 44, 56–60. http://dx.doi.org/10.1038/nature05226.

Jonas, D., Cusack, K., Forneris, C., Wilkins, T., Sonis, J., Middleton, J., Gaynes, B., 2013. Comparative Effectiveness Review: Psychological Treatments and Pharmacological Treatments for Adults with Post-Traumatic Stress Disorder (PTSD). Retrieved from, http://effectivehealthcare.ahrq.gov/ehc/products/347/1435/PTSD-adult-treatment-report-130403.pdf.

Koralek, A.C., Jin, X., Ii, J.D.L., Costa, R.M., Carmena, J.M., 2012. Corticostriatal plasticity is necessary for learning intentional neuroprosthetic skills. Nature 483 (7389), 331–335. http://dx.doi.org/10.1038/nature10845.

Kübler, A., Birbaumer, N., 2008. Brain–computer interfaces and communication in paralysis: extinction of goal directed thinking in completely paralysed patients? Clin. Neurophysiol. 119 (11), 2658–2666. http://dx.doi.org/10.1016/j.clinph.2008.06.019.

Legon, W., Sato, T.F., Opitz, A., Mueller, J., Barbour, A., Williams, A., Tyler, W.J., 2014. Transcranial focused ultrasound modulates the activity of primary somatosensory cortex in humans. Nat. Neurosci. 17 (2), 322–329. http://dx.doi.org/10.1038/nn.3620.

Likhtik, E., Paz, R., 2015. Amygdala–prefrontal interactions in (mal)adaptive learning. Trends Neurosci. 38 (3), 158–166. http://dx.doi.org/10.1016/j.tins.2014.12.007.

Little, S., Pogosyan, A., Neal, S., Zavala, B., Zrinzo, L., Hariz, M., Brown, P., 2013. Adaptive deep brain stimulation in advanced Parkinson disease. Ann. Neurol. 74 (3), 449–457. http://dx.doi.org/10.1002/ana.23951.

Makkar, S.R., Zhang, S.Q., Cranney, J., 2010. Behavioral and neural analysis of GABA in the acquisition, consolidation, reconsolidation, and extinction of fear memory. Neuropsychopharmacology 35 (8), 1625–1652. http://dx.doi.org/10.1038/npp.2010.53.

Malone, D.A., Dougherty, D.D., Rezai, A.R., Carpenter, L.L., Friehs, G.M., Eskandar, E.N., Greenberg, B.D., 2009. Deep brain stimulation of the ventral capsule/ventral striatum for treatment-resistant depression. Biol. Psychiatry 65 (4), 267–275. http://dx.doi.org/10.1016/j.biopsych.2008.08.029.

Maris, E., Oostenveld, R., 2007. Nonparametric statistical testing of EEG- and MEG-data. J. Neurosci. Methods 164 (1), 177–190. http://dx.doi.org/10.1016/j.jneumeth.2007.03.024.

Marzullo, T.C., Miller, C.R., Kipke, D.R., 2006. Suitability of the cingulate cortex for neural control. IEEE Trans. Neural Syst. Rehabil. Eng. 14 (4), 401–409. http://dx.doi.org/10.1109/TNSRE.2006.886730.

Mavoori, J., Jackson, A., Diorio, C., Fetz, E., 2005. An autonomous implantable computer for neural recording and stimulation in unrestrained primates. J. Neurosci. Methods 148 (1), 71–77.

Mayberg, H.S., Lozano, A.M., Voon, V., McNeely, H.E., Seminowicz, D., Hamani, C., Kennedy, S.H., 2005. Deep brain stimulation for treatment-resistant depression. Neuron 45 (5), 651–660. http://dx.doi.org/10.1016/j.neuron.2005.02.014.

Mehić, E., Xu, J.M., Caler, C.J., Coulson, N.K., Moritz, C.T., Mourad, P.D., 2014. Increased anatomical specificity of neuromodulation via modulated focused ultrasound. PLoS ONE 9(2). http://dx.doi.org/10.1371/journal.pone.0086939.

Milad, M.R., Rauch, S.L., 2012. Obsessive-compulsive disorder: beyond segregated cortico-striatal pathways. Trends Cogn. Sci. 16 (1), 43–51. http://dx.doi.org/10.1016/j.tics.2011.11.003.

Miranda, R.A., Casebeer, W.D., Hein, A.M., Judy, J.W., Krotkov, E.P., Laabs, T.L., Ling, G.S.F., 2015. DARPA-funded efforts in the development of novel brain–computer interface technologies. J. Neurosci. Methods 244, 52–67. http://dx.doi.org/10.1016/j.jneumeth.2014.07.019.

Mondello, S.E., Kasten, M.R., Horner, P.J., Moritz, C.T., 2014. Therapeutic intraspinal stimulation to generate activity and promote long-term recovery. Neuroprosthetics 8, 21. http://dx.doi.org/10.3389/fnins.2014.00021.

Morishita, T., Fayad, S.M., Higuchi, M., Nestor, K.A., Foote, K.D., 2014. Deep brain stimulation for treatment-resistant depression: systematic review of clinical outcomes. Neurotherapeutics 11 (3), 475–484. http://dx.doi.org/10.1007/s13311-014-0282-1.

Moritz, C.T., Perlmutter, S.I., Fetz, E.E., 2008. Direct control of paralysed muscles by cortical neurons. Nature 456, 639–643.

Morrell, M.J., 2011. Responsive cortical stimulation for the treatment of medically intractable partial epilepsy. Neurology 77 (13), 1295–1304. http://dx.doi.org/10.1212/WNL.0b013e3182302056.

Nishimura, Y., Perlmutter, S.I., Eaton, R.W., Fetz, E.E., 2013. Spike-timing-dependent plasticity in primate corticospinal connections induced during free behavior. Neuron 80 (5), 1301–1309. http://dx.doi.org/10.1016/j.neuron.2013.08.028.

Nuttin, B., Cosyns, P., Demeulemeester, H., Gybels, J., Meyerson, B., 1999. Electrical stimulation in anterior limbs of internal capsules in patients with obsessive-compulsive disorder. Lancet 354 (9189), 1526. http://dx.doi.org/10.1016/S0140-6736(99)02376-4.

Ojemann, G.A., Ojemann, J., Ramsey, N.F., 2013. Relation between functional magnetic resonance imaging (fMRI) and single neuron, local field potential (LFP) and electrocorticography (ECoG) activity in human cortex. Front. Hum. Neurosci. 7, 34. http://dx.doi.org/10.3389/fnhum.2013.00034.

Olds, J., 1958. Self-stimulation of the brain: its use to study local effects of hunger, sex, and drugs. Science 127 (3294), 315–324.

Ooms, P., Blankers, M., Figee, M., Mantione, M., van den Munckhof, P., Schuurman, P.R., Denys, D., 2014. Rebound of affective symptoms following acute cessation of deep brain stimulation in obsessive-compulsive disorder. Brain Stimul. 7 (5), 727–731. http://dx.doi.org/10.1016/j.brs.2014.06.009.

Pitman, R.K., Rasmusson, A.M., Koenen, K.C., Shin, L.M., Orr, S.P., Gilbertson, M.W., Liberzon, I., 2012. Biological studies of post-traumatic stress disorder. Nat. Rev. Neurosci. 13 (11), 769–787. http://dx.doi.org/10.1038/nrn3339.

Pittenger, C., Bloch, M.H., 2014. Pharmacological treatment of obsessive-compulsive disorder. Psychiatr. Clin. N. Am. 37 (3), 375–391. http://dx.doi.org/10.1016/j.psc.2014.05.006.

Powers, M.B., Halpern, J.M., Ferenschak, M.P., Gillihan, S.J., Foa, E.B., 2010. A meta-analytic review of prolonged exposure for posttraumatic stress disorder. Clin. Psychol. Rev. 30 (6), 635–641. http://dx.doi.org/10.1016/j.cpr.2010.04.007.

Price, J.L., Drevets, W.C., 2012. Neural circuits underlying the pathophysiology of mood disorders. Trends Cogn. Sci. 16 (1), 61–71. http://dx.doi.org/10.1016/j.tics.2011.12.011.

Regier, D.A., Narrow, W.E., Clarke, D.E., Kraemer, H.C., Kuramoto, S.J., Kuhl, E.A., Kupfer, D.J., 2013. DSM-5 field trials in the United States and Canada, part II: test-retest reliability of selected categorical diagnoses. Am. J. Psychiatr. 170 (1), 59–70. http://dx.doi.org/10.1176/appi.ajp.2012.12070999.

Riva-Posse, P., Choi, K.S., Holtzheimer, P.E., McIntyre, C.C., Gross, R.E., Chaturvedi, A., Mayberg, H.S., 2014. Defining critical white matter pathways mediating successful subcallosal cingulate deep brain stimulation for treatment-resistant depression. Biol. Psychiatry 76 (12), 963–969. http://dx.doi.org/10.1016/j.biopsych.2014.03.029.

Rolston, J.D., Gross, R.E., Potter, S.M., 2010. Closed-loop, open-source electrophysiology. Front. Neurosci. 4. http://dx.doi.org/10.3389/fnins.2010.00031.

Rosa, M.A., Lisanby, S.H., 2012. Somatic treatments for mood disorders. Neuropsychopharmacology 37 (1), 102–116. http://dx.doi.org/10.1038/npp.2011.225.

Rosin, B., Slovik, M., Mitelman, R., Rivlin-Etzion, M., Haber, S.N., Israel, Z., Bergman, H., 2011. Closed-loop deep brain stimulation is superior in ameliorating Parkinsonism. Neuron 72 (2), 370–384. http://dx.doi.org/10.1016/j.neuron.2011.08.023.

Rouse, A.G., Stanslaski, S.R., Cong, P., Jensen, R.M., Afshar, P., Ullestad, D., Denison, T.J., 2011. A chronic generalized bi-directional brain–machine interface. J. Neural Eng. 8, 036018. http://dx.doi.org/10.1088/1741-2560/8/3/036018.

Rush, A.J., Marangell, L.B., Sackeim, H.A., George, M.S., Brannan, S.K., Davis, S.M., Cooke, R.G., 2005. Vagus nerve stimulation for treatment-resistant depression: a randomized, controlled acute phase trial. Biol. Psychiatry 58 (5), 347–354. http://dx.doi.org/10.1016/j.biopsych.2005.05.025.

Samuel, M., Rodriguez-Oroz, M., Antonini, A., Brotchie, J.M., Ray Chaudhuri, K., Brown, R.G., Lang, A.E., 2015. Management of impulse control disorders in Parkinson's disease: controversies and future approaches. Mov. Disord. 30 (2), 150–159. http://dx.doi.org/10.1002/mds.26099.

Schlaepfer, T.E., Bewernick, B.H., Kayser, S., Mädler, B., Coenen, V.A., 2013. Rapid effects of deep brain stimulation for treatment-resistant major depression. Biol. Psychiatry 73 (12), 1204–1212. http://dx.doi.org/10.1016/j.biopsych.2013.01.034.

Siegle, J.H., Hale, G.J., Newman, J.P., Voigts, J., 2015. Neural ensemble communities: open-source approaches to hardware for large-scale electrophysiology. Curr. Opin. Neurobiol. 32, 53–59. http://dx.doi.org/10.1016/j.conb.2014.11.004.

Stidd, D.A., Vogelsang, K., Krahl, S.E., Langevin, J.-P., Fellous, J.-M., 2013. Amygdala deep brain stimulation is superior to paroxetine treatment in a rat model of posttraumatic stress disorder. Brain Stimul. 6 (6), 837–844. http://dx.doi.org/10.1016/j.brs.2013.05.008.

Stypulkowski, P.H., Stanslaski, S.R., Jensen, R.M., Denison, T.J., Giftakis, J.E., 2014. Brain stimulation for epilepsy—local and remote modulation of network excitability. Brain Stimul. 7 (3), 350–358. http://dx.doi.org/10.1016/j.brs.2014.02.002.

Tan, D.W., Schiefer, M.A., Keith, M.W., Anderson, J.R., Tyler, J., Tyler, D.J., 2014. A neural interface provides long-term stable natural touch perception. Sci. Transl. Med. 6 (257), 257ra138. http://dx.doi.org/10.1126/scitranslmed.3008669.

Tufail, Y., Yoshihiro, A., Pati, S., Li, M.M., Tyler, W.J., 2011. Ultrasonic neuromodulation by brain stimulation with transcranial ultrasound. Nat. Protoc. 6 (9), 1453–1470. http://dx.doi.org/10.1038/nprot.2011.371.

Wang, X.-J., Krystal, J.H., 2014. Computational psychiatry. Neuron 84 (3), 638–654. http://dx.doi.org/10.1016/j.neuron.2014.10.018.

Wheeler, J.J., Baldwin, K., Kindle, A., Guyon, D., Nugent, B., Segura, C., et al., 2015. An implantable 64-channel neural interface with reconfi-gurable recording and stimulation. In: Engineering in Medicine and Biology Society (EMBC). 2015 37th Annual International Conference of the IEEE. IEEE Retrieved from http://ieeexplore.ieee.org/xpls/abs_all.jsp?arnumber=7320208.

Whelan, R., Garavan, H., 2013. When optimism hurts: inflated predictions in psychiatric neuroimaging. Biol. Psychiatry. http://dx.doi.org/10.1016/j.biopsych.2013.05.014.

Widge, A.S., Dougherty, D.D., 2015. Managing patients with psychiatric disorders with deep brain stimulation. In: Marks Jr., W.J. (Ed.), Deep Brain Stimulation Management. second ed. Cambridge University Press, Cambridge, NY.

Widge, A.S., Moritz, C.T., 2014. Pre-frontal control of closed-loop limbic neurostimulation by rodents using a brain-computer interface. J. Neural Eng. 11 (2), 024001. http://dx.doi.org/10.1088/1741-2560/11/2/024001.

Widge, A.S., Avery, D.H., Zarkowski, P., 2013. Baseline and treatment-emergent EEG biomarkers of antidepressant medication response do not predict response to repetitive transcranial magnetic stimulation. Brain Stimul. 6 (6), 929–931. http://dx.doi.org/10.1016/j.brs.2013.05.001.

Widge, A.S., Dougherty, D.D., Moritz, C.T., 2014. Affective brain-computer interfaces as enabling technology for responsive psychiatric stimulation. Brain Comput. Interfaces 1 (2), 126–136. http://dx.doi.org/10.1080/2326263X.2014.912885.

Widge, A.S., Arulpragasam, A.R., Deckersbach, T., Dougherty, D.D., 2015a. Deep brain stimulation for psychiatric disorders. In: Scott, R.A., Kosslyn, S.M. (Eds.), Emerging Trends in the Social and Behavioral Sciences. John Wiley & Sons Inc., New York, NY. Retrieved from, http://onlinelibrary.wiley.com/doi/10.1002/9781118900772.etrds0103/abstract.

Widge, A.S., Deckersbach, T., Eskandar, E.N., Dougherty, D.D., 2015b. Deep brain stimulation for treatment-resistant psychiatric illnesses: what has gone wrong and what should we do next? Biol. Psychiatry. http://dx.doi.org/10.1016/j.biopsych.2015.06.005.

Widge, A.S., Licon, E., Zorowitz, S., Corse, A., Arulpragasam, A.R., Camprodon, J.A., Dougherty, D.D., 2015c. Predictors of hypomania during ventral capsule/ventral striatum deep brain stimulation. Journal of Neuropsychiatry and Clinical Neurosciences, Accepted.

Widge, A.S., Zorowitz, S., Tang, W., Miller, E.K., Deckersbach, T., Dougherty, D.D., 2015d. Behavioral and neural biomarkers of improved top-down control mediate clinical response to ventral capsule/ventral striatum deep brain stimulation in major depression. In: Society of Biological Psychiatry Annual Meeting. Toronto, Canada.

Wu, H., Ghekiere, H., Beeckmans, D., Tambuyzer, T., van Kuyck, K., Aerts, J.-M., Nuttin, B., 2015. Conceptualization and validation of an open-source closed-loop deep brain stimulation system in rat. Sci. Rep. 4, 9921. http://dx.doi.org/10.1038/srep09921.

Yang, J.C., Ginat, D.T., Dougherty, D.D., Makris, N., Eskandar, E.N., 2013. Lesion analysis for cingulotomy and limbic leucotomy: comparison and correlation with clinical outcomes. J. Neurosurg. http://dx.doi.org/10.3171/2013.9.JNS13839.

Zanos, S., Richardson, A.G., Shupe, L., Miles, F.P., Fetz, E.E., 2011. The neurochip-2: an autonomous head-fixed computer for recording and stimulating in freely behaving monkeys. IEEE Trans. Neural Syst. Rehabil. Eng. 19 (4), 427–435. http://dx.doi.org/10.1109/TNSRE.2011.2158007.

Chapter 18

Conscious Brain-to-Brain Communication Using Noninvasive Technologies ☆

G. Ruffini*,†

*Starlab, Barcelona, Spain, †Neuroelectrics Corporation, Cambridge, MA, United States

ABBREVIATIONS

B2B	brain-to-brain
BCI	brain-computer interface
CBI	computer-brain interface
EEG	electroencephalography
MRI	magnetic resonance imaging
tACS	transcranial alternating current stimulation
tCS	transcranial current stimulation
tDCS	transcranial direct current stimulation
tEFS	transcranial endogenous field stimulation
TMS	transcranial magnetic stimulation

As long as our brain is a mystery, the universe, the reflection of the structure of the brain will also be a mystery.

Santiago Ramón y Cajal

Everything you can imagine is real.

Pablo Picasso

1 INTRODUCTION

Information has now taken a central role in human endeavors, including science and technology. Formally, it quantifies the number of yes/no questions required to specify a physical (or abstract) state, and it defines the mathematical concept of the *bit*—the essence of dichotomy and an invention of Leibniz. Information is the currency brains use to characterize states and infer things about them (see Fig. 1). It is digital by nature and can be transmitted as a sequence of binary symbols. Because the brain is where subjective experiences are generated as a result of the exchange of information with the world outside (which for the purposes of the discussion here includes the rest of the body), computation and algorithmic information theory

(Cover and Thomas, 1991) is bound to play a central role in neuroscience as well as in physics. We can define a brain as *a semi-isolated physical/computational system capable of controlling some of its couplings/information interfaces with the rest of the universe* (Ruffini, 2007, 2009, 2016). To illustrate this, let us consider a simple *gedanken* (thought experiment). Imagine a subject—let us call her *Alice*—in an environment where a computer—using a world model or driven by external data (eg, from sensors or from another brain)—prepares and controls all her sensorial inputs via immersive audio and video 3D displays, haptic or olfactory interfaces, vestibular stimulation, etc., and in which her brain outputs are similarly intercepted by cameras, microphones, motion capture systems, etc. If this experiment is carried out successfully, that is, if the bidirectional bit-stream is managed properly, Alice will feel fully immersed in the virtual reality environment and, in behaviorally and physiologically measurable terms, act as if what she is experiencing is real (Sanchez-Vives and Slater, 2005). Note that in this quite realistic experiment, it suffices to describe Alice's universe as information (bits), because the environment is really a program in a computer managing sensorial inputs and brain outputs—see Fig. 1—and, crucially, the bidirectional interface exchanges only bits. With regard to the underlying physics, if needed, the discussion can be extended by the concepts of quantum information and quantum computation theory (using *qubits* instead of bits, see eg, Lloyd, 2013).

Humans need to interact with the outside world—which includes other humans—and since the beginning of time have endeavored to increase the bandwidth of a limited set of natural communication channels. It is for this reason that speech evolved and that new sensors (infrared, radar, micro or macroscopic, etc.), writing, the Internet, and telepresence have recently been developed. The goal of these efforts and derived technologies is to facilitate the exchange of information between brains and the universe in which they live, to increase their mutual information (in a

---

☆ The author is co-owner of Neuroelectrics and Starlab, and holds patents on multisite tCS.

Closed Loop Neuroscience. http://dx.doi.org/10.1016/B978-0-12-802452-2.00018-4

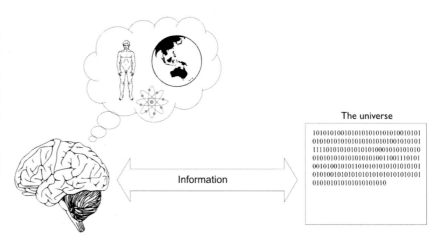

FIG. 1 The brain creates the model of reality through information exchange (in and out) with the outside world. In this case we show the full universe subdivided into a brain interacting with the rest of the universe by bidirectional (quantum) information exchange, which naturally occurs through sensors and effectors. As I discuss, the brain could be that of Alice's in our *gedanken* but in a radically immersive, virtual reality environment mediated by a direct bidirectional brain-to-brain interface. In either case, the universe could include other brains in the real world, or just a computer simulating them.

technical sense) as rapidly and efficiently as possible. The role of shared neural representations (mutual information increase) in social interaction is now being studied using modern neuroimaging techniques (eg, Hasson et al., 2008; Astolfi et al., 2010). In this chapter I discuss technologies to create such shared representations directly, which will mark a major turning point for humanity when fully realized (Nicolelis, 2010).

The ultimate vision for direct brain-to-brain (B2B) communication is often popularly referred to as *telepathy* (Telepathy, 2012):

Telepathy (from the Ancient Greek τῆλε, tele meaning "distant" and πάθος, pathos or -patheia meaning "feeling, perception, passion, affliction, experience") is the purported transmission of information from one person to another without using any of our known sensory channels or physical interaction.

In this definition, brains exchange information without the intervention of the senses or any other part of the peripheral nervous system (PNS) (sensors or effectors). In fact, it is as if no physical interaction at all would be

needed—something beyond the realm of science. We will talk instead about "hyperinteraction" here: the direct, conscious transmission of information from one cerebral cortex to another cortex through some sort of noninvasive physical interaction. Without loss of generality, we may also use this term for the noninvasive bidirectional transmission of information from brains to computers. Such a link requires three elements (Fig. 2). The first is a brain-computer interface (BCI), that is, a device that can extract information from the brain of a subject we shall call the "emitter" (Alice). The second is a medium for transmission of this information. The third is computer-brain interface (CBI) to transmit information into the brain of a subject we will call the "receiver" (Bob). A bidirectional arrangement is needed for establishing a B2B dialog, but this is a natural extension of the directional system just described.

We emphasize the terms "direct" and "conscious" in our definition. The first implies a bypass of the PNS in the process. The second requires that both emitter and receiver experience the transmission of information in a fully conscious manner—a rather restrictive requirement, since other types of direct communication between brains may be of

FIG. 2 The basic elements of a brain-to-brain (B2B) interface or hyperinteraction. On the left we have the emitter *Alice*, equipped with a BCI, and on the right the receiver *Bob*, with a CBI. Data is transmitted over the Internet. Equivalently, we can require a bidirectional brain-computer interface.

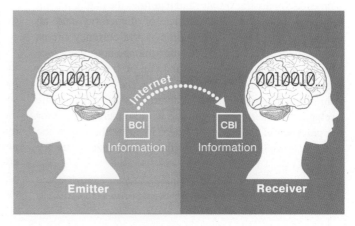

interest. For example, some recent human experiments have linked a BCI in an emitter subject with a brain stimulation system for motor control of the hand of a receiver (Rao et al., 2014). While such work provides a demonstration of B2B communication, the fact that the receiver was quite likely made aware (conscious) of the information flow by generated peripheral hand sensations (the moving hand) invalidates it as a demonstration of a direct *conscious* B2B link. Indeed, an important challenge in hyperinteraction experiments is to ensure that the conscious information transfer takes place directly via the central nervous system (CNS) and not via the PNS.

In summary, we restrict our discussion to the question of how to link two human minds directly fulfilling three important conditions, namely that the link must be (a) noninvasive, (b) cortically based, and (c) consciously driven. In this chapter I will describe a recent set of experiments carried out in close collaboration with colleagues worldwide, which demonstrated for the first time that hyperinteraction in humans is possible, and then discuss its further development.

2 NONINVASIVE ELECTROMAGNETIC INTERACTION TECHNOLOGIES

Our work relies on noninvasive electromagnetic technologies. Although invasive approaches offer great potential (see, eg, Chapin et al., 1999; O'Doherty et al., 2011; Pais-Vieira et al., 2013, 2015; Herff et al., 2015), the need for surgery and associated complications makes for an unlikely practical solution in all but extreme cases—at least in the near future. Nevertheless, invasive solutions can overcome the formidable barrier of scalp, skull, and cerebrospinal fluid to give access to the cortex, and should, in principle, be more capable of implementing broadband communication.

Electroencephalography (EEG) is a technique pioneered by Hans Berger in 1929 (Nunez and Srinivasan, 2006) to measure the tiny electrical fields generated by the living brain noninvasively from the scalp. According to our understanding of the origin of bioelectric signals in the brain (Plonsey, 1969), the potentials we observe on the scalp are due to current sources at the neuron membranes. Early research in the 1960s already demonstrated a BCI based on the voluntary control of EEG alpha rhythm power to send messages based on Morse code (Dewan, 1967), and since then the field has advanced considerably (see, eg, Wolpaw et al., 2002; Birbaumer, 2006). Modern EEG is quite capable, thanks to much improved electronics, novel mathematical algorithms, and tremendous computational power. Using EEG, it is possible today to—somewhat rudimentarily—control computers and, from them, any other devices (robots, cars) after appropriate training

(Allison et al., 2013). While the temporal resolution of EEG is excellent and well matched to the natural processes it measures (of the order of a millisecond), it currently provides rather low-resolution spatial information. However, it is now known that scalp-recorded EEG signals are mainly produced by a summation of field activities associated with large cortical pyramidal cells that are typically oriented perpendicular to the cortical surface (Nunez and Srinivasan, 2006). By determining cortical source location and orientation constraints from magnetic nuclear resonance brain imaging (MRI), the orientations of potential sources can be assumed a priori, greatly reducing the possible space of solutions in the EEG inverse problem. High-density electrode arrays (currently available with up to 512 channels), realistic, personalized head modeling, and derived cortical or tomographic mapping methods, together with signal processing, Bayesian methods, machine learning, and improved electronics will undoubtedly continue to push the spatial resolution barrier in EEG processing.

Magnetoencephalography (MEG), a closely related technique, is more sensitive to the (stronger) primary currents generated by cortical neurons and thus less affected by uncertainty in conductivity values of tissues. These currents generate magnetic fields via the Biot-Savart law, which are measured by magnetic field sensor multiarrays of about 300 channels. EEG and MEG can be measured concurrently with benefit: MEG is mainly sensitive to tangential dipoles, which are parallel to the scalp, but rather insensitive to radially oriented dipoles aligned normal to the scalp (like at the top of a gyrus or bottom of a sulcus); EEG is sensitive to the activity of both radial and tangential dipoles. MEG has also successfully been used as a BCI (see, eg, Cichy et al., 2014), also in combination with EEG (Pantazis et al., 2015).

Transcranial electrical stimulation of the brain comes in two main forms: current controlled as in transcranial current stimulation (tCS) or electromagnetically induced as in transcranial magnetic stimulation (TMS). Despite some differences, both of these techniques base their effectiveness on the interaction of their generated electric fields with neuronal populations. tCS englobes a family of noninvasive, very tolerable, and safe techniques, which include direct current (tDCS), alternating current (tACS), and random noise current stimulation. More generally, tCS is based on the delivery of arbitrary weak currents through the scalp (with electrode current intensity to area ratios of about 0.3–5 A/m) at low frequencies (typically 1 kHz) resulting in weak electric fields in the brain (with amplitudes of about 0.2–2 V/m) (Ruffini et al., 2013). Such electric fields are unlikely to produce per se action potentials, but influence the likelihood of neuronal firing by the alteration of neuronal transmembrane potentials. The sub- or suprathreshold nature of tCS depends also on its temporal characteristics—simple models of neurons already display a frequency-dependent sensitivity to tACS (resonance, Miranda et al., 2009). Other suprathreshold noninvasive techniques include transcranial electrical stimulation (TES) and electroconvulsive therapy, both of which involve much stronger (and painful) currents and electric fields than tCS.

TMS is an important noninvasive technique based on the application of strong, short (pulsed), localized electric fields to the cortex, and, unlike tCS, is it capable of directly inducing action potentials but very well tolerated—see, for example, Wassermann et al. (2008). The effects of TMS can include peripheral motor activity as well as—conveniently for us—subjective experiences of seeing light (*phosphenes*, Taylor et al., 2010) or being touched (*tactenes*, Feurra et al., 2011).

The basic mechanism in tCS (Ruffini et al., 2013 and references therein) and also TMS (Fox et al., 2004) is believed to be the coupling of electric fields to populations of ordered, elongated form-factor neurons such as pyramidal cells. Pyramidal cells account for 75% of all cortical neurons and are arranged perpendicularly to the cortical surface (Silva et al., 2008). The role of other types of neurons (eg, interneurons such as basket cells) or other brain cells like glia is not well understood. Physically, the external electric field forces the displacement of intracellular ions (which move to cancel the intracellular electric field), altering the neuron's internal charge distribution and modifying the transmembrane potential difference. For a long, straight finite fiber with space constant λ in a homogeneous electric field **E**, the transmembrane potential difference is largest at the fiber termination, with a value that can be approximated by $\delta\Phi = \lambda\mathbf{n}\cdot\mathbf{E}$, where **n** is the unit vector defining the fiber axis. In this expected first-order result, the spatial scale is provided by the space constant, and the effect is modulated by the field to fiber relative orientation (Fig. 3). These considerations imply that the greatest effect of stimulation will be determined by both field magnitude and by its direction relative to the orientation of neuronal populations.

Electrical current propagation models play a key role in EEG/MEG and in electrical stimulation of the brain. These techniques are related at the fundamental level by Poisson's equation (Miranda et al., 2009) and the associated reciprocity theorem (Plonsey, 1969). In particular, starting from a realistic forward model of electric currents, we can develop a forward model of EEG and associated inverse models to map scalp to source space, as already shown by Rush and Driscoll (1969) (see also Ruffini, 2015, for a generalization of this theorem to multiple sources and electrodes, of relevance to tCS). In Miranda et al. (2013), the authors describe advances in finite element modeling (FEM) of current propagation in an MRI-derived realistic human head and provide a realistic five-layer head model derived from magnetic resonance images applicable to tCS using multiple small electrodes. The model emphasizes the role of the normal (orthogonal) electric fields in the cortical surface, in agreement with our understanding of the role of pyramidal neurons in the cortex in tCS (and also EEG). Modeling can be used to understand the impact of tCS and, more importantly, to design optimized tCS montages (Ruffini et al., 2014)—as I discuss below. Fig. 4 displays similarities and differences of TMS and modern multichannel tCS generated electric fields from such realistic modeling.

3 NATURAL VERSUS ARTIFICIAL CODING; OTHER DIMENSIONS

Our experimental paradigm hinges on the idea of transmitting binary data—with perhaps some semantics associated, such as "right" or "left"—generated by a random (50/50) uniform distribution, and showing that transmission error is lower than randomly expected in a statistically

FIG. 3 Model for interaction of electric field with elongated neuron. On the left, pyramidal neuron population from the human cortex (from Santiago Ramon and Cajal, 1899. Comparative Study of the Sensory Areas of the Human Cortex. Wikipedia Public Domain.) and electric field. On the right, zoomed view of a single pyramidal neuron (from the 1903 edition of Sobotta's Histology, Wikipedia Public Domain.) displayed with superimposed electric field and space constant vectors pointing into orthodromic direction. The electric field displaces charges inside the cell and this alters the transmembrane potential of the neuron.

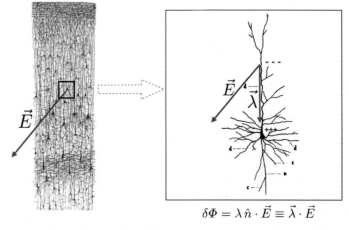

$$\delta\Phi = \lambda\,\hat{n}\cdot\vec{E} \equiv \vec{\lambda}\cdot\vec{E}$$

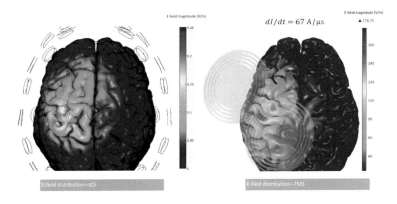

$$dI/dt = 67 \text{ A}/\mu s$$

FIG. 4 The electric fields (magnitude) generated by multichannel tCS (left) and TMS (right). Note the similar spatial distributions despite the different scales. *(Salvador, R., Miranda, P. private communication.)*

significant way. It can be argued that this paradigm (in binary form or not) will apply to any B2B experiment in one way or another, since any such experiment will aim to show that information has been correctly transmitted. For instance, we may have found a way to monitor Alice as she visualizes "house" or a "dog" to, respectively, stimulate Bob to feel a touch to the left or right hand, and use such codes to transmit binary information. The probability that a randomly generated sequence of 1's and 0's will be correctly transmitted by pure chance alone will quickly approach zero if the list of symbols is long enough. For example, the probability of guessing correctly 140 random bits with an error rate of 20% (28 errors out of 140) or less is extremely low (equivalent to obtaining 112 heads or more after 140 tosses of a fair coin, $p < 10^{-13}$). Hence, statistically robust results should be readily obtained with a working B2B system.

We can distinguish two approaches to B2B communication. In the first we use what we call *natural coding*. By natural coding we mean transmission protocols in which the source and target cortical areas and modalities resemble the ones already used by the brain. For example, we may stimulate a part of the somatosensory cortex (SMC) of Alice associated with the hand using electric fields with the frequencies or other temporal aspects of EEG associated with hand motor imagery—with the hope of eliciting tactile sensations. The idea is to use the natural neural codes used in the brain. If the emitter Alice imagines moving her right hand and the receiver Bob feels or imagines moving his right hand, using such means we will have achieved natural transmission. No training of the code is in principle needed (although may be used to improve link quality): we may hypothesize that the brain will respond naturally to the electrical stimulus used because it is being stimulated in a way it already recognizes—hence the term "natural." Pioneering work by Pais-Vieira et al. (2013, 2015) has demonstrated invasive B2B communication with a natural code in the rat by interfacing the sensory motor cortices of animals—linking the neural activity in the animals brains

electrophysiologically. Such natural codes may carry information by transmitting sensory signals, or, perhaps, others such as imagery or emotions. A recent paper describes the use of what we could call seminatural coding: anodal tDCS over the left motor cortex is shown to induce general and athletic motor imagery (Speth et al., 2015).

In contrast, *artificial coding* relies on any form of conscious CBI or BCI to transmit arbitrary data using ad hoc coding schemes. For instance, if we want to transmit the concept "yes" or "no," we may use a motor imagery code coupled to a visual phosphene code to transmit the world "red house" in binary form. The receiver Bob will know what concept was transmitted, but has to rely on the artificially generated code to decipher the message (eg, "light sensation" means "yes"). The percepts or sensations used in the BCI and/or the CBI are unrelated to the message content.

The concept of *ephaptic interaction* may provide the basis for electromagnetic natural coding. Ephaptic interactions are postulated to represent another form of communication in the brain, one that travels at fast electrical speeds (according to Plonsey, 1969, of the order of 1×10^5 m/s): the direct interaction between neurons that through the self-generated electric fields. Endogenous electric field activity in the brain typically induces extracellular fields under 5 V/m (see references in Anastassiou et al., 2011). Jefferys (1995) defined ephaptic interaction as "population field effect in which the synchronous activity of populations of neurons cause large electric fields that can affect the excitability of suitably oriented, but not closely neighboring, neurons." Pyramidal cell populations are natural candidates for this role as ephaptic antennas: in order to radiate or receive effectively, sufficient power is needed, and this can only be achieved through temporal and spatial coherence of neuronal populations. Such coherence is observed in the brain at low frequencies—precisely the EEG realm. It has been shown that neuronal circuits are surprisingly sensitive to rather weak external fields (Deans et al., 2007), and, in fact, the field of tCS rests on this phenomenon. Recent work in vitro by Fröhlich and McCormick (2010) showed that neuronal populations in the neocortex are sensitive to endogenously generated electric fields and hints at the potential role of ephaptic interactions in the cortex by showing that weak endogenous fields can enhance and entrain physiological neocortical network activity. Similarly, Anastassiou et al. (2011) stimulated and recorded from rat cortical pyramidal neurons in vitro and found that, despite the small size of extracellular fields, they could strongly entrain action potentials, particularly for slow (8 Hz) fluctuations of the extracellular field.

These observations imply that neural stimulation using tCS may be tapping into very powerful natural mechanisms of brain function, since the magnitude and frequencies of tCS and endogenous electric fields are similar. Ephaptic mechanisms may be used by brains to internally transmit information at very fast speeds to complement synaptic communication. In this sense, we may say that if ephaptic, weak field mechanisms are used in the brain to connect populations, we will expand their reach with B2B technology to achieve "tele-ephaptic" interaction—especially in the case of natural coding. Going a bit further, if EEG generated electric fields are not an epiphenomenon but a causal ephaptic agent in the brain, we could guide a noninvasive CBI with EEG. Naively, we would use a multichannel tCS device to bring together brains via such tele-ephaptic interaction: the EEG from Alice (in some form, eg, filtered to some band) being transmitted to a tCS system, which would stimulate Bob with electric fields based on this EEG pattern—and vice versa. More concretely, consider the following thought experiment: (1) Alice visualizes hand motor movements, which are picked up by a BCI system in the form of EEG from a set of electrodes, (2) the EEG is processed and perhaps simplified by software, and then (3), the associated waveform is "injected" via tCS into Bob using electrodes over the same motor area. We may call the last step "transcranial endogenous (or ephaptic) field stimulation" (tEFS). Such a setup would take ephaptic interactions up a level to the interaction across humans, in a way connecting brains using a natural mechanism. It would be very fast. I will discuss this possibility further.

Another important element to consider is the need for training of the subjects to receive signals. Training may help a subject detect faint signals, but there is the danger that he will—knowingly or not—learn to rely on faint sensory cues. Even without explicit training, if the receiver Bob detects that the transmission is based on two states of the CBI apparatus, he can make a guess as to which state corresponds to what as long as the B2B transmission's performance is better than random. This may help him to better detect the signals after a while, but then the transmission of information will partly take place via the PNS. There are several ways to handle this problem in CBI with tCS, including, for example, using families of stimulation montages, which are designed to generate the same cortical effects but with different PNS effects, or controlling by reduced performance of the PNS, that is, by working with subjects in which the afferent PNS receptors are compromised (eg, by using a local anesthetic on the skin).

Finally, we may also point out another dimension in CBI: it may either target an area associated with sensory processing (eg, the SMC, occipital cortex) or not (eg, Broca's area or the dorsolateral prefrontal cortex, or a network associated with negative emotions as in Chang et al., 2015). While we expect more immediate success in the former, the latter is possibly the more interesting aspect to study. Ultimately, brain networks throughout the cortex will be targeted, as discussed below.

4 THE FIRST TESTS (2008–12)

The first experiments were partly funded by the EU FET Open project HIVE (http://hive-eu.org), which aimed to explore noninvasive technologies for hyperinteraction. The project helped to develop a new class of noninvasive brain stimulation technologies, including hybrid EEG/multichannel tCS (Schestatsky et al., 2013) and realistic FEM of the brain for simulation of electric fields and EEG. Several hyperinteraction experiments in humans and animals were carried out. At that time, our team had ample experience with EEG-based BCIs, and in particular in the use of the BCI2000 analysis software (Schalk and Mellinger, 2010) with the EEG system (*Enobio*) we had developed. The first conscious CBI experiments (2010–12) relied on tACS with the goal of eliciting tactenes, a technique described by Feurra et al. (2011), who investigated tACS over the primary SMC to elicit tactile sensations in humans. Their results indicated that tACS in alpha (10–14 Hz) and high gamma (52–70 Hz) frequency range produced tactile sensations in the contralateral hand, with a weaker effect also observed in beta (16–20 Hz). After carrying out our tests, however, we found this approach to be insufficiently reliable for our purposes. Although we observed a statistical shift in the lateralization of perceived sensations with stimulation, we decided to test a TMS-based CBI for a faster demonstration of hyperinteraction. Nevertheless, I argue below that there is still much research to do with tCS, with a wide range of montages and waveforms to test.

Thus, we next investigated (2012) the elicitation of "tactenes" (tactile sensations) using TMS over the SMC (as also described by Feurra et al., 2011). We found this method to be more reliable for transmission of information, with error rates of about 20%. However, it appeared to be hard to elicit tactenes without generating concurrent motor activation of the hand (as monitored by EMG)—the motor and somatosensory areas of the cortex being rather close relative to the spatial resolution of TMS. Even with no motor evoked potentials present, it was unclear that the TMS pulse was not affecting the cortical motor area and initiating motor activity, and cueing the subject indirectly via PNS mediated tactile sensations. In parallel, we tested TMS over the occipital cortex to generate phosphenes (visual sensations of light), and realized it would be a reliable medium for transmission, although there remained the challenge of ensuring that subjects would not be cued by visual, tactile, or auditive sensations. Unlike in tCS, TMS phosphenes are known to be CNS mediated (Fried et al., 2011)—tCS currents can actually substantially flow through the retina (Laakso and Hirata, 2013). I note that in these first experiments we did not use robotized, neuronavigated TMS, and found it difficult to produce robust, controllable, and repeatable effects with "hand-navigation." For this reason I definitely do not discount TMS elicitation of tactenes as a potential technique for CBI.

At the same time, other members of the HIVE team were carrying out some very interesting experiments in a rabbit model. First, Márquez-Ruiz et al. (2013, 2014) showed in behaving rabbits that tDCS applied over the SMC modulates cortical processes by localized stimulation of the whisker pad or of the corresponding area of the ventroposterior medial (VPM) thalamic nucleus. Rabbits were prepared for the chronic recording of local field potentials in the SMC in response to whisker and/or VPM thalamic nucleus stimulations in the presence of tDCS and tACS. tDCS and tACS applied over the SMC-modulated cerebral cortical processes subsequent to the localized stimulation of the whisker pad or of the corresponding area of the VPM nucleus. Longer stimulation periods indicate that poststimulation effects were only observed in the SMC after cathodal tDCS. Consistently with these polarity-specific effects, the acquisition of a classical eye-blink conditioning was potentiated or depressed by the application of anodal or cathodal tDCS, respectively, when stimulation of whisker pad was used as a conditioned

stimulus (CS). Using the same animal model (Márquez-Ruiz et al., 2012a,b, 2016) the team then investigated tCS for the direct generation of perceptions and, hence, transmission of information, by transcranial stimulation of the cerebral cortex. A group of animals was prepared for classical eye-blink conditioning and simultaneous tCS. In this group, the authors showed that certain types of tACS (100 ms, 30 Hz tACS) could successfully substitute for a whisker CS during an associative learning task. That is, tACS-CS in the classically conditioned rabbits induced conditioned responses similar to those observed when a direct stimulation of the whisker pad was carried out, showing that peripheral whisker stimulation can be substituted by tACS as CS when the proper stimulation frequencies are applied. These experiments represent the first noninvasive animal CBI known to the author, although there has also been interesting work in this direction using ultrasound (Yoo et al., 2011, 2013).

5 TRANSMISSION

The first complete online computer-mediated hyperinteraction transmissions using artificial coding took place in early 2014 (concretely, Mar. 28 for transmission of the word "hola" and Apr. 7th for "ciao"), with data sent from Thiruvananthapuram (Kerala, India) (BCI site) to Strasbourg, France (CBI), as described in Grau et al. (2014). The emitter used a noninvasive BCI system based on voluntary, conscious motor imagery (of moving feet or hands) to discriminate among two states with signatures in EEG spectral power in the motor cortex (coding for the bit values of "0" and "1"). To monitor EEG we used a wireless EEG recording system (500 S/s, 24 bit, *Starstim* tCS/EEG system, Fig. 5 left). Eight Ag/AgCl electrodes were placed at F3, F4, T7, C3, Cz, C4, T8, and Fz scalp sites (10–20

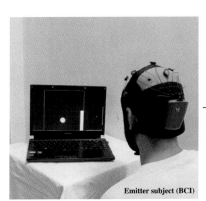

0, 1, 1, 0, 0, ...
Internet

Receiver subjects (CBI)

Emitter subject (BCI)

FIG. 5 The elements of a brain-to-brain (B2B) interface: (left) the emitter with an EEG BCI is about to correctly encode a 0 through motor imagery of moving feet. The device is actually a hybrid multichannel tCS/EEG system (*Starstim*). On the right, top, an early version (2012) of the CBI TMS element on the author (note manual coil placement). (Right, bottom) The final, robotized, neuronavigated implementation of the CBI, which allowed for controlled, repeatable stimulation effects.

EEG positioning system) and electrically referenced by a clip electrode placed in the right ear lobe. A Laplacian spatial filter was applied to the electrodes of interest (C3, Cz, and C4) by re-referencing them to the average potential of the neighboring electrodes. To transform EEG signals into binary information we used the BCI-2000 platform (Schalk and Mellinger, 2010) for the detection of anatomically localized changes in EEG related with voluntary motor imagery. The emitter was shown on a screen a sequential representation of the bits to be transmitted (the message). Each bit was represented either by a target cue in the downright part of the screen (bit value 0) or in the upright part (bit value 1) (Figs. 4–6). The emitter was to encode a bit value of 1 (or 0) through motor imagery of the hands (or feet). These motor imagery tasks controlled the vertical movement of a ball appearing on the screen from the left with a constant horizontal speed. If the BCI-controlled ball hit the displayed target, the transmitted bit was correctly encoded (and vice versa). Whatever the outcome, the BCI encoded bits were then automatically sent via email to the CBI subsystem. Following a training period, the emitter was able to regularly achieve an accuracy of well over 90% in BCI encoding.

The CBI system was based on TMS phosphene elicitation. Our initial robotized offline CBI tests (subject 1) used a position-coding strategy with one location (bit = 1) being the phosphene hotspot and another scalp location (displaced about 2 cm from the first) representing the silent condition (bit = 0). This strategy was used for CBI transmissions of 60 bit messages with a low error rate. A first hyper-interaction experiment (Barcelona to Strasbourg) was carried out offline, that is, with the BCI and CBI branches of transmission separated in time by buffering the data after BCI transmission (specifically with the BCI transmission on Sep. 23, 2013 and the CBI transmission 2 days later) and resulted in a 15% transmission error rate (5% in the BCI segment and 11% in the CBI one) using this position-dependent CBI encoding. However, we could not discount the possibility of the receiver being cued on the (active or silent) stimulation condition by PNS sensory inputs (tactile, auditory, or visual) associated with the repositioning of the coil at different scalp sites. In order to rule this out, we implemented a series of measures. First, to avoid contact related cues and taking advantage of the anisotropic response of the visual cortex to TMS (Kammer et al., 2007), we adopted the strategy of encoding bits through *rotation* of the TMS coil: the location and active orientation of the coil (producing phosphenes in most trials and coding a "1") were chosen with the condition that a 90-degree rotation of the coil on the same location did not produce phosphenes (coding a "0," Fig. 6). For each of three receiver subjects, we identified such a TMS phosphene-producing hotspot in the right visual occipital cortex (approximately 2 cm anterior and 2 cm right from inion). We thereafter achieved the required high precision in relocation and reorientation of the TMS target by using a neuronavigated (Ginhoux et al., 2013; Bashir et al., 2011; Julkunen et al., 2009), robotized TMS system (*Axilum Robotics TMS-Robot*, http://www.axilumrobotics.com, piloted by a *Localite 2.8 Neuronavigation* system using the *MagVenture MagPro R30 TMS Stimulator* with a "butterfly" coil of type *Cool-B65-RO*)—see Figs. 5–7. Subjects went through a familiarization period in which we administered several TMS pulses to the chosen hotspot site using various rotations of the coil, and identified the intensity of TMS pulses (range 57–90% of maximum intensity of the coil) that optimally discriminated *active* (ie, producing phosphenes) from *silent* (not producing phosphenes) orientations (Figs. 4–6). Subjects described the sensations of light produced by TMS pulses of the active orientation as having a strong, clear, and reliable nature, and located at the bottom of the visual field contralateral to the

FIG. 6 A more detailed view of the flow of information from the mind (conscious brain) of the emitter to the mind of the receiver with a focus on the communication channels: EEG, Bluetooth, Internet and, finally, TMS.

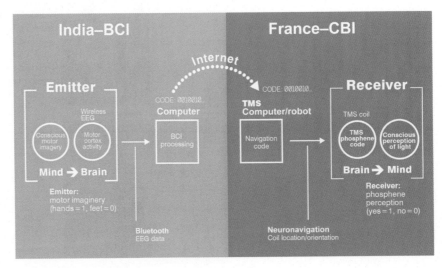

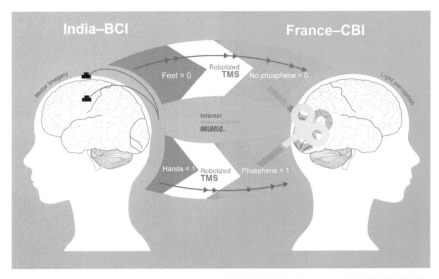

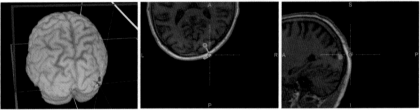

FIG. 7 (Top) Overall view of the final implementation using EEG for BCI and neuronavigated TMS with orientation encoding of information. On the left, the BCI subsystem is shown schematically, including electrodes over the motor cortex and the EEG amplifier/transmitter wireless box in the cap. Motor imagery of the feet encodes the bit value "0," of the hands the bit value "1." On the right, the CBI, highlighting the role of TMS coil orientation for encoding the two bit values. Communication between components is mediated by the Internet. (Bottom) Location and orientation of hotspot for phosphene production overlaid on MRI image of the head of subject 2. *(Reprinted from Grau, C., Ginhoux, R., Riera, A., Nguyen, T.L., Chauvat, H., Berg, M., Amengual, J., Pascual-Leone, A., Ruffini, G., 2014. Conscious brain-to-brain communication in humans using non-invasive technologies. PLoS ONE 9(8), e105225. doi: 10.1371/journal.pone.0105225.)*

stimulation site (Fried et al., 2011). They were instructed to report orally the presence of phosphenes after TMS pulse delivery. TMS pulses were administered by the robotized TMS system controlled by a researcher sitting away from the visual field of the subject, or later directly programmed into the neuronavigation computer by the BCI emailed message sequence. Sequences of two or three redundant TMS pulses were delivered 2 s apart to ensure better signal to noise ratio. The robot was programmed to move the coil away from the scalp after the delivery of each dyad/triad of TMS pulses. A force sensor on the coil surface was used to maintain a constant contact force with the scalp in all conditions. The TMS cable holder on the robot was adjusted to keep the coil's cable at a good distance from the subject's shoulders and back, preventing contact during coil rotation. To avoid auditory identification of coil orientation, subjects wore earplugs and the robot moved the coil between each pair or triad of TMS pulses toward a parking site located approximately 1 cm away from the scalp with an intermediate rotation of 45 degrees. This forced the robot to realize a movement of similar duration and with equal noise levels for all bit transmission events, irrespective of coil orientation. Lastly, we blocked visual cues on stimulation configuration by having subjects close their eyes and wear an eye mask.

To assess the effectiveness of our PNS-blocking measures, we carried out a series of control studies using the sensitivity index (or *d-prime*) statistic (Green and Swets,

1966; MacMillan and Creelman, 2005) comparing pairs of stimuli delivered either with the same or different orientations of the coil. This statistic is used in signal detection theory to measure how well a receiver can discriminate between two states in a communication channel. The first control studied TMS noise induced auditory cueing and had subjects (2 and 3) wear eye mask and earplugs and report on a sequence of 32 balanced pairs of 3 TMS stimuli randomly interspersed over silent and active conditions. We mimicked the contact of the coil, but eliminated the production of phosphenes by interposing between coil and scalp a single piece of foam slightly displacing ($\sim$1 cm) the center of the coil away from the head. After the administration of each dyad or triads of stimuli, subjects were asked if they had been delivered with the equal or different orientations. We performed a similar control experiment to evaluate tactile cues from coil skin contact, based on another sequence of 32 balanced pairs, without foam on the coil but setting the magnetic stimulator to zero intensity. Results from these tests indicated with high confidence that, after correct blinding of auditory, visual, or tactile cues, the subjects were unable to distinguish coil orientation in the absence of actual phosphene elicitation (subject 2: $d' = 0.0$ in the auditory task, $d' = -0.1$ in the skin contact task; subject 3: $d' = 0.6$ in the auditory task, $d' = 0.1$ in the skin task). That is, responses were for our purposes essentially random.

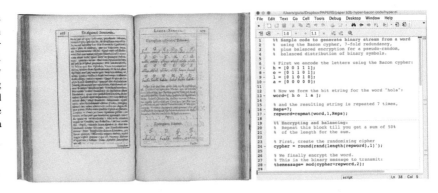

FIG. 8 (Left) The Bacon code, from Francis Bacon's *De Augmentis Scientiarum* (1623) (from the Folger Digital image collection, Folger Shakespeare Library, http://luna.folger.edu). (Right) Matlab code used to generate the binary sequence associated with the word *hola* using the Bacon code with sevenfold redundancy and further encryption using one time cipher. The symbol "a" (or "b") in Bacon becomes here a "0" (or "1").

The final set of experiments targeted the demonstration of online hyperinteraction transmission of information between remotely located subjects. On Mar. 28, 2014, 140 bits were encoded by the BCI emitter in Thiruvananthapuram and automatically emailed to Strasbourg, where a CBI receiver (subject 3) was located. There, a program parsed incoming emails to navigate the robot and deliver TMS pulses precisely over the selected site and with the appropriate coil orientation. A similar transmission with receiver subject 2 took place on Apr. 7, 2014. In both cases, the transmitted pseudorandom sequences carried encrypted messages encoding a word—"*hola*" ("*hello*" in Catalan or Spanish) in the first transmission, "*ciao*" ("hello" or "goodbye" in Italian) in the second. Words were encoded using a 5-bit Bacon cipher (Gaines, 1989) employing 20 bits with sevenfold redundancy (for a total of 140 bits). The resulting bit streams were further encrypted using a random cipher selected to produce a balanced pseudorandom sequences of 0's and 1's for the purposes of subject blinding and proper statistical analysis (Fig. 8). On reception, deciphering and majority voting from the copies of the word were used to decode the message. The BCI and CBI elements as well as the complete B2B link provided transmission of pseudorandom information with high integrity. In the first transmission the error rates were of 6%, 5%, and 11% for the BCI, CBI, and the combined B2B components, respectively, while in the second, error rates were of 2%, 1%, and 4%, respectively (after rounding). We recall that the probability of transmission of lists of 140 binary symbols having occurred with the low observed error rates or less by chance is negligible ($p < 10^{-22}$). With regard to bandwidth, BCI and CBI transmission rates were of 3 and 2 bits/min, respectively. The overall B2B transmission speed was of 2 bits/min and in practice limited by the CBI segment. We made no effort to speed up the process, since our primary goal was to transmit data faithfully. The encoded words were transmitted with full integrity by all links—BCI, CBI, and B2B.

Despite the complexity of the above setup, it is worth keeping in mind that in these experiments, for the first time

in history, a human being knew what another one was thinking ("I am moving my feet." or "I am moving my hands.") using a direct B2B link.

6 FUTURE

The current challenge for EEG and MEG (alone or combined) will be to resolve features of about 0.5 mm (Tanaka, 1997; Haynes, 2012), that is, at the level of cortical columns, which appear to be the basic functional unit of the cortex, and deploy multivariate pattern analysis tools—since the brain processes information in a spatially distributed form. Some recent results from electrophysiological signal analysis indicate this may indeed be possible (Cichy and Pantazis, 2015). As Stokes et al. (2015) eloquently state, this work suggests that both MEG and EEG contain rich spatial information in addition to unparalleled temporal resolution. Multivariate pattern analysis and machine learning can provide the tools to extract it. If this work evolves favorably, the field of BCI will experience a qualitative jump. Realistic-head modeling will certainly be needed to achieve this, as well as estimation of conductivity values using techniques such as electrical impedance tomography (Oostendorp, 2000; Gonçalves et al., 2003). Such work can leverage on advances in fMRI given the fact that blood-oxygen-level-dependent contrast imaging (BOLD) signals correlate with electrophysiological signals (Mukamel et al., 2005), and, specifically, with EEG activity in different bands (Scheeringa et al., 2011; Murta et al., 2015)— the correlation being positive in the gamma band >30 Hz, and negative in theta, alpha, and beta bands. On the other hand, fMRI multivariate pattern analysis (machine learning) has demonstrated significant potential for detection of subjective experiences in subjects (Haynes and Rees, 2006; Shinkareva et al., 2008; LaConte, 2011; Haynes, 2012), including emotions (Kassam et al., 2013; Chang et al., 2015). It is thus a natural, powerful candidate for BCI. fMRI is however, slow, bulky, and expensive. Functional near-infrared spectroscopy (fNIRS), diffuse optical spectroscopy, and other optical techniques such as diffuse correlation spectroscopy (DCS, Durduran and Yodh, 2014) are related, low-resolution noninvasive methods that allow for regional assessment of blood volume, flow, and the oxygenation state of hemoglobin in tissue (Chance et al., 1993; Edwards et al., 1993), and thus neuronal metabolism. fNIRS measures cerebral concentrations of oxyhemoglobin (HbO_2) and deoxyhemoglobin (HHb) by observing the absorption of near-infrared light, and has already been used for BCI (Naseer and Hong, 2015). DCS uses the temporal fluctuations of near-infrared light to measure cerebral blood flow noninvasively (Durduran and Yodh, 2014).

Recent advances in biophysical modeling will no doubt play an important role in the development of both BCI and CBI. Many groups—including ours—are now developing MRI-based, realistic numerical models of current propagation in the realistic human head (eg, Miranda et al., 2013) to allow to (a) improve the accuracy of source localization solutions and (b) predict tCS/TMS generated electric fields. Coupled with more capable hardware, modeling can help target tCS or TMS systems. Nevertheless, fundamental insights are needed to specify what it is to be achieved, that is, what the desired spatiotemporal electric field patterns are for encoding a given percept or thought.

It is well known that brain function is mediated by brain networks (see, eg, Sporns, 2013), which can then be targeted by tCS and TMS. Advances in neuroimaging technology, such as positron emission tomography, EEG, MEG, and resting state functional connectivity MRI are now allowing us to noninvasively visualize brain networks in humans with increasing clarity. In a parallel development, technologies now allow for multifocal stimulation. In Ruffini et al. (2014) we provided a method based on our current understanding of the interaction of tCS electric fields with the human cortex to optimize tCS targeting of both localized regions and cortical networks. Building on the hypothesis that the effects of current stimulation are to first order due to the interaction of electric fields with populations of elongated cortical neurons, we argued that the optimization problem for tCS can be defined in terms of the component of the electric field orthogonal to the cortical surface (in principle, the same methodology and logic are applicable to TMS). Using neuroimaging and other data for specification of a weighted target map on the cortical surface of excitatory, inhibitory, or neutral stimulation and a constraint on the maximal number of electrodes and currents, we showed how an optimal montage solution (electrode currents and locations) can be obtained. The present implementation of this method (called *Stimweaver*) relies on the fast calculation of multifocal tCS electric fields (including components normal and tangential to the cortical boundaries) using a five-layer finite element model of a realistic head (Miranda et al., 2013). Solutions are found using constrained least squares to optimize current intensities, with electrode number and location selected using

a genetic algorithm. As examples of the potential application of this technique for fMRI-derived network-targeting CBI, I refer, for example, to Shinkareva et al. (2008) (identification of cognitive states associated with perception of tools and dwellings), Damarla and Just (2013) (brain patterns for quantities of objects), Kassam et al. (2013) (discrimination among nine self-induced emotional states), Chang et al. (2015) (identification of a sensitive and specific brain signature that predicts the intensity of negative emotion in individuals), or Huth et al. (2016) (fMRI-derived semantic tiling of the cortex)—which also represent clear opportunities for experiments with fMRI-based BCIs of abstract concepts or emotions.

The brain is a dynamical system. The generalization of the proposed method to the case of tACS or other arbitrary temporal tCS waveforms (such as those derived from endogenous fields, tEFS, or of others such as amplitude-modulated tACS as studied by Witkowski et al. (2015)—of interest for closed-loop applications) is nontrivial, even though the process for calculation of electric fields for low frequencies (<1 kHz) is the same as for tDCS (ie, currents are simply mapped to fields independently at each time point). The difficulty lies in the definition of a physiological meaningful optimization problem. Current studies show that brain function involves orchestrated oscillatory activity at spatially separated brain regions. Phase or amplitude synchronization may relate different functional regions operating at the same or different frequencies via cross-frequency synchrony. In principle, multichannel tCS is potentially capable of coupling to such natural network rhythms through the process of resonance (tCS devices already allow for the simultaneous multisite stimulation of different cortical regions with specific frequencies and relative phases or even arbitrary temporal waveforms). Here, integrated, realistic models of EEG coupled with tCS such as the one proposed in Merlet et al. (2013) will become important tools to design montages. As I discuss now, EEG or MEG data can be used to define a weighted spatiotemporal target cortical map. In Ruffini (2015) I provided an overview on how to use the reciprocity theorem with realistic brain current propagation models to derive forward and inverse EEG models. In addition, I derived a slight generalization of the reciprocity theorem (Helmholtz, 1853; Plonsey, 1969) to the case of multiple electrode contact points and EEG dipole sources. This can be used to guide EEG-based optimization of tCS with multiple channels. The expression for the generalized reciprocity theorem for multichannel tCS with N channels is

$$\sum_{a=1}^{N} V_a I_a = -\int dV \, \vec{J}(x) \cdot \vec{E}(x),$$

where I_a is the current injected by the tCS stimulation system with a channel at point a of the scalp, $E(x)$ is the

FIG. 9 The reciprocity theorem for a multichannel tCS system using multiple current channels and interacting with a dipole source field in the brain.

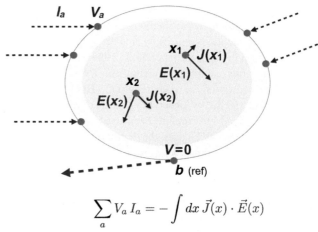

$$\sum_a V_a I_a = -\int dx\, \vec{J}(x) \cdot \vec{E}(x)$$

(1) **EEG**: dipole field generates scalp potential: $J(x) \rightarrow V_a$
(2) **tCS**: currents generate the E field: $I_a \rightarrow E(x)$

electric vector field generated by the tCS montage at brain point x, and V_a is the voltage produced at scalp point a by a set of dipole sources in the brain $J(x)$ (the last product is a vector dot product). Both currents and potentials are referenced to a common point b (see Fig. 9). This rather elegant expression says that if we wish transcranially generated electric fields and EEG dipole sources to be linearly correlated (ie, be made "parallel" and thus coupled), currents and potential arrays have to anticorrelate. This gives a simple, computationally fast way to determine optimal stimulation currents given the dynamical scalp potential target, since the goal will be to make currents and EEG potentials to be as "parallel" or "antiparallel" as possible, given some constraints. The optimization task will be to maximize (or minimize) $\sum I_a V_a$ subject to constraints on maximal current, total injected current, etc. Spontaneous EEG or event-related potential (ERP) data can be prefiltered to bands of interest or linearly processed in other ways—for example, principal component or independent component analysis as in Marco-Pallarés et al. (2005)—to refine further the optimization target, since such transformations will implicitly select the appropriate dipole generating networks. Since EEG is time varying, the resulting multichannel tCS protocols may also have complex temporal characteristics (tEFS) beyond those in the DC or AC paradigms. As a limitation, note that this approach does not allow for optimization of quantities, such as the electric field magnitude at the dipole sources. Rather, it is the specific component of the electric field *parallel* to the sources that will be optimized. Another limitation of this model-free approach is that the magnitude of generated electric fields will not be precisely known, and the difficulty in working with weighted target maps or targeting other field component, etc. Yet, this method provides a fast recipe to optimize currents to align multichannel electric fields with

EEG sources, which can be especially useful when models are not available, or in closed-loop applications such as epilepsy, where rapid adaptation of tCS parameters will be needed.

For a more generic approach, EEG/ERP/MEG data can be prefiltered to bands of interest or treated in other ways as previously described prior to tomographic inversion, for example, linearly preprocessed data can be inverted to the cortical source space using a generic or personalized realistic-head EEG model (Ruffini, 2015) to define a weighted, dynamic dipole field target map, and a tCS montage optimized as discussed before—time point by time point if needed. If the same head model is used for tCS and EEG, modeling errors will tend to cancel out. The advantage of this approach is that it allows for the full specification of electric field target and weight maps.

Similar ideas apply to the case of multichannel TMS (Ruohonen and Ilmoniemi, 1998; Roth et al., 2014). Just as with tCS, multichannel TMS could make feasible the stimulation multiple targets at functionally relevant millisecond timescales without slow (manual or robotic) repositioning of coils, synthesizing the desired electric field patterns by linear superposition under electronic control. Again, such electric field reshaping would allow modulation of large-scale brain networks with spatiotemporal precision. Work focussing on modeling the fields generated by such arrays has recently been presented by Nummenmaa et al. (2015), for example. The electric fields thus generated will be stronger than with tCS. Unlike the case of tCS, however, the pulsed nature of TMS places limits on the achievable field dynamics. Low field magnetic stimulation (Rohan et al., 2014) is another option, and more closely related to tCS.

One of the limitations of tCS or TMS stems from the underlying physics. An important result from purely mathematical considerations is that at the low frequencies considered in tCS (quasistatic approximation regime, governed by Poisson's equation), it is not possible to focalize the electric field within a constant conductivity medium (Heller and Hulsteyn, 1992). Neither the electric field magnitude nor any component can attain a local maximum or minimum within a region of constant conductivity: local maxima or minima of electric field components must occur at the boundaries of the cortex. Such a limitation does not apply to other forms of stimulation. Transcranial focussed ultrasound—already mentioned as a potential candidate for CBI—can be delivered by lenses, curved transducers, or phase arrays for precise targeting, as in its recent applications in humans for ablation of brain tissue (see, eg, Coluccia et al., 2014) or neuromodulation of electrical brain activity (Legon et al., 2014). Similarly, relatively recent efforts have been directed to study the effects of mobile phone-modulated radiofrequency (RF) radiation on EEG. For example, in Lustenberger et al. (2013), slow wave activity during sleep was increased by pulsed RF. In Roggeveen et al. (2015), significant RF radiation effects were found in the alpha, slow-beta, fast-beta, and gamma EEG bands. As with the case of ultrasound, RF waves can also be applied using antennas or phase arrays for targeting of deep brain regions: both phenomena are based on similar wave equations (as opposed to Poisson's equation).

The future evolution of hyperinteraction, or, more generally, B2B technology will undoubtedly raise ethical concerns. Some of these have recently been reviewed by Hildt (2015). Of special interest, I highlight the implications on the meaning of self and from the possibility of direct communication with other species. Are these technologies dangerous? From the point of view that the fundamental aim of hyperinteraction is to enable transfer of information consciously and voluntarily between brains, there should not be substantially novel, ethically difficult questions raised by what is—in the end—a communication technology. However, we do need to understand the basic science better to ensure that both information extraction and information injection can be controlled by a subjective conscious gatekeeper—even after subjects agree to voluntarily use a noninvasive B2B, BCI, or CBI system. On the other hand, the science and technology needed for hyperinteraction will foster our understanding of the human brain and undoubtedly spill over into clinical applications of BCI, CBI, and B2B, including open or closed-loop brain stimulation treatment of pathologies such as depression or epilepsy.

ACKNOWLEDGMENTS

This interdisciplinary work would have not been possible without the help of my collaborators—as cited in Grau et al. (2014) with regard to the first experiment—and the FET Open, HIVE, Starlab, and Neuroelectrics teams, to all of whom I am extremely grateful.

REFERENCES

Allison, B.Z., Dunne, S., Leeb, R., Millán, D.R.J., Nijholt, A. (Eds.), 2013. Towards Practical Brain-Computer Interfaces. Springer, Berlin. ISBN: 978-3-642-29746-5.

Anastassiou, C.A., Perin, R., Markram, H., Koch, C., 2011. Ephaptic coupling of cortical neurons. Nat. Neurosci. 14, 217–223.

Astolfi, L., Toppi, J., De Vico Fallani, F., Vecchiato, G., Salinari, S., Mattia, D., Cincotti, F., Babiloni, F., 2010. Neuroelectrical hyperscanning measures simultaneous brain activity in humans. Brain Topogr. 23 (3), 243–256. http://dx.doi.org/10.1007/s10548-010-0147-9.

Bacon, F., 1623. Opera Francisci Baronis de Verulamio, vice-comitis Sancti Albani, tomus primus: qui continet De dignitate & augmentis scientiarum libros IX. Ad regem suum. In officina Ioannis Haviland, London (De augmentis scientiarum). http://luna.folger.edu/luna/servlet/s/n9165h.

Bashir, S., Edwards, D., Pascual-Leone, A., 2011. Neuronavigation increases the physiologic and behavioral effects of low-frequency rTMS of primary motor cortex in healthy subjects. Brain Topogr. 24 (1), 54–64. http://dx.doi.org/10.1007/s10548-010-0165-7.

Birbaumer, N., 2006. Breaking the silence: brain-computer interfaces (BCI) for communication and motor control. Psychophysiology 43, 517–532.

Chance, B., Zhuang, Z., UnAh, C., Alter, C., Lipton, L., 1993. Cognition-activated low-frequency modulation of light absorption in human brain. Proc. Natl. Acad. Sci. U. S. A. 90, 3770–3774.

Chang, L.J., Gianaros, P.J., Manuck, S.B., Krishnan, A., Wager, T.D., 2015. A sensitive and specific neural signature for picture-induced negative affect. PLoS Biol. 13 (6), e1002180. http://dx.doi.org/10.1371/journal.pbio.1002180.

Chapin, J.K., Moxon, K.A., Markowitz, R.S., Nicolelis, M.A.L., 1999. Real-time control of a robot arm using simultaneously recorded neurons in the motor cortex. Nat. Neurosci. 2, 664–670. http://dx.doi.org/10.1038/10223.

Cichy, R.M., Pantazis, D., 2015. Can visual information encoded in cortical columns be decoded from magnetoencephalography data in humans? In: Poster Presented at OHBM 2015, Honolulu. http://arxiv.org/ftp/arxiv/papers/1503/1503.08030.pdf.

Cichy, R.M., Pantazis, D., Oliva, A., 2014. Resolving human object recognition in space and time. Nat. Neurosci. 17, 455–462. http://dx.doi.org/10.1038/nn.3635.

Coluccia, D., Fandino, J., Schwyzer, L., O'Gorman, R., Remonda, L., Anon, J., Martin, E., Werner, B., 2014. First noninvasive thermal ablation of a brain tumor with MR-guided focused ultrasound. J. Ther. Ultrasound 2, 17.

Cover, T.M., Thomas, J.A., 1991. Elements of Information Theory. Wiley-Interscience, New York, NY.

Damarla, S.R., Just, M.A., 2013. Decoding the representation of numerical values from brain activation patterns. Hum. Brain Mapp. 34, 2624–2634.

Deans, J.K., Powell, A.D., Jefferys, J.G.R., 2007. Sensitivity of coherent oscillations in rat hippocampus to AC electric fields. J. Physiol. 583 (2), 555–565.

Dewan, E.M., 1967. Occipital alpha rhythm eye position and lens accommodation. Nature 214, 975–977. http://dx.doi.org/10.1038/214975a0.

Durduran, T., Yodh, A.G., 2014. Diffuse correlation spectroscopy for noninvasive, micro-vascular cerebral blood measurement. NeuroImage 85, 5163.

Edwards, A.D., Richardson, C., van der Zee, P., Elwell, C., Wyatt, J.S., Cope, M., Delpy, D.T., Reynolds, E.O., 1993. Measurement of hemoglobin flow and blood flow by near-infrared spectroscopy. J. Appl. Physiol. 75, 1884–1889.

Feurra, M., Paulus, W., Walsh, V., Kanai, R., 2011. Frequency specific modulation of human somatosensory cortex. Front. Psychol. 2, (Article 13).

Fox, P.T., Narayana, S., Tandon, N., Sandoval, H., Fox, S.P., Kochunov, P., Lancaster, J.L., 2004. Column-based model of electric field excitation of cerebral cortex. Hum. Brain Mapp. 22 (1), 1–14.

Fox, M.D., Buckner, R.L., Liu, H., Chakravarty, M.M., Lozano, A.M., Pascual-Leone, A., 2014. Resting-state networks link invasive and noninvasive brain stimulation across diverse psychiatric and neurological diseases. Proc. Natl. Acad. Sci. U. S. A. 111 (41), E4367–E4375. http://dx.doi.org/10.1073/pnas.1405003111.

Fried, P.J., Elkin-Frankston, S., Rushmore, R.J., Hilgetag, C.C., Valero-Cabre, A., 2011. Characterization of visual percepts evoked by noninvasive stimulation of the human posterior parietal cortex. PLoS ONE 6 (11), e27204.

Fröhlich, F., McCormick, D.A., 2010. Endogenous electric fields may guide neocortical network activity. Neuron 67, 129–143.

Gaines, H.F., 1989. Cryptanalysis: A Study of Ciphers and Their Solutions. Dover Publications, New York, NY.

Ginhoux, R., Renaud, P., Zorn, L., Goffin, L., Bayle, B., Foucher, J., Lamy, J., Armspach, J.P., de Mathelin, M., 2013. A custom robot for transcranial magnetic stimulation: first assessment on healthy subjects. Conf. Proc. IEEE Eng. Med. Biol. Soc. 2013, 5352–5355.

Gonçalves, S.I., de Munck, J.C., Verbunt, J.P.A., Bijma, F., Heethaar, R.M., Lopes da Silva, F., 2003. In vivo measurement of the brain and skull resistivities using an EIT-based method and realistic models for the head. IEEE Trans. Biomed. Eng. 50 (6), 754–767.

Grau, C., Ginhoux, R., Riera, A., Nguyen, T.L., Chauvat, H., Berg, M., Amengual, J., Pascual-Leone, A., Ruffini, G., 2014. Conscious brain-to-brain communication in humans using non-invasive technologies. PLoS ONE 9 (8), e105225. http://dx.doi.org/10.1371/journal.pone.0105225.

Green, D.M., Swets, J.A., 1966. Signal Detection Theory and Psychophysics. Wiley, New York, NY.

Hasson, U., Furman, O., Clark, D., Dudai, Y., Davachi, L., 2008. Enhanced intersubject correlations during movie viewing correlate with successful episodic encoding. Neuron 57, 452–462.

Haynes, J.D., 2012. Brain reading. In: Richmond, S.D., Rees, G., Edwards, S.J.L. (Eds.), I Know What You're Thinking: Brain Imaging and Mental Privacy. Oxford University Press, Oxford (Chapter 3).

Haynes, J.D., Rees, G., 2006. Decoding mental states from brain activity in humans. Nat. Rev. Neurosci. 7, 523–534.

Heller, L., Hulsteyn, D.B., 1992. Brain stimulation using electromagnetic sources: theoretical aspects. Biophys. J. 63, 129–138.

Helmholtz, H., 1853. Uber einige Gesetz der Vertheilung elektrischer Strdme in korperlichen Leitern, mit Anwendung auf die thierisch-elektrischen Versuche. Ann. Phys. Chem. 89, 211–233. 353–377.

Herff, C., Heger, D., de Pesters, A., Telaar, D., Brunner, P., Schalk, G., Schultz, T., 2015. Brain-to-text: decoding spoken phrases from phone representations in the brain. Front. Neurosci. http://dx.doi.org/10.3389/fnins.2015.00217.

Hildt, E., 2015. What will this do to me and my brain? Ethical issues in brain-to-brain interfacing. Front. Syst. Neurosci. 9 (Article 17).

Huth, A.G., de Heer, W.A., Griffiths, T.L., Theunissen, F.E., Gallant, J.L., 2016. Natural speech reveals the semantic maps that tile human cerebral cortex. Nature 532, 453–458.

Jefferys, J.G.R., 1995. Nonsynaptic modulation of neuronal activity in the brain: electric currents and extracellular ions. Physiol. Rev. 75 (4), 1995.

Julkunen, P., Säisänen, L., Danner, N., Niskanen, E., Hukkanen, T., Mervaala, E., Könönen, M., 2009. Comparison of navigated and non-navigated transcranial magnetic stimulation for motor cortex mapping, motor threshold and motor evoked potentials. NeuroImage 44, 790–795.

Kammer, T., Vorwerg, M., Herrnberger, B., 2007. Anisotropy in the visual cortex investigated by neuronavigated transcranial magnetic stimulation. NeuroImage 36, 313–321.

Kassam, K.S., Markey, A.R., Cherkassky, V.L., Loewenstein, G., Just, M.A., 2013. Identifying emotions on the basis of neural activation. PLoS ONE 8 (6), e66032.

Laakso, I., Hirata, A., 2013. Computational analysis shows why transcranial alternating current stimulation induces retinal phosphenes. J. Neural Eng. 10 (4), 046009. http://dx.doi.org/10.1088/1741-2560/10/4/046009.

LaConte, S.M., 2011. Decoding fMRI brain states in real-time. NeuroImage 56, 440–454.

Legon, W., Sato, T.F., Opitz, A., Mueller, J., Barbour, A., Williams, A., Tyler, W.J., 2014. Transcranial focused ultrasound modulates the activity of primary somatosensory cortex in humans. Nat. Neurosci. 17, 322–329. http://dx.doi.org/10.1038/nn.3620.

Lloyd, S., 2013. The universe as quantum computer. arXiv: 1312.4455v1 [quant-ph], http://arxiv.org/abs/1312.4455v1.

Lustenberger, C., Murbach, M., Dürr, R., Schmid, M.R., Kuster, N., Achermann, P., Huber, R., 2013. Stimulation of the brain with radiofrequency electromagnetic field pulses affects sleep-dependent performance improvement. Brain Stimul. 6 (5), 805–811. http://dx.doi.org/10.1016/j.brs.2013.01.017.

MacMillan, N., Creelman, C., 2005. Detection Theory: A User's Guide. Lawrence Erlbaum Associates, Mahwah, NJ.

Marco-Pallarés, J., Grau, C., Ruffini, G., 2005. Combined ICA-LORETA analysis of mismatch negativity. NeuroImage 25 (2), 471–477.

Márquez-Ruiz, J., Leal-Campanario, R., Sánchez-Campusano, R., Molaee-Ardekani, B., Wendling, F., Miranda, P.C., Ruffini, G., Gruart, A., Delgado-García, J.M., 2012a. Transcranial direct-current stimulation modulates, synaptic mechanisms involved in associative learning in behaving rabbits. PNAS 109 (17), 6710–6715.

Márquez-Ruiz, J., Leal-Campanario, R., Sánchez-Campusano, Ammann, C., Molaee-Ardekani, B., Wendling, F., Miranda, P.C., Ruffini, G., Gruart, A., Delgado-García, J.M., 2012b. Transcranial Current Stimulation (tCS) Over Somatosensory Cortex Modulates Classical Eyeblink Conditioning and Induces Tactile Perception in Rabbits. Program No. 706.03/EEE19. 2012 Neuroscience Meeting Planner, Society for Neuroscience, New Orleans, LA.

Márquez-Ruiz, J., Ammann, C., Leal-Campanario, R., Wendling, F., Ruffini, G., Gruart, A., Delgado-García, J.M., 2013. OP 8. Modulating tactile perception learning processes by tCS in animal models: hyperinteraction viability experiments (HIVE). Clin. Neurophysiol. 1388-2457124 (10), e59–e60. http://dx.doi.org/10.1016/j.clinph.2013.04.075.

Márquez-Ruiz, J., Leal-Campanario, R., Wendling, F., Ruffini, G., Gruart, A., Delgado-García, J.M., 2014. Transcranial electrical stimulation in animals. In: Cohen Kadosh, R. (Ed.), The Stimulated Brain. Academic Press, San Diego, CA, ISBN: 9780124047044, pp. 117–144 (Chapter 5).

Márquez-Ruiz, J., Ammann, C., Leal-Campanario, R., Ruffini, G., Gruart, A., Delgado-García, J.M., 2016. Synthetic tactile perception induced by transcranial alternating-current stimulation can substitute for natural sensory stimulus in behaving rabbits. Sci. Rep 6, 1–12.

Merlet, I., Birot, G., Salvador, R., Molaee-Ardekani, B., Mekonnen, A., Soria-Frish, A., Ruffini, G., Miranda, P.C., Wendling, F., 2013. From oscillatory transcranial current stimulation to scalp EEG changes: a biophysical and physiological modeling study. PLoS ONE 8 (2), e57330. http://dx.doi.org/10.1371/journal.pone.0057330.

Miranda, P., Wendling, F., Merlet, I., Molaee-Ardekani, B., Dunne, S., Soria-Frisch, A., Whitmer, D., 2009. In: Ruffini, G. (Ed.), Brain Stimulation: Models, Experiments and Open Questions, HIVE: Hyper Interaction Viability Experiments, EU FET Open—222079, Deliverable D1.1. Available at http://wiki.neuroelectrics.com/images/4/4d/HIVE-D1.1_State-of-the-art-V1.3withcovers-small.pdf.

Miranda, P.C., Mekonnen, A., Salvador, R., Ruffini, G., 2013. The electric field in the cortex during transcranial current stimulation. NeuroImage 70, 4858.

Mukamel, R., Gelbard, H., Arieli, A., Hasson, U., Fried, I., Malach, R., 2005. Coupling between neuronal firing, field potentials, and FMRI in human auditory cortex. Science 309, 951–954.

Murta, T., Leite, M., Carmichael, D.W., Figueiredo, P., Lemieux, L., 2015. Electrophysiological correlates of the BOLD signal for EEG-informed fMRI. Hum. Brain Mapp. 36, 391–414.

Naseer, N., Hong, K.-S., 2015. fNIRS-based brain-computer interfaces: a review. Front. Hum. Neurosci. 9, 3. http://dx.doi.org/10.3389/fnhum.2015.00003.

Nicolelis, M.A., 2010. Beyond Boundaries: The New Neuroscience of Connecting Brains With Machines and How It Will Change Our Lives. St. Martin's Griffin, New York, NY.

Nummenmaa, A., Raij, T., Hämäläinen, M., Okada, Y., 2015. Targeting accuracy and efficiency of a multichannel TMS array: a simulation study. In: Presented at OHBM 2015, Honolulu.

Nunez, P.L., Srinivasan, R., 2006. The Electric Fields of the Brain: the Neurophysics of EEG, second ed. Oxford University Press, New York, NY.

O'Doherty, J.E., Lebedev, M.A., Ifft, P.J., Zhuang, K.Z., Shokur, S., Bleuler, H., Nicolelis, M.A.L., 2011. Active tactile exploration enabled by a brain-machine-brain interface. Nature 479, 228–231.

Oostendorp, T.F., 2000. The conductivity of the human skull: results of in vivo and in vitro measurements. IEEE Trans. Biomed. Eng. 47 (11), 1487–1492.

Pais-Vieira, M., Lebedev, M., Kunicki, C., Wank, J., Nicolelis, M.A.L., 2013. A brain-to-brain interface for real-time sharing of sensorimotor information. Sci. Rep. 3, 1319. http://dx.doi.org/10.1038/srep01319.

Pais-Vieira, M., Chiuffa, G., Lebedev, M., Yadav, A., Nicolelis, M.A.L., 2015. Building an organic computing device with multiple interconnected brains. Sci. Rep. 5, 11869. http://dx.doi.org/10.1038/srep11869.

Pantazis, D., Cichy, R., Chang, Y.-T., Bagherzadeh, Y., 2015. Multivariate pattern analysis of MEG and EEG reveals the emergence of human object representations. In: Abstract Presented at OHBM 2015, Honoluluj. Poster Number:1893.

Plonsey, R., 1969. Bioelectric Phenomena. McGraw-Hill, New York, NY.

Rao, R.P.N., Stocco, A., Bryan, M., Sarma, D., Youngquist, T.M., Wu, J., Prat, C.S., 2014. A direct brain-to-brain interface in humans. PLoS ONE 9 (11), e111332.

Roggeveen, S., van Os, J., Viechtbauer, W., Lousberg, R., 2015. EEG changes due to experimentally induced 3G mobile phone radiation. PLoS ONE 10 (6), e0129496. http://dx.doi.org/10.1371/journal.pone.0129496.

Rohan, M.L., Yamamoto, R.T., Ravichandran, C.T., Cayetano, K.R., Morales, O.G., Olson, D.P., Vitaliano, G., Paul, S.M., Cohen, B.M., 2014. Rapid mood-elevating effects of low field magnetic stimulation in depression. Biol. Psychiatry 76, 186–193.

Roth, Y., Levkovitz, Y., Pell, G.S., Ankry, M., Zangen, A., 2014. Safety and characterization of a novel multi-channel TMS stimulator. Brain Stimul. 7 (2), 194–205.

Ruffini, G., 2007. Information, complexity, brains and reality (Kolmogorov Manifesto). arXiv:0704.1147 [physics.gen-ph]http://arxiv.org/abs/0704.1147.

Ruffini, G., 2009. Reality as simplicity. arXiv:0903.1193 [physics.gen-ph], http://arxiv.org/abs/0903.1193.

Ruffini, G., 2015. Application of the reciprocity theorem to EEG inversion and optimization of EEG-driven tCS (tDCS, tACS and tRNS). arXiv:1506.04835 [physics.bio-ph], http://arxiv.org/abs/1506.04835.

Ruffini, G., 2016. One second of consciousness, in preparation, arxiv.org.

Ruffini, G., Wendling, F., Merlet, I., Molaee-Ardekani, B., Mekonnen, A., Salvador, R., Soria-Frisch, A., Grau, C., Dunne, S., Miranda, P.C.M., 2013. Transcranial current brain stimulation (tCS): models and technologies. IEEE Trans. Neural Syst. Rehab. Eng. 21 (3), 333–345.

Ruffini, G., Fox, M.D., Ripolles, O., Miranda, P.C., Pascual-Leone, A., 2014. Optimization of multifocal transcranial current stimulation for weighted cortical pattern targeting from realistic modeling of electric fields. NeuroImage 89, 216–225.

Ruohonen, J., Ilmoniemi, R.J., 1998. Focusing and targeting of magnetic brain stimulation using multiple coils. Med. Biol. Eng. Comput. 36 (3), 297–301.

Rush, S., Driscoll, D.A., 1969. EEG electrode sensitivity: an application of reciprocity. IEEE Trans. Biomed. Eng. BME-16, 15–22.

Sanchez-Vives, M.V., Slater, M., 2005. From presence to consciousness through virtual reality. Nat. Rev. Neurosci. 6, 332.

Schalk, G., Mellinger, J.A., 2010. Practical Guide to Brain-Computer Interfacing With BCI2000, first ed. Springer, Berlin. 264 pp.

Scheeringa, R., Fries, P., Petersson, K.-M., Oostenveld, R., Grothe, I., Norris, D.G., Hagoort, P., Bastiaansen, M.C.M., 2011. Neuronal dynamics underlying high- and low-frequency EEG oscillations contribute independently to the human BOLD signal. Neuron 69, 572–583.

Schestatsky, P., Morales-Quezada, L., Fregni, F., 2013. Simultaneous EEG monitoring during transcranial direct current stimulation. J. Vis. Exp. (76), e50426. http://dx.doi.org/10.3791/50426.

Shinkareva, S.V., Mason, R.A., Malave, V.L., Wang, W., Mitchell, T.M., Just, M.A., 2008. Using fMRI brain activation to identify cognitive states associated with perception of tools and dwellings. PLoS ONE 3 (1), e1394. http://dx.doi.org/10.1371/journal.pone.0001394.

Silva, S., Basser, P.J., Miranda, P.C., 2008. Elucidating the mechanisms and loci of neuronal excitation by transcranial magnetic stimulation using a finite element model of a cortical sulcus. Clin. Neurophysiol. 119, 2405–2413.

Speth, J., Speth, C., Harley, T.A., 2015. Transcranial direct current stimulation of the motor cortex in waking resting state induces motor imagery. Conscious. Cogn. 36, 298–305.

Sporns, O., 2013. Structure and function of complex brain networks. Dialogues Clin. Neurosci. 15 (3), 247–262.

Stokes, M.G., Wolff, M.J., Spaak, E., 2015. Decoding rich spatial information with high temporal resolution. Trends Cogn. Sci. 19 (11), 636–638.

Tanaka, K., 1997. Mechanisms of visual object recognition: monkey and human studies. Curr. Opin. Neurobiol. 7 (4), 523–529.

Taylor, P.C.J., Walsh, V., Eimer, M., 2010. The neural signature of phosphene perception. Hum. Brain Mapp. 31 (9), 1408–1417. http://dx.doi.org/10.1002/hbm.20941.

Telepathy, 2012. In Wikipedia. https://en.wikipedia.org/wiki/Telepathy (retrieved 20.07.15).

Wassermann, E.M., Epstein, C.M., Ziemann, U., Walsh, V., Paus, T., Lisanby, S.H., 2008. The Oxford Handbook of Transcranial Stimulation. Oxford University Press, New York, NY.

Witkowski, M., Garcia-Cassio, E., Chander, B.S., Braun, C., Birbaumer, N., Robinson, S.E., Soekadar, S.R., 2015. Mapping entrained brain oscillations during transcranial alternating current stimulation (tACS). Neuroimage 17. http://dx.doi.org/10.1016/j.neuroimage.2015.10.024. pii: S1053-8119(15)00934-9.

Wolpaw, J.R., Birbaumer, N., McFarland, D.J., Pfurtscheller, G., Vaughan, T.M., 2002. Brain-computer interfaces for communication and control (Review). Clin. Neurophysiol. 113 (6), 767–791.

Yoo, S.-S., Bystritsky, A., Lee, J.-H., Zhang, Y., Fischer, K., Min, B.-K., McDannold, N.J., Pascual-Leone, A., Jolesz, F.A., 2011. Focused ultrasound modulates region-specific brain activity. NeuroImage 56, 1267–1275.

Yoo, S.-S., Kim, H., Filandrianos, E., Taghados, S.J., Park, S., 2013. Non-invasive brain-to-brain interface (BBI): establishing functional links between two brains. PLoS ONE 8 (4), e60410. http://dx.doi.org/10.1371/journal.pone.0060410.

Part III

Philosophical Axis

Chapter 19

Philosophical Aspects of Closed-Loop Neuroscience

W. Glannon* and C. Ineichen†

*University of Calgary, Calgary, AB, Canada, †University of Zurich, Zurich, Switzerland

1 INTRODUCTION

Experimental and clinical neuroscience has advanced to the point where implantable stimulators can alter a range of neural functions. One of the most significant applications of the science is therapeutic deep brain stimulation (DBS) to modulate dysregulated neural circuits implicated in neurological and psychiatric disorders. In DBS, electrical stimulation of subcortical brain circuits at certain frequencies (generally >100 Hz) can restore normal physiological oscillations in these circuits and alleviate symptoms (Lozano and Lipsman, 2013). A DBS system consists of one or more electrodes implanted unilaterally or bilaterally in a targeted brain region using MRI-guided stereotactic techniques. The electrodes are connected to leads and stimulated by a pulse generator implanted subcutaneously below the clavicle or abdomen. Activation of the generator and the level of current transmitted to the electrodes are controlled by a manually operated programmable device. DBS is FDA-approved for movement disorders such as Parkinson's disease (PD), essential tremor, and dystonia (Abramowicz et al., 2014). The technique has also been used to treat seizure disorders. It was granted a humanitarian device exemption for obsessive-compulsive disorder (OCD) in 2009, but is still considered experimental and investigational for treatment-refractory OCD, major depressive disorder (MDD), and other psychiatric disorders (Holtzheimer and Mayberg, 2011).

Although DBS has improved motor, cognitive, affective, and volitional functions for many patients with these disorders, questions remain about its efficacy. This is especially the case in psychiatry. For example, the BROADEN (BROdmann Area 25 DEep brain Neuromodulation) Study of DBS for depression sponsored by device manufacturer St. Jude Medical was discontinued in 2013 after disappointing initial results (Underwood, 2013). Also, DBS has resulted in a number of adverse events. These include effects associated with intracranial surgery, such as intracerebral hemorrhage, edema, and infection, which are within the range of typical neurosurgical complications. They also include effects associated with stimulation, such as hypomania, mania, and compulsive behaviors such as gambling and hypersexuality (Muller and Christen, 2011; Christen et al., 2012). These sequelae may result from not stimulating targeted circuits with the requisite precision or frequency, overstimulation or from expanding effects to other circuits. Focussed stimulation is challenging because many brain functions involve distributed and interacting neural pathways that send projections to and receive projections from each other. An area of the brain targeted by DBS may involve nuclei in closely related circuits regulating different neural processes. For example, the subthalamic nucleus in the basal ganglia is one of the areas stimulated to restore motor control in PD. Yet the basal ganglia consist of a complex network involving not only a motor circuit, but also associative and limbic circuits mediating cognitive and emotional processes. The compulsive behavior of some PD patients receiving DBS may be explained by unintended excitatory effects on the limbic circuit (Castrioto et al., 2014).

Many adverse events may be attributed to a technical feature in the open-loop devices (OLDs) currently used in DBS. There is no information feedback from the neural output of stimulation to stimulator input and no mechanism for the frequency to adjust to changes in the brain. In closed-loop devices (CLDs), information is fed back from changes in neural circuitry to the stimulator in real-time, and the stimulator can adjust the electrical frequency accordingly. This makes CLDs preferable to OLDs by ensuring that neural circuits are neither constantly overstimulated nor understimulated. They appear to be a safer and more effective means of neuromodulation by providing a greater degree of precision in activating electrophysiological mechanisms in neural pathways. By overcoming the design flaws in open-loop systems, closed-loop systems are more likely to maximize benefit and minimize harm for people suffering from diseases of the brain.

But the prospect of CLDs offering a greater degree of control at the neural-circuit level raises questions about whether they would impair or undermine control at the psychological level of the subject. It generates concern about whether people with these devices implanted in their brains would have conscious control of their behavior and retain their agency, autonomy, and identity. The source of their thoughts and actions and the mechanism that ensures the unity and integrity of their psychological properties would appear to be an artificial device operating outside of their awareness. For patients undergoing DBS, it seems that something alien to them is doing most of the work in regulating their physical and mental functions. Intuitively, CLDs cause more concern about these issues than open-loop ones because CLDs operate entirely on objective quantifiable measures and do not rely on subjective reports of symptoms from patients. After describing the neuroscientific and ethical respects in which CLDs are superior to OLDs, we argue that CLDs would not threaten but would restore and maintain agency, autonomy, and identity by restoring neural and mental functions to normal levels. These devices are also consistent with a nonreductive materialist conception of the mind-brain relation.

2 OPEN-LOOP VERSUS CLOSED-LOOP SYSTEMS: TECHNICAL DIFFERENCES

In open-loop DBS systems, the output quantity has no effect on input. Electrical impulses are delivered unidirectionally and independently of any feedback. "Open-loop" in this sense means that the implanted pulse generator constantly delivers invariant preprogrammed stimulatory pulses without adjustment of these pulses. They are not sensitive and do not respond to the unique and dynamically fluctuating characteristics of the disease state. Stimulation parameters are programmed into the device and held constant until the next programming session, regardless of changes in the neural environment (Foltynie and Hariz, 2010).

The main technical disadvantage of OLDs is that the lack of sensing feedback signals prevents the system from automatically correcting any errors in the parameters of stimulation frequency, intensity, and pulse duration. These errors can interfere with the intended physiological modulation. Without the necessary feedback, the device cannot adjust to alterations in the brain and thus is limited in its therapeutic output. The present way of stimulation by OLDs involves a very crude form of stimulus application reflected in a static approach to what is an inherently dynamic system. The only way of knowing of device malfunction

or suboptimal function is through a subjective report from a patient when symptoms appear or reappear. Device programmers and clinicians can then confirm that the device is the source of the problem and adjust the parameters in a clinical setting. This approach is far from optimal because it is time-intensive, observer-dependent, and limits battery life. Any adverse effects may be immediately detectable in movement disorders, where the symptoms are physical. But cognitive or affective symptoms from circuit dysregulation due to device malfunction in a psychiatric disorder may take weeks or months to appear. Aberrant neural activity may be more difficult to modulate the longer the aberration goes uncorrected, thus preventing patients from receiving maximal clinical benefit. These problems highlight the importance of developing more technologically sophisticated neuromodulating devices (Ineichen et al., 2014).

A closed-loop system can prevent neurological and psychiatric sequelae from an open-loop system stimulating at frequencies that are too high or too low, and which cannot be detected by the device itself. By detecting errors through feedback of this information, theoretically a closed-loop system can correct these errors on its own (Abbott, 2006; Rolston et al., 2010; Rosin et al., 2011; Santos et al., 2011; Potter et al., 2014). CLDs can provide reliable bidirectional sensing and responding to either neurochemical or electrophysiological biomarkers to update stimulation as the device is operating. Still, how well the feedback loop functions depends on precise programming of the system by teams of bioengineers, neurologists, neurosurgeons, and psychiatrists. It also depends on understanding the relationship between a patient's clinical state and how neuronal signals are influenced by external stimulation (Hebb et al., 2014).

Automatic parameter adjustment would not rule out the possibility of expanding effects of stimulation on nontargeted normal circuits. These effects might not be attributable to stimulation as such, but to the interconnectedness of pathways in the brain and the complex ways in which they project to and from each other. In this regard, CLDs might not be any better than OLDs in preventing these effects. Electrophysiological and neurochemical recording of brain activity, combined with functional imaging and reprogramming, may be necessary to reduce or prevent the risk of unintended stimulation. It may require interventions from practitioners beyond the initial programming and implantation of the device in the brain. So, a closed-loop system would not be entirely problem-free. In addition, even though CLDs might be functionally superior to OLDs in restoring physiological oscillations in circuits affected by disease, damage in these circuits from disease may entail the irreversible loss of highly specialized information processing in them. They may fail to respond to any form of stimulation. CLDs might not be therapeutically superior to OLDs in this regard either.

Yet by allowing feedback from neural outputs to stimulation inputs, a closed-loop neurostimulating system can restore a relatively stable equilibrium by modulating pathological oscillatory activity in the brain when this has become dysregulated from malfunctioning natural feedback mechanisms. In a healthy and stable neural environment, cortical and subcortical circuits are neither constantly overactive nor underactive. This balance is disrupted in neurological and psychiatric disorders due to a complex combination of genetics, environmental factors, and neurodevelopmental and neurodegenerative processes. In PD, stimulation of the subthalamic nucleus can downregulate an overactive motor circuit of the basal ganglia, modulate a cortico-basal ganglia-thalamic loop, and improve motor control (Castrioto et al., 2014). In the subtype of depression characterized by anhedonia and avolition, stimulation of the nucleus accumbens (NAcc) can upregulate an underactive reward circuit, modulate a fronto-striato-limbic loop, and improve mood and motivation (Schlaepfer et al., 2014). A different dysregulated, in general, loop has been implicated in OCD. Stimulating circuits in this loop can downregulate hyperactivity in them and ameliorate the obsessions and compulsions (Greenberg et al., 2010a; Melloni et al., 2012). Overall, closed-loop systems can do more to restore normal neural activity and promote equilibrium among circuits and pathways in the neural environment more effectively than open-loop systems. As "molecular prostheses" that automatically detect disease states and respond with appropriate counter-measures, CLDs in general are better tailored to the individual patient and more likely to produce greater clinical benefit and fewer side effects (Rosin et al., 2011; Santos et al., 2011).

3 BENEFIT-COST ANALYSIS

The constant frequency at which OLDs operate may not be optimal in producing neuromodulatory effects. Electrical current may continue to be delivered to circuits at the same intensity regardless of a patient's symptoms, due to the lack of feedback from the circuits to the stimulator. OLDs do not respond to variability in neural activity among different patients. Real-time adjustment of stimulation parameters, where information about this activity is fed back to the stimulator, can reduce the incidence of adverse effects of DBS. These effects are partly caused by unnecessary or excessive stimulation. By responding to particular alterations in a patient's brain as they occur, rather than by following a predetermined program, CLDs can contribute to "precision medicine" for neurological and psychiatric disorders. This consists of "treatments targeted to the needs of individual patients on the basis of genetics, biomarkers, phenotypic, or psychosocial characteristics that distinguish a given patient from other patients with similar clinical presentations. Inherent in this definition is the goal of improving

clinical outcomes for individual patients and minimizing unnecessary side effects for those less likely to have a response to a particular treatment" (Jameson and Longo, 2015, p. 1). In precision or "smart" DBS, neurostimulation would be tailored to individual patients. Researchers and clinicians could use neurochemical biomarkers identified through neuroimaging and electrophysiological recording to predict how patients might respond to different frequencies, intensities, and durations of stimulation and confirm whether they do in fact respond favorably to it (Grahn et al., 2014, p. 4).

In addition to minimizing harm to patients by reducing the incidence of side effects, the more efficient stimulation of CLDs can extend the battery life of the pulse generator. It may also reduce the probability of lead fractures, which can result in open or short circuits in the device. Most commercially available DBS systems using OLDs are nonrechargeable, and battery depletion requires replacement. Longer battery and lead life with closed-looped devices would benefit patients by significantly reducing the frequency of costly and time-consuming travel to clinics to replace batteries or leads. This is related to the main shortcoming of OLDs, which Grahn et al. summarize: "Although many DBS patients require minimal stimulation adjustment following surgery, many more require several months of regular parameter adjustments before optimal therapeutic results can be achieved." This may involve "adjustment of stimulation parameters every few months" (Grahn et al., 2014, p. 3; see also Okun et al., 2005; Mayberg et al., 2005).

The burden on patients of having to return to a clinic every few months for parameter adjustment or lead or battery replacement could contribute to nonadherence to device monitoring. Some may fail to keep these appointments, which would defeat the therapeutic purpose of the technique. Obviously, it would not be in a patient's best interests to fail to adhere to a treatment plan. But the psychological and financial cost of having to travel regularly to and from a medical clinic for adjustment and the additional financial cost of absence from work, for example, to keep these appointments could be a disincentive against adhering to such a schedule. The intermittent or discontinued stimulation associated with OLD design could result in incomplete or failed neuromodulation. Focussing mainly on movement disorders, Grahn et al. point out that: "existing clinical programming and stimulation paradigms are poorly suited to cope with the dynamic and comorbid nature of most neurological disorders. This, in turn, highlights the need for dynamic feedback systems that can continually and automatically adjust stimulation parameters in response to changes within the environment of the brain." (Grahn et al., 2014, p. 2). The need is even more acute in psychiatric disorders, where the effects of dysregulation on cognition, mood, and motivation are often not immediate, but delayed. An extended period of dysregulation due to imperfections in OLDs and/or patient nonadherence may allow a return of symptoms and possibly result in permanent deleterious changes to neural circuits. All of these considerations underscore the limitations of OLDs.

Closed-loop systems shed new light on the debate between the comparative benefits and risks between neurostimulation and ablation for neurological and psychiatric disorders. In the late 1980s, Benabid and colleagues used high-frequency stimulation as therapy for PD without any neural tissue ablation (Benabid et al., 1987). This showed promise of treating movement disorders without having to destroy tissue through lesioning. The most noteworthy advantage of DBS over lesioning is that adverse events from neurostimulation are generally reversible. The stimulation parameters can be adjusted, depleted batteries in the pulse generator or fractured leads can be replaced, or the electrodes can be surgically removed from the brain. This last procedure entails risks from intracranial surgery. But the potential for long-term harmful neurological and psychological effects from the device would be reduced, if not eliminated. Techniques such as MRI-guided stereotactic Gamma Knife radiosurgery are much safer and effective than the crude lesioning procedures of the past. But the fact that the effects of ablated neural tissue and any adverse events associated with it cannot be reversed is significant and provides medical and ethical reasons against it.

If ablation could be performed at a level of precision that would prevent all or most sequelae, then the benefits of neurostimulation over ablation would not be so clear. Ablative procedures are generally faster, more cost-effective, and require considerably less postoperative management than DBS. These are all relevant considerations regarding patient adherence to device monitoring and follow-up. They are also important in light of the mandate to control health care costs. A recent review of the literature on these two procedures shows that anterior capsulotomy for treatment-refractory OCD may be as effective as DBS (Pepper et al., 2015; Greenberg et al., 2010b). The authors of the review claim that a preference for neurostimulation over ablation in treating this condition is not due to actual differences in efficacy. Instead, it is due to the perception of patients and clinicians that the first procedure is more acceptable than the second because it does not involve destruction of any neural tissue. In spite of this claim, the irreversibility of the effects of lesioning, and the fact that these effects include apathy, cognitive impairment, and other chronic mentally disabling conditions make this a persistent problem that may not be resolved (Ruck et al., 2008).

If CLDs reduced the time and costs associated with post-implantation monitoring of open-looped devices, then this, combined with the fact that it does not destroy neural tissue, would make closed-loop DBS clearly preferable to lesioning. Devices that automatically adjusted to changes in the brain and obviated the need for external monitoring would eliminate the burden of having to undergo regular parameter adjustment in the clinical setting. Patients and medical practitioners programming and operating these devices would perceive them as more acceptable because they would, in fact, be more objectively acceptable given their neurophysiological properties, cost-effectiveness, and therapeutic potential.

Closed-loop systems raise some concerns, however. Focussing exclusively on objective electrophysiological and neurochemical measures of brain activity may ignore the phenomenological aspect of having a neurological or psychiatric disorder. It could fail to appreciate the patient's unique experience of having such a disorder and the effects of the device. It is persons, rather than neural circuits that have and suffer from diseases of the brain and mind. The medical, psychological, and social needs generated by these diseases cannot be met entirely by exclusive attention to the mechanical properties of the device. Nor is there always a strict correlation between measurable neural dysfunction and symptoms. Subtle changes in neural circuits may not be detectable by imaging or other objective measures, but may manifest in the behavioral characteristics of a neuropsychiatric disorder. Subjective reports of motor, cognitive, affective, or volitional impairment may be reliable indicators of abnormalities in brain regions mediating these physical and mental capacities. Feedback from outputs to inputs at a brain-systems level should supplement, rather than supplant feedback from patients to clinicians at an interpersonal level. How a patient reports symptoms, how a clinician responds to the report, and how the patient responds in turn to the clinician are components of the therapeutic process. Patient reactions to clinicians may influence whether symptoms persist or are ameliorated. This is especially the case in psychiatric disorders characterized by impaired cognition, mood, and motivation. CLDs may enable therapeutic responses by modulating aberrant neural activity underlying the impairment. But the device cannot completely account for these responses.

Moreover, CLDs are not immune to malfunction. Imprecise or incorrect programming of the device, lack of sensitivity to changes in neural activity, or technical failures may result in inappropriate responses to alterations in brain circuitry from electrical stimulation. The device might incorrectly read the information it receives from circuits and increase the frequency of the electrical current when it should be decreased or decrease it when it should be increased. Because of design flaws or programming errors, the device could become "stuck" in a feedforward loop that could maintain or exacerbate, rather than resolve circuit dysregulation. These problems may be less likely to occur than with an OLD; but they would be possible nonetheless. Since CLDs run more automatically, such effects might be observed later than in the case of OLDs and allow a longer period of dysregulation. Constant monitoring would be needed to inform the clinician or researcher of any technical problems.

There are cases in which patients with neurological and psychiatric disorders are allowed to operate closed- and open-loop stimulators on their own outside of a clinical setting and without constant monitoring by investigators and clinicians. The rationale is to enable them to have greater control of their symptoms by allowing them to alter the frequency depending on their experience of symptoms. Although these patients are deemed cognitively capable of understanding the potential negative consequences of overstimulation and giving informed consent to using the device, allowing them to do this entails some risk of self-harm. The potential for self-harm can be seen in a slightly different actual case. A patient with generalized anxiety and OCD became less anxious and experienced improved mood following DBS. He then asked his psychiatrist to increase the voltage of the stimulator so that he could feel even better. This caused him to feel "unrealistically good" and "overwhelmed by a feeling of happiness and ease" (Synofzik et al., 2012, p. 32). Yet he retained insight into his condition and the possibility of losing control of his behavior, as reflected in his comment about fearing that his euphoria would "tilt over" and that his anxiety would return. Accordingly, he agreed to have the voltage reduced. In cases where patients can control the voltage and intensity of the electrical current on their own, they would have to retain the capacity to know how to maintain optimal levels of mood and motivation within therapeutic stimulation parameters. These remarks cast doubt on the idea of neuroenhancement. It suggests that neurostimulation benefits patients as therapy to treat pathologies due to dysregulated neural circuits, but does not benefit and may even harm those with optimal levels of neural and mental functions when it is used to raise them above these levels.

More importantly, insofar as CLDs could achieve optimal levels of neural activity and therapeutic outputs without any contribution from the patient, there would be no neurophysiological reason for a patient to turn an OLD on or off on his own. The only reasons for doing this would be practical, such as passing through airport security without triggering alarms. Indeed, the risk of hypomania and compulsive behavior from increasing voltage beyond an optimal level would provide a compelling reason against allowing him to stimulate his brain. A CLD would preclude this risk, as well as the risk of sequelae from mistakes of a patient adjusting the stimulator or electrodes in an open-loop system.

An equally important ethical concern about research involving both OLDs and CLDs is that manufacturers of the devices can determine which experiments are worthwhile and which should be conducted. This is an obstacle for researchers intending to determine the therapeutic potential of neurostimulation for a particular disorder. It can also be unfair to subjects participating in clinical trials testing these devices because of the costs. "In the United States, companies and institutions sponsoring research are rarely, if ever, required to pay medical costs that trial subjects incur as a result of their participation" (Underwood, 2015, pp. 1186–1187). In addition, if a device manufacturer goes out of business and neurostimulation is no longer available, then any benefit a subject received from the technique would be for only a limited time. If battery and lead replacement or the stimulating system as a whole was not an option because of a lack of this equipment, then symptoms could reappear and the condition could return to an uncontrolled state. This is a more serious concern than the potential of nonadherence from frequent trips to clinic for parameter adjustment. President Barack Obama's Brain Research through Advanced Innovative Neurotechnologies (BRAIN) Initiative may lead to adequate funding from the National Institute of Mental Health (NIMH) to overcome these obstacles. But it is still too early to predict what the outcome might be.

At a more philosophical level, the idea that a device can control a person's behavior completely outside of his conscious awareness suggests that he is not the author or source of his thoughts and actions. This issue seems more disturbing for CLDs than for OLDs because the former provide a greater degree of mechanistic and automatic control of brain processes than the latter. Who or what is the agent behind our behavior? Are our actions really our own, or are they entirely the products of neural circuits or devices operating to ensure that these circuits function properly? Do CLDs support the idea that we are nothing more than the neural architecture of our brains? It is to these more fundamental philosophical questions about agency, autonomy, and identity that we now turn.

4 AGENCY, AUTONOMY, CONTROL

Agency consists in the executive ability to translate mental states such as desires, beliefs, and intentions into actions. This ability has sensorimotor, cognitive, affective, and volitional components. One or more of these components is impaired in different neurological and psychiatric disorders. In PD, dysfunction in the basal ganglia can impair the motor capacity to perform voluntary bodily movements. In generalized anxiety disorder or MDD, dysfunction in cortical-limbic pathways can impair the cognitive, affective, and volitional capacity to form and carry out action plans. These and other conditions can impair agency by impairing these capacities. DBS can restore some degree of agency by modulating neural circuits implicated in these disorders, ameliorating the relevant capacities and thereby ameliorating control of thought and action.

Having voluntary control of one's behavior suggests that conscious mental processes have a causal role in decision-making and acting. The fact that a CLD can operate automatically outside of awareness and without any apparent conscious contribution from the subject seems to threaten this control. It is not enough for persons to be agents. They must be autonomous agents to control their behavior. Yet the function of these devices appears to undermine autonomous agency, with the person's actions traceable to an artificial source. It seems that the device rather than the person is in control of her behavior (Klaming and Haselager, 2013).

Autonomy consists of two general capacities: competency and authenticity (Kant, 1785/1983; Dworkin, 1988; Frankfurt, 1988; Taylor, 1991; Mele, 1995). The first involves the cognitive and affective capacity to critically reflect on the mental states that issue in one's actions. The second involves the cognitive and affective capacity to identify with or endorse these mental states following a period of reflection. The process of critically reflecting on and identifying with one's mental states and actions is what makes them one's *own*. This reflects the original meaning of "autonomy": *autos* = self; *nomos* = rule or law. Mental states with which one does not identify or endorse may be considered "alien" to the agent. In ego-dystonic psychiatric disorders such as MDD and OCD, there is incongruence between the mental states one wants to have and those one has with the pathology. Patients retain insight into their condition, interpret their symptoms as imposed upon them, and yearn to be rid of them (Meynen, 2010). Through reflection and deliberation, an

autonomous agent is able to regulate the set of motor and mental springs of her actions. Autonomy is synonymous with this self-regulating process and being in control of one's behavior.

There are limits to this reflective capacity, however, which can become pathological beyond a certain level. To illustrate, persons with OCD engage in excessive conscious deliberation or rumination about how to act. Exerting too much conscious control over their thoughts and actions impairs their ability to act (De Haan et al., 2015; Glannon, 2015). Hyperreflectivity interferes with unconscious automatic behavior that enables one to perform a range of cognitive tasks and motor skills without having to think about performing them (Fuchs, 2011). Autonomy requires a certain degree of conscious reflection on the reasons or motivation for action. Yet OCD shows that too much reflection can undermine autonomy. Autonomous behavior must, to some extent, be automatic. It requires a balance between deliberative and automatic processes mediated by interacting fronto-striato-thalamic pathways. Dysregulation in these pathways prevents individuals with OCD from performing basic actions they would ordinarily perform as a matter of course. The pathological need for control is symptomatic of a loss of control and a form of mental paralysis. The mental states that move one to act in this and other psychiatric disorders are not the sorts of states one would endorse as the springs of one's actions and therefore are not autonomous (Fig. 1).

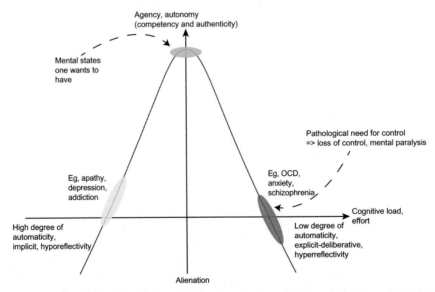

FIG. 1 Depicts the concepts of hypo-hyperreflectivity and alienation-autonomy along two axes. The inverted U curve highlights the relationship between a high degree of automaticity and a high degree of reflectivity. Both decrease autonomy as evidenced in pathological states such as MDD and OCD. Hyperreflectivity interferes with normal automatic behavior that ordinarily enables one to perform common motor and cognitive tasks as evidenced in OCD, schizophrenia, and anxiety disorders. The pathological need for control may induce loss of control and mental paralysis. On the other spectrum, disorders such as apathy, MDD, or addiction can impair the capacity to carry out action plans due to conditions such as mental fatigue or compulsive drug-seeking. Hence, autonomy to some extent has to be automatic and concomitantly requires a moderate degree of conscious reflection. It requires a healthy balance between deliberative and automatic processes mediated by interacting fronto-striato-limbic and fronto-striato-thalamic pathways. The degree of symptom severity impairing agency and/or autonomy in different neuropsychiatric disorders is reflected by the vertical position on the curve. The mental states that move one to act in different psychiatric disorders are not states one would endorse as the springs of one's action and therefore are not autonomous either.

The fact that an OLD or CLD operates outside of a person's awareness does not undermine but instead supports behavior control by modulating dysregulated neural circuits that generate and sustain thought and action. Electrical stimulation of circuits in the basal ganglia in PD and circuits in fronto-striato-thalamic pathways in MDD and OCD can restore the phenomenology or feeling of being in control of motor and mental functions. The subjects' implicit knowledge that electrodes are implanted and activated in the brain does not figure in the explicit content of their awareness. Most normal brain processes are not transparent to us. We have no direct access to our efferent system, for example, and only experience the sensorimotor consequences of our unconscious motor plans. We carry out these plans by performing actions without having to think about them. It does not matter whether these consequences are produced by a natural or artificial system. Provided that an artificial system such as DBS connects in the right way with the neural inputs and outputs that regulate behavior, it allows the subject to initiate and execute action plans (Glannon, 2014a,b). Insofar as the device ensures that the subject has the motor and mental capacities necessary to perform actions she wants to perform, she can identify it as her own, as an expanded feature of her brain–mind.

To further clarify the idea of control, it is instructive to compare the feedback in closed-loop stimulating systems with different versions of the same process in brain-computer interfaces (BCIs) and neurofeedback (NFB). There are two types of feedback in BCIs. The first type involves direct feedback about the level of brain activity itself. The second type involves information about the outcome of a self-initiated, BCI-mediated action, such as moving a computer cursor or robotic arm. It provides only indirect feedback about brain activity. But it is relevant to the subject's experience of control in that he can perceive the success or failure of his mental act (Wolpaw and Wolpaw, 2012). In NFB, the visual feedback a subject receives from information about the brain displayed on EEG or fMRI enables her to modulate aberrant brain activity (Weiskopf, 2012).

Among these techniques, NFB appears to do the most to promote behavior control and autonomy because the modulation of neural circuitry results from the active participation of the subject (Schermer, 2015; Glannon, 2015). Activity at conscious (expectation) and unconscious (conditioning) mental levels induces salutary changes at the neural level, which in turn improves symptoms. Like cognitive behavioral therapy (CBT), NFB is an example of mind-brain and brain-mind interaction. BCIs can also restore some degree of control of mental and physical actions, though the subject has a more limited causal role in these systems than in NFB. NFB does not require any devices on the head or in the brain to cause changes in neural networks. Imaging techniques record and display neural activity but do not have a direct causal role in modulating it. The mental states of the subject do this without any aid from an artificial device. BCIs rely on electrodes in a cap placed on the scalp or microelectrode arrays implanted in the motor cortex to enable the subject to translate cortical signals into intended actions. Still, there is a mental aspect to behavior control in BCIs because the subject has to consciously formulate and, with the help of a computer algorithm, execute an action plan (Glannon, 2014d). The mental acts that issue in physical acts of the sort we have mentioned in these techniques are enabled by and thus depend on certain neural functions. But their content involves more than mechanisms at the neural-circuit level. In contrast, there is no conscious contribution from the subject in CLDs. The device alone produces the neuromodulatory effects at a neural-systems level without any involvement of processes at a mental-systems level. The subject does not appear to have any causal role in the effects of the CLD.

This does not mean that the person with a CLD implanted in her brain is less autonomous and has less control of her behavior than a subject using a BCI or NFB. In fact, it is the disease rather than the technique that undermines autonomy. All behavior is regulated by a balance of interacting conscious and unconscious mental and neural processes. The fact that the subject's mental states can induce changes in the brain in BCIs or NFB, but not in CLDs is not the key issue for control. When operating effectively, CLDs regulate neural functions and the mental and physical capacities impaired by diseases of the brain. Being a "passive recipient" of the effects of a CLD does not imply that the subject has no control of his behavior. Although it operates outside of conscious awareness, the device does not replace him as the source or author of his actions but enables voluntary and effective agency by restoring the functional integrity of the neural circuits that mediate the relevant capacities (Lipsman and Glannon, 2013). Recall that in OCD, too much conscious reflection on cognitive and motor tasks impairs agency by impairing the ability to perform these tasks automatically. The etiology and pathophysiology of this condition illustrates that more activity at the conscious mental level does not necessarily guarantee a greater degree of behavior control and indeed may reduce it (see Fig. 1). Also, a CLD can reduce the subject's cognitive load and improve decision-making by combining its neuromodulatory action with the endogenous action of functionally intact neural circuits. Nor does the subject have to bear the burden of regular clinical visits for monitoring. Subjects engaged in NFB need the clinical setting of the EEG or fMRI in order to induce the positive changes in their brains and experience symptom relief.

One who successfully uses NFB as therapy for pain or depression exercises conscious control in restoring normal brain function. Still, what matters for behavior control is not whether modulation results from an artificial device or psychological responses to information about the brain. Rather, what matters is that the critical neural circuits are modulated and that the relevant mental and physical capacities mediated by these circuits are restored. In this way, CLDs can liberate the subject from constraints imposed on these capacities by diseases of the brain and restore and maintain autonomous agency to the same degree as other neuromodulating techniques. These devices do not threaten a person's control of behavior any more than OLDs do. The shared behavior control between the conscious person and the artificial device is not fundamentally different from the shared behavior control between a conscious person's mental states and naturally occurring unconscious processes in her normally functioning brain.

5   NEUROSTIMULATION AND IDENTITY

Personal identity is defined in terms of the connectedness and continuity of psychological states necessary for a person to persist through time as the same individual (Parfit, 1984, Part III; Schechtmann, 1997). Connectedness and continuity provide the integrity and unity of these states implied by the notion of persistence. As with the concern about autonomous agency, if the CLD restores and sustains the critical links between psychological states that define who we are, then some might ask whether the device that maintains these links undermines the psychological sense of identity.

When it modulates neural circuits, a subject need not perceive the stimulation system as something that undermines his identity. Again, there is shared behavior control between the subject and the device, which restores abilities that have been disabled by neural-circuit dysfunction. Ideally, a CLD operates as an enabling device that compensates for impaired mental or motor functions while complementing functions that are intact. It does not supplant these functions, but supplements them. The device and its neural and mental representation become integrated into the subject's brain and mind. It can be perceived as a form of expanded or extended embodiment that becomes part of his identity (Glannon, 2014c).

Some patients may feel uneasy about continued dependence on a mechanical device to maintain normal neural and mental processes. Yet there are no rational grounds for this attitude if the device produces a therapeutic response. If neurostimulation ameliorates a patient's symptoms and improves her well-being, then it would be mistaken to describe its therapeutic effect as part of a dependence relation, at least not in the sense of an addiction. The technique aims to resolve, rather than create pathology. There may be a stronger sense of identification of the subject with a CLD than an OLD. The ability of CLDs to provide more continuous, self-adjusting long-term

modulatory changes on a molecular level than OLDs may be more likely to "cause nerve cells to change their spontaneous firing patterns or by making different proteins. In this way, they can form cellular 'memories' of the stimulation. As a result, the chronic brain stimulation may become very much a part of the patients' normal neural network activity" (Linden, 2014, p. 111). This is not problematic for the psychological sense of identity if the device enables the patient to have the mental states he wants to have and restores the connections between these states that obtained before the disorder affected him. Symptom fluctuation and disruption in continuity of care from having to undergo periodic parameter adjustment or battery and lead replacement may cause the patient to become more aware of an OLD and its operational imperfections as a threat to his identity. Insofar as CLDs avoid these problems, the subject's identification with such a device at the neural level can complement her identification with the capacities it regulates at the mental level.

Adverse effects of neurostimulation can disrupt the unity and continuity of the psychological properties on the basis of which one experiences oneself persisting through time. But a properly functioning device can resolve pathological states and restore the unity and continuity of the psychological properties that defined one's premorbid self (Witt et al., 2013; Glannon, 2014c). When one has experienced depression for many years, one may gradually come to identify with the symptoms of the disorder as forming the core of one's self. However, in this and other ego-dystonic disorders, most patients want to rid themselves of these symptoms and reclaim the phenomenology and content of the mental states they had before the onset of the disorder. This can motivate patients to seek treatment and adhere to a treatment regimen. Stimulation sequelae such as hypomania or suicidal ideation could preclude insight into the disorder and disrupt identity as much as the dysregulated circuits because they would be equally incongruent with the psychological properties a healthy and rational individual would persistently endorse over time. This underscores the need for careful use of neurostimulation in targeting the right circuits with the right frequency, intensity, and pulse duration to restore optimal levels of mental and physical functions.

The sometimes rapid and substantial changes in personality and other forms of behavior from DBS can result in a difficult period of psychological adjustment for both patients and caregivers who have become accustomed to a characteristic pattern of symptoms. This pertains to both neurological and psychiatric disorders because it is not improvement in motor or cognitive–affective functions as such that is the issue, but the emotional response of patients and caregivers to the changes in the patient's behavior. One's relational identity to others may change in undesirable ways (Baylis, 2014; Levy and Clausen, 2014).

Nevertheless, provided that the changes are beneficial and improve the patient's quality of life, it clearly would be preferable to adjust to what is essentially a return of the patient's real self than to continue dealing with the greater challenges associated with living with a mentally and physically disabling disease. When the goal of neuromodulation is to alter one's affective states, it may be difficult to differentiate desired changes in the psyche from undesired ones and predict how these changes might alter identity (Muller and Christen, 2011). But potentially adverse changes in identity have to be weighed against the main goal of therapy, which is to resolve pathology, relieve suffering, and restore normal brain function.

6 NEUROSTIMULATION AND THE MIND-BRAIN RELATION

The ability of open- and closed-loop DBS to probe and modulate neural circuits and alter the mental states these circuits mediate supports a materialist view of mind. It validates the view that all normal and abnormal states of mind have a neurobiological underpinning. It invalidates the dualist view that there is an immaterial soul and that brain and mind are distinct substances. Yet closed-looped stimulation does not imply that everything about the mind can be explained in terms of the brain and that conscious mental states have no causal role in modulating dysregulated neural circuits. Mental events are not reducible to events and processes in the brain because they cannot be completely explained in neurobiological terms. The content of mental states, the events in the external world to which they are directed, are outside of the brain. Nor can the subjective quality, or "qualia," of experiencing a neurological or psychiatric disorder, or relief from their symptoms be accounted for entirely in terms of dysfunctional or functional neural circuits.

These considerations indicate the need to distinguish between reductive and nonreductive forms of materialism to elucidate the brain-mind relation and the implications of neuromodulation for this relation. This is especially significant for the use of DBS for psychiatric disorders. According to reductive materialism, phenomena at one level can be completely explained in terms of more basic elements at a different level (Kim, 1998). On this view, normal and abnormal mental states can be completely explained in terms of brain function and dysfunction. According to nonreductive materialism, the brain necessarily generates and sustains mental states, but cannot account for all of their properties (Baker, 2009). By themselves, probing and altering neural circuits of people with diseases of the brain-mind fail to capture the negative and positive effects of the mind on the brain in the development of and therapy for these diseases. Proponents of

nonreductive materialism hold that mental properties are part of the material world. They also hold that mental properties can be causally efficacious in altering material or physical properties of the brain without being reducible to these properties. Critics of this position argue that if mental events and processes are not reducible to physical events and processes, then they are epiphenomenal. Mental events are the effects of material or physical causes in the brain, but cannot cause any material or physical neural events to occur. If mental processes associated with beliefs, desires, emotions, and intentions are not reducible to their neural correlates and are epiphenomenal, then presumably they cannot influence the etiology of neuropsychiatric disorders nor have any role in therapeutic interventions for them.

But there are many examples in clinical neuroscience where patients' mental states, *qua* mental, can be causally efficacious in disrupting and modulating neural pathways. Chronic stress or fear can cause the adrenal medulla to release high circulating levels of adrenaline and noradrenaline, which can disrupt the balance among the components of the hypothalamic-pituitary-adrenal axis and between cortical-limbic pathways. Fear that persists beyond the duration of a threat or is out of proportion to the real nature of a stimulus can be a factor in the etiology of depression and anxiety. This is evidenced by hyperreactivity of the amygdala. On the positive side, patients' reframing their beliefs in CBT can re-wire cortical circuits and modulate their projections to and from limbic circuits (Goldapple et al., 2004). This is one example of how mental states can induce salutary changes in the brain. In NFB, patients' cognitive and affective responses to information about their brains through EEG or fMRI can downregulate hyperactivity in the rostral anterior cingulate cortex associated with chronic pain and thereby reduce their perception of pain (de Charms et al., 2005). Similarly, patients with anhedonia and avolition associated with depression can use information about their brains in NFB to upregulate activity in the reward system and improve motivation and mood (Linden et al., 2012). These examples show that mental and neural processes are not separate, but interacting and interdependent.

In the psychiatric disorders for which it has been used, neurostimulation alone does not always completely restore normal neural and mental functions. Additional interventions may be necessary to achieve the therapeutic goal. DBS can modulate the neural circuits implicated in MDD and OCD to make them amenable to CBT. The bottom-up effects of DBS on subcortical circuits and their projections to limbic and cortical circuits enable the top-down effects of subjects' cognitive and affective responses on cortical activity. Moreover, exclusive focus on a reductive neural-circuit model of these disorders and therapies can overlook variability in patients' responses to DBS. Two depressed patients suffering from anhedonia and avolition may improve from electrical stimulation of the NAcc. Even if fMRI or other imaging techniques show the same general pattern of neural activity during and after stimulation, one of these patients may be more successful than the other in forming and executing action plans. It is plausible to assume that the difference between these patients may be attributed to greater mental effort that one of them puts into performing actions. While the modulating effects of stimulation restore the same basic volitional capacity in both patients, the greater degree of trying that goes into the more successful agency of the first patient cannot be accounted for entirely by recording alterations in brain activity through quantitative measures. Something more than neural-circuit level function is necessary to account for it. OLDs and CLDs restore and sustain only the physiological basis on which the individual performs intended motor and mental actions.

The clinician or researcher relies on objective data about electrical stimulation of neural circuits before setting the critical parameters and feedback mechanisms when programming, implanting, and activating a CLD. This decision and action rely on more than objective measures. They are made only after discussing the potential benefits and risks of the technique with the patient. This includes a psychological assessment of the patient's attitudes about and responses to the chronicity and severity of her symptoms, as well as the patient's understanding and expectations about how neurostimulation would affect her. Applying the technique would also be preceded by consideration of how potential changes in the patient's personality might affect caregivers or others living with the patient. Neural and psychosocial factors inside and outside of the brain would be involved in using CLDs.

These examples suggest that the design and effects of CLDs are consistent with a nonreductive materialist conception of the mind-brain/brain-mind relation. They also support what Linden et al. (2012, p. 8) describe as "a holistic approach that overcomes bio-psychological dualisms" regarding the understanding of and therapies for many psychiatric and some neurological disorders. In neurological disorders, these devices can be a critical component in normalizing predominantly motor functions. We say "predominantly" because patients with a movement disorder such as PD suffer from affective and cognitive states associated with it as well. This is why an increasing number of clinical neuroscientists categorize both neurological and psychiatric disorders as "neuropsychiatric disorders." In psychiatric conditions such as MDD and OCD, CLDs can be a critical component in normalizing bidirectional brain-mind and mind-brain interactions and how they mediate cognitive and affective functions.

The fact that a device regulating neural circuits involves a closed loop does not mean that the device and its effects on circuits are "closed" to mental and environmental influences. Although neurological and psychiatric disorders are caused by neural-circuit dysfunction, it is not neural circuits, but persons who are affected by and suffer from them. Persons are constituted by their brains but are not identical to them. They are also constituted by genes, immune, endocrine, and cardiovascular systems in which they are embodied and the natural and social environment in which they are embedded. CLDs may be necessary to modulate dysregulated neural circuits and restore the motor, cognitive, affective, and volitional capacities impaired by or lost from diseases of the brain-mind. But because of the multifactorial etiology and pathophysiology of these diseases, the effects of CLDs on neural circuits will not yield a complete account of why they occur and will not be sufficient in achieving all therapeutic goals in treating them.

7 CONCLUSION

CLDs are mechanically superior to OLDs in modulating neural circuits because of their ability to automatically adjust themselves in response to changes in the brain. Their built-in feedback mechanism is more likely to normalize pathological oscillations and strike the right balance between excitatory and inhibitory neural activity. In this way, they can reduce the incidence of adverse effects from suboptimal levels of stimulation frequency, intensity, and pulse duration. They can also reduce the number of clinical interventions to correct problems associated with battery depletion, lead fracture, and unintended neurochemical and physiological events resulting from the lack of feedback from circuits to the electrodes. CLDs can be a safer and more effective form of therapy for neurological and psychiatric disorders than OLDs if equipped with a safety-monitoring device and more likely to maximize benefit and minimize harm for patients who have them. The ability of CLDs to adjust themselves to brain changes in real-time, combined with the identification of biomarkers, such as particular structural and functional features of the brain by neuroimaging, biochemical sensing, or electrophysiological recording, can provide "smart" DBS that may overcome many of the limitations of OLDs.

CLDs would not threaten the agency, autonomy, and identity of persons. An artificial device implanted in the brain modulating neural circuits and the motor and mental capacities they mediate would not undermine, but restore and sustain them. There is shared control between the device and the subject in regulating her behavior. As an enabling device, a CLD could integrate into the brain and mind of the subject, who could identify with it as a form of expanded or extended embodiment. The changes the device induced in the brain would not disrupt, but reestablish the psychological connectedness and continuity of their premorbid self and thus reestablish their identity.

Nevertheless, questions remain about the long-term effects of neurostimulation on neural circuits and pathways.

Theoretically, CLDs could provide a greater degree of behavior control for subjects affected by neurological and psychiatric disorders. However, as Grahn et al. note: "Before such control strategies can be implemented, it is necessary to improve the understanding of the cellular mechanisms for the network effects of DBS" (Grahn et al., 2014, p. 3). Very little is known about how implantable devices work. This highlights the importance of conducting foundational research to understand the underlying mechanisms and their therapeutic indications. There is more uncertainty about these effects from CLDs than from OLDs, since the first type has been used in fewer human trials (Morrell, 2011; Almeida et al., 2015). One issue not addressed here is the possibility of "brain hacking," where third parties could disrupt the operation of the device and adversely affect brain function. Security measures would have to be implemented and enforced to prevent this from occurring (Pycroft et al., 2016).

There is no definitive evidence that DBS can reverse the pathophysiology and progression of neurodegenerative diseases. This, more so than symptom control, would be the ideal outcome of the technique. Results of at least one study of DBS for PD indicate that initiating the technique in younger patients at an earlier stage of the disease produces greater benefit than it does in older patients at a later stage (Schuepbach et al., 2013). Replication of these results could generate an ethical obligation to administer DBS just (soon) after the onset of symptoms. This could apply to motor, seizure, and psychiatric disorders. Administering DBS before an advanced disease state and in younger brains may promote neuroplasticity and the release of trophic and other factors inducing neurogenesis. This could generate an even stronger ethical obligation to use the technique because the therapeutic benefit to those harmed by these diseases could be substantial. One can only speculate at this point whether the ability of CLDs to adjust to changes in the brain could produce these neurogenerative effects. But it is an intriguing hypothesis that warrants further investigation.

REFERENCES

Abbott, A., 2006. Neuroprosthetics: in search of the sixth sense. Nature 442, 125–127. http://dx.doi.org/10.1038/442125a.

Abramowicz, M., Zuccotti, G., Pflomm, J.-M., 2014. Deep brain stimulation for Parkinson's disease with early motor complications. J. Am. Med. Assoc. 311, 1686–1687. http://dx.doi.org/10.1001/jama.2014.3323.

Almeida, L., Martinez-Ramirez, D., Rossi, P., Peng, Z., Gunduz, A., Okun, M., 2015. Chasing tics in the human brain: development of open, scheduled and closed-loop responsive approaches to deep brain stimulation for Tourette syndrome. J. Clin. Neurol. 11, 122–131. http://dx.doi.org/10.3988/jcn.2015.11.2.122.

Baker, L., 2009. Non-reductive materialism. In: McLaughlin, B., Beckerman, A. (Eds.), The Oxford Handbook of Philosophy of Mind. Oxford University Press, Oxford, pp. 109–120.

Baylis, F., 2014. Neuroethics and identity. In: Levy, N., Clausen, J. (Eds.), Handbook of Neuroethics. Springer, Dordrecht, pp. 367–372.

Benabid, A.-L., Pollak, P., Louveau, A., Henry, S., de Rougemont, J., 1987. Combined (thalamotomy and stimulation) stereotactic surgery of the VIM thalamic nucleus for bilateral Parkinson disease. Appl. Neurophysiol. 50, 344–346.

Castrioto, A., Lhommee, E., Moro, E., Krack, P., 2014. Mood and behavioral effects of subthalamic stimulation in Parkinson's disease. Lancet Neurol. 13, 287–305. http://dx.doi.org/10.1016/S1474-4422(13)70294-1.

Christen, M., Bittlinger, M., Walter, H., Brugger, P., Muller, S., 2012. Dealing with side effects of deep brain stimulation: lessons learned from stimulating the STN. Am. J. Bioeth. Neurosci. 3, 37–43. http://dx.doi.org/10.1080/21507740.2011.636627.

De Charms, R., Maeda, F., Glover, G., Ludlow, D., Pauly, J., Soneji, D., et al., 2005. Control over brain activation and pain learned by using real-time functional MRI. Proc. Natl. Acad. Sci. 102, 18626–18631. http://dx.doi.org/10.1073/pnas.0505210102.

De Haan, S., Rietveld, E., Denys, D., 2015. Being free by losing control: what obsessive-compulsive disorder can tell us about free will. In: Glannon, W. (Ed.), Free Will and the Brain: Neuroscientific, Philosophical, and Legal Perspectives. Cambridge University Press, Cambridge, pp. 83–102.

Dworkin, R., 1988. The Theory and Practice of Autonomy. Cambridge University Press, New York, NY.

Foltynie, T., Hariz, M., 2010. Surgical management of Parkinson's disease. Expert. Rev. Neurother. 10, 903–914. http://dx.doi.org/10.1586/ern.10.68.

Frankfurt, H., 1988. Identification and externality. In: Frankfurt (Ed.), The Importance of What We Care About. Cambridge University Press, New York, NY, pp. 58–72.

Fuchs, T., 2011. The psychopathology of hyperreflexivity. J. Specul. Philos. 24, 239–255. http://dx.doi.org/10.1353/jsp.2010.0010.

Glannon, W., 2014a. Neuromodulation, agency and autonomy. Brain Topogr. 27, 46–54. http://dx.doi.org/10.1007/s10548-012-0269-3.

Glannon, W., 2014b. Prostheses for the will. Front. Syst. Neurosci. 8, 1–3. http://dx.doi.org/10.3389/fnsys.2014.00079.

Glannon, W., 2014c. Philosophical reflections on therapeutic brain stimulation. Front. Comput. Neurosci. 8, 1–3. http://dx.doi.org/10.3389/fncom.2014.00054.

Glannon, W., 2014d. Ethical issues with brain-computer interfaces. Front. Syst. Neurosci. 8, 1–3. http://dx.doi.org/10.3389/fnsys.2014.00136.

Glannon, W. (Ed.), 2015. Free Will and the Brain: Neuroscientific, Philosophical and Legal Perspectives. Cambridge University Press, Cambridge.

Goldapple, K., Segal, Z., Garson, C., Lau, M., Bieling, P., Kennedy, S., et al., 2004. Modulation of cortical-limbic pathways in major depression: treatment-specific effects of cognitive behavior therapy. Arch. Gen. Psychiatry 61, 34–41. http://dx.doi.org/10.1001/archpsyc.61.1.34.

Grahn, P., Mallory, G., Khurram, O., Berry, B.M., Hachmann, J., Bieber, A., et al., 2014. A neurochemical closed-loop controller for deep brain stimulation: toward individualized smart neuromodulation therapies. Front. Neurosci. 8, 1–11. http://dx.doi.org/10.3389/fnins.2014.00169.

Greenberg, B., Gabriels, L., Malone, D., Rezai, A., Friehs, G., Okun, M., et al., 2010a. Deep brain stimulation of the ventral internal capsule/ventral striatum for obsessive-compulsive disorder: worldwide experience. Mol. Psychiatry 15, 64–79. http://dx.doi.org/10.1038/mp.2008.55.

Greenberg, B., Rauch, S., Haber, S., 2010b. Invasive circuitry-based neurotherapeutics: stereotactic ablation and deep brain stimulation for OCD. Neuropsychopharmacology 35, 317–336. http://dx.doi.org/10.1038/npp.2009.128.

Hebb, A., Zhang, J., Mahoor, M., Tsiokos, C., Matlack, C., Chizeck, H., et al., 2014. Creating the feedback loop: closed-loop neurostimulation. Neurosurg. Clin. N. Am. 25, 187–204. http://dx.doi.org/10.1016/j.nec.2013.08.006.

Holtzheimer, P., Mayberg, H., 2011. Deep brain stimulation for psychiatric disorders. Annu. Rev. Neurosci. 34, 289–307. http://dx.doi.org/10.1146/annurev-neuro-061010-113638.

Ineichen, C., Glannon, W., Temel, Y., Baumann, C., Surucu, O., 2014. A critical reflection on the technological development of deep brain stimulation (DBS). Front. Hum. Neurosci. 8, 1–7. http://dx.doi.org/10.3389/fnhum.2014.00730.

Jameson, J.L., Longo, D., 2015. Precision medicine—personalized, problematic, and promising. N. Engl. J. Med. 387, 1–6. http://dx.doi.org/10.1056/NEJMsb1503104.

Kant, I., 1785/1983. Grounding of the Metaphysics of Morals (J. Ellington, Trans.). Hackett, Indianapolis, IN.

Kim, J., 1998. Mind in a Physical World: An Essay on the Mind-Body Problem and Mental Causation. MIT Press, Cambridge, MA.

Klaming, L., Haselager, P., 2013. Did my brain implant make me do it? Questions raised by DBS regarding psychological continuity, responsibility for actions and mental competence. Neuroethics 6, 527–539. http://dx.doi.org/10.1007/s12152-010-9093-1.

Levy, N., Clausen, J. (Eds.), 2014. Springer Handbook of Neuroethics. Springer, Berlin.

Linden, D., 2014. Brain Control: Developments in Therapy and Implications for Society. Palgrave Macmillan, Basingstoke.

Linden, D., Habes, L., Johnston, S., Linden, S., Tatineni, R., Subramanian, L., et al., 2012. Real-time self-regulation of emotion networks in patients with depression. PLoS ONE 7. http://dx.doi.org/10.1371/journal.pone.0038115.

Lipsman, N., Glannon, W., 2013. Brain, mind and machine: what are the implications of deep brain stimulation for perceptions of personal identity, agency and free will? Bioethics 27, 465–470. http://dx.doi.org/10.1111/j.1467-8519.2012.01978.x.

Lozano, A., Lipsman, N., 2013. Probing and regulating dysfunctional circuits using deep brain stimulation. Neuron 77, 406–424. http://dx.doi.org/10.1016/j.neuron.2013.01.020.

Mayberg, H., Lozano, A., Voon, V., McNeely, H., Seminowicz, D., Hamani, C., et al., 2005. Deep brain stimulation for treatment-resistant depression. Neuron 45, 651–660. http://dx.doi.org/10.1016/j.neuron.2005.02.014.

Mele, A., 1995. Autonomous Agents: From Self-Control to Autonomy. Oxford University Press, New York, NY.

Melloni, M., Urbistando, C., Sedeno, L., Gelormini, C., Kichic, R., Ibanez, A., 2012. The extended frontal-striatal model of obsessive-compulsive disorder: convergence from event-related potentials, neurophysiology and neuroimaging. Front. Hum. Neurosci. 6, 1–24. http://dx.doi.org/10.3389/fnhum.2012.00259.

Meynen, G., 2010. Free will and mental disorder: exploring the relationship. Theor. Med. Bioeth. 31, 429–443. http://dx.doi.org/10.1007/s11017-010-9158-5.

Morrell, M., 2011. Responsive cortical stimulation for the treatment of medically intractable partial epilepsy. Neurology 77, 1295–1304. http://dx.doi.org/10.1212/WNL.0b0113e3182302056.

Muller, S., Christen, M., 2011. Deep brain stimulation in Parkinsonian patients—ethical evaluation of cognitive, affective and behavioral sequelae. Am. J. Bioeth. Neurosci. 2, 3–13. http://dx.doi.org/10.1080/21507740.2010.533151.

Okun, M., Tagliati, M., Pourfar, M., Fernandez, H., Rodriguez, R., Alterman, R., et al., 2005. Management of referred deep brain stimulation failures: retrospective analysis from 2 movement disorders

centers. Arch. Neurol. 62, 1250–1255. http://dx.doi.org/10.1001/archneur.62.8.noc40425.

Parfit, D., 1984. Reasons and Persons. Clarendon Press, Oxford.

Pepper, J., Hariz, M., Zrinzo, L., 2015. Deep brain stimulation versus anterior capsulotomy for obsessive-compulsive disorder: a new review of the literature. J. Neurosurg. 59, 1–3. http://dx.doi.org/10.3171/2014.11.JNS132618.

Potter, S., El Hady, A., Fetz, E., 2014. Closed-loop neuroscience and neuroengineering. Front. Neural Circ. 8, 1–3. http://dx.doi.org/10.3389/fncir.2014.00115.

Pycroft, L., Boccard, S., Owen, S., Stein, J., Fitzgerald, J., Green, A., Aziz, T., 2016. Brainjacking: implant security issues in invasive neuromodulation. World Neurosurg. http://dx.doi.org/10.1016/j.wneu.2016.05.010.

Rolston, J., Gross, R., Potter, S., 2010. Closed-loop, open-source electrophysiology. Front. Neurosci. 4, 1–8. http://dx.doi.org/10.3389/fnins.2010.00031.

Rosin, B., Slovik, M., Mitelman, R., Rivlin-Etzion, M., Haber, S., Israel, Z., et al., 2011. Closed-loop deep brain stimulation is superior in ameliorating Parkinsonism. Neuron 72, 370–384. http://dx.doi.org/10.1016/j.neuron.2011.08.023.

Ruck, C., Karlsson, A., Steele, D., Edman, G., Meyerson, B., Ericson, K., et al., 2008. Capsulotomy for obsessive-compulsive disorder: long-term follow-up of 25 patients. Arch. Gen. Psychiatry 65, 914–922. http://dx.doi.org/10.1001/archpsyc.65.8.914.

Santos, F., Costa, R., Tecuapetla, F., 2011. Stimulation on demand: closing the loop on deep brain stimulation. Neuron 72, 197–198. http://dx.doi.org/10.1016/j.neuron.2011.10.004.

Schechtmann, M., 1997. The Constitution of Selves. Cornell University Press, Ithaca, NY.

Schermer, M., 2015. Reducing, restoring or enhancing autonomy with neuromodulation techniques. In: Glannon (Ed.), Free Will and the Brain. Cambridge University Press, Cambridge, pp. 205–227.

Schlaepfer, T., Bewernick, B., Kayser, S., Hurlemann, R., Coenen, V., 2014. Deep brain stimulation of the human reward system for major depression—rationale, outcomes and outlook. Neuropsychopharmacology 39, 1303–1314. http://dx.doi.org/10.1038/npp.2014.28.

Schuepbach, W., Rau, J., Knudsen, K., Volkmann, J., Krack, P., Timmermann, L., et al., 2013. Neurostimulation for Parkinson's disease with early motor complications. N. Engl. J. Med. 368, 610–622. http://dx.doi.org/10.1056/NEJMoa1205158.

Synofzik, M., Schlaepfer, T., Fins, J., 2012. How happy is too happy? Euphoria, neuroethics and deep brain stimulation of the nucleus accumbens. Am. J. Bioeth. Neurosci. 3, 30–36. http://dx.doi.org/10.1080/21507740.2011.635633.

Taylor, C., 1991. The Ethics of Authenticity. Harvard University Press, Cambridge, MA.

Underwood, E., 2013. Short-circuiting depression. Science 342, 548–551. http://dx.doi.org/10.1126/science.342.6158.548.

Underwood, E., 2015. Brain implant trials raise ethical concerns. Science 348, 1186–1187. http://dx.doi.org/10.1126/science.348.6240.1186.

Weiskopf, N., 2012. Real-time fMRI and its application to neurofeedback. Neuroimage 62, 682–692. http://dx.doi.org/10.1016/j.neuroimage.2011.10.009.

Witt, K., Kuhn, L., Timmermann, L., Zurowski, M., Woopen, C., 2013. Deep brain stimulation and the search for identity. Neuroethics 6, 499–511. http://dx.doi.org/10.1007/s12152-011-9100-1.

Wolpaw, J., Wolpaw, E., 2012. Brain-Computer Interfaces: Principles and Practice. Oxford University Press, New York, NY.

Chapter 20

Closed Loops in Neuroscience and Computation: What It Means and Why It Matters

C.J. Maley* and G. Piccinini[†]

*University of Kansas, Lawrence, KS, United States, [†]University of Missouri—St. Louis, St. Louis, MO, United States

1 INTRODUCTION

Different kinds of computational systems have different computational power, depending on their structure and organization, including whether they contain closed loops. Thus, understanding neural systems as computational systems sheds light on their computational power and the role that closed loops contribute to such power. Here, we analyze some interesting parallels between neural systems and computational systems as we consider each in increasing complexity. Recent developments in closed-loop approaches in neuroscience provide us an opportunity to illustrate a model of computation that is better suited to reflect this novel take on how neural systems should be conceptualized.

First, we discuss how neural systems incorporate feedback, although much research in neuroscience has ignored this important phenomenon. We demonstrate how the closed-loop approach constitutes the next step in understanding—and taking seriously—feedback in neural systems.

Next, we explain how to apply the abstract, mathematical theory of computation to concrete physical objects, such as neural systems. We introduce various automata and computational devices of increasing complexity, showing how feedback and memory contribute to the computational power of a system.

Finally, we argue that a new computational model is needed to understand a phenomenon that the closed-loop approach brings out forcefully. Specifically, closed-loop approaches emphasize the tight coupling between input to a single computation, processing of that computation, and output to a subsequent computation, which may be a different computation than the first. While traditional Turing Machines (TMs) are an awkward fit at best, a simple modification to the TM fits much better.

2 CLOSED LOOPS IN NEURAL SYSTEMS

Typical neural circuits are recurrent rather than purely feedforward. A purely feedforward circuit contains one-directional connections from upstream to downstream neurons, and no other connections. By contrast, in addition to feedforward connections, typical neural circuits also contain either sideways connections between neurons belonging to the same level of processing, or backward connections from downstream layers to upstream layers, or both.

Typical neural systems are sensitive to feedback. That is, typical neural systems respond in real time to the consequences of their outputs. There are two reasons for this. One is that typical neural systems contain sideways and recurrent connections, which send outputs back toward same-layer neurons or upstream neurons within a circuit, respectively. The other is that neural systems' outputs affect the organism and its environment continuously, and those consequences are picked up by the organism's sensors in real time and fed back into the same neural systems.[1]

For example, the vast majority of the mammalian retina's output projects to the lateral geniculate nucleus (LGN), which then has projections to the visual cortex. Viewed as a stream of information, visual stimuli go from the retina to the LGN to areas in the visual cortex. But the visual cortex also projects back to the LGN. Thus, the LGN also receives information from the visual cortex, and this feedback is used in processing incoming retinal information (Murphy et al., 1999). In addition, the retina, LGN, and

---

1. A third reason is that neural processes are sensitive to local feedback (other than action potentials)—for instance, whether a downstream neuron fires immediately after an upstream neuron affects the strength of a synapse. More on this below.

Closed Loop Neuroscience. http://dx.doi.org/10.1016/B978-0-12-802452-2.00020-2

visual cortex's inputs are constantly changing as a function of the movements of the eyes, head, and body of the organism, which, in turn, are affected by the retina, LGN, and visual cortex's outputs. Thus, the visual system is constantly receiving feedback from within and from without.

These two facts—recurrent circuits and sensitivity to feedback—have been known for a long time. The importance of feedback within the nervous system was emphasized already by Cannon (1932), while the importance of closed loops was emphasized by Lorente de Nó (1933). Both facts were front and center within the cybernetics movement, which was hugely influential on contemporary neuroscience (McCulloch and Pitts, 1990; Wiener, 1948).

Despite all of this, these same two facts have often been ignored, underappreciated, and understudied within mainstream experimental neuroscience. This is understandable, however: it is both conceptually and experimentally easier to treat neurons and neural systems as systems that simply turn an input into an output and send the output downstream.

Beginning with the early experiments of Hodgkin and Huxley (1952), neuroscientists have made great strides in understanding the electrophysiology of the neuron by recording the responses that neurons generate from particular inputs or input patterns. The simplest example is the generation of a single action potential: when the voltage at the cell body of the neuron rises to a certain threshold, the neuron generates an action potential, a wave of rapidly rising voltage significantly higher than the original voltage of the neuron. This action potential travels along the axon, ultimately leading to the release of various types of neurotransmitters, which will then go on to have an effect on other neurons. Thus, the output from one neuron becomes part of the input to another neuron, and we have a series of feedforward devices.

More complex behavior can result from the repeated firing of single neurons, still treated as feedforward devices. For example, some neurons will generate pulses of action potentials in well-defined frequencies in response to continued stimulation; others will generate bursts of action potentials. These patterns, in turn, have various effects on the neurons that receive these outputs. Plasticity effects such as long-term potentiation (LTP) and long-term depression (LTD) can be explained at the level of the single neuron and provide good examples of the kinds of feedforward devices that neurons can be.

Even at this level, it is not quite correct to think of neurons as *merely* feedforward devices. When neuronal connections undergo LTP and LTD, neurons change their response to the same input over time. In LTP, neurons produce more outputs given the same inputs over time, and the opposite happens in LTD. These phenomena are thought to provide the neurophysiological basis for certain kinds of memory. In order to produce these phenomena, neurons modify themselves in response to certain inputs

so that, when later inputs of the same kind arrive, a neuron's output will be different.

Allowing for a neuron to dynamically modify its responses to future inputs based on past inputs is a basic form of feedback. Treated as a feedback device, an individual neuron can be conceptualized as a device whose outputs will depend, not only on particular inputs, but also on the present state of the neuron as determined by past inputs.

As already mentioned, feedback in neurons and neural circuits has been acknowledged for some time, even though it has been underappreciated. Recent advances in experimental techniques have taken this idea in a new direction. To understand this new, closed-loop approach, we must first take a moment to more fully articulate what it contrasts with. Many traditional techniques involve creating a well-defined set of stimuli, presenting them to the system or organism under investigation, and then recording responses to those stimuli. At the level of neurons and neural circuits, patterns of inputs are presented at an input source, and responses are recorded at an output source. At the level of entire organisms (for example, cognitive neuroscience studies in humans), participants are presented with, say, visual stimuli of a certain kind, and differential changes in brain areas of interest are recorded. In these kinds of investigations, researchers determine ahead of time what inputs will be used, and the relationship of those inputs to the outputs that are recorded reveals information about the structure or function (or both) of the system under investigation.

The new closed-loop approach operates differently. Rather than determining the full set of inputs that will be used in an experiment ahead of time, inputs are partially determined by how the system responds in real time. Thus, an initial stimulus is provided to the system, and the system responds. That response is recorded and, importantly, is *also* used to generate the next stimulus. An example will help to illustrate the novelty of this technique.

A recent study used the closed-loop approach to study whether human attention could be improved with immediate feedback (deBettencourt et al., 2015). In this study, participants were presented with images that were a blend of a human face (male or female) and a visual scene (indoor or outdoor).[2] The degree of blending could differ, so that one image might be 20% face and 80% scene, which would make the face more difficult to see. Participants were asked to determine whether the face was male or female and, as they did so, their brain was continuously scanned. Using real-time analysis of each participant's brain data, researchers were able to determine the degree to which

---

2. We will simplify and omit some of the details, including different experimental variations, as these are irrelevant to illustrating the basic principle of the closed-loop approach.

participants were attending to the task at hand, which they then used to alter subsequent stimuli. For example, if a participant was attending to the task, subsequent images had a higher percentage of the "face" component, relative to the "scene" component; in essence, attending to the task made the task become easier. But if a participant's attention to the task began to decrease, images would begin to have a lower "face" component, making the task more difficult.

The participants who received feedback (in the form of easier or more difficult images) as a function of the amount of attention measured from their brain were able to perform the task better on a subsequent day than those who did not. In other words, a session of this kind of feedback training improved how well they did on the task. This finding exemplifies the closed-loop methodology: the set of stimuli that each participant received was *not* determined ahead of time, but was generated in real-time, according to the brain state of each participant. Thus, what a participant does at one point in time immediately effects what she experiences at the next point in time.

What is striking about this experimental methodology is that it better reflects how organisms perceive and respond to the world around them. Even in simple visual perception, what we see is a function of whatever behavior we have just engaged in. When we look for cars while crossing the street, what we see depends on how we move our head, and how we move our head depends on what we see. Inputs and outputs are tightly coupled and thoroughly interdependent.

3   PHYSICAL COMPUTATION

There is much to be gained by using theoretical concepts from computation and computer science to study and understand neural systems. One preliminary issue is that theoretical computer science concerns computations in the abstract, which may or may not be physically implemented. In order to compare neural systems to computing systems, we need to get clear on what a *physical* computing system is; in other words, we need an account of what it means for a computational system to be physically implemented (as opposed to merely *simulated* computationally, which is possible for virtually all physical systems). We adopt the mechanistic account of physical computation developed by one of us specifically for this kind of analysis. The account is defended at greater length elsewhere (Piccinini, 2015).

A physical computing system is a mechanism whose function is performing physical computations. A physical computation, in turn, is the processing of physical variables according to rules that are sensitive solely to differences between spatiotemporal portions of the variables along relevant dimensions of variation (or degrees of freedom). The physical variables in question may be implemented in different physical media (electrical, electromechanical,

magnetic, etc.) so long as those media possess the relevant dimensions of variation (degrees of freedom) and the mechanism is organized so as to process them according to the relevant rules. In this sense, physical computations are medium-independent: the same computation may be implemented in physically different media.

To illustrate, consider a logic gate, which is the simplest physical computing device. More specifically, consider an OR gate. A physical system is an OR gate so long as it satisfies the following conditions: (1) it takes two inputs during appropriate finite time intervals, (2) it yields an output during the same time intervals, (3) the inputs and outputs are physical variables that stabilize around two different macrostates—usually called 0 and 1—during the relevant time intervals, (4) the mechanism distinguishes between the two macrostates—0 and 1—reliably, and (5) the output is equal to 1 if and only if at least one of the inputs is equal to 1. Condition (5) states the rule followed by the device.

An OR gate is a primitive computing device—it cannot be decomposed into simpler computing devices. But it can be combined with other logic gates to form Boolean circuits, which can be combined together (with appropriate feedback signals) to form memory registers, control circuits, and processing circuits, which in turn can be combined together to form full-blown digital computers. In addition, a logic gate can be physically implemented in different sorts of physical devices using different sorts of physical components that manipulate different sorts of physical media, so long as they satisfy the conditions stated above.

Ordinary logic gates, and combinations thereof, are examples of *digital* computing systems. There are also nondigital computing systems. The nondigital computing systems that are historically contrasted with digital systems are *analog* computing systems. The main difference is the following. Digital computing systems manipulate strings (or sequences) of digits. Digits are states that, during relevant time intervals, stabilize around finitely many macrostates (typically two, called 0 and 1). By contrast, analog computing systems manipulate real (ie, continuous) variables, which vary continuously in time and are assumed to be able to take any real value within appropriate intervals.[3]

---

3. Thus, in this essay we use "digital" and "analog" for types of variables, without any commitment about whether they represent and, if they do, whether they represent in a digital or analog fashion. With respect to *digital* and *analog types of representation*, one of us argues elsewhere that they do not exhaust the types of representation available to computing systems, nor should they be thought of as synonymous with *discrete* and *continuous*, respectively (Maley, 2011).

At this point, two questions arise. First, are neural systems computational? Second, are they digital, analog, or neither?

The answer to the first question appears to be positive, at least to a good approximation. This is because the main function of neural systems appears to be to process spike trains based on their firing rate and, in some special cases, the exact timing of specific spikes. These are medium-independent properties of neural signals, that is, these properties could be implemented using media that are physically different from action potentials (eg, in electronic circuits, as some simulations do). Because neural systems process variables that are medium-independent and processing medium-independent variables is the defining feature of physical computing systems, neural systems are physical computing systems (Piccinini and Bahar, 2013).

The answer to the second question appears to be that neural computations are *sui generis*—neither digital nor analog—at least in the general case. This is because there are significant differences between the variables that neural computations manipulate and both digital and analog variables. The difference with analog variables is that the most functionally relevant properties of spike trains appear to be (average) spike rates and spike timing, not the precise physical values taken by the variable during a time interval. The difference with digital variables is that there don't seem to be any time intervals during which digits could be defined (Piccinini and Bahar, 2013).

Although neural computations appear to be *sui generis*, in what follows we will compare neural systems to digital computing systems. The reason is straightforward. Digital computing systems are easily understood and there is a well-developed mathematical theory of their computational power. On that basis, digital computing systems can be divided into classes of increasing computational power—and their computational power depends, to an important extent, on whether they include feedback loops. This hierarchy between digital computing systems of increased power, and the role that feedback loops play in that hierarchy, will help illuminate a corresponding hierarchy between neural systems—even though neural systems may ultimately have to be analyzed using the mathematical apparatus of *sui generis* neural computation rather than digital computation. In light of this, looking at the role of closed loops in digital computational systems will shed light on closed loops in neural systems.

4 CLOSED LOOPS AND COMPUTATIONAL SYSTEMS

An important aspect of the study of computation involves analyzing how simpler, comparatively less powerful devices can be used together to create more complex and thus more powerful ones. For example, in digital logic design, one needs to determine how to create something as complex as an adder (a circuit that outputs the sum of two numbers given as input in a binary representation) out of logic gates that perform only basic operations, such as logical AND, OR, and NOT. The simplest class of devices of this sort is Boolean circuits, also known as combinational logic devices. The functions they compute can be described using Boolean algebra or truth tables.

Combinational logic devices are purely feedforward: they take an input of bounded length and produce an output of bounded length without the use of any kind of memory. For example, one can create a circuit that is a series of eight AND gates. With the right connections, this device will output TRUE if all eight inputs are TRUE, and false otherwise. A similarly connected device using OR gates will output TRUE as long as at least one of its inputs is TRUE. But without a memory, combinational logic devices will never respond differently to inputs of the same type.

Moving to an increasing level of complexity, we can consider sequential logic devices. Multiplication and division circuits are examples of sequential logic devices. Before they yield their output (eg, the product of two numbers in a binary representation), sequential logic devices must go through a number of operations, where each operation depends in part on the results of previous operations. In order to operate this way, sequential logic devices must include memory registers, where they can store internal states that result from previous operations. The functions computed by sequential logic devices cannot be captured using Boolean algebra or truth tables, but they can be defined using the formalism of finite-state automata (FSA).

An FSA produces an output string of digits from an input string. An everyday example of an FSA is a vending machine. Beginning from an initial state (in which no money is in the machine), the machine keeps track of how much money is in so far. At any point, the machine does not record precisely how that money got there, just what the "running total" is. If a certain amount is reached, then the machine outputs an item; if more than that is reached, the machine outputs an item and some change. At that point, the machine returns to its initial state, and is ready to be run again.

Herein lays a limitation of combinational logic devices and their formal counterpart, FSAs: the functions they compute can only require a finite memory store of bounded size. Because of this, combinational logic devices cannot even compute the full multiplication and division functions, defined over all natural numbers. At most they can compute multiplication and division over finitely many inputs of bounded length, where the bound on the input size depends on the size of the device's internal memory.

Still more complex is the well-known TM, which *can* compute multiplication or division over all natural numbers. Like FSAs, TMs have various internal states that serve as one kind of memory. However, unlike the FSA, TMs also have access to an unbounded "tape," upon which they can both read and write digits. By writing digits on the tape while they perform a computation, TMs can keep track of what can be called intermediate results of unbounded length. Because of their unbounded tape, TMs can compute functions such as multiplication and division for all possible inputs.

Two striking features of TMs are well known, but worth repeating here. First, it is widely believed that everything that is computable by following an algorithm is also computable by a TM. This is the Church-Turing thesis. Second, there are so-called Universal Turing Machines (UTMs) that can compute *anything* computable by any other TM. UTMs take as input both a description of some TM and an input i to that TM, then output what that TM would output if given input i. In essence, contemporary computers are just this sort of machine: given a program, the computer runs that program, and the output produced will depend on what program the computer is given to run.

TMs have been thought of as the standard model for computation throughout many disciplines, and for good reason. Given the widespread acceptance of the Church-Turing thesis as well as the relatively clarity of the TM model of computation, TMs have found their way into many disciplines outside of computer science. For example, cognitive scientists have wondered whether the human mind is equivalent to a TM, and if so, what kind of TM it is. And even if the mind itself is not equivalent to a specific TM, cognitive scientists have *modeled* various mental capacities in terms of computation. While there are some exotic models of computation that can, in an abstract or theoretical sense, do more than TMs (such as completing infinitely many steps in finite time, taking advantage of relativistic effects of spacetime in extreme conditions, or using infinitely-precise real numbers), there is no evidence that they can be physically implemented (Piccinini, 2015, Chapters 15 and 16). The standard TM is here to stay. However, as we will explain below, when we compare closed loops in neuroscience with closed loops in computation, an enhancement of the standard TM might well be in order to help us model the closed-loop approach.

5 CLOSED LOOPS IN NEUROSCIENCE AND COMPUTATION

A close connection between neural systems and computing systems has been realized since at least the seminal work of McCulloch and Pitts (Piccinini, 2004). The ways in which neurons and neural systems have been studied both *as* computational devices and *using* computational techniques are

many; here, we only aim to show how advances in thinking about neural systems, from input/output devices to feedback devices to closed-loop systems, have important parallels to thinking about computational systems in similar terms.

Frank Rosenblatt was one of the early neural network researchers who made McCulloch and Pitts networks modifiable (Rosenblatt, 1962). He designed perceptrons, which are trainable feedforward neural networks that initially had only two layers of neurons. Perceptrons are equivalent to relatively simple combinational logic circuits. Quite interesting logical functions can be performed by constructing simple networks of logic gates. Nevertheless, the computational power of simple networks such as these is limited, and with the publication of an influential monograph proving some such limitations (Minsky and Papert, 1969), the utility of further research along these lines was temporarily put in doubt.

The project of investigating the connection between neural circuits and networks of computational "nodes" underwent a resurgence in the 1970s after it was discovered that multilayered networks of artificial neurons could be effectively trained to perform more complex computations than two-layered perceptrons (Rumelhart et al., 1986). While the algorithms that alter the connections within artificial neural networks may be more or less biologically realistic, the fact that connections among nodes *are* plastic reflects actual neural networks.

Still more complexity is gained when we allow for recurrent connections in neural networks. By allowing outputs to feed back as inputs to earlier, "upstream" nodes in a network, neural networks can alter their future output as a function of both their past outputs and future inputs. This allows for more powerful devices, insofar as they are able to respond differentially to inputs over time. An analogous feature is found in sequential logic circuits, as opposed to combinational logic circuits. The simplest such device is a "latch" or "flip-flop," which stores a single bit. The outputs produced by a flip-flop depend on the input given, plus the state that the flip-flop is already in; combinational logic devices have no such dependencies.

More generally, sequential logic devices can be modeled as FSAs, discussed above. Both recurrent artificial neural networks and FSAs use an internal state to function as a kind of memory, although they do so in different ways. Nevertheless, neither of these is as powerful as the TM, which, as mentioned above, can compute anything that is computable by following an algorithm. A TM may be compared to a neural system that includes a processing network connected to a memory network, where the memory network is large enough to approximate the unboundedness of a TM's tape for the purposes of the needed computation. There are some difficulties with comparing these various automata to neural systems, however, and recent work in closed-loop neuroscience illustrates precisely what those difficulties are.

Consider the difference between two kinds of feedback or closed-loop activity. During a single neural process, a feedback loop may be activated "in real time," meaning that some particular circuit is activated in a particular pattern for the duration of that process. Output is fed back into an earlier, upstream part of the circuit, and thus functions as part of subsequent input. Yet this feedback is not self-sustaining, creating an endless cycle of activation. Instead, the feedback alters the immediate effect of subsequent inputs, and in the absence of further inputs, the circuit settles back into a resting state. We can call this short-term feedback.

There is also long-term feedback, in which activity does not simply feed back as input to an upstream part of the neural circuit, but alters the strength of connections between neurons. This results in numerous phenomena, but in general, allows for changes in how subsequent inputs are processed long after the particular neural process that causes the change has ended. This feedback is thus not any kind of reverberating activity, but a longer-lasting alteration in how future inputs will be processed.

It is straightforward to characterize short-term feedback in computational terms. Like reverberating networks, automata, including TMs, take advantage of sequential logic so that the effects of inputs can be used to process subsequent inputs within a computation. Thus, if we think of reverberating networks as performing a single computation in which outputs have effects on subsequent inputs, this is precisely analogous to an FSA or a TM operating on the individual digits of an input string. The effect that a single digit will have within a computation depends not just on that digit, but also on what digits have come before. And this is true whether the digits are symbols on a TM tape or action potentials in a neural circuit.

It is not so straightforward to characterize long-term feedback in computational terms. To see this, we must remember that the formal study of automata is about the characteristics of computable functions. Assuming the Church-Turing Thesis, for every computable function, there is a TM that computes that function (and in some cases, a less powerful automaton might suffice). Within the computation of a function, TMs (and to some degree other automata) have a memory of what they have done. But once a computation is over, TMs and other automata reset to their initial state and lose all memory of what they have thus far computed.

One special exception is UTMs. UTMs devote a portion of their tape to the program, and they may be set up so that they alter the program as a result of the computation. So a UTM that modifies its own tape can keep records of its previous computations, in a sense. This is probably not the most useful model of closed-loop neural systems—for starters, because we don't want to assume from the start that such systems are universal computing devices. Maybe some are, but probably the typical ones are not. Additionally, this requires that a UTM compute not *just* the program and the input to that program that are on its tape, but also what *new*

program will be rewritten on its tape. From now on, we will set this exception aside and continue to focus on ordinary (nonuniversal) TMs.

We thus have a problem with thinking of neural computation in terms of ordinary TMs: TMs do not have a stable memory across multiple computations. The power of TMs—relative to FSAs—is their tape. Once a computation is complete, however, a TM begins anew in its starting state and carries no information about previous computations to subsequent computations. This is because the tape actually serves triple-duty: inputs are written on the tape, the TM reads and writes individual symbols on the tape (which functions as a memory), and the final output is written on the tape (assuming that the TM halts on the given input). Thus, the TM carries no information about previous computations. It is possible to remedy this problem by allowing a TM to produce as its output enough information such that this can then be read back as input to subsequent computations. This solution, however, creates additional problems. To name just one, this means that output strings will become increasingly longer as more computations are performed.

A more elegant solution is to enhance TMs so that they have memory *across* computations, and not just *within* a single computation. There are a variety of ways one might accomplish this. The clearest, known as a Persistent Turing Machine (PTM), is due to Goldin et al. (2004). The main idea is quite simple: instead of a single tape that serves three separate purposes (input, working memory, and output), a PTM divides these three tasks into three separate tapes. Thus, a PTM has: (1) an input tape from which it can only read; (2) an output tape upon which it can only write; and (3) a "working memory" tape which it can write to and read from. Crucially, the contents of the working memory tape remain unaltered between computations. This allows a PTM to operate both on its input and whatever has been recorded on its working memory tape. Computations are defined on PTMs similarly to TMs: when the PTM halts, its output is simply whatever has been written on its output tape. It may also have written symbols on its working memory tape, but those need not be interpreted as output.

As an example, consider the following PTM that emulates an answering machine with commands to: (1) record a symbol X (or string of symbols); (2) erase all symbols in memory; and (3) play all symbols in memory. As output to "record X," the PTM prints "ok"; as output to "erase," the PTM prints "done"; as output to "play," the PTM prints all symbols in memory. So, given the input "record A," the PTM prints "ok," and the working memory tape has an A. Input "erase" prints "done," and all symbols on the memory tape are erased. Input "record B" prints "ok," "record CD" prints "ok," and "play" prints "BCD." Each step is a single computation, but unlike a TM, this PTM is able to have, in memory, results from prior computations.

When thinking about the connection between computation and neural systems, it is clear that memory must be preserved across individual computations: this is made especially clear in the case of closed-loop approaches in neuroscience. PTMs are a natural way to allow for this capacity. For example, Egelhaaf et al. (2012) describe the dynamic process by which insects shape their visual input in response to rapid changes in their behavior. Flying insects are able to fly in highly complex ways very quickly. Part of this is due to the insects' ability to alter what visual information comes in as they fly, which is a function of their immediately prior visual information as well as their own recent behavior. Understanding this tight feedback process in terms of multiple TM computations would be both difficult and unenlightening: it is natural to think of insect flying behavior as a quick success of computations, and without the ability to carry memory across computations, TMs are not a good fit. But a PTM allows us to think of this process quite naturally: visual inputs come in and the precise way in which those inputs are processed depends on the contents of the working memory tape, which in turn has been affected by prior computations.

Whether thinking of neural systems as PTMs will prove fruitful remains to be seen. However, reflecting upon the ways in which neuroscientists conceptualize neural systems in computational terms allows us opportunities to continuously reexamine what *kind* of computational power neural systems possess and how that power is a function of neurocognitive architecture including the presence of closed loops. That, in turn, requires reexamining the extent to which available computational models are a good fit for thinking about neural systems. Thinking of neural systems as computational systems has been enormously fruitful for the past several decades, so until there is strong evidence to the contrary, we should continue to see how to best think of neural systems in computational terms as both our understanding of neural systems and computational systems continue to evolve.

6 CONCLUSION

Closed-loop approaches in neuroscience advance the view that thinking of neural systems as mere feedforward devices is insufficient. Furthermore, while adding flexibility in neural responses by way of internal memory goes some way toward understanding the complexity of their behavior, it is not enough. Input, output, and processing are tightly coupled. While thinking of the connection between different models of computation and the functioning of neurons and neural systems has been incredibly fruitful, an important difference between traditional models of computation and neural systems is that the former are ultimately models of particular, individual, computable functions. It is increasingly clear that the neural operations of functioning organisms cannot be seen simply as the repeated application of the same computable functions. Rather, insofar as it is appropriate to think of organisms as performing computations—and indeed, we think it is quite

appropriate—we need a new way of understanding the kind of computation organisms are performing. Closed-loop approaches in neuroscience show that inputs, outputs, and internal states are connected so closely that it is difficult to understand neural processing as a sequence of TM operations: organisms rely on complex information transfer *across* and *between* computations, as well as within a single computation; TMs alone cannot handle the former. But the simple addition of a persistent memory across computations—as used in PTMs—does just that. As closed-loop approaches in neuroscience illustrate how fruitful it is to think of the tightly coupled nature of inputs, outputs, and their processing, so, too, do they require us to revise our ideas of how to best model neural systems as computational systems.

REFERENCES

Cannon, W., 1932. The Wisdom of the Body. W.W. Norton and Company, New York, NY.

deBettencourt, M.T., Cohen, J.D., Lee, R.F., Norman, K.A., Turk-Browne, N.B., 2015. Closed-loop training of attention with real-time brain imaging. Nat. Neurosci. 18, 470–475. http://dx.doi.org/10.1038/nn.3940.

Egelhaaf, M., Böddeker, N., Kern, R., Kurtz, R., Lindemann, J.P., 2012. Spatial vision in insects is facilitated by shaping the dynamics of visual input through behavioral action. Front. Neural Circuits 6, http://doi.org/10.3389/fncir.2012.00108.

Goldin, D.Q., Smolka, S.A., Attie, P.C., Sonderegger, E.L., 2004. Turing machines, transition systems, and interaction. Inf. Comput. 194 (2), 101–128.

Hodgkin, A.L., Huxley, A.F., 1952. A quantitative description of membrane current and its application to conduction and excitation in nerve. J. Physiol. 117 (1–2), 500–544.

Lorente de Nó, R., 1933. Vestibulo-ocular reflex arc. Arch. Neurol. Psychiatry 30, 245–291.

Maley, C.J., 2011. Analog and digital, continuous and discrete. Philos. Stud. 155 (1), 117–131.

McCulloch, W.S., Pitts, W., 1990. A logical calculus of ideas immanent in nervous activity (reprinted from Bulletin of Mathematical Biophysics, Vol. 5, 115–133, 1943). Bull. Math. Biol. 52, 99–115.

Minsky, M.L., Papert, S.A., 1969. Perceptrons. MIT Press, Cambridge, MA.

Murphy, P.C., Duckett, S.G., Sillito, A.M., 1999. Feedback connections to the lateral geniculate nucleus and cortical response properties. Science 286 (5444), 1552–1554. http://dx.doi.org/10.1126/science.286.5444.1552.

Piccinini, G., 2004. The first computational theory of mind and brain: a close look at McCulloch and Pitts's 'logical calculus of ideas immanent in nervous activity'. Synthese 141 (2), 175–215. http://dx.doi.org/10.1023/B:SYNT.0000043018.52445.3e.

Piccinini, G., 2015. Physical Computation: A Mechanistic Account. Oxford University Press, Oxford.

Piccinini, G., Bahar, S., 2013. Neural computation and the computational theory of cognition. Cognit. Sci. 34, 453–488.

Rosenblatt, F., 1962. Principles of Neurodynamics: Perceptrons and the Theory of Brain Mechanisms. Spartan, Washington, DC.

Rumelhart, D.E., McClelland, J.L., PDP Research Group, 1986. Parallel Distributed Processing. MIT Press, Cambridge, MA.

Wiener, N., 1948. Cybernetics: Or Control and Communication in the Animal and the Machine. MIT Press, Cambridge, MA.

Subject Index

Note: Page numbers followed by *f* indicate figures and *t* indicate tables.

Printed in the United States
By Bookmasters